Probability and Statistics for Engineering and Sciences

Probability and Statistics for Engineering and Sciences

Editor: Derek Beave

NY RESEARCH PRESS
New York

Published by NY Research Press
118-35 Queens Blvd., Suite 400,
Forest Hills, NY 11375, USA
www.nyresearchpress.com

Probability and Statistics for Engineering and Sciences
Edited by Derek Beaven

International Standard Book Number: 978-1-63238-880-3 (Hardback)

loging-in-Publication Data

ability and statistics for engineering and sciences / edited by Derek Beaven.
cm.
es bibliographical references and index.
78-1-63238-880-3
bilities. 2. Mathematical statistics. 3. Engineering--Statistical methods.
QA--Statistical methods. I. Beaven, Derek.
519 2022

Contents

Preface

The purpose of the book is to provide a glimpse into the dynamics and to present opinions and studies of some of the scientists engaged in the development of new ideas in the field from very different standpoints. This book will prove useful to students and researchers owing to its high content quality.

The chance or likelihood of an event occurring is referred to as probability. It can be quantified as a number between 0 and 1, where 0 represents impossibility and 1 represents certainty. Probability theory is widely used in the area of mathematics, science (particularly physics) statistics, finance, computer science, etc. Statistics, on the other hand, is a mathematical body of science which deals with data collection, analysis, organization interpretation and presentation. Statistics and probability are two related as well as separated academic branches which are studied together. This book contains some path-breaking studies in the field of probability and statistics for engineering and sciences. The book studies, analyzes and upholds the pillars of Probability and Statistics and its utmost significance in Engineering and Sciences. This book, with its detailed analyses and data, will prove immensely beneficial to professionals and students involved in this area at various levels.

At the end, I would like to appreciate all the efforts made by the authors in completing their chapters professionally. I express my deepest gratitude to all of them for contributing to this book by sharing their valuable works. A special thanks to my family and friends for their constant support in this journey.

Editor

1

Bootstrap Order Determination for ARMA Models: A Comparison between Different Model Selection Criteria

Livio Fenga

University of California San Diego (UCSD), San Diego, CA, USA

Correspondence should be addressed to Livio Fenga; lfenga@ucsd.edu

Academic Editor: Ramón M. Rodríguez-Dagnino

The present paper deals with the order selection of models of the class for autoregressive moving average. A novel method—previously designed to enhance the selection capabilities of the Akaike Information Criterion and successfully tested—is now extended to the other three popular selectors commonly used by both theoretical statisticians and practitioners. They are the final prediction error, the Bayesian information criterion, and the Hannan-Quinn information criterion which are employed in conjunction with a semiparametric bootstrap scheme of the type sieve.

1. Introduction

Autoregressive moving average (ARMA) models [1] are a popular choice for the analysis of stochastic processes in many fields of applied and theoretical research. They are mathematical tools employed to model the persistence, over time and space, of a given time series. They can be used for a variety of purposes, for example, the generation of predictions of future values, to remove the autocorrelation structure from a time series (prewhitening) or to achieve a better understanding of a physical system. As it is well known, performances of an ARMA model are critically affected by the determination of its order: once properly built and tested, such models can be successfully employed to describe the reality, for example, trend patterns of economic variables and temperature oscillations in a given area, or to build futures scenarios through simulation exercises. Model order choice plays a key role not only for the validity of the inference procedures but also, from a more general point of view, for the fulfillment of the fundamental principle of parsimony [2, 3]. Ideally, the observation of this principle leads to choosing models showing simple structures on one hand but able to provide an effective description of the data set under investigation on the other hand. Less parsimonious models tend to extract idiosyncratic information and therefore are prone to introduce high variability in the estimated parameters. Such a variability determines for the model a lack of generalization capabilities (e.g., when new data become available), even though, by adding more and more parameters, an excellent fit of data is usually obtained [4]. Overfitting is more likely to occur when the system under investigation is affected by different sources of noise, for example, related to changes in survey methodologies, time evolving processes, and missing observations. These phenomena, very common and in many cases simply unavoidable in "real life" data, might have a significant impact on the quality of the data set at hand. Under noisy conditions, a too complex model is likely to fit the noise components embedded in the time series and not just the signal and therefore it is bound to yield poor future values' predictions. On the other hand, bias in the estimation process arises when underfitted models are selected, so that only a suboptimal reconstruction of the underlying Data Generating Process (DGP) can be provided. As it will be seen, bias also arises as a result of the uncertainty conveyed by the process *itself* of model selection. ARMA model order selection is a difficult step in time series analysis. This issue has attracted a lot of attention so that, according to different philosophies, theoretical and practical assumptions as well as several methods, both parametric and nonparametric, have been proposed over the years as a result. Among them, bootstrap strategies [5–9] are gaining more and more acceptance among researchers and practitioners.

In particular, in [6] bootstrap-based procedures applied to the Akaike Information Criterion (AIC) [10, 11] in the case of ARMA models, called b-MAICE (bootstrap-Minimum AIC Estimate), has proven to enhance the small sample performances of this selector. The aim of this work is to extend such a procedure to different selectors, that is, final prediction error (FPE) [12] and two information based criteria, that is, Bayesian information criterion (BIC) [13, 14] and Hannan-Quinn criterion (HQC) [15, 16]. In particular, the present paper is aimed at giving empirical evidences of the quality of the bootstrap approach in model selection, by comparing it with the standard procedure, which, as it is well known, is based on the minimization of a selection criterion. In particular, the empirical study (presented in Section 4) has been designed to contrast the performances of each of the considered selectors both in nonbootstrap and bootstrap world. The validity of the proposed method is assessed not only in the case of pure ARMA processes, but also when real life phenomena are simulated and embedded in the artificial data. In practice, the problem of order determination is considered also when the observed series is contaminated with outliers and additive Gaussian noise. The last type of contamination has been employed, for example, in [17], for testing a model selection approach driven by information criteria in the autoregressive fractionally integrated moving average (ARFIMA) and ARMA cases. Such a source of disturbance has been employed here in order to test the degree of robustness of the method proposed against overfitting. As it will be seen, computer simulations show that the addition of white noise generates a number of incorrect specifications comparable to those resulting from the contamination of the process with outliers of the type innovation. Outliers are a common phenomenon in time series, considering the fact that real life time series from many fields, for example, economic, sociology, and climatology, can be subjected and severely influenced by interruptive events, such as strikes, outbreaks of war, unexpected heat or cold waves, and natural disasters [18, 19]. The issue is absolutely nontrivial, given that outliers can impact virtually all the stages of the analysis of a given time series. In particular, model identification can be heavily affected by additive outliers, as they can induce the selection of underfitted models as a result of the bias elements introduced into the inference procedures. In the simulation study (Section 4), outliers of the type additive (i.e., added to some observations) and innovative (i.e., embedded in the innovation sequence driving the process) [19] will be considered.

The remainder of the paper is organized as follows: in Section 2, after introducing the problem of order identification for time series, the considered selectors are illustrated along with the related ARMA identification procedure. In Section 3 the employed bootstrap selection method is illustrated and the bootstrap scheme briefly recalled. Finally, small sample performances of the proposed method will be assessed via Monte Carlo simulations in Section 4.

2. Order Selection for Time Series Models

A key concept underlying the present paper is that, in general, "reality" generates complex structures, possibly ∞-dimensional, so that a model can at best capture only the main features of the system under investigation in order to reconstruct a simplified version of a given phenomenon. Models are just approximations of a given (nontrivial) phenomenon and the related identification procedures could never lead to the determination of the "true" model. In general, there is no true model in a finite world. What we can do is to find the one giving the best representation of the underlying DGP, according to a predefined rule. In this section, after highlighting the role played by model selection procedures in generating uncertainty, we briefly introduce the models belonging to the class ARMA along with the order selectors considered. Finally, the information criterion-based standard selection procedure is illustrated.

2.1. Uncertainty in Model Selection. Uncertainty is an unfortunate, pervasive, and inescapable feature characterizing real life data which has to be faced continually by both researchers and practitioners. The framework dealt with here is clearly no exception: if the true model structure is an unattainable goal, approximation strategies have to be employed. Such strategies are generally designed on iterative basis and provide an estimate of the model structure which embodies, by definition, a certain amount of uncertainty. Common sources of uncertainty are those induced by the lack of discriminating power of the employed selector and by the so-called model selection bias [20, 21], which arises when a model is specified and fitted on the *same* data set. Unfortunately, not only are these two types of uncertainty not mutually exclusive but also statistical theory provides little guidance to quantify their effect in terms of bias introduced in the model as a result [22]. Particularly dangerous is this last form of uncertainty, as it is based upon the strong and unrealistic assumption of making correct inference *as if* a model is known to be true, while its determination has been made on the same set of data. On the other hand, the first source of uncertainty is somehow less serious, given its direct relationship with the size of the competition set, which is usually included in the design of the experiment. In practice, it is related to the fact that very close SC minima can be found in the model selection process, so that even small variations in the data set can cause the identification of different model structures. In general, trying to explain only in part the complexity conveyed in the observed process by means of as simple as possible structures is a way to minimize uncertainty in the model selection, as it is likely to lead to the definition of a smaller set of candidate models. This approach can be seen as an extension of the principle of parsimony to the competition set. In the sequel, how the proposed procedure, being aimed at replicating both the original process and the related selection procedure, has a positive effect in reducing both the considered sources of uncertainty will be emphasized [23].

2.2. The Employed Identification Criteria. Perhaps the most well-known model order selection criteria (SC), among those considered, are the AIC and the FPE, whose asymptotic equivalence to the F-test has been proved in [24]. AIC has been designed on information-theoretic basis as an asymptotically unbiased estimate of the Kullback-Leibler divergence [25] of the fitted model relative to the true model.

Assuming X_T, T being the sample size, to be randomly drawn from an unknown distribution $H(x)$ with density $h(x)$, the estimation of h is done by means of a parametric family of distributions, with densities $[f(x \mid \theta; \theta \in \Theta)]$, θ being the unknown parameters' vector. Denoting $f(z \mid \hat{\theta})$ as the predictive density function, f as the true model, and h as the approximating one, Kullback-Leibler discrepancy can be expressed as follows:

$$I\left(h(z); f\left(z \mid \hat{\theta}\right)\right) = \int h(z) \log h(z)\, dz - \int h(z) \log f\left(z \mid \hat{\theta}\right) dz. \quad (1)$$

As the first term on the right hand side of (1) does not depend on the model, it can be neglected so that we can rewrite the distance in terms of the expected log likelihood, $L(X_T; H)$; that is,

$$L(X_T; H) = \int h(z) \log f\left(z \mid \hat{\theta}\right) dz = \int \log f\left(z \mid \hat{\theta}\right) dH(z). \quad (2)$$

This quantity can be estimated by replacing H with its empirical distribution $\hat{H}$, so that we have that $L(X_T; \hat{H}) = (1/T)\sum_{\alpha=1}^{T} \log f(X_\alpha \mid \hat{\theta})$. This is an overestimated quantity of the expected log likelihood, given that $\hat{H}$ is closer to $\hat{\theta}$ than H. The related bias can be written as follows:

$$b(H) = E_H\left\{L\left(X_T; \hat{H}\right) - L(X_T; H)\right\}, \quad (3)$$

and therefore an information criterion can be derived from the bias-corrected log likelihood; that is, $(1/T)\sum_{\alpha}^{T} \log f(X_\alpha \mid \hat{\theta}) - B(\hat{H})$

Denoting by k and T the number of estimated parameters and the sample size, respectively, Akaike proved that $b(H)$ is asymptotically equal to k/T, so that the information based criterion takes the form $L(X_T; \hat{H}) + k/T$. By multiplying this quantity by -2, finally AIC is defined as $-2\log L(X_T; \hat{H}) + 2k$. In such a theoretical framework, AIC can be seen as a way to solve the Akaike Prediction Problem [6], that is, to find a model M_0 producing estimation of density $\hat{f}$ minimizing *Kullback-Leibler* discrepancy (1). Originally conceived for AR process, extended to the ARMA case by Soderstrom and Stoica [24], FPE was designed as the minimizer of the one-step-ahead mean square forecast error, after taking in account the inflating effect of the estimated parameter. FPE statistic is defined as $\text{FPE}(k) = [(1 + k/T)/(1 - k/T)]\hat{\sigma}_\varepsilon^2(k)$, where $\hat{\sigma}_\varepsilon^2$ is the estimated variance of the residuals and k is the model's size. A different perspective has led to the construction of BIC-type criteria, which are grounded on the maximization of the model posterior probability [14]. In more detail, they envision the specification of the prior distribution on parameter values and the models, respectively, denoted by $P(\theta \mid k)$ and $P(k)$, and their introduction into the analysis through the joint probability function $P(\theta, k) = P(k)P(\theta \mid k)$. Posterior probabilities for (θ, k) are then obtained through Bayes theorem, so that the value of k maximizing (4), that is,

$$P(k \mid X_t) \propto P(k) \int_{\theta \in \Theta} f(X_t; \theta, k) P(\theta \mid k)\, d\theta, \quad (4)$$

is found. With $f(X_T; \theta, k)$ being the likelihood function associated with both the data X_T and the model M_k, the selected order will be $\hat{k} = \arg\max_k P(k \mid X_T)$. By assuming all the models equally probable, that is, $p(k) = 1/(k_{\max} + 1)$, the BIC criterion is hence defined by $-2\log L(\hat{\theta}) + 2k\log(T)$. The last criterion considered—constructed from the law of iterated algorithm—is the BIC, in which the penalty function grows at a very slow rate as the samples size increases. It is defined as follows: $\text{HQC} = \log L(\hat{\theta}) + 2k\log(\log(T))$.

All these selectors can be divided into two groups: one achieving asymptotic optimality [26] and one selection consistency. AIC and FPE fall in the first group, in the sense that the selected model asymptotically tends to reach the smallest average squared error [27, 28], if the true DGP is not included in the competition set. On the other hand, BIC and HQ are dimension consistent [29], in that the probability of selection of the "true" model approaches 1 as the sample size goes to infinity. However, it should be pointed out that such an asymptotic property holds only if the true density is in the set of the candidate models. In this regard, AIC and FPE as well as the other Shibata efficient criteria (e.g., Mallows C_p [30]) fail to select the "true" model asymptotically. As pointed out earlier, ∞-dimensionality of the "truth" implies for all the models being "wrong" to some extent—except in trivial cases—so that no set of competition models will ever encompass the true DGP. As long as this approach is held true, asymptotic efficient criteria might be preferred. In this case, one may argue a lack of significance in comparing any finite list of candidate models when we rule out the existence of a true one. Such an approach is justified in that, even if no model can ever represent the truth, we can achieve the goal to find the one being approximately correct. Conversely, if one does believe that the true density belongs to the model space, hence dimension consistent selection criteria can be preferred.

2.3. ARMA Model Selection through Minimization of Selection Criteria. In what follows, it is assumed that the observed time series $\{X_t\}_{t\in\mathbb{Z}^+}$ is a realization of a real valued, 0–mean, second-order stationary process, admitting an autoregressive moving average representation of orders p and q; that is, $x_t \sim \text{ARMA}(p, q)$, with $(p, q) \in \mathbb{Z}^+$. Its mathematical expression is as follows:

$$\phi(B)(x_t) = \theta(B)\varepsilon_t; \quad t \in \mathbb{Z}^+, \quad (5)$$

with $\phi(z) = (1 - \sum_{i=1}^{p} \phi_i z^i)$ and $\theta(z) = (1 - \sum_{i=1}^{q} \theta_i z^i)$, being $\phi_i \in \mathbb{R}$ and $\theta_i \in \mathbb{R}$, AR polynomial, and MA polynomial, respectively. With B the backward shift operator, such that $B^k Y_t = Y_{t-k}$, is denoted whereas ε_t is assumed to be sequence of centered, uncorrelated variables with common variance σ^2. The parameters vector is denoted by $\mathbf{\Gamma}$. Standard assumptions of stationarity and invertibility, respectively, of AR and MA polynomials, that is,

$$|\phi(z)| \neq 0 \quad |z| \leq 1, \tag{6}$$

$$|\theta(z)| \neq 0 \quad |z| \leq 1, \tag{7}$$

are supposed to be satisfied. Finally, the ARMA parameters of the true underlying DGP (5) are denoted by (p°, q°) (i.e., $\{X\}_t^T \sim \text{ARMA}(p^{\circ}, q^{\circ})$) and the related model by $M^0(\Gamma)$.

Identification procedures of the best approximating model for M^0 is carried out on a priori specified set Λ of plausible candidate models M_g; that is,

$$\{\Lambda \supseteq M_g(\widehat{\Gamma}) \ g = 1, 2, \ldots, G\}, \tag{8}$$

where the chosen model, say $M_0(\widehat{\Gamma}) = (\widehat{p}_0, \widehat{q}_0)$, is selected from (i.e., $[M_0(\widehat{\Gamma}) \equiv (\widehat{p}_0, \widehat{q}_0) \subset \Lambda] \approx M^0(\Gamma)$). In the ARMA case, each model $M_g \in \Lambda$ represents a specific combination of autoregressive and moving average parameters (p, q). The set Λ is upper bounded by the two integers P and Q for the AR and MA part, respectively; that is,

$$\Lambda = \{(p, q) : 0 \leq p_0 \leq P, \ 0 \leq q_0 \leq Q\}. \tag{9}$$

This assumption is a necessary condition for the above-mentioned Shibata efficiency and dimension consistency properties to hold other than for the practical implementation of the procedure (the model space needs to be bounded). From an operational point of view, the four SC considered in this work, when applied to models of the class ARMA, take the following form:

$$\text{AIC}(p, q) = T \ln \widehat{\sigma}_{p,q}^2 + 2(p + q + 1), \tag{10}$$

$$\text{FPE}(p, q) = \widehat{\sigma}_{p,q}^2 \left\{ \frac{T - (p + q + 1)}{T + (p + q + 1)} \right\}, \tag{11}$$

$$\text{BIC}(p, q) = T \ln \left[\widehat{\sigma}_{p,q}^2\right] + \left[(p + q + 1)\ln(T)\right], \tag{12}$$

$$\begin{aligned} \text{HQC}(p, q) &= T \ln \left[\widehat{\sigma}_{p,q}^2\right] \\ &\quad + \left[2(p + q + 1)\ln(\ln(T))\right], \end{aligned} \tag{13}$$

where $\widehat{\sigma}_{p,q}$ is an estimate of the Gaussian pseudo-maximum likelihood residual variance when fitting ARMA (p, q) models; that is,

$$\begin{aligned} \widehat{\sigma}_{p,q}^2 &= \frac{1}{T - (p + q + 1)} \\ &\quad \cdot \sum_{\max(p,q)}^{T} \left[y_t - \left(\sum_{j=0}^{P} \phi_j y_{t-j} - \sum_{i=0}^{Q} \theta_i y_{t-i} \right) \right]^2. \end{aligned} \tag{14}$$

Equations (10)–(13) can be synthetically expressed as follows:

$$\text{SC}(\widehat{p}, \widehat{q}) = f\left(\widehat{\sigma}_{p,q}^2, \xi_{pq}\right), \tag{15}$$

where $\widehat{\sigma}_{p,q}^2$ is defined in Section 3 and ξ is the penalty term as a function of model complexity.

The standard identification procedure, here called for convenience Minimum Selection Criterion Estimation (MSCE), is based on the minimization of the SC. In practice, the model M_0 minimizing a given SC is the winner; that is,

$$M_0 : (\widehat{p}_0, \widehat{q}_0) = \arg \min_{p<P, q<Q} \text{SC}(p, q). \tag{16}$$

3. The Bootstrap Method

As already pointed out, in [6] a bootstrap selection method has been proposed to perform AIC-based ARMA structure identification. The comparative Monte Carlo experiment with its nonbootstrap counterpart, commonly referred to as MAICE (Minimum Akaike Information Criterion Expectation) procedure, gave empirical evidences in favor of *b*-MAICE procedure. Such results motivated us to extend this approach to other selectors (see (11), (12), and (13)). For convenience, the proposed generalized version of *b*-MAICE procedure has been called bMSE (Bootstrap Minimum Selector Expectation) procedure. Finally, in order to keep the paper as self-contained as possible, and to reduce uncertainty in the experimental outcomes, AIC has also been included in the experiment.

3.1. The Bootstrapped Selection Criteria. The proposed bMSE method relies on the bootstrapped version of a given SC, obtained by bootstrapping both the residual variance term $\widehat{\sigma}_{\varepsilon}^2$ and the penalty term, so that (15) becomes

$$\text{SC}^* = f\left[\left(\widehat{\sigma}_{p,q}^2\right)^*, \xi^*\right]. \tag{17}$$

The particularization of (17) to the criteria object of this study is straightforward and yields their bootstrapped versions; that is,

$$\begin{aligned} \text{AIC}^*(p, q) &= T \ln \left(\widehat{\sigma}_{p,q}^2\right)^* + 2(p + q + 1)^*, \\ \text{FPE}^*(p, q) &= \left(\widehat{\sigma}_{p,q}^2\right)^* \left\{ \frac{T - (p + q + 1)}{T + (p + q + 1)} \right\}^*, \\ \text{BIC}^*(p, q) &= T \ln \left[\widehat{\sigma}_{p,q}^2\right]^* + \left[(p + q + 1)\ln(T)\right]^*, \\ \text{HQC}^*(p, q) &= T \ln \left[\widehat{\sigma}_{p,q}^2\right]^* \\ &\quad + \left[2(p + q + 1)\ln(\ln(T))\right]^*, \end{aligned} \tag{18}$$

with T, p, q being as above defined and $\sigma_{p,q}^2$ being the residual variance of the residuals from the fitting of the bootstrapped series y_t^* with its ARMA estimate $\widehat{y}_t^*$. In symbols,

$$\begin{aligned} \left(\widehat{\sigma}_{p,q}^2\right)^* &= \left[\frac{1}{T - (p + q + 1)}\right]^* \\ &\quad \cdot \sum_{[\max(p,q)]^*}^{T} \left[y_t^* - \left(\sum_{j=0}^{P^*} \phi_j y_{t-j}^* - \sum_{i=0}^{Q^*} \theta_i y_{t-i}^* \right) \right]^2. \end{aligned} \tag{19}$$

In essence, bMSE method works as follows: MSCE procedure is applied iteratively on each X_b^* bootstrap replication $b = 1, \ldots, B$ of the observed series. A winner model M_g is selected at each iteration on the basis of a given SC, which in turns works exploiting the bootstrap estimated variances of the residuals. The final model is chosen on the basis of its relative frequency over the B bootstrap replication.

3.2. The Applied Bootstrap Scheme. Sieve [31] [32, 33] is the bootstrap scheme employed here. It is an effective and conceptually simple tool to borrow randomness from white noise residuals, generated by the fitting procedure of a "long" autoregression to the observed time series. This autoregression, here supposed to be 0–*mean*, is of the type $y_t = \sum_{j=1}^{p} a_t(y_{t-j}) + \varepsilon_t$, $t \in \mathbb{Z}$, under the stationarity conditions as in (6). Its use is here motivated by the AR(∞) representation of process of type (5); that is,

$$X_t = \sum_{j=1}^{\infty} a_j \left(X_{t-j}\right) + \varepsilon_t \quad t = 1, 2, \ldots, T, \tag{20}$$

with $(\varepsilon_t)_{t\in\mathbb{Z}}$ being a sequence of iid variables with $E[\varepsilon_t] = 0$ and $\sum_{j=0}^{\infty} a_j^2 < \infty$. In essence, *sieve* bootstrap approximates a given process by a finite autoregressive process, whose order $\widehat{p} = p(T)$ increases with the sample size T such that $p(T) \rightarrow \infty$, $p(T) = o(T)$, $T \rightarrow \infty$. In this regard, in the empirical study the estimation of the *p-vector* of coefficients $(\widehat{a}_1, \ldots, \widehat{a}_{\widehat{p}})$ has been carried out through the Yule-Walker equations. The residuals $\widehat{\varepsilon}_t = \sum_{j=1}^{\widehat{p}} a_j X_{t-j} + \varepsilon_t$ $t = 1, 2, \ldots, T$ obtained from the fitting procedure of this autoregression to the original data are then employed to build up the centered empirical distribution function, which is defined as

$$\widehat{F}_{\varepsilon}(x) = \widehat{P}\left[\varepsilon_t \le x\right] = (T - p)^{-1} \sum_{t=\widehat{p}+1}^{n} \mathbf{1}_{[S_t - \bar{S} \le x]}, \tag{21}$$

where $S_t = X_t - \sum_{j=1}^{p} \widehat{a}_j X_{t-j}$, with $\bar{S}$ being the mean value of the available residuals, that is, S_t, $t = \widehat{p} + 1, \ldots, T$. From $\widehat{F}_{\varepsilon}$ bootstrap samples $\mathbf{X}_T^* = (X_{1-\widehat{p}}^*, \ldots, X_T^*)$ are generated by the recursion

$$\sum_{j=1}^{\widehat{p}} \widehat{a}_j \left(X_{t-j}^* - \bar{X}\right) = \varepsilon_t^* \quad t \in \left(\widehat{p}, \ldots, T\right), \tag{22}$$

with starting values $X_t^* = 0$, $\varepsilon_t^* = 0$ for $t \le -\max(p, q)$, $t = T + 1, \ldots, 2T$.

3.3. The Proposed bMSE Procedure. Let $\{x_t\}_t^T$ be the observed time series realization of ARIMA (p, q) DGP (5), from which B bootstrap replications $\{x_{b,t}^*;\ b = 1, 2, \ldots, B\}_t^T$ are generated via *sieve* method (Section 3). Our B-MSCE procedure is based on the minimization, over all the combinations of ARMA structures, of a given SC by applying MSCE procedure to each bootstrap replication $x_{t,b}^*$ of the original time series x_t.

In what follows the proposed procedure is summarized in a step-by-step fashion.

(1) A maximum ARMA order (P, Q) is arbitrarily chosen, so that exhaustive set Λ of tentative ARMA models, that is, $\{M_g,\ g = 0, 1, \ldots, G\}$, with $p \le P$, $q \le Q$, of size $((P = Q) + 1)^2$, is defined.

(2) The number B of bootstrap replications is chosen.

(3) A bootstrap replication, x_b^*, of the original time series x_t is generated via *sieve* method.

(4) The competition set Λ is iteratively fitted to x_b^* so that G values (one for each of the models in Λ) of the SC* are computed and stored in the G-dimension vector v_G.

(5) Minimum SC* value is extracted from v_G so that a winner model, $M_{0,b}^*$, is selected; that is,

$$M_{0,b}^* : \left(\widehat{p}^*, \widehat{q}^*\right) = \arg \min_{p<P,\, q<Q} \mathrm{SC}^*\left(p, q\right). \tag{23}$$

(6) By repeating B times steps (3) to (5), the final model M_0^* is chosen according to a mode-based criterion, that is, on the basis of its more frequent occurrence in the set of the bootstrap replications. In practice, the selected model is chosen according to the following rule:

$$\# \left[\mathrm{SC}^*\left(M_{0,b}\right) < \mathrm{SC}^*\left(M_{g,b}\right)\right], \quad g = 0, \ldots, G - 1,\ b = 1, 2, \ldots, B, \tag{24}$$

with the symbol # being used as a counter of the number of the cases satisfying the inequality condition expressed in (24).

The order p_0^{sieve} of the *sieve* autoregression is chosen by iteratively computing the Ljung-Box statistic [34] on the residuals resulting from the fitting of tentative autoregression on the original time series with sample size T_0. Further orders, say p_i^{sieve}, $i = 1, 2, \ldots$, for increasing sample sizes, T_i, $i = 1, 2, \ldots$, are selected according to the relation $p_i^{sieve} = c(T_i)^{1/3}$, where $c = p_0^{sieve}/T_0^{1/3}$ (in [6] p_0^{sieve} is chosen by iteratively computing the spectral density on the residuals resulting from the fitting of tentative autoregression on the original time series; the order $\widehat{p}$ for which the spectral density is approximately constant is then selected).

The presented method is exhaustive and then highly computer intensive, as for all the $(p + 1) * (q + 1)$ possible pairs (in the attempt to reduce such a burden, sometimes, see, e.g., [35], the set of the ARMA orders under investigation is restricted to $\Lambda = \{(\psi, \psi - 1){:}\ 0 \le \psi \le \Psi\}$; i.e., the competition set is made up of ARMA $(\psi, \psi - 1)$; however, the fact that such an approach entails the obvious drawback of not being able to identify common processes, such as ARMA $(2, 0)$, has appeared to be a too strong limitation; therefore, in spite of its ability to drastically reduce the computational time, such an approach has not been followed here), the values of the given SC* must be computed for each of the B bootstrap replications.

4. Empirical Study

In this section, the outcomes of a simulation study will be reported. It has been designed with the twofold purpose of

Table 1: ARMA DGPs.

Set	Parameters	DGP 1	DGP 2	DGP 3	DGP 4	DGP 5	DGP 6	DGP 7	DGP 8	DGP 9	DGP 10
	ϕ_1	−.85	−.75	−.65	−.60	−.45	.50	.65	.70	.60	.7
$\mathcal{J}_1$	ϕ_2	—	—	—	—	—	—	—	—	—	
	θ_1	.50	.40	−.40	.40	−.65	−.25	−.50	−.40	−.40	−.5
	θ_2	—	—	—	—	—	—	—	—	—	
	ϕ_1	−1.2	−1.1	−0.9	−0.8	−0.6	−.5	− 0.4	0.35	0.7	0.4
$\mathcal{J}_2$	ϕ_2	−.9	−0.9	−0.9	−0.8	−0.5	−4	−0.35	−0.6	−0.8	0.65
	θ_1	.5	.5	0.5	0.5	0.5	.35	0.35	.50	−.5	−.5
	θ_2	—	—	—	—	—	—	—	—	—	—
	ϕ_1	−0.9	−0.8	−0.7	−0.60	−0.5	−0.4	0.4	0.5	0.7	0.80
$\mathcal{J}_3$	ϕ_2	—	—	—	—	—	—	—	—	—	—
	θ_1	−0.50	−0.35	−0.35	0.5	0.	0.35	0.70	−0.9	−0.5	−0.5
	θ_2	0.35	0.50	0.35	0.35	−0.5	0.35	0.25	−0.5	− 0.5	0.35

(i) evaluating bMSE procedure's small sample performances and (ii) giving some evidences of its behavior for increasing sample sizes. As a measure of performances, the percentage frequency of selection of the true order (p^0, q^0), in the sequel denoted as g and g^* for MSCE and bMSE procedure, respectively, has been adopted; that is,

$$g = 100 * \left[\frac{\#\ \text{time series correctly identified}}{S}\right], \qquad (25)$$

with S denoting the number of the artificial time series employed in the experiment and # the quantifier symbol, expressing the number of times the statement "time series correctly identified" is true. Its extension to the bootstrap case g^* is straightforward.

Aspect (i) consists of a series of Monte Carlo experiments carried out on three different sets of time series, 10 for each set, detailed in Table 1, which (1) are realization of three prespecified ARMA orders, that is, (1, 1), (2, 1), and (1, 2) (one order for each set), and (2) differ from each other, within the same set, only for the coefficients' values, but not for the order (p, q). Two sample sizes will be considered, that is, $T = 100, 200$. Formally, these sets are, respectively, denoted as $\{(\mathcal{J}_1, \mathcal{J}_2, \mathcal{J}_3)\}$ and supposed to belong to the order subspace $\mathcal{S}$: $\{\mathcal{S} \supseteq \mathcal{J}_j;\ j = 1, 2, 3\}$. For each DGP $\in \mathcal{S}$, 10 different coefficient vectors are specified, that is, $\{\mathcal{S} \supseteq \mathcal{J} \equiv (p_1, q_1), \ldots, (p_{10}, q_{10})\}$. The validity of the presented method is assessed on comparative basis, using as benchmark the standard MSCE procedure. For the sake of concision, the values g and g^* will be computed averaging over all the DGPs belonging to either the same set $\mathcal{J}_j$ or $\mathcal{S}$. In practice, two indicators, that is, the Percentage Average Discrepancy (PAD) and the Overall Percentage Average Discrepancy (OPAD), depending on weather only one set $\mathcal{J}_j$ or the whole order subspace $\mathcal{S}_j$ is considered, will be employed. They are formalized as follows:

$$\text{PAD (SC)} = \left[g^*_{[\mathcal{J},\text{SC}]} - g_{[\mathcal{J},\text{SC}]}\right] \quad \forall \mathcal{J} \subseteq \mathcal{S}, \qquad (26)$$

$$\text{OPAD (SC)} = \frac{1}{|\mathcal{S}|} \sum_{\mathcal{J}=1}^{|\mathcal{S}|} \left[g^*_{[\mathcal{J},\text{SC}]} - g_{[\mathcal{J},\text{SC}]}\right] \quad \forall \mathcal{J} \subseteq \mathcal{S}, \qquad (27)$$

where with the symbol $|\cdot|$ being the cardinality of a set is denoted. In other words, the average percentage differences in the frequency of selection of the true model is used as a measure of the gain/loss generated by bMSE procedure with regard to a single $\mathcal{J}$ (26) or by averaging over the sets $\mathcal{S}$ (27). As already outlined, in analyzing aspect (ii) the attention is focused on the behavior of the proposed method for increasing sample sizes, that is, $T = 100, 200, 500, 1000$. In Table 4, the results obtained for the case of 4 DGPs—detailed in the same table—will be given. In both (i) and (ii), for each DGP $\in (\mathcal{J}_1, \mathcal{J}_2, \mathcal{J}_3)$, a set of $S = 500$ time series has been generated. Each time series s_i $(i = 1, 2, \ldots, S)$ has been artificially replicated $B = 125$ times using the bootstrap scheme outlined in Section 3.2 (the simulations have been implemented using the software R (8.1 version) and performed using the hardware resources of the University of California, San Diego; in particular, the computer server EULER (maintained by the Mathematical Department) and the supercomputer IBM-TERAGRID have been employed). The number of bootstrap replications B employed has been chosen on empirical basis, as the best compromise between performances yielded by the method and computational time.

The parameter space of all the DGPs considered always satisfies the invertibility and stationarity conditions (see (6), (7)), whereas the maximum order P and Q investigated has been kept fixed and low throughout the whole experiment $(P = Q = 3)$ mainly to keep the overall computational time reasonably low. However, such an arbitrary choice seems to be able to reflect time series usually encountered in practice in a number of fields, such as economy, ecology, or hydrology. However, it should be emphasized that in many other contexts (e.g., signal processing) higher orders must be considered.

4.1. The Experiments. Other than on the pure ARMA signal, aspect (i) has been investigated in terms of the robustness shown against outliers and noisy conditions. In practice, the simulated DGPs are assumed to be

a: a pure process (no contamination),

b: contaminated with outliers of the type IO (experiments $\mathbf{b}_1, \mathbf{b}_2$) and AO (experiment $\mathbf{b}_3$),

c: contaminated with Gaussian additive noise.

The first set of simulations (experiment **a**) is designed to give empirical evidences for the case of noise-free, uncontaminated ARMA process of type (5). Experiment **b** is aimed at mimicking a situation where a given dynamic system is perturbed by shocks resulting in aberrant data, commonly referred to as outliers. As already pointed out, such abnormal observations might be generated by unpredictable phenomena (e.g., sudden events related to strikes, wars, and exceptional meteorological conditions) or noise components which have the ability to lead to an inappropriate model identification, other than to biased inference, low quality forecast performances, and, if seasonality is present in the data, poor decomposition. Without any doubt, outliers represent a serious issue in time series analysis; therefore testing the degree of robustness of any procedure against such potentially disruptive source of distortion is an important task. This topic has attracted much attention from both theoretical statisticians and practitioners. Detection of time series outliers was first studied by Fox [19], whose results have been extended to ARIMA models by Chang et al. [36]. Other references include [37–39]. In addition, more and more often outlier detection algorithms are provided in the form of stand-alone efficient routines—for example, the library TSO of the software "R," based on the procedure of Chen and Liu (1993) [37]—or included in automatic model identification procedures provided by many software packages, as in the case of the statistical program TRAMO (Time series Regression with ARIMA noise, Missing observations, and Outliers [40]) or SCA (Scientific Computing Associates [41]). Following [19], two common types of outliers, that is, additive (AO) and innovational (IO), will be considered. As it will be illustrated, unfortunately the proposed identification procedure shows sensitivity to outliers, as they are liable, even though to different extents, to noticeable deterioration of the selecting performances.

In more detail, the observed time series x_t is considered as being affected by a certain number ρ of deterministic shocks at different time $t = \tau_1, \ldots, \tau_\rho$; that is,

$$x_t = \sum_{j=1}^{n} h_j \xi_j(B) I_t^{(\tau_j)} + z_t, \tag{28}$$

where z_t is the uncontaminated one of type (5), h_j measures the impact of the outlier at time $t = \tau_j$, and $I_t^{(\tau_j)}$ is an indicator variable taking the value 1 for $t = \tau_j$ and 0 otherwise. Outlier-induced dynamics are described by the function $\xi(B)$ which takes the form

$$\xi_j(B) = \begin{cases} 1, & \text{for AO} \\ \dfrac{\theta(B)}{\phi(B)} & \text{for IO.} \end{cases} \tag{29}$$

As the onset of an external cause, outliers of the type IO have the ability to affect the level of the series at the time they occur until a lag τ_j, whose localization depends on the memory mechanism encoded in the ARMA model. Their effect can be even temporally unbounded, for example, under ARIMA DGPs with nonzero integrating stationary inducing constant I. Conversely, AOs affect only the level of the observations at the time of its occurrence (in this regard, typical examples are errors related to the recording process or to the measurement device employed). They are liable to corrupting the spectral representation of a process, which tends to be of the type *white noise* and in general the autocorrelations are pulled towards zero (their effect on the Autocorrelation Function (ACF) and the spectral density level has been discussed in the literature (see, e.g., [42] and the references therein)), so that meaningful conclusion based on these functions—depending on their location, magnitude, and probability of occurrence—might be severely compromised. On the other hand, the effects produced by IOs are usually less dramatic as the ACF tends to maintain the pattern of the uncontaminated process z_t and the spectral density $G_x(\omega)$, ω being the frequency, roughly shows a shape consistent with the one computed on z_t (i.e., $G_x(\omega) \propto G_z(\omega)$). The outcomes of the simulations conducted are consistent with the above.

In the present study, IOs have been randomized and introduced according to a Bernoulli (BER) distribution with parameter $\pi = .04$. In order to better assess the sensitivity of the proposed procedure to outlying observations, experiment **b** has been conducted considering two different levels of standard errors, that is, $\sigma = 3$ (experiment $\mathbf{b}_1$) and $\sigma = 4$ (experiment $\mathbf{b}_2$); in symbols, recalling (5), we have

$$\begin{gathered} \varepsilon_t = (1 - Q_t)\varepsilon_{1,t} + Q_t \varepsilon_{2,t} \quad Q_t \sim \text{BER}(\pi) \quad \pi = 0.04, \\ \varepsilon_{1,t} \sim \text{NID}(0, \sigma^2) \quad \varepsilon_{2,t} \sim \text{NID}(0, k\sigma^2) \quad k = \sqrt{3}, 2. \end{gathered} \tag{30}$$

In $\mathbf{b}_3$, AOs have been placed according to the following scheme:

$$I_t^{(\tau_j)} = \begin{cases} 1 & j = 25, 35, 65, 75 \quad T = 100 \\ 1, & j = 125, 135, 165, 175 \quad T = 200 \\ 0, & \text{Otherwise.} \end{cases} \tag{31}$$

The last experiment, that is, **c**, has been designed to mimic a situation characterized by low quality data, induced, for example, by phenomena like changes in survey methodologies (e.g., sampling design or data collecting procedures) or in the imputation techniques. Practically, a Gaussian-type noise ν_t is added to the output signal, so that $x_t = z_t + \nu_t$, z_t being the pure ARMA process. Using (5), we have $x_t = [\phi(B)/\theta(B)]\varepsilon_t + \nu_t$, where $\varepsilon_t \sim \text{nid}(0, \sigma^2)$ and $\nu_t \sim \text{nid}(0, w^2)$ is additive noise, independent of z_t. The variance of ν_t, say w^2, has been chosen according to the relation $w^2 = (1/10)\sigma^2(x_t)$.

4.2. Results. The empirical results pertaining to aspect (i) are summarized in Tables 2 and 3 for the sample sizes $T = 100$ and $T = 200$, respectively. By inspecting these tables it is possible to notice that, with the exception of experiment b_3, in all the other cases bMSE procedure gives no negligible

TABLE 2: Frequency of selection of the true model in the nonbootstrap (nb) and bootstrap (b) world for $T = 100$.

Model	Test	AIC		FPE		BIC		HQC	
		g	g^*	g	g^*	g	g^*	g	g^*
ARMA (1, 1) (SET $\mathcal{J}_1$)	a	42.3	52.4	41.5	50.2	52.4	62.7	47.7	60.2
	b_1	37.9	44.7	36.3	42.5	42.1	50.3	44.2	52.7
	b_2	36.3	42.4	32.7	38.1	38.4	45.4	42.0	49.3
	b_3	23.1	20.3	21.9	18.5	29.8	26.7	25.8	24.5
	c	31.2	38.3	26.1	32.5	39.9	48.8	43.1	52.3
ARMA (2, 1) (SET $\mathcal{J}_2$)	a	49.0	59.6	47.1	55.5	59.2	71.3	52.4	65.5
	b_1	47.5	53.4	41.0	46.6	51.8	59.4	53.5	62.7
	b_2	42.0	47.2	39.2	43.6	46.0	53.5	47.6	55.6
	b_3	28.2	26.3	24.2	21.5	27.6	30.2	31.5	33.4
	c	38.3	44.1	34.3	39.4	43.5	51.8	45.9	57.0
ARMA (1, 2) (SET $\mathcal{J}_3$)	a	47.5	57.9	43.6	53.2	55.4	67.0	48.5	61.3
	b_1	32.4	38.4	29.3	34.6	45.3	52.8	48.1	56.3
	b_2	31.6	36.7	28.1	32.4	34.7	41.3	38.5	45.7
	b_3	25.8	23.3	19.6	16.5	24.6	23.7	29.4	28.3
	c	35.1	41.4	32.2	37.6	46.4	54.2	44.7	54.4

TABLE 3: Frequency of selection of the true model in the nonbootstrap (g^*) and bootstrap (g^*) world for $T = 200$.

Model	Test	AIC		FPE		BIC		HQC	
		g	g^*	g	g^*	g	g^*	g	g^*
ARMA (1, 1) (SET I)	a	49.6	58.7	46.3	54.1	56.7	66.8	53.2	64.4
	b_1	45.5	51.0	39.5	44.7	47.9	54.8	51.9	59.4
	b_2	39.0	43.4	37.1	41.2	45.4	51.2	48.6	54.7
	b_3	29.5	26.4	26.8	23.6	33.4	30.7	30.9	27.4
	c	42.4	48.4	39.5	44.7	48.3	55.6	50.6	58.3
ARMA (2, 1) (SET II)	a	58.7	68.9	53.5	62.3	65.4	77.2	60.4	73.1
	b_1	51.9	57.6	48.1	53.4	55.7	62.8	57.6	66.5
	b_2	49.0	53.3	44.6	48.8	51.5	57.2	53.0	59.8
	b_3	33.8	30.7	22.7	25.3	31.4	33.6	36.3	37.4
	c	50.3	56.4	43.0	48.6	57.2	64.6	58.1	66.3
ARMA (1, 2) (SET III)	a	52.4	62.1	50.5	58.6	60.3	71.3	57.6	69.7
	b_1	48.9	54.2	42.5	47.4	51.1	57.2	54.5	61.7
	b_2	44.2	48.3	35.4	38.6	47.2	52.3	50.8	56.2
	b_3	24.7	22.6	21.0	18.4	26.1	24.2	32.0	29.8
	c	44.4	50.6	36.2	41.3	55.0	61.2	56.3	62.9

improvements over the standard procedure. In particular, it proves to perform particularly well in the pure ARMA case (experiment a) where, for $T = 100$, it brings the frequency of selection of the true model (g^*), averaging over all the three sets of orders, from 45.2% and 52.1% to 54.8% and 64.7% in the case of Shibata efficient and dimension consistent criteria, respectively. Considering the latter, the PAD values recorded are between 11.3 (BIC) and 13.8 (HQC). On the other hand, the bootstrapped version of AIC still shows good improvements (PAD over 10) whereas FPE provides the smaller gains (PAD between 8.4 for $\mathcal{J}_2$ and 9.6 for $\mathcal{J}_3$). As expected, for $T = 200$ both the methods show an increasing average frequency of selection of the correct model for all the SC: averaging over $\mathcal{J}$ and all the SC the values of 55.4% and 65.6% have been recorded for g and g^*, respectively. Regarding the gains over the standard procedure, now BIC and HQC show PAD values above 10 (with a spike of 12.7 of HQC in the case of $\mathcal{J}_2$), whereas the performances for the AIC (PAD above 9) are still good. Less satisfactory job is done by the FPE (PAD = 8.2). Finally, it is worth mentioning that the greatest gains pertain to the HQC, with PAD($\mathcal{J}_1$) = 15.5 for $T = 100$ and PAD($\mathcal{J}_2$) = 12.7 for $T = 200$.

Even though to different extents, both the procedures are affected by the presence of outliers, especially in the case of the smaller sample size. However, as long as IOs (experiments b_1 and b_2) are involved, bMSE seems to do a good job in

TABLE 4: Frequency of selection of the true model in the nonbootstrap (g) and bootstrap (g^*) world, for different sample sizes.

DGP	Γ	SC	$T = 100$		$T = 200$		$T = 500$		$T = 1000$	
			g	g^*	g	g^*	g	g^*	g	g^*
A	$\phi_1 = -.9$, $\phi_2 = -.9$, $\theta = .6$	AIC	59.2	72.0	71.5	80.5	79.1	84.3	81.8	85.5
		FPE	56.2	65.6	68.5	76.1	75.4	80.2	78.5	82.3
		BIC	65.5	79.7	78.7	89.8	87.3	93.8	92.6	95.5
		HQC	63.3	77.2	76.4	85.3	86.6	91.2	91.5	93.4
B	$\phi_1 = -.8$, $\phi_2 = -.8$, $\theta = .5$	AIC	52.5	61.2	65.7	73.6	74.3	80.3	79.6	82.8
		FPE	51.4	59.1	66.0	71.5	72.9	78.7	75.5	80.4
		BIC	57.3	67.5	72.4	79.7	80.7	84.6	85.2	87.1
		HQC	50.8	63.5	67.4	74.2	76.3	79.4	81.5	83.7
C	$\phi_1 = -.6$, $\phi_2 = -.5$, $\theta = .5$	AIC	45.3	55.7	60.5	69.3	70.2	75.6	73.7	77.2
		FPE	42.5	54.1	57.0	64.9	64.4	70.1	67.6	71.7
		BIC	53.1	65.8	67.3	77.6	76.5	83.1	82.7	84.9
		HQC	50.7	65.4	67.5	75.6	76.0	80.6	81.3	83.1
D	$\phi_1 = -.8$, $\phi_2 = -.6$, $\theta = .4$	AIC	47.8	58.1	61.2	68.4	68.3	72.6	69.7	73.7
		FPE	44.2	53.8	56.5	62.7	64.7	68.6	66.6	70.8
		BIC	51.9	64.2	65.3	71.8	72.5	76.6	74.7	77.5
		HQC	49.7	62.3	66.4	75.0	73.3	79.4	78.0	80.6

counteracting their adverse effects. In fact, for $T = 200$, this procedure, applied to dimension consistent criteria, selects the right model always more than 50% (experiment b_2) and approximately 55% of the times in experiment b_1. For this type of criteria, the average gain over the standard procedure is noticeable, especially in the case of experiment b_1 (OPAD = 6.7 for BIC and 7.9 for HQC). On the other hand, Shibata efficient criteria achieve less remarkable results: with PAD values ranging from 4.9 for the FPE (PAD($\mathcal{J}_3$)) to 5.7 for the AIC (PAD($\mathcal{J}_2$)). As expected, for $T = 100$ the impact of the IOs is stronger: applied to Shibata consistent criteria, bMSE procedure selects the right model in average approximately 43.4% of the time with a minimum of 34.6% recorded for FPE in the case of $\mathcal{J}_2$, whereas dimension consistent criteria show a g^* value in average equal to 55.7%. Selecting performances granted by the proposed method, even though still acceptable, tended to deteriorate to a greater extent considering the experiment b_2, especially with $T = 100$: here the frequency of selection of the true model for Shibata efficient SC is around 40.1% versus 35% of the standard procedure, for a recorded OPAD amounting to 5.5 for the AIC and 4.7 for the FPE. Slightly better results for $T = 200$ are recorded, where the correct model has been identified by dimension consistent criteria 55.2% (OPAD = 5.8%) of the times versus 49.4% of the standard procedure.

Experiment b_3 is where the proposed procedure crashes and offers little or no improvements over the standard one. The most seriously affected selector is the FPE, which shows an ability to select the correct model in average only 18.9% and 22.4% of the times, versus 21% and 25.2% recorded for the nonbootstrap counterpart, respectively, for $T = 100$ and 200. Finally the effect of the injection of a Gaussian noise to the output signal (experiment c) is commented on. Here, the performances of the method appear to be adequate: averaging over $\mathcal{J}$, the value recorded for g^* is 61.8% ($g = 54.6$) for dimension consistent criteria ($T = 200$) with particularly interesting improvements over the standard procedure yielded by HQC, which shows OPAD values amounting to 10% and 7.5% for $T = 100$ and 200, respectively. The bootstrapped version of HQC performs consistently better than the other criteria: in fact it chooses the correct model in average 63.2% and 56.6% of the times for $T = 100$ and $T = 200$, respectively. On the other hand, FPE detects the true model with the smallest probability by reaching the average frequency of selection of the true model of 39.8 ($T = 100$) and 45.1 ($T = 200$). Shibata consistent criteria show also the smallest gains over the standard procedure; for example, for $T = 100$ the maximum PAD is equal to 7.1 and 6.4 for AIC and FPE, respectively (both values' recorder for $\mathcal{J}_2$), whereas dimension consistent criteria, for the same sample sizes, show a maximum PAD of 8.9 and 11.1, in the case of BIC ($\mathcal{J}_1$) and HQC ($\mathcal{J}_2$), respectively.

In the analysis of aspect (ii), the performances yielded by the two procedures, in terms of frequency of selection of the correct model, are considered for increasing sample sizes ($T = 100, 200, 500, 1000$). The results for four different ARMA (2, 1) models, along with their details, are presented in Table 4. As possibly seen by inspecting this table, all the SC under test exhibit roughly a similar pattern: for the small sample size, remarkable disclosures in selecting performances between the two methods are noticeable whereas such discrepancies become less pronounced for $T = 500$ and very small for $T = 1000$. For example, considering all the 4 DGPs, BIC shows a PAD ranging from 12.6 (series D) to 14.7 (series C) with sample size $T = 100$, whereas for $T = 1000$, PAD is in the range 1.9–2.9 for the series B and A, respectively. For this sample size, the smallest PAD has been recorded

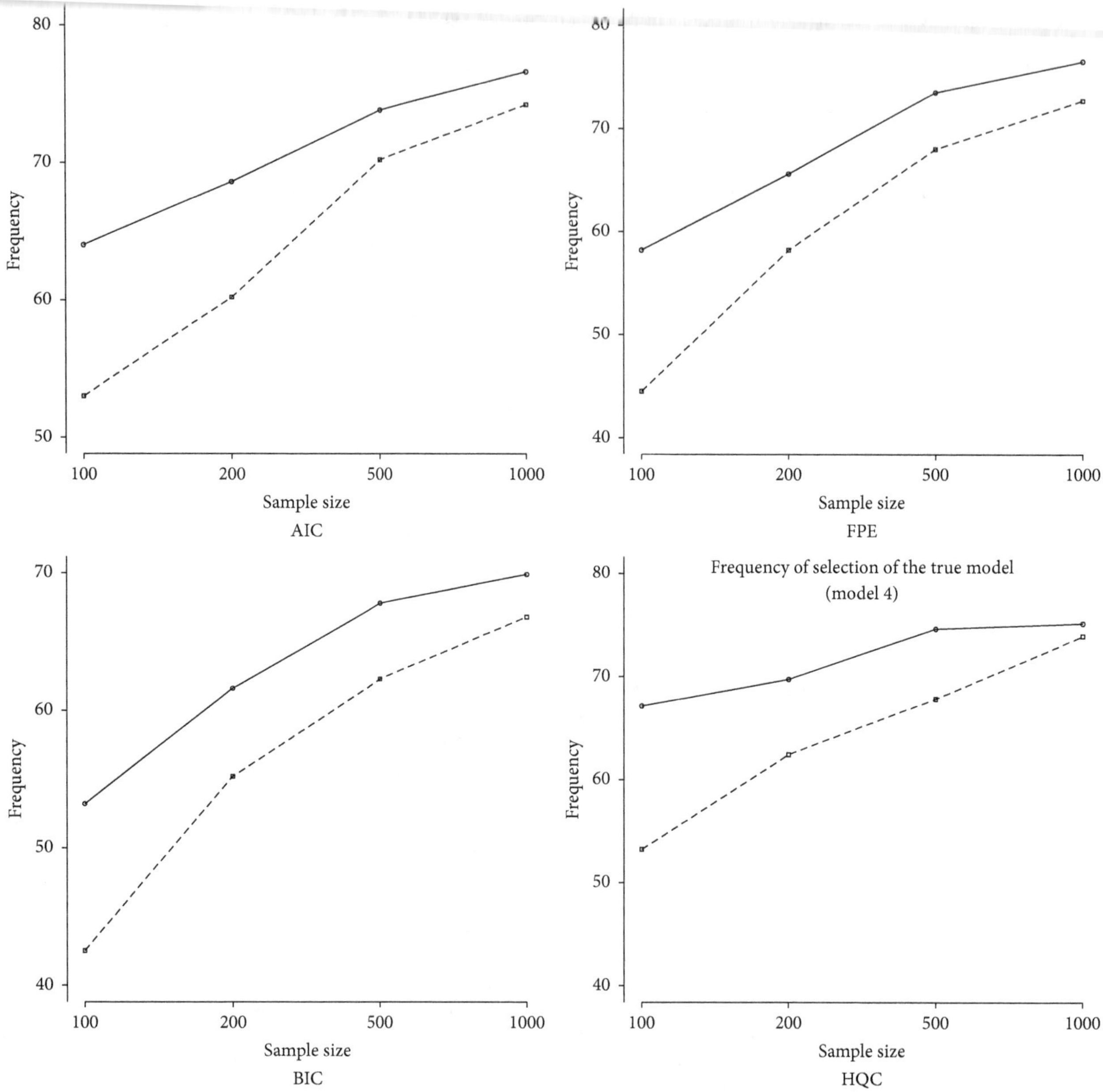

FIGURE 1: Frequency of selection of the true model, DGP C, Table 4: SC (continuous line) and SC* (dashed line).

for the HQC, whose values range from 1.8 (series C) to 2.6 (series D). Graphical evidence of such a behavior is finally given in Figure 1, where the results recorded for series C are considered.

Among the many factors influencing the performances of the method, the sample size is clearly one of the most crucial ones. Unreported simulations conducted with a sample size $T = 60$ confirm a still acceptable selection rate of the right model, provided that the time series is either noise- and outlier-free and generated by a robust parametric DGP structure. Under such conditions—met, for instance, in the case of DGPs 1-2 (set $\mathcal{J}_2$)—no substantial reductions in the selection performances have been noticed between the smallest sample size considered, that is, $T = 60$ and $T = 100$. For instance, for $T = 60$ and averaging between DGPs 1 and 2 (set $\mathcal{J}_2$), the bootstrap procedure applied to the dimension consistent criteria identifies the right model approx 64% and 61% of the times, versus 49.6% and 42.3% achieved by their nonbootstrap counterpart, for the BIC and HQC, receptively. However, with time series subjected to disturbances and characterized by weaker ARMA structures, as in the case of DGP 6 (set $\mathcal{J}_2$), the proposed method tends to perform poorly and to select the correct model a number of times not far from the standard MAICE procedure.

It is worth pointing out that, unlike artificial setups, in real life data set the true model order is generally unknown; therefore the optimality of the bootstrap method can be inferred on empirical basis, that is, by using the relative frequency of selection of the different tentative ARMA models. In practice, the B winning models generated at each and every bootstrap replication are ranked according to their relative frequency of selection of the true model. In this way, our confidence in the bootstrap selection procedure is linked to the difference in the relative frequency of selection of the winner model (with the highest selection rate) compared to the ones achieved by its closest competitors. Ideal situations are characterized by high rate of choices of the winner model, which drops sharply considering the rest of the competition set. In such a case, we can reasonably be sure that the selected model is closer to the true order than the one found by using the standard MAICE procedure (clearly if different models are selected). On the other hand, slight discrepancies (say 3-4%) between the winning model and the others should be regarded with suspicion and carefully evaluated on a case-by-case basis.

5. Final Remarks and Future Directions

In this paper two pairs of selectors, differing for their derivation and properties, have been brought in a bootstrap framework with the purpose of enhancing their selecting capabilities. A set of Monte Carlo-type experiments has been employed in order to assess the magnitude of the improvements achieved. These encouraging results obtained can be explained in terms of the reduction of uncertainty induced by the bootstrap approach. Identification procedures of the type MSCE, in fact, base the choice of the final model on the minimum value attained by a given SC, no matter how small the differences in the values showed by other competing models might be. When they are actually small, standard MSCE procedures are likely to introduce significant amount of uncertainty in the selection procedure; that is, different order choices can be determined by small variations in the data set. The proposed procedure accounts for such a source of uncertainty, by reestimating the competing models and recomputing the related SC value B times (one for each bootstrap replication). In doing so the identification procedure is based on B different data replications each of them embodying random variations. Also the improvements achieved by the proposed method in the case of IOs can also be explained in the light of reduction of uncertainty. Basically, what the procedure does is to reallocate these outliers B times, so that the related selection procedure can control for such anomalous observations. On the other hand, bMSE procedure breaks down in the case of AOs, probably because of the fact that the employed maximum likelihood estimation procedure is carried out on the residuals, which are severely affected by these types of outliers. Consistently with other Monte Carlo experiments, in the proposed simulations the best results are achieved by dimension consistent criteria, especially by BIC. However, two drawbacks affect this criterion: tendency in the selection of underfitted models and consistency achieved only in case of very large sample [4], under the condition that the true model *is included* in the competition set. The last assumption implies the existence of a model able to provide full explanation of the reality and the "existence" of an analyst able to include it in the competition set. Unfortunately, even assuming finite dimensionality of real life problems, reality is still very complex so that a large number of models are likely to be included in the competition set. As a result of that, selection uncertainty will rise. Superiority of BIC should also be reconsidered in the light of different empirical framework, as Monte Carlo experiment cannot capture the aforementioned problems. It is in fact characterized by the presence of the true model in the portfolio of candidate models. This appears unfair if we consider that criteria of the types AIC and FPE are designed to relax such a strong, in practice unverifiable, assumption and that they enjoy the nice Shibata efficiency property. In addition, in order to keep the computational time acceptable, in Monte Carlo experiments the true DGP is generally of low order, so that BIC underestimation tendency is likely to be masked or, at least, to appear less serious. For these reasons, from a more operational point of view, it can be advisable to consider the indications provided by both AIC^* and BIC^*, which are the best selectors in their respective categories, according to the simulation experiment. This is particularly true when the sample size is "small" and the information criteria, either considered in their standard or bootstrap form, tend to yield values close to each other for closer models. As a result of that, significant amount of uncertainty can be introduced in the selection process. Finally, as a future direction, it might be worth emphasizing that the purpose of a given model is built and thus identified, which can be usefully considered to assess the selector's performances. For instance, in many cases computational time is a critical factor, so that one might be willing to accept less accurate model outcomes by reducing the number of bootstrap replications. In fact, global fitting is not necessarily the only interesting feature one wants to look at, as a model might be also evaluated on the basis of the *potential* ability in solving the specific problems it has been built for. In this regard, selection procedures optimized on a case-by-case basis and implemented in the bootstrap world might result in a more efficient tool for a better understanding of the reality.

Conflicts of Interest

The author declares that he has no conflicts of interest.

References

[1] G. E. Box and G. M. Jenkins, *Times Series Analysis. Forecasting and Control*, Holden-Day, San Francisco, Calif, USA, 1970.

[2] M. R. Forster, "Key concepts in model selection: performance and generalizability," *Journal of Mathematical Psychology*, vol. 44, no. 1, pp. 205–231, 2000.

[3] E. E. Leamer, *Specification Searches: Ad Hoc Inference with Experimental Data*, John Wiley & Sons, New York, NY, USA, 1978.

[4] K. P. Burnham and D. Anderson, *Model Selection and Multimodal Inference: A Practical Information-Theoretic Approach*, Springer, New York, NY, USA, 2nd edition, 2002.

[5] H.-Y. Chung, K.-W. Lee, and J.-Y. Koo, "A note on bootstrap model selection criterion," *Statistics & Probability Letters*, vol. 26, no. 1, pp. 35–41, 1996.

[6] L. Fenga and D. N. Politis, "Bootstrap-based ARMA order selection," *Journal of Statistical Computation and Simulation*, vol. 81, no. 7, pp. 799–814, 2011.

[7] D. A. Freedman, "Bootstrapping regression models," *The Annals of Statistics*, vol. 9, no. 6, pp. 1218–1228, 1981.

[8] J. S. Raho and R. Tibshirani, "Bootstrap model selection via the cost complexity parameter in regression," Tech. Rep., University of Toronto, 1993.

[9] J. Shao, "Bootstrap model selection," *Journal of the American Statistical Association*, vol. 91, no. 434, pp. 655–665, 1996.

[10] H. Akaike, "A new look at the statistical model identification," *IEEE Transactions on Automatic Control*, vol. 19, pp. 716–723, 1974.

[11] H. Akaike, "Statistical predictor identification," *Annals of the Institute of Statistical Mathematics*, vol. 22, pp. 203–217, 1970.

[12] H. Akaike, "Fitting autoregressive models for prediction," *Annals of the Institute of Statistical Mathematics*, vol. 21, pp. 243–247, 1969.

[13] G. Schwarz, "Estimating the dimension of a model," *The Annals of Statistics*, vol. 6, no. 2, pp. 461–464, 1978.

[14] R. E. Kass and A. E. Raftery, "Bayes factors," *Journal of the American Statistical Association*, vol. 90, no. 430, pp. 773–795, 1995.

[15] E. J. Hannan, "The estimation of the order of an ARMA process," *The Annals of Statistics*, vol. 8, no. 5, pp. 1071–1081, 1980.

[16] E. J. Hannan and B. G. Quinn, "The determination of the order of an autoregression," *Journal of the Royal Statistical Society, Series B*, vol. 41, pp. 190–195, 1979.

[17] K. Torre, D. Delignie, and L. c. Lemoine, "Detection of long-range dependence and estimation of fractal exponents through ARFIMA modelling," *The British Journal of Mathematical and Statistical Psychology*, vol. 60, no. 1, pp. 85–106, 2007.

[18] A. D. McQuarrie and C.-L. Tsai, "Outlier detections in autoregressive models," *Journal of Computational and Graphical Statistics*, vol. 12, no. 2, pp. 450–471, 2003.

[19] A. J. Fox, "Outliers in time series," *Journal of the Royal Statistical Society. Series B. Methodological*, vol. 34, pp. 350–363, 1972.

[20] P. Zhang, "Inference after variable selection in linear regression models," *Biometrika*, vol. 79, no. 4, pp. 741–746, 1992.

[21] L. Breiman, "The little bootstrap and other methods for dimensionality selection in regression: X-fixed prediction error," *Journal of the American Statistical Association*, vol. 87, no. 419, pp. 738–754, 1992.

[22] C. Chatfield, "Model uncertainty, data mining and statistical inference," *Journal of the Royal Statistical Society, Series A (Statistics in Society)*, vol. 158, no. 3, pp. 419–466, 1995.

[23] J. S. Hjorth, *Computer Intensive Statistical Methods—Validation Model Selection and Bootstrap*, Chapman & Hall, London, UK, 1994.

[24] T. Soderstrom and P. Stoica, *System Identification*, Prentice-Hall, Englewood Cliffs, NJ, USA, 1989.

[25] S. Kullback and R. A. Leibler, "On information and sufficiency," *Annals of Mathematical Statistics*, vol. 22, no. 1, pp. 79–86, 1951.

[26] R. Shibata, "Asymptotically efficient selection of the order of the model for estimating parameters of a linear process," *The Annals of Statistics*, vol. 8, no. 1, pp. 147–164, 1980.

[27] J. Shao, "An asymptotic theory for linear model selection," *Statistica Sinica*, vol. 7, no. 2, pp. 221–264, 1997.

[28] Y. Yang, "Can the strengths of AIC and BIC be shared? A conflict between model indentification and regression estimation," *Biometrika*, vol. 92, no. 4, pp. 937–950, 2005.

[29] H. Bozdogan, "Model selection and Akaike's information criterion (AIC): the general theory and its analytical extensions," *Psychometrika*, vol. 52, no. 3, pp. 345–370, 1987.

[30] C. L. Mallows, "Some comments on Cp," *Technometrics*, vol. 15, no. 4, pp. 661–675, 1973.

[31] P. Bühlmann, "Bootstraps for time series," *Statistical Science*, vol. 17, no. 1, pp. 52–72, 2002.

[32] P. Bühlmann, "Sieve bootstrap for time series," *Bernoulli*, vol. 3, no. 2, pp. 123–148, 1997.

[33] J. P. Kreiss, "Bootstrap procedures for AR (∞)—processes," in *Bootstrapping and Related Techniques*, K. H. Jockel, G. Rothe, and W. Sendler, Eds., vol. 376 of *Lecture Notes in Economics and Mathematical Systems*, pp. 107–113, Springer, Berlin, Germany, 1992.

[34] G. M. Ljung and G. E. P. Box, "On a measure of lack of fit in time series models," *Biometrika*, vol. 65, no. 2, pp. 297–303, 1978.

[35] P. M. T. Broersen, "Automatic spectral analysis with time series models," *IEEE Transactions on Instrumentation and Measurement*, vol. 51, no. 2, pp. 211–216, 2002.

[36] I. Chang, G. C. Tiao, and C. Chen, "Estimation of time series parameters in the presence of outliers," *Technometrics*, vol. 30, no. 2, pp. 193–204, 1988.

[37] C. Chen and L. Liu, "Joint estimation of model parameters and outlier effects in time series," *Journal of the American Statistical Association*, vol. 88, no. 421, pp. 284–297, 1993.

[38] D. Peña, "Influential observations in time series," *Journal of Business and Economic Statistics*, vol. 8, no. 2, pp. 235–241, 1990.

[39] A. G. Bruce and D. Martin, "Leave-k-out diagnostics for time series (with discussion)," *Journal of the Royal Statistical Society, Series B*, vol. 51, no. 3, pp. 363–424, 1989.

[40] V. c. Gomez and A. n. Maravall, "Estimation, prediction, and interpolation for nonstationary series with the Kalman filter," *Journal of the American Statistical Association*, vol. 89, no. 426, pp. 611–624, 1994.

[41] L. M. Liu, G. Hudak, G. E. P. Box, M. E. Muller, and G. C. Tiao, *The SCA Statistical System: Reference Manual for Forecasting and Time Series Analysis*, Scientific Computing Associates, DeKalb, Ill, USA, 1986.

[42] M. A. Carnero, D. Peña, and E. Ruiz, "Outliers and conditional autoregressive heteroscedasticity in time series," *Estadística*, vol. 53, pp. 143–213, 2001.

2

Estimation of the Parameters of a Chirp Type Model with Stationary Residuals

K. Perera

Department of Engineering Mathematics, Faculty of Engineering, University of Peradeniya, Peradeniya, Sri Lanka

Correspondence should be addressed to K. Perera; kanthip@pdn.ac.lk

Academic Editor: Aera Thavaneswaran

Let $X_{n1}, \dots, X_{nn}$ be the observations from a chirp type statistical model $\{X_{nt}\}$, $X_{nt} = A\cos(\omega t + (\Delta/n)t^2) + B\sin(\omega t + (\Delta/n)t^2) + \epsilon_t$, where $\{\epsilon_t\}$ is a stationary noise. We consider a method of estimation of parameters, A, B, ω, Δ, and ν, (where ν is the variance of ϵ_t's) which is basically an approximate least-squares method. The main advantage of the proposed approach is that no assumptions are required. We make use of the three theorems which were established associated with the kernel $\sum_{t=1}^{n} e^{i(ut+vt^2)}$ and then use them to prove, under certain conditions, the consistency of the estimators.

1. Introduction

In [1] 1973 Walker considered the problem of estimating the parameters of a sine wave,

$$X_t = A\cos\omega t + B\sin\omega t + \epsilon_t, \tag{1}$$

where X_t's are the observations and ϵ_t's are independent, identically distributed random variables with mean zero and finite but unknown variance ν. The parameters A, B, ω, and ν are assumed unknown and are to be estimated. He showed that, as $n \to \infty$, the estimators $\widehat{A}_n$, $\widehat{B}_n$, $\widehat{\omega}_n$, and $\widehat{\nu}_n$ converge in probability to the actual values A_0, B_0, ω_0, and ν_0, respectively. That is, he showed that the estimators are consistent. He then showed that the differences between the estimators and the actual values they estimate have a joint normal distribution.

Suppose now that the frequency of the above sine wave changes with time. If ω is the initial frequency, the frequency at time t may be written as

$$\omega_t = \omega + \alpha_t. \tag{2}$$

The simplest case is one in which α_t is linear; that is, $\alpha_t = \alpha t$. Then $\omega_t t = \omega t + \alpha t^2$. This leads to the model

$$X_t = A\cos\left(\omega t + \alpha t^2\right) + B\sin\left(\omega t + \alpha t^2\right) + \epsilon_t. \tag{3}$$

The parameters for this model are A, B, ω, ν, and α. This is sometimes called the "chirp" model (see [2, 3]). Several authors have considered the parameter estimation of the chirp model and are found in [4–7]. Different approaches to the estimation of chirp parameters in similar kinds of models are found in [8–12].

But our approach is entirely different from them. We do not make any assumptions. The method used to estimate the parameters and to prove the consistency of the estimators is similar to that by Walker in [13]. We consider not only estimates of the parameters of A, B, ω, and ν but also an estimate $\widehat{\alpha}_n$ of α. Although this leads to an interesting problem in estimation, the model is somewhat unrealistic. Over the course of n observations, the frequency changes from ω to $\omega + \alpha n$. As n goes to ∞, the sine wave oscillates faster and faster (unless $\alpha = 0$). Its frequency becomes infinite, and its period approaches 0. We therefore change the model by assuming that the change in frequency over the course of the n observations is a number Δ independent of n. We also assume, as in (3), that the change in frequency is linear. This leads to the model or, more precisely, sequence of models

$$X_{nt} = A\cos\left(\omega t + \frac{\Delta}{n}t^2\right) + B\sin\left(\omega t + \frac{\Delta}{n}t^2\right) + \epsilon_t, \quad 1 \le t \le n. \tag{4}$$

We assume, as Walker does in his 1971 paper, that ϵ_t's are independent and identically distributed with mean zero and variance ν. Thus the parameters are A, B, ω, ν, and Δ and the estimators of the parameters are $\widehat{A}$, $\widehat{B}$, $\widehat{\omega}$, $\widehat{\nu}$, and $\widehat{\Delta}$. Our objective is to show that these estimators are consistent.

We make use of the following three theorems which we have established.

Theorem 1 (see [14]). *Let $S_n(\omega,\alpha) = \sum_{t=1}^{n} t^{\beta} e^{i(\omega t+\alpha t^2)}$, where β is a nonnegative real number. Then $S_n(\omega,\alpha) = o(n^{\beta+1})$, for α not a rational multiple of π, uniformly in ω.*

Theorem 2 (see [15]). *Let (ϵ_t) be a sequence of independent random variables such that $E(\epsilon_t) = 0$, $E(\epsilon_t^2) = \nu < \infty$, and $E(\epsilon_t^4) = \sigma < \infty$. Then*

$$\max_{\substack{0\le\omega<\pi \\ 0\le\alpha<\pi}} \left|\sum_{t=1}^{n} t^{\beta}\epsilon_t e^{i(\omega t+\alpha t^2)}\right|^2 = O_p\left(n^{\beta+7/8}\right). \quad (5)$$

Theorem 3 (see [16]). *For any sufficiently small $\delta_1 > 0$ and $\delta_2 > 0$*

$$\overline{\lim} \max_{\substack{\{n^{-1}\delta_1\le|u|\le\pi,\ |v|\le\pi-n^{-2}\delta_2\} \\ or \\ \{|u|\le\pi-n^{-1}\delta_1,\ n^{-2}\delta_2\le|v|\le\pi\}}} \frac{1}{n}\left|\sum_{t=1}^{n} e^{i(ut+vt^2)}dx\right| < 1. \quad (6)$$

2. Estimation of the Parameters

If ϵ_t's are normally distributed, the likelihood function of the observations $X_{n1},\ldots,X_{nn}$ is

$$f(x_{n1},\ldots,x_{nn}) = \frac{1}{\left(\sqrt{2\pi\nu}\right)^n} \cdot e^{-(1/2\pi)\sum_{t=1}^{n}[x_{nt}-A\cos(\omega t+(\Delta/n)t^2)-B\sin(\omega t+(\Delta/n)t^2)]^2}. \quad (7)$$

Then the log likelihood function of the observations $X_{n1}, \ldots, X_{nn}$ is

$$L_n(A,B,\omega,\Delta,\nu) = -\frac{1}{2}n\log(2\pi\nu) - \frac{(1/2)\,S_n(A,B,\omega,\Delta)}{\nu}, \quad (8)$$

where

$$S_n(A,B,\omega,\Delta) = \sum_{t=1}^{n}\left[x_{nt} - A\cos\left(\omega t+\frac{\Delta}{n}t^2\right) - B\sin\left(\omega t+\frac{\Delta}{n}t^2\right)\right]^2. \quad (9)$$

Now consider $S_n(A,B,\omega,\Delta)$:

$$\begin{aligned} S_n(A,B,\omega,\Delta) &= \sum_{t=1}^{n}\left[X_{nt} - A\cos\left(\omega t+\frac{\Delta}{n}t^2\right) - B\sin\left(\omega t+\frac{\Delta}{n}t^2\right)\right]^2 = \sum_{t=1}^{n}X_{nt}^2 \\ &- 2\sum_{t=1}^{n}X_{nt}\left[A\cos\left(\omega t+\frac{\Delta}{n}t^2\right) + B\sin\left(\omega t+\frac{\Delta}{n}t^2\right)\right] + \sum_{t=1}^{n}\left[A\cos\left(\omega t+\frac{\Delta}{n}t^2\right) + B\sin\left(\omega t+\frac{\Delta}{n}t^2\right)\right]^2 = \sum_{t=1}^{n}X_{nt}^2 \\ &- 2\sum_{t=1}^{n}X_{nt}\left[A\cos\left(\omega t+\frac{\Delta}{n}t^2\right) + B\sin\left(\omega t+\frac{\Delta}{n}t^2\right)\right] + \sum_{t=1}^{n}A^2\cos^2\left(\omega t+\frac{\Delta}{n}t^2\right) \\ &+ B^2\sin^2\left(\omega t+\frac{\Delta}{n}t^2\right) + \sum_{t=1}^{n}2AB\cos\left(\omega t+\frac{\Delta}{n}t^2\right) \cdot \sin\left(\omega t+\frac{\Delta}{n}t^2\right). \end{aligned} \quad (10)$$

By using the identities $\cos 2\theta = 2\cos^2\theta - 1$, $\cos 2\theta = 1 - 2\sin^2\theta$, and $\sin 2\theta = 2\sin\theta\cos\theta$ we obtain

$$\begin{aligned} S_n(A,B,\omega,\Delta) &= \sum_{t=1}^{n}X_{nt}^2 - 2\sum_{t=1}^{n}X_{nt}\left[A\cos\left(\omega t+\frac{\Delta}{n}t^2\right) + B\sin\left(\omega t+\frac{\Delta}{n}t^2\right)\right] \\ &+ \sum_{t=1}^{n}A^2\frac{\left[\cos 2\left(\omega t+(\Delta/n)t^2\right)+1\right]}{2} + \sum_{t=1}^{n}B^2\frac{\left[1-\cos 2\left(\omega t+(\Delta/n)t^2\right)\right]}{2} + AB\sin 2\left(\omega t+\frac{\Delta}{n}t^2\right) \\ &= \sum_{t=1}^{n}X_{nt}^2 - 2\sum_{t=1}^{n}X_{nt}\left[A\cos\left(\omega t+\frac{\Delta}{n}t^2\right) + B\sin\left(\omega t+\frac{\Delta}{n}t^2\right)\right] \end{aligned}$$

$$+\frac{n}{2}\left(A^2+B^2\right)+\frac{1}{2}\left(A^2-B^2\right)\cos 2\left(\omega t+\frac{\Delta}{n}t^2\right)$$
$$+AB\sum_{t=1}^{n}\sin 2\left(\omega t+\frac{\Delta}{n}t^2\right). \tag{11}$$

From Theorem 1, the last two terms in (11) are of order $o(n)$. Therefore if we let

$$U_n(A,B,\omega,\Delta)=\sum_{t=1}^{n}X_{nt}^2 - 2\sum_{t=1}^{n}X_{nt}\left[A\cos\left(\omega t+\frac{\Delta}{n}t^2\right)+B\sin\left(\omega t+\frac{\Delta}{n}t^2\right)\right] + \frac{n}{2}\left(A^2+B^2\right), \tag{12}$$

then U_n is an approximation of S_n. More precisely, $U_n - S_n = o(n)$. Thus

$$L_n^*(A,B,\omega,\Delta,\nu)=-\frac{1}{2}n\log(2\pi\nu) - \frac{(1/2)U_n(A,B,\omega,\Delta)}{\nu} \tag{13}$$

is an approximation of the log likelihood function $L_n(A, B, \omega, \Delta, \nu)$. Now fix ν in (13). Then we maximize $L_n^*(A, B, \omega, \Delta, \nu)$ to obtain the estimates for A, B, ω, and Δ. That is, we minimize $U_n(A, B, \omega, \Delta)$ over the region $R \times R \times [0, \pi/2] \times [-\pi/4, \pi/2]$. Now fix ω and Δ. If $|A| \to \infty$ or $|B| \to \infty$ then $U_n(A, B, \omega, \Delta) \to \infty$. Since $U_n(A, B, \omega, \Delta)$ is continuous in (A, B), the minimum is achieved at a point $(\widehat{A}_n(\omega, \Delta), \widehat{B}_n(\omega, \Delta))$. The partial derivatives $\partial U_n/\partial A$ and $\partial U_n/\partial B$ must vanish at $(\widehat{A}_n(\omega, \Delta), \widehat{B}_n(\omega, \Delta))$. Thus the estimators of A and B are solutions of the equations

$$\begin{aligned}\frac{\partial U_n}{\partial A}&=-2\sum_{t=1}^{n}X_{nt}\cos\left(\omega t+\frac{\Delta}{n}t^2\right)+n\widehat{A}=0,\\ \frac{\partial U_n}{\partial B}&=-2\sum_{t=1}^{n}X_{nt}\sin\left(\omega t+\frac{\Delta}{n}t^2\right)+n\widehat{B}=0.\end{aligned} \tag{14}$$

The solution to these equations is given by

$$\widehat{A}_n(\omega,\Delta)=\frac{2}{n}\sum_{t=1}^{n}X_{nt}\cos\left(\omega t+\frac{\Delta}{n}t^2\right), \tag{15}$$

$$\widehat{B}_n(\omega,\Delta)=\frac{2}{n}\sum_{t=1}^{n}X_{nt}\sin\left(\omega t+\frac{\Delta}{n}t^2\right). \tag{16}$$

To obtain estimates for ω and Δ we substitute $\widehat{A}_n(\omega, \Delta)$ and $\widehat{B}_n(\omega, \Delta)$ for A and B, respectively, in (12) and minimize $U_n(\widehat{A}_n(\omega, \Delta), \widehat{B}_n(\omega, \Delta), \omega, \Delta)$ as a function of ω and Δ. Since this last expression is equal to

$$\begin{aligned}&\sum_{t=1}^{n}X_{nt}^2-2\sum_{t=1}^{n}X_{nt}\left[\frac{2}{n}\cdot\sum_{t=1}^{n}X_{nt}\cos\left(\omega t+\frac{\Delta}{n}t^2\right)\cos\left(\omega t+\frac{\Delta}{n}t^2\right)+\frac{n}{2}\cdot\sum_{t=1}^{n}X_{nt}\sin\left(\omega t+\frac{\Delta}{n}t^2\right)\sin\left(\omega t+\frac{\Delta}{n}t^2\right)\right]\\ &+\frac{n}{2}\left[\left(\frac{2}{n}\sum_{t=1}^{n}X_{nt}\cos\left(\omega t+\frac{\Delta}{n}t^2\right)\right)^2+\left(\frac{2}{n}\cdot\sum_{t=1}^{n}X_{nt}\sin\left(\omega t+\frac{\Delta}{n}t^2\right)\right)^2\right]=\sum_{t=1}^{n}X_{nt}^2-2\\ &\cdot\frac{2}{n}\left|\sum_{t=1}^{n}X_{nt}e^{i(\omega t+(\Delta/n)t^2)}\right|^2+\frac{2}{n}\left|\sum_{t=1}^{n}X_{nt}e^{i(\omega t+(\Delta/n)t^2)}\right|^2\\ &=\sum_{t=1}^{n}X_{nt}^2-\frac{2}{n}\left|\sum_{t=1}^{n}X_{nt}e^{i(\omega t+(\Delta/n)t^2)}\right|^2,\end{aligned} \tag{17}$$

with respect to ω and Δ; this is same as maximizing

$$I_n\left(\omega,\frac{\Delta}{n}\right)=\frac{2}{n}\left|\sum_{t=1}^{n}X_{nt}e^{i(\omega t+(\Delta/n)t^2)}\right|^2. \tag{18}$$

Since (ω, Δ) varies over a compact set, there is a point $(\widehat{\omega}_n, \widehat{\Delta}_n)$ which maximizes this expression. We then estimate A and B by

$$\begin{aligned}\widehat{A}_n&=\widehat{A}_n\left(\widehat{\omega}_n,\widehat{\Delta}_n\right),\\ \widehat{B}_n&=\widehat{B}_n\left(\widehat{\omega}_n,\widehat{\Delta}_n\right).\end{aligned} \tag{19}$$

The minimum value of $U_n(A, B, \omega, \Delta)$ is assumed at $(\widehat{A}_n, \widehat{B}_n, \widehat{\omega}_n, \widehat{\Delta}_n)$.

Now we obtain an estimator $\widehat{\nu}_n$ of ν. If ν is zero then, since $E(\epsilon_t) = 0$ and $E(\epsilon_t^2) = 0$, $\epsilon_t = 0$ almost surely. Thus there is no randomness in the model. We assume, then, that $0 < \nu < \infty$. To obtain $\widehat{\nu}_n$ we substitute $\widehat{A}_n$, $\widehat{B}_n$, $\widehat{\omega}_n$, and $\widehat{\Delta}_n$ in (13) and maximize $L_n^*(\widehat{A}_n, \widehat{B}_n, \widehat{\omega}_n, \widehat{\Delta}_n, \nu)$. From (13),

$$L_n^*\left(\widehat{A}_n,\widehat{B}_n,\widehat{\omega}_n,\widehat{\Delta}_n,\nu\right)=-\frac{n}{2}\left[\log 2\pi+\log\nu\right]-\frac{(1/2)U_n\left(\widehat{A}_n,\widehat{B}_n,\widehat{\omega}_n,\widehat{\Delta}_n\right)}{\nu}. \tag{20}$$

We will show that $L_n^*(\widehat{A}_n, \widehat{B}_n, \widehat{\omega}_n, \widehat{\Delta}_n)$, as a function of ν, achieves its positive maximum provided $U_n(\widehat{A}_n, \widehat{B}_n, \widehat{\omega}_n, \widehat{\Delta}_n) >$

0, which we show is positive with high probability for large values of n. From (4) and (9) we have

$$S_n\left(\widehat{A}_n,\widehat{B}_n,\widehat{\omega}_n,\widehat{\Delta}_n\right)=\sum_{t=1}^{n}\epsilon_t^2. \tag{21}$$

Using (11) and (12) we have

$$\begin{aligned}S_n&\left(\widehat{A}_n,\widehat{B}_n,\widehat{\omega}_n,\widehat{\Delta}_n\right)\\&=U_n\left(\widehat{A}_n,\widehat{B}_n,\widehat{\omega}_n,\widehat{\Delta}_n\right)\\&\quad+\frac{1}{2}\left(\widehat{A}_n^2-\widehat{B}_n^2\right)\sum_{t=1}^{n}\cos 2\left(\widehat{\omega}_nt+\frac{\widehat{\Delta}_n}{n}t^2\right)\\&\quad+\widehat{A}_n\widehat{B}_n\sum_{t=1}^{n}\sin 2\left(\widehat{\omega}_nt+\frac{\widehat{\Delta}_n}{n}t^2\right).\end{aligned} \tag{22}$$

Using the mean value theorem,

$$\begin{aligned}\frac{1}{n}&\sum_{t=1}^{n}\cos 2\left(\widehat{\omega}_nt+\frac{\widehat{\Delta}_n}{n}t^2\right)-\frac{1}{n}\sum_{t=1}^{n}\cos 2\left(\omega_0t+\frac{\Delta_0}{n}t^2\right)\\&=-\frac{1}{n}\sum_{t=1}^{n}2t\sin 2\left(\widetilde{\omega}_nt+\frac{\widetilde{\Delta}_n}{n}t^2\right)2\left(\widehat{\omega}_n-\omega_0\right)\\&\quad+\frac{1}{n}\sum_{t=1}^{n}4\frac{t}{n}\sin 2\left(\widetilde{\omega}_nt+\frac{\widetilde{\Delta}_n}{n}t^2\right)2\frac{t^2}{n}\left(\widehat{\Delta}_n-\Delta_0\right),\end{aligned} \tag{23}$$

where $(\widetilde{\omega},\widetilde{\Delta}/n)$ is a point on the line segment connecting $(\omega_0,\Delta_0/n)$ and $(\widehat{\omega}_n,\widehat{\Delta}_n/n)$.

Thus

$$\begin{aligned}&\left|\frac{1}{n}\sum_{t=1}^{n}\cos 2\left(\widehat{\omega}_nt+\frac{\widehat{\Delta}_n}{n}t^2\right)\right.\\&\left.\quad-\frac{1}{n}\sum_{t=1}^{n}\cos 2\left(\omega_0t+\frac{\Delta_0}{n}t^2\right)\right|\le\frac{4}{n}\left|\left(\widehat{\omega}_n-\omega_0\right)\right|\sum_{t=1}^{n}t\\&\quad+\frac{8}{n^3}\sum_{t=1}^{n}t^3\left|\left(\widehat{\Delta}_n-\Delta_0\right)\right|\le 4n\left(\widehat{\omega}_n-\omega_0\right)+8n\left|\widehat{\Delta}_n\right.\\&\quad\left.-\Delta_0\right|.\end{aligned} \tag{24}$$

In Section 3 we prove that $\widehat{\omega}_n-\omega_0=o_p(n^{-1})$ and $\widehat{\Delta}_n-\Delta_0=o_p(n^{-1})$. Therefore we have

$$\begin{aligned}\frac{1}{n}&\sum_{t=1}^{n}\cos 2\left(\widehat{\omega}_nt+\frac{\widehat{\Delta}_n}{n}t^2\right)-\frac{1}{n}\sum_{t=1}^{n}\cos 2\left(\omega_0t+\frac{\Delta_0}{n}t^2\right)\\&=4no_p\left(n^{-1}\right)+8no_p\left(n^{-1}\right)=o_p(1).\end{aligned} \tag{25}$$

Also in Section 3 we show that

$$\sum_{t=1}^{n}\cos 2\left(\omega_0t+\frac{\Delta_0}{n}t^2\right)=o(n). \tag{26}$$

Thus (25) implies that

$$\frac{1}{n}\sum_{t=1}^{n}\cos 2\left(\widehat{\omega}_nt+\frac{\widehat{\Delta}_n}{n}t^2\right)=o_p(1). \tag{27}$$

Similarly

$$\frac{1}{n}\sum_{t=1}^{n}\cos 2\left(\widehat{\omega}_nt+\frac{\widehat{\Delta}_n}{n}t^2\right)=o_p(1). \tag{28}$$

In Section 3 we also show that $\widehat{A}_n\xrightarrow{P}A_0$ and $\widehat{B}_n\xrightarrow{P}B_0$. Thus (22) implies that

$$S_n\left(\widehat{A}_n,\widehat{B}_n,\widehat{\omega}_n,\widehat{\Delta}_n\right)-U_n\left(\widehat{A}_n,\widehat{B}_n,\widehat{\omega}_n,\widehat{\Delta}_n\right)=o_p(1). \tag{29}$$

Using (21) in the above equation we obtain

$$\frac{1}{n}U_n\left(\widehat{A}_n,\widehat{B}_n,\widehat{\omega}_n,\widehat{\Delta}_n\right)=\frac{1}{n}\sum_{t=1}^{n}\epsilon_t^2-o_p(1). \tag{30}$$

By the weak law of large numbers

$$\frac{1}{n}\sum_{t=1}^{n}\epsilon_t^2\xrightarrow{P}\nu>0. \tag{31}$$

Using this in (30) we obtain that $U_n(\widehat{A}_n,\widehat{B}_n,\widehat{\omega}_n,\widehat{\Delta}_n)>0$ with probability arbitrarily close to 1 for sufficiently large values of n. Suppose, then, that $U_n(\widehat{A}_n,\widehat{B}_n,\widehat{\omega}_n,\widehat{\Delta}_n)>0$. Since

$$\begin{aligned}L_n^*\left(\widehat{A}_n,\widehat{B}_n,\widehat{\omega}_n,\widehat{\Delta}_n\right)&=-\frac{n}{2}\left[\log 2\pi+\log\nu\right]\\&\quad-\frac{(1/2)\,U_n\left(\widehat{A}_n,\widehat{B}_n,\widehat{\omega}_n,\widehat{\Delta}_n\right)}{\nu}\end{aligned} \tag{32}$$

and $0<\nu<\infty$, $L_n^*(\widehat{A}_n,\widehat{B}_n,\widehat{\omega}_n,\widehat{\Delta}_n)\to-\infty$ as $\nu\to 0$ or $\nu\to\infty$. Thus since $L_n^*(\widehat{A}_n,\widehat{B}_n,\widehat{\omega}_n,\widehat{\Delta}_n)$ is a continuous function in ν the maximum is achieved when $\nu=\widehat{\nu}_n$. The partial derivative $\partial L_n^*/\partial\nu$ vanishes at $\widehat{\nu}_n$. Therefore $\widehat{\nu}_n$ is the solution to equation

$$\frac{\partial L_n^*}{\partial\nu}=-\frac{1}{2}\frac{n}{\nu}+\frac{(1/2)\,U_n\left(\widehat{A}_n,\widehat{B}_n,\widehat{\omega}_n,\widehat{\Delta}_n\right)}{\nu^2}=0. \tag{33}$$

So

$$\widehat{\nu}_n=\frac{1}{2}U_n\left(\widehat{A}_n,\widehat{B}_n,\widehat{\omega}_n,\widehat{\Delta}_n\right). \tag{34}$$

Note that $\widehat{\nu}_n$ exits only with high probability for large n. Using (12) and substituting for $\widehat{A}_n$ and $\widehat{B}_n$ we obtain

$$U_n\left(\widehat{A}_n,\widehat{B}_n,\widehat{\omega}_n,\widehat{\Delta}_n\right)=\sum_{t=1}^{n}X_{nt}^2-2\cdot\frac{2}{n}\left|\sum_{t=1}^{n}X_t e^{i(\widehat{\omega}_n t+(\widehat{\Delta}_n/n)t^2)}\right|^2+\frac{2}{n}\left\{\sum_{t=1}^{n}\left[\cos\left(\widehat{\omega}_n t+\frac{\widehat{\Delta}_n}{n}t^2\right)\right]^2+\sum_{t=1}^{n}\left[\sin\left(\widehat{\omega}_n t+\frac{\widehat{\Delta}_n}{n}t^2\right)\right]^2\right\}=\sum_{t=1}^{n}X_{nt}^2-2\cdot\frac{2}{n}\left|\sum_{t=1}^{n}X_{nt}e^{i(\widehat{\omega}_n t+(\widehat{\Delta}_n/n)t^2)}\right|^2+\frac{2}{n}\left|\sum_{t=1}^{n}X_{nt}e^{i(\widehat{\omega}_n t+(\widehat{\Delta}_n/n)t^2)}\right|^2=\sum_{t=1}^{n}X_{nt}^2-\frac{2}{n}\left|\sum_{t=1}^{n}X_{nt}e^{i(\widehat{\omega}_n t+(\widehat{\Delta}_n/n)t^2)}\right|^2. \tag{35}$$

Now, using (18) in the above equation, we obtain

$$U_n\left(\widehat{A}_n,\widehat{B}_n,\widehat{\omega}_n,\widehat{\Delta}_n\right)=\sum_{t=1}^{n}X_{nt}^2-I_n\left(\widehat{\omega}_n,\frac{\widehat{\Delta}}{n}\right). \tag{36}$$

Therefore, substituting in (34) we have

$$\widehat{\nu}_n=\frac{1}{n}\left\{\sum_{t=1}^{n}X_{nt}^2-I_n\left(\widehat{\omega}_n,\frac{\widehat{\Delta}}{n}\right)\right\}. \tag{37}$$

3. The Consistency of the Estimators

Now we will establish under certain conditions the consistency of the estimators $\widehat{A}_n$, $\widehat{B}_n$, $\widehat{\omega}_n$, $\widehat{\Delta}_n$, and $\widehat{\nu}_n$.

Theorem 4. *Let $\{\epsilon_t\}_{t=1}^{\infty}$ be a sequence of independent random variables with $E(\epsilon_t)=0$, $E(\epsilon_t^2)=\nu<\infty$, and $E(\epsilon_t^4)=\sigma<\infty$. Let A_0, B_0 be any real numbers and $0<\omega<\pi/2$. Let $|\Delta_0|\le\pi/4$ and let $\alpha_n=\Delta_0/n$. For each $n=1,2,\ldots$, let*

$$X_{nt}=A_0\cos\left(\omega_0 t+\frac{\Delta_0}{n}t^2\right)+B_0\sin\left(\omega_0 t+\frac{\Delta_0}{n}t^2\right)+\epsilon_t,\quad 1\le t\le n. \tag{38}$$

The estimators $\widehat{A}_n$, $\widehat{B}_n$, $\widehat{\omega}_n$, $\widehat{\Delta}_n$, and $\widehat{\nu}_n$ of A_0, B_0, ω_0, Δ_0, and ν_0 given in (12), (13), (15), and (17), respectively, are consistent estimators of A_0, B_0, ω_0, Δ_0, and ν_0, respectively. Furthermore $\widehat{\omega}_n-\omega_0=o_p(n^{-1})$ and $\widehat{\Delta}_n-\Delta_0=o_p(n^{-1})$.

Proof. From (15)

$$\frac{n}{2}I_n\left(\omega,\frac{\Delta}{n}\right)=\left|\sum_{t=1}^{n}X_t e^{i(\omega t+(\Delta/n)t^2)}\right|^2. \tag{39}$$

But

$$X_{nt}=A_0\cos\left(\omega_0 t+\frac{\Delta_0}{n}t^2\right)+B_0\sin\left(\omega_0 t+\frac{\Delta_0}{n}t^2\right)+\epsilon_t. \tag{40}$$

Thus

$$\frac{n}{2}I_n\left(\omega,\frac{\Delta}{n}\right)=\left|\sum_{t=1}^{n}\left[A_0\cos\left(\omega_0 t+\frac{\Delta_0}{n}t^2\right)+B_0\sin\left(\omega_0 t+\frac{\Delta_0}{n}t^2\right)+\epsilon_t\right]e^{i(\omega t+(\Delta/n)t^2)}\right|^2. \tag{41}$$

If we use complex exponentials instead of sines and cosines we have

$$X_{nt}=A_0\cos\left(\omega_0 t+\frac{\Delta_0}{n}t^2\right)+B_0\sin\left(\omega_0 t+\frac{\Delta_0}{n}t^2\right)=D_0e^{i(\omega_0 t+(\Delta_0/n)t^2)}+D_0^*e^{-i(\omega_0 t+(\Delta_0/n)t^2)}, \tag{42}$$

where $D_0=(A_0-iB_0)/2$ and $D_0^*=(A_0+iB_0)/2$, and then

$$\frac{n}{2}I_n\left(\omega,\frac{\Delta}{n}\right)=\left|\sum_{t=1}^{n}\left[D_0e^{i[(\omega+\omega_0)t+(\Delta/n+\Delta_0/n)t^2]}+D_0^*e^{-i[(\omega_0-\omega)t+(\Delta_0/n-\Delta/n)t^2]}+\epsilon_t e^{i(\omega t+(\Delta/n)t^2)}\right]\right|^2. \tag{43}$$

Let

$$M_n(u,v)=\sum_{t=1}^{n}e^{i(ut+vt^2)}. \tag{44}$$

Then

$$\frac{n}{2}I_n\left(\omega,\frac{\Delta}{n}\right)=\left|D_0M_n\left(\omega+\omega_0,\frac{\Delta}{n}+\frac{\Delta_0}{n}\right)+D_0^*M_n\left(\omega-\omega_0,\frac{\Delta}{n}-\frac{\Delta_0}{n}\right)+\sum_{t=1}^{n}\epsilon_t e^{i(\omega t+(\Delta/n)t^2)}\right|^2=\left|\sum_{t=1}^{n}\epsilon_t e^{i(\omega t+(\Delta/n)t^2)}\right|^2+2\Re\sum_{t=1}^{n}\epsilon_t e^{-i(\omega t+(\Delta/n)t^2)}\left[D_0M_n\left(\omega+\omega_0,\frac{\Delta}{n}+\frac{\Delta_0}{n}\right)+D_0^*M_n\left(\omega-\omega_0,\frac{\Delta}{n}-\frac{\Delta_0}{n}\right)\right]+\left|D_0M_n\left(\omega+\omega_0,\frac{\Delta}{n}+\frac{\Delta_0}{n}\right)+D_0^*M_n\left(\omega-\omega_0,\frac{\Delta}{n}-\frac{\Delta_0}{n}\right)\right|^2. \tag{45}$$

Expanding the third term on the right, we get

$$\frac{n}{2}I_n\left(\omega,\frac{\Delta}{n}\right)=\left|\sum_{t=1}^{n}\epsilon_t e^{i(\omega t+(\Delta/n)t^2)}\right|^2 + 2\Re\sum_{t=1}^{n}\epsilon_t e^{-i(\omega t+(\Delta/n)t^2)}\left[D_0M_n\left(\omega+\omega_0,\frac{\Delta}{n}+\frac{\Delta_0}{n}\right)+D_0^*M_n\left(\omega-\omega_0,\frac{\Delta}{n}-\frac{\Delta_0}{n}\right)\right]+\left|D_0\right|^2\cdot\left|M_n\left(\omega+\omega_0,\frac{\Delta}{n}+\frac{\Delta_0}{n}\right)\right|^2+2\Re D_0^2M_n\left(\omega+\omega_0,\frac{\Delta}{n}+\frac{\Delta_0}{n}\right)M_n\left(\omega-\omega_0,\frac{\Delta_0}{n}-\frac{\Delta}{n}\right)+\left|D_0\right|^2\left|M_n\left(\omega-\omega_0,\frac{\Delta}{n}-\frac{\Delta_0}{n}\right)\right|^2. \tag{46}$$

Lemma 5.

$$\frac{1}{n}I_n\left(\omega_0,\frac{\Delta_0}{n}\right)=\frac{A_0^2+B_0^2}{2}+o_p(1). \tag{47}$$

Proof. From (46) we have

$$\frac{n}{2}I_n\left(\omega_0,\frac{\Delta_0}{n}\right)=\left|\sum_{t=1}^{n}\epsilon_t e^{i(\omega_0 t+(\Delta_0/n)t^2)}\right|^2 + 2\Re\sum_{t=1}^{n}\epsilon_t e^{-i(\omega_0 t+(\Delta_0/n)t^2)}\left[D_0M_n\left(2\omega_0,2\frac{\Delta_0}{n}\right)+D_0^*M_n(0,0)\right]+\left|D_0\right|^2\left|M_n\left(2\omega_0,2\frac{\Delta_0}{n}\right)\right|^2+2\Re D_0^2M_n\left(2\omega_0,2\frac{\Delta_0}{n}\right)M_n(0,0)+\left|D_0^*\right|^2\cdot\left|M_n(0,0)\right|^2. \tag{48}$$

By virtue of Theorem 2

$$\max_{\substack{0\le\omega\le\pi\\0\le\alpha\le\pi}}\left|\sum_{t=1}^{n}\epsilon_t e^{i(\omega t+\alpha t^2)}\right|=O_p\left(n^{7/8}\right). \tag{49}$$

Hence

$$\max_{\substack{0\le\omega\le\pi\\0\le\Delta/n\le\pi}}\left|\sum_{t=1}^{n}\epsilon_t e^{i(\omega t+(\Delta/n)t^2)}\right|=O_p\left(n^{7/8}\right). \tag{50}$$

Thus we obtain the following estimates for the first and second terms in (48):

$$\left|\sum_{t=1}^{n}\epsilon_t e^{i(\omega_0 t+(\Delta_0/n)t^2)}\right|^2=O_p\left(n^{7/4}\right),$$
$$2\Re\sum_{t=1}^{n}\epsilon_t e^{-i(\omega_0 t+(\Delta_0/n)t^2)}\left[D_0M_n\left(2\omega_0,2\frac{\Delta_0}{n}\right)+D_0^*M_n(0,0)\right]=O_p\left(n\cdot n^{7/8}\right)=O_p\left(n^{15/8}\right). \tag{51}$$

Consider the term $|M_n(2\omega_0,2(\Delta_0/n))|^2$ in (48). Using (44) we have

$$M_n\left(2\omega_0,2\frac{\Delta_0}{n}\right)=\sum_{t=1}^{n}e^{i(2\omega_0 t+2(\Delta_0/n)t^2)}. \tag{52}$$

Now we can write

$$2\omega_0 t+2\frac{\Delta_0}{n}t^2=2\pi\frac{\left[2\omega_0 t+2(\Delta_0/n)t^2\right]}{2\pi}=2\pi f(t), \tag{53}$$

where $f(t)=(2\omega_0 t+2(\Delta_0/n)t^2)/2\pi$.

Also

$$\left|f'(t)\right|=\frac{2\omega_0}{2\pi}+\frac{4\left(\left|\Delta_0\right|t/n\right)}{2\pi}\le\frac{\omega_0}{\pi}+2\frac{\left|\Delta_0\right|t}{n\pi}\le\frac{\omega_0}{\pi}+2\frac{\left|\Delta_0\right|n}{n\pi}<\frac{1}{2}+\frac{2\left|\Delta_0\right|}{\pi}\quad\text{since }\omega_0<\frac{\pi}{2}$$
$$<1\quad\text{since }\left|\Delta_0\right|\le\frac{\pi}{4}. \tag{54}$$

It follows by virtue of Lemma [16] that there is $A_1>0$ for which

$$\left|\sum_{t=1}^{n}e^{i(2\omega_0 t+(2\Delta_0/n)t^2)}-\int_0^n e^{i(2\omega_0 t+(2\Delta_0/n)t^2)}dt\right|\le A_1. \tag{55}$$

Thus we have

$$\frac{1}{n}M_n\left(2\omega_0,\frac{2\Delta_0}{n}\right)=\frac{1}{n}\int_0^n e^{i(2\omega_0 n(t/n)+2\Delta_0 n(t^2/n^2))}dt+O\left(\frac{1}{n}\right). \tag{56}$$

Let $x=t/n$. Then we obtain

$$\frac{1}{n}M_n\left(2\omega_0,2\frac{\Delta_0}{n}\right)=\int_0^1 e^{i(2\omega_0 nx+2\Delta_0 nx^2)}dx+o(1). \tag{57}$$

In the proof of Theorem 3 [16, p. 65, eq 1.9], we showed that if at least one of s_n or t_n tends to ∞ then

$$\int_0^1 e^{i(2s_n x+2t_n x^2)}dx=o(1). \tag{58}$$

Therefore, since $\omega_0>0$, we have

$$\int_0^1 e^{i(2\omega_0 nx+2\Delta_0 nx^2)}dx=o(1). \tag{59}$$

This together with (57) implies that

$$M_n\left(2\omega_0, \frac{2\Delta_0}{n}\right) = o(n). \tag{60}$$

Thus using this equation (51), and (60) in (48) we obtain

$$\frac{n}{2}I_n\left(\omega_0, \frac{\Delta_0}{n}\right) = O_p\left(n^{7/4}\right) + O_p\left(n^{15/8}\right) + o\left(n^2\right) + o(n)\,n + \frac{A_0^2 + B_0^2}{4}n^2. \tag{61}$$

Thus

$$n^{-1}I_n\left(\omega_0, \frac{\Delta_0}{n}\right) = \frac{A_0^2 + B_0^2}{2} + o_p(1). \tag{62}$$

This completes the proof of Lemma 5.

Lemma 6. *Let $K(n, \delta_1, \delta_2)$ be defined by*

$$K(n, \delta_1, \delta_2) = \max_{\substack{|\omega-\omega_0|\ge n^{-1}\delta_1 \\ or \\ |\Delta-\Delta_0|\ge n^{-1}\delta_2}} I_n\left(\omega, \frac{\Delta}{n}\right), \tag{63}$$

where $0 < \omega < \pi/2$ and $|\Delta| \le \pi/4$. Then

$$n^{-1}K(n, \delta_1, \delta_2) \le \frac{A_0^2 + B_0^2}{2n^2} \max_{\substack{|\omega-\omega_0|\ge n^{-1}\delta_1 \\ or \\ |\Delta-\Delta_0|\ge n^{-1}\delta_2}} \left|\int_0^n e^{i[(\omega-\omega_0)t+(\Delta-\Delta_0)t^2]}dt\right|^2 + o_p(1). \tag{64}$$

Proof. Throughout the proof it is assumed that $0 < \omega < \pi/2$ and $|\Delta| \le \pi/4$. From (46) we have

$$\begin{aligned}\frac{n}{2}K(n, \delta_1, \delta_2) \le{}& \max_{\substack{|\omega-\omega_0|\ge n^{-1}\delta_1 \\ \text{or} \\ |\Delta-\Delta_0|\ge n^{-1}\delta_2}} \left|\sum_{t=1}^n \epsilon_t e^{i(\omega t+(\Delta/n)t^2)}\right|^2 \\ &+ \max_{\substack{|\omega-\omega_0|\ge n^{-1}\delta_1 \\ \text{or} \\ |\Delta-\Delta_0|\ge n^{-1}\delta_2}} 2\Re\sum_{t=1}^n \epsilon_t e^{-i(\omega t+\Delta t^2)}\left[D_0 M_n\left(\omega+\omega_0, \frac{\Delta_0}{n}+\frac{\Delta}{n}\right) + D_0^* M_n\left(\omega-\omega_0, \frac{\Delta}{n}-\frac{\Delta_0}{n}\right)\right] \\ &+ \max_{\substack{|\omega-\omega_0|\ge n^{-1}\delta_1 \\ \text{or} \\ |\Delta-\Delta_0|\ge n^{-1}\delta_2}} |D_0|^2\left|M_n\left(\omega+\omega_0, \frac{\Delta_0}{n}+\frac{\Delta}{n}\right)\right|^2 + \max_{\substack{|\omega-\omega_0|\ge n^{-1}\delta_1 \\ \text{or} \\ |\Delta-\Delta_0|\ge n^{-1}\delta_2}} 2\Re D_0^2 M_n\left(\omega+\omega_0, \frac{\Delta}{n}+\frac{\Delta_0}{n}\right)M_n\left(\omega_0-\omega, \frac{\Delta_0}{n}-\frac{\Delta}{n}\right) \\ &+ \max_{\substack{|\omega-\omega_0|\ge n^{-1}\delta_1 \\ \text{or} \\ |\Delta-\Delta_0|\ge n^{-1}\delta_2}} |D_0|^2\left|M_n\left(\omega-\omega_0, \frac{\Delta}{n}-\frac{\Delta_0}{n}\right)\right|^2.\end{aligned} \tag{65}$$

From Theorem 2

$$\max_{0\le\omega<\pi}\left|\sum_{t=1}^n \epsilon_t e^{i(\omega t+\alpha t^2)}\right| = O_p\left(n^{7/8}\right). \tag{66}$$

This implies that

$$\max_{0\le\omega<\pi}\left|\sum_{t=1}^n \epsilon_t e^{i(\omega t+(\Delta/n)t^2)}\right| = O_p\left(n^{7/8}\right). \tag{67}$$

Thus we obtain the following estimates for the first and second terms in (65):

$$\begin{aligned}&\max_{\substack{|\omega-\omega_0|\ge n^{-1}\delta_1 \\ \text{or} \\ |\Delta-\Delta_0|\ge n^{-1}\delta_2}} \left|\sum_{t=1}^n \epsilon_t e^{i(\omega t+(\Delta/n)t^2)}\right|^2 = O_p\left(n^{7/4}\right), \\ &\max_{\substack{|\omega-\omega_0|\ge n^{-1}\delta_1 \\ \text{or} \\ |\Delta-\Delta_0|\ge n^{-1}\delta_2}} 2\Re\sum_{t=1}^n \epsilon_t e^{-i(\omega t+(\Delta/n)t^2)}\left[D_0 M_n\left(\omega+\omega_0, \frac{\Delta_0}{n}+\frac{\Delta}{n}\right) + D_0^* M_n\left(\omega-\omega_0, \frac{\Delta}{n}-\frac{\Delta_0}{n}\right)\right] = O_p\left(n\cdot n^{7/8}\right) = O_p\left(n^{15/8}\right).\end{aligned} \tag{68}$$

Consider the term $|M_n(\omega+\omega_0, (\Delta/n)+(\Delta_0/n))|^2$. Using (44) we have

$$M_n\left(\omega+\omega_0, \frac{\Delta}{n}+\frac{\Delta_0}{n}\right) = \sum_{t=1}^n e^{i(\omega+\omega_0)t+(\Delta/n+\Delta_0/n)t^2}. \tag{69}$$

We can write

$$\begin{aligned}&(\omega+\omega_0)t + \left(\frac{\Delta}{n}+\frac{\Delta_0}{n}t^2\right) \\ &= 2\pi\frac{\left[(\omega+\omega_0)t + \left(\Delta/n + (\Delta_0/n)t^2\right)\right]}{2\pi} = 2\pi f(t),\end{aligned} \tag{70}$$

where $f(t) = [(\omega+\omega_0)t + (\Delta/n + \Delta_0/n)t^2]/2\pi$.

Then

$$\begin{aligned}\left|f'(t)\right| &\le \frac{(\omega+\omega_0)}{2\pi} + 2t\frac{|(\Delta/n+\Delta_0/n)|}{2\pi} \\ &\le \frac{(\omega+\omega_0)}{2\pi} + 2n\frac{|\Delta+\Delta_0|}{\pi n} < \frac{1}{2} + \frac{|\Delta+\Delta_0|}{\pi} \\ &\qquad \text{since } 0<\omega_0<\frac{\pi}{2},\ \sup_{0<\omega_0<\pi/2}(\omega+\omega_0) < \pi \\ &< \frac{1}{2} + \frac{\pi}{2\pi} \quad \text{since } |\Delta| \le \frac{\pi}{4},\ |\Delta_0| \le \frac{\pi}{4} \\ &= 1.\end{aligned} \tag{71}$$

It follows by virtue of Lemma [9, p. 1] that there is $A_2 > 0$ for which

$$\left|\sum_{t=1}^n e^{i(\omega+\omega_0)t+(\Delta/n+\Delta_0/n)t^2} - \int_0^n e^{i(\omega+\omega_0)t+(\Delta/n+\Delta_0/n)t^2}\right| \le A_2. \tag{72}$$

Similarly we can show that

$$\left|\sum_{t=1}^{n} e^{i[(\omega-\omega_0)t+(\Delta/n-\Delta_0/n)t^2]} - \int_0^n e^{i[(\omega-\omega_0)t+(\Delta/n-\Delta_0/n)t^2]}dt\right| \le A_3. \tag{73}$$

Thus we have

$$\begin{aligned}\frac{1}{n}M_n\left(\omega+\omega_0, \frac{\Delta}{n}+\frac{\Delta_0}{n}\right) \\ = \frac{1}{n}\int_0^n e^{i[(\omega+\omega_0)t+(\Delta+\Delta_0)(t^2/n)]}dt + O\left(\frac{1}{n}\right),\end{aligned} \tag{74}$$

$$\begin{aligned}\frac{1}{n}M_n\left(\omega-\omega_0, \frac{\Delta}{n}-\frac{\Delta_0}{n}\right) \\ = \frac{1}{n}\int_0^n e^{i[(\omega-\omega_0)t+(\Delta-\Delta_0)(t^2/n)]}dt + O\left(\frac{1}{n}\right).\end{aligned} \tag{75}$$

Let $x = t/n$ in (74). Then we have

$$\begin{aligned}\frac{1}{n}M_n\left(\omega+\omega_0, \frac{\Delta}{n}+\frac{\Delta_0}{n}\right) \\ = \int_0^1 e^{i[(\omega+\omega_0)nx+(\Delta+\Delta_0)nx^2]}dx + O\left(\frac{1}{n}\right).\end{aligned} \tag{76}$$

Let $R = \{(\Delta, \omega) : |\Delta| \le \pi/4,\ 0 \le \omega \le \pi/2\}$. We will show that

$$\max_R \left|\int_0^1 e^{i[(\omega+\omega_0)nx+(\Delta+\Delta_0)nx^2]}dx\right| = o(1). \tag{77}$$

Suppose

$$\max_R \left|\int_0^1 e^{i[(\omega+\omega_0)nx+(\Delta+\Delta_0)nx^2]}dx\right| \ne o(1). \tag{78}$$

Then there is a sequence n_k and $\delta > 0$ for which

$$\max_R \left|\int_0^1 e^{i[(\omega+\omega_0)n_k x+(\Delta+\Delta_0)n_k x^2]}dx\right| > \delta, \tag{79}$$

for each k. Then we can find $(\omega_k, \Delta_k) \in R$ for which

$$\max_R \left|\int_0^1 e^{i[(\omega_k+\omega_0)n_k x+(\Delta+\Delta_0)n_k x^2]}dx\right| > \frac{\delta}{2}, \tag{80}$$

for each k. Let $s_k = (\omega_k + \omega_0)n_k$, $t_k = (\Delta_k + \Delta_0)n_k$. Since $(\omega_k, \Delta_k) \in R$, $\omega_k \ge 0$, $\omega_k + \omega_0 \ge \omega_0 > 0$; hence $s_k = (\omega_k + \omega_0)n_k \ge \omega_0 n_k \to \infty$. It follows from equation 1.9 [16, p. 65] in the proof of Theorem 3 that

$$\int_0^1 e^{i[(\omega_k+\omega_0)n_k x+(\Delta+\Delta_0)n_k x^2]}dx \longrightarrow 0. \tag{81}$$

Thus we have a contradiction. Therefore

$$\max_R \left|\int_0^1 e^{i[(\omega+\omega_0)nx+(\Delta+\Delta_0)nx^2]}dx\right| = o(1). \tag{82}$$

Hence by (76)

$$\max_R M_n\left(\omega+\omega_0, \frac{\Delta}{n}+\frac{\Delta_0}{n}\right) = o(n). \tag{83}$$

This implies that

$$\max_{\substack{|\omega-\omega_0|\ge n^{-1}\delta_1 \\ \text{or} \\ |\Delta-\Delta_0|\ge n^{-1}\delta_2}} M_n\left(\omega+\omega_0, \frac{\Delta}{n}+\frac{\Delta_0}{n}\right) = o(n). \tag{84}$$

Substituting (68) and (84) in (65) we obtain

$$\begin{aligned}\frac{n}{2}K(n, \delta_1, \delta_2) \le O_p\left(n^{7/4}\right) + O_p\left(n^{15/8}\right) + o\left(n^2\right) \\ + o(n)\,n + |D_0|^2\frac{A_0^2+B_0^2}{4} \\ \cdot \max_{\substack{|\omega-\omega_0|\ge n^{-1}\delta_1 \\ \text{or} \\ |\Delta-\Delta_0|\ge n^{-1}\delta_2}} \left|\int_0^n e^{i[(\omega-\omega_0)t+(\Delta/n-\Delta_0/n)t^2]}dt\right| + A_3.\end{aligned} \tag{85}$$

Thus

$$\begin{aligned}n^{-1}K(n, \delta_1, \delta_2) \le n^{-2}\frac{A_0^2+B_0^2}{2} \\ \cdot \max_{\substack{|\omega-\omega_0|\ge n^{-1}\delta_1 \\ \text{or} \\ |\Delta-\Delta_0|\ge n^{-1}\delta_2}} \left|\int_0^n e^{i[(\omega-\omega_0)t+(\Delta/n-\Delta_0/n)t^2]}dt\right|^2 \\ + o_p(1).\end{aligned} \tag{86}$$

This completes the proof of Lemma 6.

Now combining Lemmas 5 and 6 we obtain

$$\begin{aligned}n^{-1}K(n, \delta_1, \delta_2) - n^{-1}I_n\left(\omega_0 + \frac{\Delta_0}{n}\right) \\ \le \frac{\left(A_0^2+B_0^2\right)}{2}\Bigg[n^{-2} \\ \cdot \max_{\substack{|\omega-\omega_0|\ge n^{-1}\delta_1 \\ \text{or} \\ |\Delta-\Delta_0|\ge n^{-1}\delta_2}} \left|\int_0^n e^{i[(\omega-\omega_0)t+(\Delta/n-\Delta_0/n)t^2]}dt\right|^2 - 1\Bigg] \\ + o_p(1).\end{aligned} \tag{87}$$

Thus

$$n^{-1}K(n, \delta_1, \delta_2) - n^{-1}I_n\left(\omega_0, \frac{\Delta_0}{n}\right) \le c_n + z_n, \tag{88}$$

where

$$c_n = \frac{\left(A_0^2 + B_0^2\right)}{2}\left[n^{-2} \cdot \max_{\substack{|\omega-\omega_0|\geq n^{-1}\delta_1 \\ \text{or} \\ |\Delta-\Delta_0|\geq n^{-1}\delta_2}} \left|\int_0^n e^{i[(\omega-\omega_0)t+(\Delta/n-\Delta_0/n)t^2]}dt\right|^2 - 1\right], \quad (89)$$

and $z_n = o_p(1)$.

From Theorem 3 we have

$$\overline{\lim} \max_{\substack{\{n^{-1}\delta_1\leq|u|\leq\pi,\, 0\leq|v|\leq\pi-n^{-2}\delta_2\} \\ \text{or} \\ \{|u|\leq\pi-n^{-1}\delta_1,\, n^{-2}\delta_2\leq|v|\leq\pi\}}} \left|\frac{1}{n}\int_0^n e^{i(ut+vt^2)}\right| < 1. \quad (90)$$

Substituting $u = \omega - \omega_0$ and $v = \Delta/n - \Delta_0/n$ in the above inequality we obtain

$$\overline{\lim} \max_{\substack{\{n^{-1}\delta_1\leq|\omega-\omega_0|\leq\pi,\, 0\leq|\Delta-\Delta_0|\leq n\pi-n^{-1}\delta_2\} \\ \text{or} \\ \{|\omega-\omega_0|\leq\pi-n^{-1}\delta_1,\, n^{-1}\delta_2\leq|\Delta-\Delta_0|\leq n\pi\}}} \left|\frac{1}{n} \cdot \int_0^n e^{i[(\omega-\omega_0)t+(\Delta/n-\Delta_0/n)t^2]}\right| < 1. \quad (91)$$

Therefore since $0 < \omega < \pi/2$ and $|\Delta| \leq \pi/4$, we have

$$\overline{\lim} \max_{\substack{|\omega-\omega_0|\geq n^{-1}\delta_1 \\ \text{or} \\ |\Delta-\Delta_0|\geq n^{-1}\delta_2}} \left|\frac{1}{n}\int_0^n e^{i[(\omega-\omega_0)t+(\Delta/n-\Delta_0/n)t^2]}\right| < 1. \quad (92)$$

Therefore $\overline{\lim}\, c_n = c < 0$.

Let $\epsilon > 0$. Since $\lim_{n\to\infty} z_n \overset{P}{=} 0$, there exits N_1 such that

$$P\left\{|z_n| < -\frac{c}{4}\right\} \geq 1 - \epsilon \quad \forall n > N_1. \quad (93)$$

Since $\overline{\lim}_{c_n} = c\ (<0)$, there exits $N_2(\delta_1, \delta_2)$ such that

$$c_n < \frac{c}{2}, \quad \forall n > N_2(\delta_1, \delta_2). \quad (94)$$

Now let $N'(\delta_1, \delta_2) = \max\{N_1, N_2(\delta_1, \delta_2)\}$. If $n > N'$, then

$$\left\{|z_n| < -\frac{c}{4}\right\} \subseteq \left\{c_n + z_n < \frac{c}{2} + -\frac{c}{4}\right\} = \left\{c_n + z_n < \frac{c}{4}\right\} \subseteq \{c_n + z_n < 0\}. \quad (95)$$

Hence

$$P(\{c_n + z_n < 0\}) \geq P\left(\left\{|z_n| < -\frac{c}{4}\right\}\right) \geq 1 - \epsilon. \quad \text{(by (93))}. \quad (96)$$

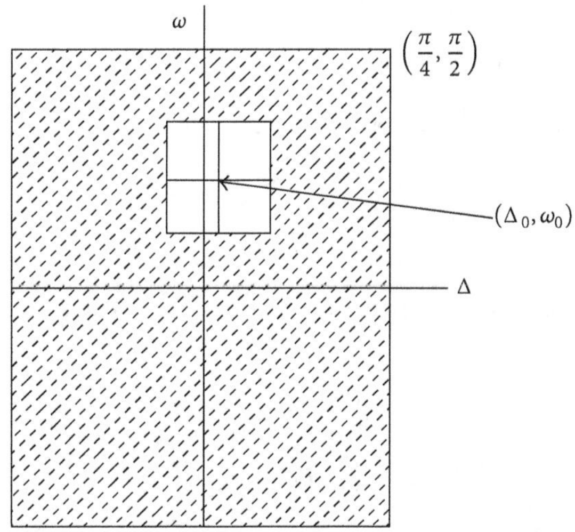

FIGURE 1

Now, from (88),

$$\{c_n + z_n < 0\} \subseteq \left\{\left[n^{-1}K(n,\delta_1,\delta_2) - n^{-1}I_n\left(\omega_0, \frac{\Delta_0}{n}\right)\right] < 0\right\}. \quad (97)$$

Therefore

$$P\left\{\left[n^{-1}K(n,\delta_1,\delta_2) - n^{-1}I_n\left(\omega_0, \frac{\Delta_0}{n}\right)\right] < 0\right\} \geq 1 - \epsilon, \quad \forall n > N'. \quad (98)$$

Since $\epsilon > 0$ is arbitrary,

$$\lim_{n\to\infty} P\left\{\left[K(n,\delta_1,\delta_2) - I_n\left(\omega_0, \frac{\Delta_0}{n}\right)\right] < 0\right\} = 1. \quad (99)$$

In other words,

$$\lim_{n\to\infty} P\left\{K(n,\delta_1,\delta_2) < I_n\left(\omega_0, \frac{\Delta_0}{n}\right)\right\} = 1, \quad (100)$$

where

$$K(n,\delta_1,\delta_2) = \max_{\substack{|\omega-\omega_0|\geq n^{-1}\delta_1 \\ \text{or} \\ |\Delta-\Delta_0|\geq n^{-1}\delta_2}} I_n\left(\omega, \frac{\Delta}{n}\right). \quad (101)$$

In the last statement the maximum is taken over the region shown in Figure 1.

Suppose (ω', Δ') belongs to the shaded region of the figure. Then $I_n(\omega', \Delta'/n)$ does not exceed $K(n, \delta_1, \delta_2)$. If $I_n(\omega_0, \Delta_0/n) > K(n, \delta_1, \delta_2)$ this implies that (ω', Δ') is not the point $(\widehat{\Delta}_n, \widehat{\omega}_n)$ which maximizes $I_n(\omega, \Delta/n)$ which in turn implies that $(\widehat{\Delta}_n, \widehat{\omega}_n)$ lies in the unshaded rectangle, namely, the region $\{(\Delta, \omega) : |\Delta - \Delta_0| < n^{-1}\delta_2,\ |\omega - \omega_0| < n^{-1}\delta_1\}$. Thus (100) implies that, for large n, the probability is high that the maximum value $(\widehat{\Delta}_n, \widehat{\omega}_n)$ lies in that region, that is,

$$\left|\Delta - \Delta_0\right| < n^{-1}\delta_2, \quad \left|\widehat{\omega}_n - \omega_0\right| < n^{-1}\delta_1. \quad (102)$$

In other words

$$\lim_n P\left(\left|\widehat{\omega}_n - \omega_0\right| < n^{-1}\delta_1, \left|\widehat{\Delta}_n - \Delta_0\right| < n^{-1}\delta_2\right) = 1. \quad (103)$$

Since δ_1 and δ_2 are arbitrary positive numbers, we have

$$\widehat{\omega}_n - \omega_0 = o_P\left(n^{-1}\right),$$
$$\widehat{\omega}_n - \Delta_0 = o_P\left(n^{-1}\right). \quad (104)$$

Next we prove the consistency of $\widehat{A}_n$ and $\widehat{B}_n$. That is, we show that

$$\lim_{n\to\infty} \widehat{A}_n \overset{p}{=} A_0,$$
$$\lim_{n\to\infty} \widehat{B}_n \overset{p}{=} B_0. \quad (105)$$

Recall from (15) and (16)

$$\widehat{A}_n = \frac{2}{n}\sum_{t=1}^{n} X_{nt}\cos\left(\widehat{\omega}_n t + \frac{\widehat{\Delta}_n}{n}t^2\right),$$
$$\widehat{B}_n = \frac{2}{n}\sum_{t=1}^{n} X_{nt}\sin\left(\widehat{\omega}_n t + \frac{\widehat{\Delta}_n}{n}t^2\right). \quad (106)$$

Thus

$$\widehat{A}_n + i\widehat{B}_n = \frac{2}{n}\sum_{t=1}^{n} X_{nt} e^{(\widehat{\omega}_n t + (\widehat{\Delta}_n/n)t^2)}. \quad (107)$$

But

$$X_{nt} = A_0\cos\left(\omega_0 t + \frac{\Delta_0}{n}t^2\right) + B_0\sin\left(\omega_0 t + \frac{\Delta_0}{n}t^2\right) + \epsilon_t; \quad (108)$$

so

$$\widehat{A}_n + i\widehat{B}_n = \frac{2}{n}\sum_{t=1}^{n}\left[A_0\cos\left(\omega_0 t + \frac{\Delta_0}{n}t^2\right) + B_0\sin\left(\omega_0 t + \frac{\Delta_0}{n}t^2\right) + \epsilon_t\right] e^{i(\widehat{\omega}_n t + (\widehat{\Delta}_n/n)t^2)}. \quad (109)$$

Letting

$$A_0\cos\left(\omega_0 t + \frac{\Delta_0}{n}t^2\right) + B_0\sin\left(\omega_0 t + \frac{\Delta_0}{n}t^2\right) = D_0 e^{(\omega_0 t + (\Delta_0/n)t^2)} + D_0^* e^{i(\omega_0 t + (\Delta_0/n)t^2)}, \quad (110)$$

where $D_0 = (A_0 - iB_0)/2$ and $D_0^* = (A_0 + iB_0)/2$, we have

$$\widehat{A}_n + i\widehat{B}_n = \frac{2}{n}\sum_{t=1}^{n}\left[D_0 e^{i(\omega_0 t + (\Delta_0/n)t^2)} + D_0^* e^{(\omega_0 t + (\Delta_0/n)t^2)} + \epsilon_t\right] e^{i(\widehat{\omega}_n t + (\widehat{\Delta}_n/n)t^2)} = \frac{2}{n} \cdot \sum_{t=1}^{n}\left[D_0 e^{i[(\omega_0 + \widehat{\omega}_n)t + (\Delta_0/n + \widehat{\Delta}_n/n)t^2]} + D_0^* e^{i[(\widehat{\omega}_n - \omega_0)t + (\widehat{\Delta}_n/n - \Delta_0/n)t^2]} + \epsilon_t e^{i(\widehat{\omega}_n t + (\widehat{\Delta}_n/n)t^2)}\right]. \quad (111)$$

Since $A_0 + iB_0 = 2D_0^*$ we obtain, using (44),

$$\left(\widehat{A}_n - A_0\right) + i\left(\widehat{B}_n - B_0\right) = \frac{2}{n}\left\{D_0 M_n\left[\left(\omega_0 + \widehat{\omega}_n, \frac{\Delta_0}{n} + \frac{\widehat{\Delta}_n}{n}\right) + D_0^* M_n\left(\widehat{\omega}_n - \omega_0, \frac{\widehat{\Delta}_n}{n} - \frac{\Delta_0}{n}\right) - n\right]\right\} + \frac{2}{n} \cdot \sum_{t=1}^{n}\epsilon_t e^{i(\widehat{\omega}_n t + (\widehat{\Delta}_n/n)t^2)}. \quad (112)$$

Using the triangle inequality we get

$$\left|\left(\widehat{A}_n - A_0\right) + i\left(\widehat{B}_n - B_0\right)\right| \le 2\frac{|D_0|}{n}\left|M_n\left(\widehat{\omega}_n + \omega_0, \frac{\widehat{\Delta}_n}{n} + \frac{\Delta_0}{n}\right)\right| + 2\frac{|D_0^*|}{n}\left|M_n\left(\widehat{\omega}_n - \omega_0, \frac{\widehat{\Delta}_n}{n} - \frac{\Delta_0}{n}\right) - n\right| + \frac{2}{n}\left|\sum_{t=1}^{n}\epsilon_t e^{i(\widehat{\omega}_n t + (\widehat{\Delta}_n/n)t^2)}\right|. \quad (113)$$

Now we estimate each of the terms on the right hand side of the above inequality.

Let $F(\omega, \Delta/n) = \Re M_n(\omega, \Delta/n)$.

By the mean value theorem

$$\Re M_n\left(\omega, \frac{\Delta}{n}\right) - \Re M_n(0,0) = \frac{\partial F}{\partial\omega}\left(\widetilde{\omega} + \frac{\widetilde{\Delta}}{n}\right)\omega + \frac{\partial F}{\partial\Delta}\left(\widetilde{\omega} + \frac{\widetilde{\Delta}}{n}\right)\Delta, \quad (114)$$

where $(\widetilde{\omega} + \widetilde{\Delta}/n)$ is a point on the open line segment connecting $(0,0)$ and $(\omega, \Delta/n)$.

Now

$$\frac{\partial F}{\partial\omega}\left(\omega, \frac{\Delta}{n}\right) = \frac{\partial}{\partial\omega}\Re M_n\left(\omega, \frac{\Delta}{n}\right) = \frac{\partial}{\partial\omega}\sum_{t=1}^{n}\cos\left(\omega t + \frac{\Delta}{n}t^2\right) = -\sum_{t=1}^{n} t\sin\left(\omega t + \frac{\Delta}{n}t^2\right), \quad (115)$$

and similarly

$$\frac{\partial F}{\partial\Delta}\left(\omega, \frac{\Delta}{n}\right) = \frac{\partial}{\partial\Delta}\Re M_n\left(\omega, \frac{\Delta}{n}\right) = -\frac{1}{n}\sum_{t=1}^{n} t^2\sin\left(\omega t + \frac{\Delta}{n}t^2\right). \quad (116)$$

Thus

$$\Re M_n\left(\omega,\frac{\Delta}{n}\right)-\Re M_n(0,0) = -\omega\sum_{t=1}^{n} t\sin\left(\widetilde{\omega}t+\frac{\widetilde{\Delta}}{n}t^2\right) - \frac{\Delta}{n}\sum_{t=1}^{n} t^2\sin\left(\widetilde{\omega}t+\frac{\widetilde{\Delta}}{n}t^2\right). \quad (117)$$

Using the triangle inequality and the fact that $M_n(0,0)=n$, we obtain

$$\left|\Re M_n\left(\omega,\frac{\Delta}{n}\right)-n\right| \le |\omega|\sum_{t=1}^{n} t+\left|\frac{\Delta}{n}\right|\sum_{t=1}^{n} t^2. \quad (118)$$

Substituting $|\widehat{\omega}_n-\omega_n|$ and $|\widehat{\Delta}_n/n-\Delta_0/n|$ for ω and Δ, respectively, we get

$$\begin{aligned}&\left|\Re M_n\left(\widehat{\omega}_n-\omega_n,\frac{1}{n}\left(\frac{\widehat{\Delta}_n}{n}-\frac{\Delta_0}{n}\right)\right)-n\right| \\ &\quad\le \left|\widehat{\omega}-\omega_0\right|\sum_{t=1}^{n} t+\left|\frac{1}{n}\left(\frac{\widehat{\Delta}_n}{n}-\frac{\Delta_0}{n}\right)\right|\sum_{t=1}^{n} t^2 \\ &\quad< \left|\widehat{\omega}_n-\omega_0\right|n^2+\frac{1}{n}\left|\frac{\widehat{\Delta}_n}{n}-\frac{\Delta_0}{n}\right|n^3,\end{aligned} \quad (119)$$

so that

$$\frac{1}{n}\left|\Re M_n\left(\widehat{\omega}_n-\omega_n,\frac{\widehat{\Delta}_n}{n}-\frac{\Delta_0}{n}\right)-n\right| < \left|\widehat{\omega}_n-\omega_n\right|n+\left|\widehat{\Delta}_n-\Delta_0\right|n. \quad (120)$$

It follows from (104) that

$$\lim_{n\to\infty}\left|\Re M_n\left(\widehat{\omega}_n-\omega_n,\frac{\widehat{\Delta}_n}{n}-\frac{\Delta_0}{n}\right)-n\right| \overset{P}{=} 0. \quad (121)$$

Similarly

$$\lim_{n\to\infty}\frac{1}{n}\left|\Im M_n\left(\widehat{\omega}_n-\omega_n,\frac{\widehat{\Delta}_n}{n}-\frac{\Delta_0}{n}\right)-n\right| \overset{P}{=} 0. \quad (122)$$

Thus

$$\lim_{n\to\infty}\frac{1}{n}\left|M_n\left(\widehat{\omega}_n-\omega_n,\frac{\widehat{\Delta}_n}{n}-\frac{\Delta_0}{n}\right)-n\right| \overset{P}{=} 0. \quad (123)$$

Now consider the first term on the right in inequality (113). Again by the mean value theorem,

$$\begin{aligned}&\Re M_n\left(\omega,\frac{\Delta}{n}\right)-\Re M_n\left(2\omega_0,2\frac{\Delta_0}{n}\right) \\ &\quad=\frac{\partial F}{\partial\omega}\left(\widetilde{\omega},\frac{\widetilde{\Delta}}{n}\right)(\omega-2\omega_0) +\frac{\partial F}{\partial\Delta}\left(\widetilde{\omega},\frac{\widetilde{\Delta}}{n}\right)\left(\frac{\Delta}{n}-2\frac{\Delta_0}{n}\right),\end{aligned} \quad (124)$$

where $(\widetilde{\omega},\widetilde{\Delta}/n)$ is a point on the open line segment connecting $(2\omega_0,2(\Delta_0/n))$ and $(\omega,\Delta/n)$. Using the above expressions for $\partial F/\partial\omega$ and $\partial F/\partial\Delta$, we have as before

$$\begin{aligned}&\Re M_n\left(\omega,\frac{\Delta}{n}\right)-\Re M_n\left(2\omega_0,2\frac{\Delta_0}{n}\right) \\ &\quad=-(\omega-2\omega_0)\sum_{t=1}^{n} t\sin\left(\widetilde{\omega}t+\frac{\widetilde{\Delta}}{n}t^2\right) -\left(\frac{\Delta}{n}-2\frac{\Delta_0}{n}\right)\sum_{t=1}^{n} t^2\sin\left(\widetilde{\omega}t+\frac{\widetilde{\Delta}}{n}t^2\right).\end{aligned} \quad (125)$$

Thus

$$\begin{aligned}&\Re M_n\left(\omega,\frac{\Delta}{n}\right) \\ &\quad=-(\omega-2\omega_0)\sum_{t=1}^{n} t\sin\left(\widetilde{\omega}t+\frac{\widetilde{\Delta}}{n}t^2\right) -\left(\frac{\Delta}{n}-2\frac{\Delta_0}{n}\right)\sum_{t=1}^{n} t^2\sin\left(\widetilde{\omega}t+\frac{\widehat{\Delta}}{n}t^2\right) \\ &\qquad+\sum_{t=1}^{n}\cos\left(2\omega_0t+\frac{2\Delta_0}{n}t^2\right).\end{aligned} \quad (126)$$

Therefore

$$\begin{aligned}\left|\Re M_n\left(\omega,\frac{\Delta}{n}\right)\right| &\le |\omega-2\omega_0|\sum_{t=1}^{n} t+\left|\frac{\Delta}{n}-2\frac{\Delta_0}{n}\right|\sum_{t=1}^{n} t^2 \\ &\quad+\left|\sum_{t=1}^{n}\cos\left(2\omega_0t+2\frac{\Delta_0}{n}t^2\right)\right|.\end{aligned} \quad (127)$$

Substituting $\widehat{\omega}_n+\omega_n$ and $\widehat{\Delta}_n/n+\Delta_0/n$ for ω and Δ, respectively, we get

$$\begin{aligned}&\left|\Re M_n\left(\widehat{\omega}_n+\omega_n,\frac{\widehat{\Delta}_n}{n}+\frac{\Delta_0}{n}\right)\right| \\ &\quad\le\left|\widehat{\omega}-\omega_0\right|\sum_{t=1}^{n} t+\left|\frac{\widehat{\Delta}_n}{n}-\frac{\Delta_0}{n}\right|\sum_{t=1}^{n} t^2 +\left|\sum_{t=1}^{n}\cos\left(2\omega_0t+\frac{2\Delta_0}{n}t^2\right)\right| \\ &\quad<\left|\widehat{\omega}_n-\omega_0\right|n^2+\left|\widehat{\Delta}_n-\Delta_0\right|n^2 +\left|\sum_{t=1}^{n}\cos\left(2\omega_0t+2\frac{\Delta_0}{n}t^2\right)\right|,\end{aligned} \quad (128)$$

so that

$$\begin{aligned}&\frac{1}{n}\left|\Re M_n\left(\widehat{\omega}_n+\omega_0,\frac{\widehat{\Delta}_n}{n}+\frac{\Delta}{n_0}\right)\right| \\ &\quad<\left|\widehat{\omega}_n-\omega_0\right|n+\left|\widehat{\Delta}_n-\Delta_0\right|n +\frac{1}{n}\left|\sum_{t=1}^{n}\cos\left(2\omega_0t+\frac{2\Delta_0}{n}t^2\right)\right|.\end{aligned} \quad (129)$$

Using (60) and (104) we obtain

$$\lim_{n\to\infty}\frac{1}{n}\left|\Re M_n\left(\widehat{\omega}_n+\omega_0,\frac{\widehat{\Delta}_n}{n}+\frac{\Delta_0}{n}\right)\right|\overset{P}{=}0. \quad (130)$$

Similarly

$$\lim_{n\to\infty}\frac{1}{n}\left|\Im M_n\left(\widehat{\omega}_n+\omega_0,\frac{\widehat{\Delta}_n}{n}+\frac{\Delta_0}{n}\right)\right|\overset{P}{=}0. \quad (131)$$

Thus

$$\lim_{n\to\infty}\frac{1}{n}\left|M_n\left(\widehat{\omega}_n+\omega_n,\frac{\widehat{\Delta}_n}{n}+\frac{\Delta_0}{n}\right)\right|\overset{P}{=}0. \quad (132)$$

Now consider the last term in inequality (113). As a consequence of Theorem 2, we have

$$\lim_{n\to\infty}\frac{1}{n}\left|\sum_{t=1}^{n}\epsilon_t e^{i(\widehat{\omega}_n t+(\widehat{\Delta}_0/n)t^2)}\right|\overset{P}{=}0. \quad (133)$$

By using (123), (132), and (133) in inequality (113), we obtain

$$\lim_{n\to\infty}\left|\left(\widehat{A}_n-A_0\right)\right|+i\left|\left(\widehat{B}_n-B_0\right)\right|\overset{P}{=}0. \quad (134)$$

Therefore

$$\begin{aligned}&\lim_{n\to\infty}\left|\widehat{A}_n-A_0\right|\overset{P}{=}0,\\&\lim_{n\to\infty}\left|\widehat{B}_n-B_0\right|\overset{P}{=}0.\end{aligned} \quad (135)$$

That is,

$$\begin{aligned}&\lim_{n\to\infty}\widehat{A}_n\overset{P}{=}A,\\&\lim_{n\to\infty}\widehat{B}_n\overset{P}{=}B_0.\end{aligned} \quad (136)$$

Now we prove the consistency of $\widehat{\nu}_n$. That is, we show that $\lim_{n\to\infty}\widehat{\nu}_n\overset{P}{=}\nu$.

Recall from (37) that

$$\widehat{\nu}_n=\frac{1}{n}\left[\sum_{t=1}^{n}X_{nt}^2-I_n\left(\widehat{\omega}_n+\frac{\widehat{\Delta}_n}{n}\right)\right]. \quad (137)$$

From the definition of $I_n(\omega,\Delta/n)$, we have

$$\begin{aligned}I_n\left(\widehat{\omega}_n,\frac{\widehat{\Delta}_n}{n}\right)&=\frac{2}{n}\left|\sum_{t=1}^{n}X_{nt}e^{i(\widehat{\omega}_n t+(\widehat{\Delta}_n/n)t^2)}\right|^2\\&=\frac{2}{n}\left[\sum_{t=1}^{n}X_{nt}\cos\left(\widehat{\omega}_n t+\frac{\widehat{\Delta}_n}{n}t^2\right)\right]^2\\&\quad+\frac{2}{n}\left[\sum_{t=1}^{n}X_{nt}\sin\left(\widehat{\omega}_n t+\frac{\widehat{\Delta}_n}{n}t^2\right)\right]^2.\end{aligned} \quad (138)$$

But from (15) and (16) we have

$$\begin{aligned}\widehat{A}_n&=\frac{2}{n}\sum_{t=1}^{n}X_{nt}\cos\left(\widehat{\omega}_n t+\frac{\widehat{\Delta}_n}{n}t^2\right),\\\widehat{B}_n&=\frac{2}{n}\sum_{t=1}^{n}X_{nt}\sin\left(\widehat{\omega}_n t+\frac{\widehat{\Delta}_n}{n}t^2\right).\end{aligned} \quad (139)$$

Hence

$$I_n\left(\widehat{\omega}_n+\frac{\widehat{\Delta}_n}{n}\right)=\frac{2}{n}\left[\widehat{A}_n^2+\widehat{B}_n^2\right]. \quad (140)$$

Therefore

$$\widehat{\nu}_n=\frac{1}{n}\left[\sum_{t=1}^{n}X_{nt}^2-\frac{n}{2}\left(\widehat{A}_n^2+\widehat{B}_n^2\right)\right]. \quad (141)$$

Since

$$X_{nt}=A_0\cos\left(\omega_0 t+\frac{\Delta_0}{n}t^2\right)+B_0\sin\left(\omega_0 t+\frac{\Delta_0}{n}t^2\right)+\epsilon_t, \quad (142)$$

we have

$$\begin{aligned}\widehat{\nu}_n&=\frac{1}{n}\sum_{t=1}^{n}\left[A_0\cos\left(\omega_0 t+\frac{\Delta_0}{n}t^2\right)+B_0\sin\left(\omega_0 t+\frac{\Delta_0}{n}t^2\right)+\epsilon_t\right]^2-\frac{1}{2}\left(\widehat{A}_n^2+\widehat{B}_n^2\right)\\&=\frac{1}{n}\sum_{t=1}^{n}\left[A_0\cos\left(\omega_0 t+\frac{\Delta_0}{n}t^2\right)+B_0\sin\left(\omega_0 t+\frac{\Delta_0}{n}t^2\right)\right]^2+\frac{2}{n}\\&\quad\cdot\sum_{t=1}^{n}\epsilon_t\left[A_0\cos\left(\omega_0 t+\frac{\Delta_0}{n}t^2\right)+B_0\sin\left(\omega_0 t+\frac{\Delta_0}{n}t^2\right)\right]+\frac{1}{n}\sum_{t=1}^{n}\epsilon_t^2-\frac{1}{2}\left(\widehat{A}_n^2+\widehat{B}_n^2\right)\\&=\frac{1}{n}\sum_{t=1}^{n}\left[A_0^2\cos\left(\omega_0 t+\frac{\Delta_0}{n}t^2\right)^2+B_0^2\sin\left(\omega_0 t+\frac{\Delta_0}{n}t^2\right)^2\right.\\&\quad\left.+2A_0B_0\cos\left(\omega_0 t+\frac{\Delta_0}{n}t^2\right)\sin\left(\omega_0 t+\frac{\Delta_0}{n}t^2\right)\right]\\&\quad+\frac{2}{n}\sum_{t=1}^{n}\epsilon_t\left[A_0\cos\left(\omega_0 t+\frac{\Delta_0}{n}t^2\right)+B_0\sin\left(\omega_0 t+\frac{\Delta_0}{n}t^2\right)\right]+\frac{1}{n}\epsilon_t^2-\frac{1}{2}\left(\widehat{A}_n^2+\widehat{B}_n^2\right).\end{aligned} \quad (143)$$

Using the identities $\cos 2\theta = 2\cos^2\theta - 1$, $\cos 2\theta = 1 - 2\sin^2\theta$, and $\sin 2\theta = 2\sin\theta\cos\theta$ we obtain

$$\begin{aligned}\widehat{\nu}_n &= \frac{1}{n}\sum_{t=1}^{n}\frac{A_0^2}{2}\left[\cos 2\left(\omega_0 t + \frac{\Delta_0}{n}t^2\right) + 1\right] + \frac{1}{n} \\ &\cdot \sum_{t=1}^{n}\frac{B_0^2}{2}\left[1 - \cos 2\left(\omega_0 t + \frac{\Delta_0}{n}t^2\right)\right] + \frac{1}{n}\sum_{t=1}^{n}A_0B_0 \\ &\cdot \sin 2\left(\omega_0 t + \frac{\Delta_0}{n}t^2\right) + \frac{2}{n} \\ &\cdot \sum_{t=1}^{n}\epsilon_t\left[A_0\cos\left(\omega_0 t + \frac{\Delta_0}{n}t^2\right)\right. \\ &\left.+ B_0\sin\left(\omega_0 t + \frac{\Delta_0}{n}t^2\right)\right] + \frac{1}{n}\sum_{t=1}^{n}\epsilon_t^2 - \frac{1}{2}\left(\widehat{A}_n^2 + \widehat{B}_n^2\right).\end{aligned} \tag{144}$$

From (60) we have

$$M_n\left(2\omega_0, 2\frac{\Delta_0}{n}\right) = \sum_{t=1}^{n} e^{i(2\omega_0 t + 2(\Delta_0/n)t^2)} = o(n). \tag{145}$$

Thus

$$\begin{aligned}\sum_{t=1}^{n}\cos 2\left(\omega_0 t + \frac{\Delta_0}{n}t^2\right) &= o(n), \\ \sum_{t=1}^{n}\sin 2\left(\omega_0 t + \frac{\Delta_0}{n}t^2\right) &= o(n).\end{aligned} \tag{146}$$

Therefore

$$\begin{aligned}\widehat{\nu}_n &= \frac{A_0^2}{2} + \frac{B_0^2}{2} + \frac{2A_0}{n}\sum_{t=1}^{n}\epsilon_t\cos\left(\omega_0 t + \frac{\Delta_0}{n}t^2\right) \\ &+ \frac{2B_0}{n}\sum_{t=1}^{n}\epsilon_t\sin\left(\omega_0 t + \frac{\Delta_0}{n}t^2\right) + \frac{1}{n}\sum_{t=1}^{n}\epsilon_t^2 \\ &- \frac{1}{2}\left(\widehat{A}_n^2 + \widehat{B}_n^2\right) + o(1) \\ &= \frac{1}{n}\sum_{t=1}^{n}\epsilon_t^2 + \frac{2A_0}{n}\sum_{t=1}^{n}\epsilon_t\cos\left(\omega_0 t + \frac{\Delta_0}{n}t^2\right) \\ &+ \frac{2B_0}{n}\sum_{t=1}^{n}\epsilon_t\sin\left(\omega_0 t + \frac{\Delta_0}{n}t^2\right) \\ &+ \frac{1}{2}\left[A_0^2 - \widehat{A}_n^2 + B_0^2 - \widehat{B}_n^2\right] + o(1).\end{aligned} \tag{147}$$

Now we estimate each term on the right in (147). First consider the term $(1/n)\sum_{t=1}^{n}\epsilon_t^2$. ϵ_t's are independent and identically distributed with $E(\epsilon_t^2) = \nu$. Therefore by the Weak Law of Large Numbers

$$\lim_{n\to\infty}\frac{1}{n}\sum_{t=1}^{n}\epsilon_t^2 \overset{P}{=} \nu. \tag{148}$$

Next consider the second and third terms in the right hand side of (147).

Using Theorem 2 we have

$$\begin{aligned}\frac{1}{n}\sum_{t=1}^{n}\epsilon_t\cos\left(\omega_0 t + \frac{\Delta_0}{n}t^2\right) &= o_p(1), \\ \frac{1}{n}\sum_{t=1}^{n}\epsilon_t\sin\left(\omega_0 t + \frac{\Delta_0}{n}t^2\right) &= o_p(1).\end{aligned} \tag{149}$$

Now consider the last term $[A_0^2 - \widehat{A}_n^2 + B_0^2 - \widehat{B}_n^2]$ in (136). From (132) we have

$$\begin{aligned}\widehat{A}_n &\xrightarrow{P} A_0, \\ \widehat{B}_n &\xrightarrow{P} B_0.\end{aligned} \tag{150}$$

Thus

$$\begin{aligned}\widehat{A}_n^2 &\xrightarrow{P} A_0^2, \\ \widehat{B}_n^2 &\xrightarrow{P} B_0^2.\end{aligned} \tag{151}$$

Therefore

$$\lim_{n\to\infty}\left[A_0^2 - \widehat{A}_n^2 + B_0^2 - \widehat{B}_n^2\right] \overset{P}{=} 0. \tag{152}$$

By using (148), (149), and (152) in (147) we obtain

$$\lim_{n\to\infty}\widehat{\nu}_n \overset{P}{=} \nu. \tag{153}$$

This completes the proof of the theorem.

Competing Interests

The author declares that there is no conflict of interests regarding the publication of this paper.

References

[1] A. M. Walker, "On the estimation of a harmonic component in a time series with stationary dependent residuals," *Advances in Applied Probability*, vol. 5, pp. 217–241, 1973.

[2] G. Larry Bretthorst, *Lecture Notes in Statistics*, Springer, 1965.

[3] C. Ray Smith and J. Erickson Gray, *Maximum-Entropy and Bayesian Spectral Analysis and Estimation Problems*, D. Reidel Publishing Company, 1987.

[4] R. Kumaresan and S. Verma, "On estimating the parameters of chirp signals using rank reduction techniques," in *Proceedings of the 21st Asilomar Conference on Signals Systems and Computers*, pp. 555–558, Pacific Grove, Calif, USA, March 1987.

[5] P. M. Djuric and S. M. Kay, "Parameter estimation of chirp signals," *IEEE Transactions on Acoustics, Speech, and Signal Processing*, vol. 38, no. 12, pp. 2118–2126, 1990.

[6] F. Gini, M. Montanari, and L. Verrazzani, "Estimation of chirp radar signals in compound-Gaussian clutter: a cyclostationary approach," *IEEE Transactions on Signal Processing*, vol. 48, no. 4, pp. 1029–1039, 2000.

[7] S. Nandi and D. Kundu, "Asymptotic properties of the least squares estimators of the parameters of the chirp signals," *Annals of the Institute of Statistical Mathematics*, vol. 56, no. 3, pp. 529–544, 2004.

[8] G. B. Giannakis and G. Zhou, "Harmonics in multiplicative and additive noise: parameter estimation using cyclic statistics," *IEEE Transactions on Signal Processing*, vol. 43, no. 9, pp. 2217–2221, 1995.

[9] G. Zhou, G. B. Giannakis, and A. Swami, "On polynomial phase signals with time-varying amplitudes," *IEEE Transactions on Signal Processing*, vol. 44, no. 4, pp. 848–861, 1996.

[10] S. Shamsunder, G. B. Giannakis, and B. Friedlander, "Estimating random amplitude polynomial phase signals: a cyclostationary approach," *IEEE Transactions on Signal Processing*, vol. 43, no. 2, pp. 492–505, 1995.

[11] A. Swami, "Cramer-Rao bounds for deterministic signals in additive and multiplicative noise," *Signal Processing*, vol. 53, no. 2-3, pp. 231–244, 1996.

[12] G. Zhou and G. B. Giannakis, "Harmonics in multiplicative and additive noise: performance analysis of cyclic estimators," *IEEE Transactions on Signal Processing*, vol. 43, no. 6, pp. 1445–1460, 1995.

[13] A. M. Walker, "On the estimation of a harmonic component in a time series with stationary independent residuals," *Biometrika*, vol. 58, pp. 21–36, 1971.

[14] K. Perera, "An application of Vander Corput's inequality," *Journal of Inequalities in Pure and Applied Mathematics*, vol. 2, no. 1, article 8, 2001.

[15] K. Perera, "Probability inequality concerning a chirp-type signal," *Journal of Inequalities and Applications*, vol. 2013, article 131, 8 pages, 2013.

[16] K. Perera, "An inequality concerning the kernal $\sum_{t=1}^{n} e^{i(ut+vt^2)}$," *Journal of Inequalities and Special Functions*, vol. 3, no. 2, pp. 59–70, 2012.

3

Odds Ratios Estimation of Rare Event in Binomial Distribution

Kobkun Raweesawat,[1] Yupaporn Areepong,[1] Katechan Jampachaisri,[2] and Saowanit Sukparungsee[1]

[1] *Department of Applied Statistics, Faculty of Applied Science, King Mongkut's University of Technology North Bangkok, Bangkok 10800, Thailand*
[2] *Department of Mathematics, Faculty of Science, Naresuan University, Phitsanulok 65000, Thailand*

Correspondence should be addressed to Saowanit Sukparungsee; saowanit.s@sci.kmutnb.ac.th

Academic Editor: Shein-chung Chow

We introduce the new estimator of odds ratios in rare events using Empirical Bayes method in two independent binomial distributions. We compare the proposed estimates of odds ratios with two estimators, modified maximum likelihood estimator (MMLE) and modified median unbiased estimator (MMUE), using the Estimated Relative Error (ERE) as a criterion of comparison. It is found that the new estimator is more efficient when compared to the other methods.

1. Introduction

The odds ratio is a measure of association between two independent groups on a categorical response with two possible outcomes, success and failure. The two independent groups can be two treatment groups or treatment and control groups. The odds ratio is widely used in many fields of medical and social science research. It is most commonly used in epidemiology to express the results of some clinical trials, such as in case-control studies.

A number of subjects in each group falling in each category can be summarized in a two-way contingency table. Total numbers of subjects in group 1 and group 2 are n_1 and n_2, which are assumed to be fixed. Numbers of successes in group 1 and group 2 are Y_1 and Y_2, which are considered as independent binomial random variables. Let π_1 and π_2 be probabilities of success in group 1 and group 2, respectively. The odds of success in group 1 are defined to be $\text{odds}_1 = \pi_1/(1-\pi_1)$, similar to group 2. The usual maximum likelihood estimator of odds ratio is defined as

$$\widehat{\text{OR}}_{\text{MLE}} = \frac{\text{odds}_1}{\text{odds}_2} = \frac{\pi_1/(1-\pi_1)}{\pi_2/(1-\pi_2)} = \frac{Y_1/(n_1-Y_1)}{Y_2/(n_2-Y_2)} = \frac{Y_1(n_2-Y_2)}{Y_2(n_1-Y_1)}. \quad (1)$$

Odds ratio is nonnegative real value. When successes are similar in both groups, the odds ratio is equal to 1, meaning that groups are independent of response. When the odds of a positive response are higher in group 1 than in group 2, the odds ratio is greater than 1 and vice versa for the value less than 1. The father of odds ratio from 1 in a given direction represents stronger association. In addition, its sampling distribution is highly skewed. Sample natural logarithm of odds ratio, which is less skewed, is often utilized for inference. However, odds ratio can be zero (if zero cell count appears in numerator of (1)) or infinity (if zero cell count is in denominator of (1)) or undefined (if there are zero cell counts in both the numerator and denominator of (1)). Haldane [1] and Gart and Zweifel [2] suggested to add a correction term 0.5 to each cell, when having zero cell count, which gives the modified maximum likelihood estimator (MMLE) as

$$\widehat{\text{OR}}_{\text{MMLE}} = \frac{(Y_1+0.5)(n_2-Y_2+0.5)}{(Y_2+0.5)(n_1-Y_1+0.5)}. \quad (2)$$

Even though $\widehat{\text{OR}}_{\text{MMLE}}$ still laid between 0 and infinity, some investigators discouraged adding 0.5 to each cell because of the appearance of adding "fake data"; see Bishop et al. [3] and Agresti and Yang [4]. Among controversy, several similar alternatives to this modified maximum likelihood estimator have been proposed. Hirji et al. [5] proposed the median unbiased estimator (MUE) of the odds ratio, obtained from the conditional noncentral hypergeometric distribution. However, the median unbiased estimator of the odds ratio still caused a problem when $Y_1 = n_1$ and $Y_2 = n_2$ or $Y_1 = Y_2 = 0$, and then the MUE was undefined. Parzen et al. [6] proposed an estimator of the odds ratio based on MUE called the modified median unbiased estimator (MMUE) of which the estimated probability of success was always in the interval (0, 1), even if there were 0 or n successes in each group. Consequently, the estimated odds ratio always laid between 0 and infinity. Additionally, this method performed well with respect to bias in small sample and was an alternative to adding "fake data."

In this paper, we focus on "rare events" which occasionally observed zero or small counts of interesting events which happened within a given time period or a given sample such as natural disasters or some diseases. As aforementioned, rare events caused difficulty in estimation of odds ratio due to the occurrence of zeros or small observed counts in numerator or in denominator or in both, resulting in the large standard error and therefore less precise confidence interval. Only a rough estimate of the odds ratio is thus obtained. Researches involving association between categorical variables in contingency table have long been studied, using both classical and Bayesian approaches. Good [7] studied association factor, at early stage, in large contingency table with small entries, assuming log-normal and Pearson type III distribution. The author also mentioned that these assumptions may be less accurate but easy to handle. Fisher [8] estimated the odds ratio based on hypergeometric distribution utilizing exact method in a 2 × 2 table. Thomas and Gart [9] constructed a table for 95% confidence limits of differences and ratio of two proportions, including odds ratio and one-tailed p value for Fisher-Irwin Exact test in various types of 2 × 2 table. Altham [10] studied association and exact p value in a 2 × 2 contingency table based on the cumulative posterior probabilities which was not easy to extract. Nurminen and Mutanen [11] proposed Bayesian approach for the estimation of difference between two proportions, risk ratio and odds ratio, using independent beta prior and provided integral expressions for the cumulative posterior distribution. They also applied the proposed method to real data regarding malignant lymphoma and colon cancer cases exposed to phenoxy acids and chlorophenols in agriculture. Nouri et al. [12] presented the estimation of the odds ratio in 2 × 2 × J tables when exposure was misclassified. They compared the matrix and inverse matrix methods to the MLE method using simulation study and found that the inverse matrix method having a closed form was more efficient than the matrix method.

As previously mentioned, the estimates of association measure in two-way contingency table can be carried out based on classical and Bayesian approaches. The exact distribution using classical approach is, however, rather difficult for mathematical tractability. In Bayesian approach, where prior belief is incorporated into derivation of posterior density, the hyperparameters, characterizing the prior density, are often unknown to researchers and need to be assessed irrespective of current data. However, controversy still exists. Alternatively, the estimation of hyperparameters is plausibly carried out with the notion of Empirical Bayes method using current data to estimate the unknown hyperparameters, contrary to Bayesian approach. As a consequence, we focus on the utilization of Empirical Bayes method to estimate the odds ratio in a two-way contingency table, focusing on small proportions of success. Our purposed estimation tends to outperform the traditional estimator, MMLE, and MMUE without interference in the original data.

The rest of this paper is organized in the following sequence. In the next section, we discuss the median unbiased estimator. The third section describes the odds ratio estimation using EB method. The forth section illustrates simulated results, and the efficiency of EB is compared with MMLE and MUE. The fifth section displays the application of our method to real data. Our conclusion is drawn in the final section.

2. The Modified Median Unbiased Estimator of Odds Ratio

Parzen et al. [6] suggested the modified median unbiased estimator (MMUE) in two independent binomial distributions. Let $\tilde{p}$ be the estimator of success probability which satisfies

$$\begin{aligned} P(\tilde{p} \le p) &\ge 0.5, \\ P(\tilde{p} \ge p) &\ge 0.5. \end{aligned} \tag{3}$$

To obtain $\tilde{p}_t$, they use the binomial distribution, $Y_t \sim B(n_t, p_t)$, where Y_t denotes random variable representing success in the tth group ($t = 1, 2$). Let y_t be the observed value of Y_t.

$$P(Y_t = y_t \mid p_t) = \binom{n_t}{y_t} p_t^{y_t} (1 - p_t)^{n_t - y_t}. \tag{4}$$

The MMUE can be computed from the distribution of sufficient statistics for binomial data.

Compute the values of p_t^L and p_t^U to be those value of p_t for which

$$\begin{aligned} P\left(Y_t \ge y_t \mid p_t = \tilde{p}_t^L\right) &\ge 0.5, \\ P\left(Y_t \le y_t \mid p_t = \tilde{p}_t^U\right) &\ge 0.5, \end{aligned} \tag{5}$$

where p_t^L and p_t^U are the smallest and largest values of p_t, respectively. Then, the MMUE is defined as

$$\tilde{p}_t = \frac{\left(\tilde{p}_t^L + \tilde{p}_t^U\right)}{2}. \tag{6}$$

When $0 < Y_t \leq n_t$, we can find values of $\tilde{p}_t^L$ and $\tilde{p}_t^U$ which satisfy

$$P\left(Y_t \geq y_t \mid p_t = \tilde{p}_t^L\right) = P\left(Y_t \leq y_t \mid p_t = \tilde{p}_t^U\right) = 0.5. \quad (7)$$

Then, solve from

$$\begin{aligned} 0.5 &= P\left(Y_t \geq y_t \mid p_t = \tilde{p}_t^L\right) \\ &= \sum_{i=y_t}^{n_t} \binom{n_t}{i} \left(\tilde{p}_t^L\right)^i \left(1 - \tilde{p}_t^L\right)^{n_t - i}, \end{aligned} \quad (8)$$

and solve $\tilde{p}_t^U$ from

$$\begin{aligned} 0.5 &= P\left(Y_t \leq y_t \mid p_t = \tilde{p}_t^U\right) \\ &= \sum_{i=0}^{y_t} \binom{n_t}{i} \left(\tilde{p}_t^U\right)^i \left(1 - \tilde{p}_t^U\right)^{n_t - i}. \end{aligned} \quad (9)$$

The values of $\tilde{p}_t^L$ and $\tilde{p}_t^U$ can then actually be obtained by using the relationship between the cumulative beta distribution and the cumulative binomial distribution function as follows (Daly [13] and Johnson et al. [14]).

Let $U \sim \text{Beta}(\alpha, \beta)$:

$$\begin{aligned} F(p_t, \alpha, \beta) &= \int_0^{p_t} \frac{\Gamma(\alpha + \beta)}{\Gamma(\alpha)\Gamma(\beta)} U^{\alpha - 1} (1 - U)^{\beta - 1} dU \\ &= \sum_{i=\alpha}^{\alpha+\beta-1} \binom{\alpha + \beta - 1}{i} p_t^i (1 - p_t)^{(\alpha+\beta-1)-i} \\ &= \sum_{i=\alpha}^{n_t} \binom{n_t}{i} p_t^i (1 - p_t)^{n_t - i}. \end{aligned} \quad (10)$$

We need to find $\tilde{p}_t^L$ and $\tilde{p}_t^U$ such that

$$\begin{aligned} F\left(\tilde{p}_t^L \mid \alpha = y_t,\ \beta = n_t - y_t + 1\right) &= 0.5, \\ F\left(\tilde{p}_t^U \mid \alpha = y_t + 1,\ \beta = n_t - y_t + 2\right) &= 0.5. \end{aligned} \quad (11)$$

In particular,

$$\begin{aligned} \tilde{p}_t^L &= F^{-1}\left(0.5 \mid \alpha = y_t,\ \beta = n_t - y_t + 1\right), \\ \tilde{p}_t^U &= F^{-1}\left(0.5 \mid \alpha = y_t + 1,\ \beta = n_t - y_t + 2\right), \end{aligned} \quad (12)$$

where $F^{-1}(Q \mid \alpha, \beta)$ is the Qth quantile of the beta-distribution with parameters α and β.

TABLE 1: The estimated values of odds ratio for $(n_1, n_2) = (10, 10)$.

(p_1, p_2)	OR	$\widehat{\text{OR}}_{\text{EB}}$	$\widehat{\text{OR}}_{\text{MMLE}}$	$\widehat{\text{OR}}_{\text{MMUE}}$
(0.01, 0.01)	1.0000	1.3665	1.1514	1.2043
(0.01, 0.03)	0.3266	0.3931	1.0029	1.0377
(0.01, 0.05)	0.1919	0.2248	0.8723	0.8935
(0.01, 0.10)	0.0909	0.1040	0.6219	0.6184
(0.01, 0.15)	0.0572	0.0650	0.4481	0.4307
(0.03, 0.01)	3.0619	3.8746	1.6008	1.7848
(0.03, 0.03)	1.0000	1.1119	1.3933	1.5383
(0.03, 0.05)	0.5876	0.6363	1.2128	1.3244
(0.03, 0.10)	0.2784	0.2942	0.8640	0.9156
(0.03, 0.15)	0.1753	0.1838	0.6227	0.6378
(0.05, 0.01)	5.2105	6.5657	2.0724	2.4094
(0.05, 0.03)	1.7018	1.8851	1.8036	2.0763
(0.05, 0.05)	1.0000	1.0787	1.5693	1.7868
(0.05, 0.10)	0.4737	0.4989	1.1181	1.2356
(0.05, 0.15)	0.2982	0.3116	0.8059	0.8609
(0.10, 0.01)	11.0000	13.7434	3.3472	4.1489
(0.10, 0.03)	3.5926	3.9471	2.9135	3.5759
(0.10, 0.05)	2.1111	2.2585	2.5352	3.0777
(0.10, 0.10)	1.0000	1.0445	1.8068	2.1288
(0.10, 0.15)	0.6296	0.6523	1.3027	1.4839
(0.15, 0.01)	17.4706	21.7299	4.7827	6.1533
(0.15, 0.03)	5.7059	6.2349	4.1625	5.3026
(0.15, 0.05)	3.3529	3.5678	3.6225	4.5648
(0.15, 0.10)	1.5882	1.6498	2.5812	3.1568
(0.15, 0.15)	1.0000	1.0303	1.8602	2.1990

Now suppose $y_t = 0$, and then

$$\begin{gathered} P\left(Y_t \geq y_t \mid p_t = \tilde{p}_t^L\right) = P\left(Y_t \geq 0 \mid p_t = \tilde{p}_t^L\right) = 1, \\ \sum_{i=0}^{n_t} \binom{n_t}{i} \left(\tilde{p}_t^L\right)^i \left(1 - \tilde{p}_t^L\right)^{n_t - i} = 1. \end{gathered} \quad (13)$$

Any value of $\tilde{p}_t^L$ in the interval [0, 1] satisfies

$$P\left(Y_t \geq 0 \mid p_t = \tilde{p}_t^L\right) \geq 0.5, \quad (14)$$

where $\tilde{p}_t^L = 0$ is the smallest possible value of $\tilde{p}_t^L$.

Similarly, when $y_t = 0$, $\tilde{p}_t^U$ satisfies

$$\begin{gathered} P\left(Y_t \leq y_t \mid p_t = \tilde{p}_t^U\right) = P\left(Y_t = 0 \mid p_t = \tilde{p}_t^U\right) = 0.5, \\ \binom{n_t}{0} \left(\tilde{p}_t^U\right)^0 \left(1 - \tilde{p}_t^U\right)^{n_t - 0} = 0.5, \\ \tilde{p}_t^U = 1 - 0.5^{(1/n_t)}. \end{gathered} \quad (15)$$

Consequently, $y_t = 0$; $\tilde{p}_t$ equals

$$\tilde{p}_t = \frac{\left(\tilde{p}_t^L + \tilde{p}_t^U\right)}{2} = \frac{1 - 0.5^{(1/n)}}{2}. \quad (16)$$

Similarly, when $\tilde{p}_t^U = 1$ is the largest possible value of $\tilde{p}_t^U$, then $\tilde{p}_t^L$ satisfies

$$\binom{n}{n}\left(\tilde{p}_t^L\right)^n\left(1-\tilde{p}_t^L\right)^{n-n} = 0.5, \qquad \tilde{p}_t^L = (0.5)^{1/n} \tag{17}$$

when $y_t = n$ and $\tilde{p}_t = (\tilde{p}_t^L + \tilde{p}_t^U)/2$.

Then, the MMUE of odds ratio estimation is defined as

$$\widehat{\text{OR}}_{\text{MMUE}} = \frac{\tilde{p}_1/(1-\tilde{p}_1)}{\tilde{p}_2/(1-\tilde{p}_2)}, \tag{18}$$

where $\tilde{p}_1$ and $\tilde{p}_2$ denote success probability estimators in groups 1 and 2, respectively.

3. Proposed Estimation of Odds Ratio

In this section, we proposed a new method for odds ratio estimation using Empirical Bayes method in two independent binomial distributions. Let Y_1 and Y_2 be random variables, distributed as binomial with equal and unequal sample sizes and unknown probability, $Y_1 \sim \text{Bin}(n_1, p_1)$ and $Y_2 \sim \text{Bin}(n_2, p_2)$, where n_1, n_2 and p_1, p_2 denote two sample sizes and two unknown success probabilities. Adopt information priors on p_i, $p_i \sim \text{beta}(\alpha_i, \beta_i)$, $i = 1, 2$, where α_i and β_i denote unknown hyperparameters. The estimation of hyperparameters can be obtained from the posterior marginal distribution function as follows:

$$m(\mathbf{y} \mid \alpha, \beta) = \int_0^1 f(\mathbf{y} \mid p)\,\pi(p)\,dp = \binom{n}{y}\frac{\Gamma(\alpha+\beta)}{\Gamma(\alpha)\Gamma(\beta)}\frac{\Gamma(y+\alpha)\Gamma(n-y+\beta)}{\Gamma(n+\alpha+\beta)}. \tag{19}$$

Consequently, the posterior marginal distribution function of y is the beta-binomial distribution (BBD).

Then, both hyperparameters in each group can be estimated using maximum likelihood method. The likelihood function of posterior marginal distribution function is then written as

$$\begin{aligned} L(\mathbf{y} \mid \alpha, \beta) &= \prod_{i=1}^{n}\binom{n}{y_i}\frac{\Gamma(\alpha+\beta)}{\Gamma(\alpha)\Gamma(\beta)}\frac{\Gamma(y_i+\alpha)\Gamma(n-y_i+\beta)}{\Gamma(n+\alpha+\beta)} \\ &= \prod_{i=1}^{n}\binom{n}{y_i}\frac{(\alpha+y_i-1)(\alpha+y_i-2)\cdots\alpha(\beta+n-y_i-1)(\beta+n-y_i-2)\cdots\beta}{(\alpha+\beta+n-1)(\alpha+\beta+n-2)\cdots(\alpha+\beta)}. \end{aligned} \tag{20}$$

Applying Newton-Raphson method to solve a nonlinear equation, the $(r+1)$th maximum likelihood estimator of hyperparameters $(r = 1, 2, \ldots)$ can be obtained from

$$\begin{bmatrix}\hat{\alpha}^{(r+1)} \\ \hat{\beta}^{(r+1)}\end{bmatrix} = \begin{bmatrix}\hat{\alpha}^{(r)} \\ \hat{\beta}^{(r)}\end{bmatrix} - \begin{bmatrix}\dfrac{\partial^2 L^{(r)}}{\partial\alpha^2} & \dfrac{\partial^2 L^{(r)}}{\partial\alpha\partial\beta} \\ \dfrac{\partial^2 L^{(r)}}{\partial\beta\partial\alpha} & \dfrac{\partial^2 L^{(r)}}{\partial\beta^2}\end{bmatrix}^{-1}\begin{bmatrix}\dfrac{\partial L^{(r)}}{\partial\alpha} \\ \dfrac{\partial L^{(r)}}{\partial\beta}\end{bmatrix}, \tag{21}$$

where

$$\begin{aligned} \frac{\partial^2 L^{(r)}}{\partial\alpha^2} &= -\sum_{i=1}^{n}\left[\frac{1}{(\alpha^{(r)}+y_i-1)^2} + \frac{1}{(\alpha^{(r)}+y_i-2)^2} + \cdots + \frac{1}{(\alpha^{(r)})^2}\right] + \sum_{i=1}^{n}\left[\frac{1}{(\alpha^{(r)}+\beta^{(r)}+n-1)^2}\right. \\ &\quad \left. + \frac{1}{(\alpha^{(r)}+\beta^{(r)}+n-2)^2} + \cdots + \frac{1}{(\alpha^{(r)}+\beta^{(r)})^2}\right], \\ \frac{\partial^2 L^{(r)}}{\partial\alpha\partial\beta} &= \sum_{i=1}^{n}\left[\frac{1}{(\alpha^{(r)}+\beta^{(r)}+n-1)^2} + \frac{1}{(\alpha^{(r)}+\beta^{(r)}+n-2)^2} + \cdots + \frac{1}{(\alpha^{(r)}+\beta^{(r)})^2}\right], \\ \frac{\partial^2 L^{(r)}}{\partial\beta^2} &= \sum_{i=1}^{n}\left[\frac{1}{(\beta^{(r)}+n+y_i-1)^2} + \frac{1}{(\beta^{(r)}+n+y_i-2)^2} + \cdots + \frac{1}{(\beta^{(r)})^2}\right] + \sum_{i=1}^{n}\left[\frac{1}{(\alpha^{(r)}+\beta^{(r)}+n-1)^2}\right. \\ &\quad \left. + \frac{1}{(\alpha^{(r)}+\beta^{(r)}+n-2)^2} + \cdots + \frac{1}{(\alpha^{(r)}+\beta^{(r)})^2}\right], \\ \frac{\partial L^{(r)}}{\partial\alpha} &= \sum_{i=1}^{n}\left[\frac{1}{\alpha^{(r)}+y_i-1} + \frac{1}{\alpha^{(r)}+y_i-2} + \cdots + \frac{1}{\alpha^{(r)}}\right] - \sum_{i=1}^{n}\left[\frac{1}{\alpha^{(r)}+\beta^{(r)}+n-1}\right. \\ &\quad \left. + \frac{1}{\alpha^{(r)}+\beta^{(r)}+n-2} + \cdots + \frac{1}{\alpha^{(r)}+\beta^{(r)}}\right], \end{aligned}$$

TABLE 2: The estimated values of odds ratio for $(n_1, n_2) = (10, 30)$.

(p_1, p_2)	OR	$\widehat{\mathrm{OR}}_{\mathrm{EB}}$	$\widehat{\mathrm{OR}}_{\mathrm{MMLE}}$	$\widehat{\mathrm{OR}}_{\mathrm{MMUE}}$
(0.01, 0.01)	1.0000	1.3656	2.9352	2.7925
(0.01, 0.03)	0.3266	0.4023	2.0063	1.8490
(0.01, 0.05)	0.1919	0.2279	1.4146	1.2576
(0.01, 0.10)	0.0909	0.4043	0.6802	0.5486
(0.01, 0.15)	0.0572	0.0652	0.3981	0.2971
(0.03, 0.01)	3.0619	3.8640	4.0816	4.1395
(0.03, 0.03)	1.0000	1.1385	2.7872	2.7376
(0.03, 0.05)	0.5876	0.6452	1.9663	1.8632
(0.03, 0.10)	0.2784	1.1466	0.9457	0.8131
(0.03, 0.15)	0.1753	0.1845	0.5536	0.4406
(0.05, 0.01)	5.2105	6.5508	5.2833	5.5870
(0.05, 0.03)	1.7018	1.9307	3.6077	3.6950
(0.05, 0.05)	1.0000	1.0940	2.5446	2.5144
(0.05, 0.10)	0.4737	1.9430	1.2236	1.0969
(0.05, 0.15)	0.2982	0.3128	0.7163	0.5944
(0.10, 0.01)	11.0000	13.7159	8.5346	9.6223
(0.10, 0.03)	3.5926	4.0427	5.8289	6.3651
(0.10, 0.05)	2.1111	2.2907	4.1132	4.3339
(0.10, 0.10)	1.0000	1.0327	1.6556	1.5221
(0.10, 0.15)	0.6296	0.6549	1.1578	1.0247
(0.15, 0.01)	17.4706	21.6662	12.1932	14.2687
(0.15, 0.03)	5.7059	6.3850	8.3278	9.4395
(0.15, 0.05)	3.3529	3.6181	5.8732	6.4225
(0.15, 0.10)	1.5882	6.4257	2.8237	2.8019
(0.15, 0.15)	1.0000	1.0345	1.6529	1.5181

TABLE 3: The estimated values of odds ratio for $(n_1, n_2) = (10, 50)$.

(p_1, p_2)	OR	$\widehat{\mathrm{OR}}_{\mathrm{EB}}$	$\widehat{\mathrm{OR}}_{\mathrm{MMLE}}$	$\widehat{\mathrm{OR}}_{\mathrm{MMUE}}$
(0.01, 0.01)	1.0000	1.5096	4.2803	3.9580
(0.01, 0.03)	0.3266	0.4210	2.3790	2.0799
(0.01, 0.05)	0.1919	0.6975	1.4513	1.1915
(0.01, 0.10)	0.0909	0.1057	0.6167	0.4475
(0.01, 0.15)	0.0572	0.0656	0.3656	0.2515
(0.03, 0.01)	3.0619	4.2760	5.9515	5.8667
(0.03, 0.03)	1.0000	1.1926	3.3063	3.0810
(0.03, 0.05)	0.5876	1.9756	2.0173	1.7653
(0.03, 0.10)	0.2784	0.2992	0.8578	0.6636
(0.03, 0.15)	0.1753	0.1856	0.5083	0.3728
(0.05, 0.01)	5.2105	7.2486	7.7012	7.9149
(0.05, 0.03)	1.7018	2.0224	4.2796	4.1586
(0.05, 0.05)	1.0000	3.3494	2.6114	2.3833
(0.05, 0.10)	0.4737	0.5073	1.1100	0.8954
(0.05, 0.15)	0.2982	0.3147	0.6579	0.5031
(0.10, 0.01)	11.0000	15.1780	12.4415	13.6337
(0.10, 0.03)	3.5926	4.2347	6.9165	7.1666
(0.10, 0.05)	2.1111	7.0132	4.2213	4.1087
(0.10, 0.10)	1.0000	1.0621	1.7938	1.5433
(0.10, 0.15)	0.6296	0.6589	1.0631	0.8669
(0.15, 0.01)	17.4706	23.9776	17.7763	20.2195
(0.15, 0.03)	5.7059	6.6884	9.8777	10.6225
(0.15, 0.05)	3.3529	11.0773	6.0256	6.0860
(0.15, 0.10)	1.5882	1.6775	2.5614	2.2869
(0.15, 0.15)	1.0000	1.0408	1.5181	1.2848

$$\frac{\partial L^{(r)}}{\partial \beta} = \sum_{i=1}^{n}\left[\frac{1}{\beta^{(r)}+n+y_i-1}+\frac{1}{\beta^{(r)}+n+y_i-2}+\cdots+\frac{1}{\beta^{(r)}}\right]-\sum_{i=1}^{n}\left[\frac{1}{\alpha^{(r)}+\beta^{(r)}+n-1}+\frac{1}{\alpha^{(r)}+\beta^{(r)}+n-2}+\cdots+\frac{1}{\alpha^{(r)}+\beta^{(r)}}\right], \tag{22}$$

where the moment estimators of hyperparameters in beta-binomial distribution are used as initial values; see Minka [15].

The posterior distribution function of p is thus calculated, yielding

$$\pi(p \mid \mathbf{y}, \alpha, \beta) = p^{(y+\alpha-1)}(1-p)^{(n-y+\beta-1)}\frac{\Gamma(n+\alpha+\beta)}{\Gamma(y+\alpha)\Gamma(n-y+\beta)}. \tag{23}$$

Substituting the estimators of α and β, we obtain

$$p \mid \mathbf{y}, \alpha, \beta \sim \mathrm{Beta}\left(y+\widehat{\alpha}, n-y+\widehat{\beta}\right). \tag{24}$$

Let p_1' and p_2' be estimators of p_1 and p_2, respectively, where

$$p_1' = \frac{y_1+\widehat{\alpha}_1}{n_1+\widehat{\alpha}_1+\widehat{\beta}_1}, \qquad p_2' = \frac{y_2+\widehat{\alpha}_2}{n_2+\widehat{\alpha}_2+\widehat{\beta}_2}. \tag{25}$$

Thus, the EB estimator of odds ratio can be obtained as follows:

$$\widehat{\mathrm{OR}}_{\mathrm{EB}} = \frac{p_1'/(1-p_1')}{p_2'/(1-p_2')}, \tag{26}$$

where p_1' and p_2' denote success probability estimators in groups 1 and 2, respectively.

4. Simulation Study for MMLE, MMUE, and EB Method

Simulation studies have been carried out using R program (version 3.2.0) [16] to assess the efficiency of the EB method in comparison with two existing methods. Binomial data are generated with equal and unequal sample sizes: $(n_1, n_2) = (10, 10), (10, 30)$ with $(10, 50)$ probabilities of success in group 1: $p_1 = 0.01, 0.03, 0.05, 0.1$, and 0.15. For each value

TABLE 4: The percentages of the Estimated Relative Error of odds ratio estimation for $(n_1, n_2) = (10, 10)$.

(p_1, p_2)	ERE_{EB}	ERE_{MMLE}	ERE_{MMUE}
(0.01, 0.01)	36.6535	15.1385	20.4316
(0.01, 0.03)	20.3505	206.7910	217.7388
(0.01, 0.05)	17.1281	354.5204	365.6116
(0.01, 0.10)	14.3630	584.0771	580.2372
(0.01, 0.15)	13.4914	682.8616	652.4367
(0.03, 0.01)	26.5436	47.7187	41.7082
(0.03, 0.03)	11.1895	39.3332	53.8303
(0.03, 0.05)	8.2767	106.3838	125.3783
(0.03, 0.10)	5.6857	210.3971	228.9552
(0.03, 0.15)	4.8556	255.2977	263.9451
(0.05, 0.01)	26.0092	60.2273	53.7592
(0.05, 0.03)	10.7763	5.9852	22.0098
(0.05, 0.05)	7.8661	56.9330	78.6812
(0.05, 0.10)	5.3167	136.0498	160.8427
(0.05, 0.15)	4.4721	170.2257	188.6564
(0.10, 0.01)	24.9404	69.5705	62.2831
(0.10, 0.03)	9.8685	18.9032	0.4659
(0.10, 0.05)	6.9818	20.0893	45.7877
(0.10, 0.10)	4.4467	80.6791	112.8820
(0.10, 0.15)	3.6016	106.9073	135.6854
(0.15, 0.01)	24.3801	72.6244	64.7792
(0.15, 0.03)	9.2720	27.0495	7.0675
(0.15, 0.05)	6.4085	8.0392	36.1422
(0.15, 0.10)	3.8773	62.5224	98.7588
(0.15, 0.15)	3.0338	86.0169	119.9015

TABLE 5: The percentages of the Estimated Relative Error of odds ratio estimation for $(n_1, n_2) = (10, 30)$.

(p_1, p_2)	ERE_{EB}	ERE_{MMLE}	ERE_{MMUE}
(0.01, 0.01)	36.5558	193.5166	179.2467
(0.01, 0.03)	23.1721	514.2925	466.1248
(0.01, 0.05)	18.7270	637.0581	555.2608
(0.01, 0.10)	344.7753	648.1943	503.4379
(0.01, 0.15)	13.8933	595.4561	419.0998
(0.03, 0.01)	26.1981	33.3063	35.1969
(0.03, 0.03)	13.8541	178.7227	173.7624
(0.03, 0.05)	9.7978	234.6089	217.0772
(0.03, 0.10)	311.9138	239.7364	192.0982
(0.03, 0.15)	5.2578	215.8678	151.3923
(0.05, 0.01)	25.7219	1.3963	7.2250
(0.05, 0.03)	13.4545	112.0004	117.1259
(0.05, 0.05)	9.3998	154.4641	151.4381
(0.05, 0.10)	310.1875	158.3124	131.5756
(0.05, 0.15)	4.8844	140.1677	99.2831
(0.10, 0.01)	24.6897	22.4126	12.5246
(0.10, 0.03)	12.5282	62.2482	77.1730
(0.10, 0.05)	8.5055	94.8364	105.2921
(0.10, 0.10)	3.2693	65.5584	52.2104
(0.10, 0.15)	4.0198	83.8863	62.7515
(0.15, 0.01)	24.0150	30.2071	18.3274
(0.15, 0.03)	11.9013	45.9518	65.4338
(0.15, 0.05)	7.9089	75.1670	91.5471
(0.15, 0.10)	304.5808	77.7868	76.4163
(0.15, 0.15)	3.4464	65.2920	51.8093

of p_1, p_2 is varied to 0.01, 0.03, 0.05, 0.1, and 0.15. Each situation is repeated 5,000 times after removing the first 1,000 iterations (1,000 burn-ins). The efficiency of proposed estimator is evaluated using Estimated Relative Error (ERE), defined as

$$\text{ERE} = \left[\frac{\left|\text{OR} - \widehat{\text{OR}}_i\right|}{\text{OR}} \right], \tag{27}$$

where OR denotes the usual maximum likelihood estimator of odds ratio and $\widehat{\text{OR}}_i$ denotes the estimate of odds ratio using EB, MMLE, and MMUE ($i = 1, 2, 3, \ldots$), respectively.

The simulation results with odds ratio estimates for sample sizes $(n_1, n_2) = (10, 10), (10, 30)$ and $(10, 50)$ are given in Tables 1–3. The performance of estimation uses ERE given in Tables 4–6 and compares this result with graph in Figure 1; the other case provides similar results. It is found that the odds ratio estimation using EB method mostly yields smallest ERE with 78.67%, while those using MMLE and MMUE methods result in smallest ERE with only 6.67% and 14.66%, respectively.

5. Illustrative Examples Using Real Data

Our first example is taken from the studies of Good [7] and Hardell [17]. As shown in Table 7, subjects with malignant lymphoma and colon cancer cases and controls who are shortly exposed to phenoxy acids in agriculture and forestry were observed, including the true odds ratios and their estimates using EB, MMLE, and MMUE. For outcome in which $(Y_1, Y_2) = (8, 9)$ out of $(n_1, n_2) = (24, 53)$ for cases and control, respectively, the estimate of the odds ratio using EB method yields the least ERE with 0.5523, while those using MMLE and MMUE methods result in ERE with 1.2805 and 4.1483, respectively.

The second example is taken from the study of Perondi et al. [18], as shown in Table 7, which compared high-dose epinephrine and standard-dose epinephrine in children with cardiac arrest with 34 children in each treatment, including the true odds ratios and their estimates using EB, MMLE, and MMUE. For outcome measure was survival at 24 hours in which $(Y_1, Y_2) = (1, 7)$ out of $(n_1, n_2) = (34, 34)$ for high dose and standard dose, respectively. The estimate of the odds ratio using EB method yields the least ERE with 5.2097, while those using MMUE and MMLE methods result in ERE with 15.5305 and 40.4643, respectively.

6. Conclusion

Based on simulated study for odds ratio estimation in rare events with two independent binomial data, the result indicates that the proposed method performs rather well.

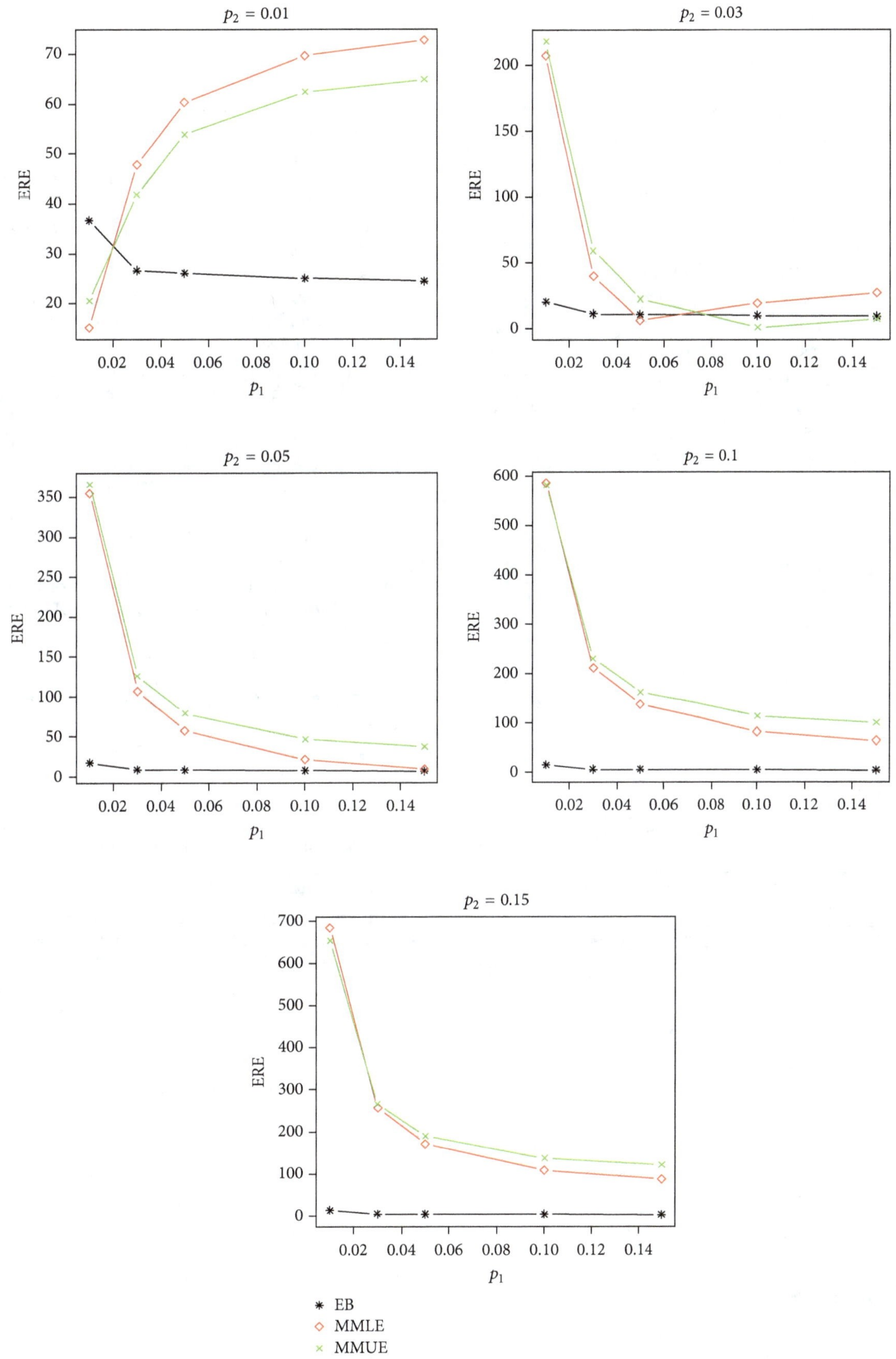

FIGURE 1: The percentages of ERE for odds ratio estimation using EB, MMLE, and MMUE when $(n_1, n_2) = (10, 10)$.

TABLE 6: The percentages of the Estimated Relative Error of odds ratio estimation for $(n_1, n_2) = (10, 50)$.

(p_1, p_2)	ERE_{EB}	ERE_{MMLE}	ERE_{MMUE}
(0.01, 0.01)	50.9628	328.0343	295.7966
(0.01, 0.03)	28.9076	628.4030	536.8268
(0.01, 0.05)	263.4419	656.1777	520.8249
(0.01, 0.10)	16.2619	578.3828	392.2232
(0.01, 0.15)	14.5797	538.7736	339.4512
(0.03, 0.01)	39.6535	94.3767	91.6069
(0.03, 0.03)	19.2635	230.6324	208.1003
(0.03, 0.05)	236.1914	243.2889	200.4045
(0.03, 0.10)	7.4747	208.1561	138.4178
(0.03, 0.15)	5.8990	190.0357	112.7091
(0.05, 0.01)	39.1153	47.8002	51.9027
(0.05, 0.03)	18.8401	151.4810	144.3693
(0.05, 0.05)	234.9386	161.1356	138.3292
(0.05, 0.10)	7.0955	134.3384	89.0341
(0.05, 0.15)	5.5239	120.5896	68.6865
(0.10, 0.01)	37.9815	13.1041	23.9423
(0.10, 0.03)	17.8718	92.5224	99.4826
(0.10, 0.05)	232.2048	99.9578	94.6234
(0.10, 0.10)	6.2116	79.3838	54.3286
(0.10, 0.15)	4.6546	68.8420	37.6836
(0.15, 0.01)	37.2457	1.7496	15.7345
(0.15, 0.03)	17.2201	73.1152	86.1673
(0.15, 0.05)	230.3755	79.7100	81.5120
(0.15, 0.10)	5.6227	61.2754	43.9902
(0.15, 0.15)	4.0763	51.8150	28.4826

TABLE 7: True odds ratios and their estimates using EB, MMLE, and MMUE, with corresponding percentages of ERE.

			Methods		
		True	EB	MMLE	MMUE
1st example	OR	2.4444	2.4309	2.4131	2.3430
	ERE	—	0.5523	1.2805	4.1483
2nd example	OR	0.1169	0.1230	0.1642	0.1350
	ERE	—	5.2097	40.4643	15.5305

The EB estimator of odds ratio is also more efficient than the other two estimators, MMLE and MMUE. In addition, our purposed estimator is an alternative method for odds ratio estimation to the MMLE method without disturbing the original data.

Competing Interests

The authors declare that they have no competing interests.

Acknowledgments

The authors are grateful to the Graduate College, King Mongkut's University of Technology North Bangkok, for the financial support.

References

[1] J. B. S. Haldane, "The estimation and significance of the logarithm of a ratio of frequencies," *Annals of Human Genetics*, vol. 20, no. 4, pp. 309–311, 1956.

[2] J. J. Gart and J. R. Zweifel, "On the bias of various estimators of the logit and its variance with application of quantal bioassay," *Biometrika*, vol. 54, no. 1-2, pp. 181–187, 1967.

[3] Y. M. Bishop, S. E. Fienberg, and P. W. Holland, *Discrete Multivariate Analysis: Theory and Practice*, Springer, New York, NY, USA, 2007.

[4] A. Agresti and M.-C. Yang, "An empirical investigation of some effects of sparseness in contingency tables," *Computational Statistics and Data Analysis*, vol. 5, no. 1, pp. 9–21, 1987.

[5] K. F. Hirji, A. A. Tsiatis, and C. R. Mehta, "Median unbiased estimation for binary data," *The American Statistician*, vol. 43, no. 1, pp. 7–11, 1989.

[6] M. Parzen, S. Lipsitz, J. Ibrahim, and N. Klar, "An estimate of the odds ratio that always exists," *Journal of Computational and Graphical Statistics*, vol. 11, no. 2, pp. 420–436, 2002.

[7] I. J. Good, "On the estimations of small frequencies in contingency tables," *Journal of the Royal Statistical Society. Series B. Methodological*, vol. 18, pp. 113–124, 1956.

[8] R. A. Fisher, "The logic of inductive inference," *Journal of the Royal Statistical Society*, vol. 98, no. 1, pp. 39–82, 1935.

[9] D. G. Thomas and J. J. Gart, "A table of exact confidence limits for differences and ratios of two proportions and their odds ratios," *Journal of the American Statistical Association*, vol. 72, no. 357, pp. 73–76, 1977.

[10] P. M. E. Altham, "Exact Bayesian analysis of a 2 × 2 contingency table, and Fisher's 'exact' significance test," *Journal of the Royal Statistical Society. Series B (Methodological)*, vol. 31, pp. 261–269, 1969.

[11] M. Nurminen and P. Mutanen, "Exact Bayesian analysis of two proportions," *Scandinavian Journal of Statistics*, vol. 14, no. 1, pp. 67–77, 1987.

[12] B. Nouri, N. Zare, and S. M. T. Ayatollahi, "Validation study methods for estimating odds ratio in 2 × 2 × *J* tables when exposure is misclassified," *Computational and Mathematical Methods in Medicine*, vol. 2013, Article ID 170120, 8 pages, 2013.

[13] L. Daly, "Simple SAS macros for the calculation of exact binomial and Poisson confidence limits," *Computers in Biology and Medicine*, vol. 22, no. 5, pp. 351–361, 1992.

[14] N. L. Johnson, A. W. Kemp, and S. Kotz, *Univariate Discrete Distribution*, John Wiley & Sons, New Jersey, NJ, USA, 2005.

[15] T. P. Minka, "Estimating a Dirichlet distribution," Tech. Rep., The MIT Press, London, UK, 2000.

[16] The R Development Core Team, An introduction to R, Vienna, 2010, http://R-project.org.

[17] L. Hardell, "Relation of soft-tissue sarcoma, malignant lymphoma and colon cancer to phenoxy acids, chlorophenols and other agents," *Scandinavian Journal of Work, Environment and Health*, vol. 7, no. 2, pp. 119–130, 1981.

[18] M. B. M. Perondi, A. G. Reis, E. F. Paiva, V. M. Nadkarni, and R. A. Berg, "A comparison of high-dose and standard-dose epinephrine in children with cardiac arrest," *The New England Journal of Medicine*, vol. 350, no. 17, pp. 1722–1730, 2004.

4

Maximum Likelihood and Bayes Estimation in Randomly Censored Geometric Distribution

Hare Krishna and Neha Goel

Department of Statistics, Ch. Charan Singh University, Meerut, India

Correspondence should be addressed to Hare Krishna; hkrishnastats@yahoo.com

Academic Editor: Hyungjun Cho

In this article, we study the geometric distribution under randomly censored data. Maximum likelihood estimators and confidence intervals based on Fisher information matrix are derived for the unknown parameters with randomly censored data. Bayes estimators are also developed using beta priors under generalized entropy and LINEX loss functions. Also, Bayesian credible and highest posterior density (HPD) credible intervals are obtained for the parameters. Expected time on test and reliability characteristics are also analyzed in this article. To compare various estimates developed in the article, a Monte Carlo simulation study is carried out. Finally, for illustration purpose, a randomly censored real data set is discussed.

1. Introduction

Lifetime experiments are conducted to collect data on items under study. The data are used for fitting a suitable lifetime model and then inferring about the statistical properties and survival/reliability characteristics of the items. These experiments may be expensive in terms of both cost and time. To save the cost and time, lifetime experiments may be censored intentionally or censoring may occur in an experiment naturally. Many types of censoring schemes have been studied in literature such as type-I, type-II, progressive, hybrid, and random censoring schemes.

Random censoring is a situation when an item under study is lost or removed randomly from the experiment before its failure. In other words, some subjects in the study have not experienced the event of interest at the end of the study. For example, in a clinical trial or a medical study, some patients may still be untreated and leave the course of treatment before its completion. In a social study, some subjects are lost for the follow-up in the middle of the survey. In reliability engineering, an electrical or electronic device such as bulb on test may break before its failure. In such cases, the exact survival time (or time to event of interest) of the subjects is unknown; therefore they are called randomly censored observations.

The random censoring was introduced in literature by Gilbert [1]. Thereafter, Breslow and Crowley [2], Koziol and Green [3], and Csörgő and Horváth [4] also discussed randomly censored data in their work. Kim [5] did chi-square goodness of fit tests for randomly censored data. In the last decade, the recent studies on randomly censored data from exponential distribution include Friesl and Hurt [6] and Saleem and Raza [7]. Rayleigh model with randomly censored data was analyzed by Ghitany [8] and Saleem and Aslam [9]; Burr Type XII was analyzed by Ghitany and Al-Awadhi [10]; generalized exponential and Weibull models were analyzed, respectively, by Danish and Aslam [11, 12]. Krishna et al. [13] studied Maxwell distribution with randomly censored samples.

In all the above cases, survival time or failure time has been assumed to be a continuous variable. However, sometimes it is impossible or inconvenient to measure the life length of a device on a continuous scale.

In real life experiments we come across situations, where failure time data is discrete either through the grouping of continuous data due to imprecise measurement or because time itself is discrete, for example, days, weeks, or months. In such circumstances, one measures the life of a device on a discrete scale. A discrete lifetime model may also consider the number of successful cycles, trials, or operations before failure of a device. In discrete lifetime models, the one parameter geometric distribution has an important position. The geometric distribution can be used as a discrete failure to investigate the ability of electronic tubes to withstand successive voltage overloads and performance of electric switches, which are repeatedly turned on and off.

In reliability theory, geometric distribution has been considered as a lifetime model by Yaqub and Khan [14], Bhattacharya and Kumar [15], Maiti [16], Krishna and Jain [17], Sarhan and Kundu [18], and so forth. In most of above studies, a complete sample or right censoring is considered. No literature is available on random censoring in any discrete distribution.

In view of the above, this paper considers classical and Bayes estimation of the unknown parameters with some reliability characteristics for geometric distribution under randomly censored data. In Section 2, mathematical formulation for randomly censored data with failure and censoring times following geometric distributions is given. Section 3 deals with the maximum likelihood estimation (MLE) for the unknown parameters along with their variances and confidence intervals. Section 4 describes the expected time on test and observed time on test. In Section 5 we obtain the Bayes estimators for the unknown parameters under generalized entropy and LINEX loss functions using beta priors. In Section 6, we consider a Monte Carlo simulation study to explore the properties of various estimates developed in the above sections. Finally, Section 7 deals with a real data example to study the applications of random censoring in geometric distribution. It is essential to mention here that we have used statistical software R [19] for computation purposes throughout the paper.

2. The Model and Its Assumptions

In a life testing experiment or a clinical trial, n items or patients are subjected to test. Let $X_1, X_2, \ldots, X_n$ be their discrete failure or survival times. Assume that X_i's are independently and identically distributed (i.i.d.) random variables (r.v.) with probability mass function (p.m.f.) $p_X(x)$ and cumulative distribution function (c.d.f.) $F_X(x)$. Further, suppose that these items may be censored before their failure at times $T_1, T_2, \ldots, T_n$. Again, assume T_i's to be i.i.d. discrete random variables with p.m.f. $p_T(t)$ and c.d.f. $F_T(t)$. It is also assumed that X_i's and T_i's are independent. In a randomly censored experiment, minimum of X_i's and T_i's, that is, $Y_i = \min(X_i, T_i)$, will actually be observed for $i = 1, 2, \ldots, n$. Let D_i be an indicator variable defined by $D_i = 1$ (0) if $X_i \le T_i$ ($X_i > T_i$). Here, D_i is a Bernoulli random variable with probability $p = P[X_i \le T_i]$.

In the present article, we consider survival and censoring time variables to follow geometric distributions Geo(θ) & Geo(λ), respectively, with p.m.f.,

$$p_X(x) = (1-\theta)\,\theta^x; \quad x = 0, 1, \ldots; \ 0 < \theta < 1,$$
$$p_T(t) = (1-\lambda)\,\lambda^t; \quad t = 0, 1, \ldots; \ 0 < \lambda < 1. \tag{1}$$

The probability of failure of an item on test before its censoring is given by

$$p = P[\text{An item fails}] = P[D_i = 1] = P[X_i \le T_i]$$
$$= \sum_{t=0}^{\infty} P[X \le t \mid T = t]\, p_T(t) = \frac{\lambda(1-\theta)}{(1-\lambda\theta)}. \tag{2}$$

The above probability of failure is tabulated in Table 1 for various values of θ and λ. From Table 1 we observe that on increasing the value of lifetime parameter θ, the probability of failure decreases. However, as we increase the value of censoring parameter λ, the probability of failure increases.

Now, for $i = 1, 2, \ldots, n$, the Bernoulli p.m.f. of D_i is

$$P(D_i = d_i) = \left(\frac{\lambda(1-\theta)}{1-\lambda\theta}\right)^{d_i} \left(\frac{1-\lambda}{1-\lambda\theta}\right)^{1-d_i}; \quad d_i = 0, 1. \tag{3}$$

Also, the c.d.f. of $Y_i = \min(X_i, T_i)$ is

$$P(Y_i \le y) = 1 - P[Y_i > y]$$
$$= 1 - P[X_i > y]\, P[T_i > y] = 1 - (\theta\lambda)^{y+1}; \quad y = 0, 1, 2, \ldots \tag{4}$$

which is a geometric distribution with parameter $(\theta\lambda)$. The independence of X_i and T_i implies the independence of Y_i and D_i. Therefore, the joint p.m.f. of (Y_i, D_i) is given by

$$p(y_i, d_i \mid \theta, \lambda) = p[Y_i = y_i]\, p[D_i = d_i]$$
$$= \theta^{y_i}(1-\theta)^{d_i}\lambda^{y_i+d_i}(1-\lambda)^{1-d_i}; \quad d_i = 0, 1; \ y_i = 0, 1, 2, \ldots; \ 0 < \theta, \lambda < 1 \tag{5}$$

with $E(Y_i) = \theta\lambda/(1-\theta\lambda)$ and $E(D_i) = p = \lambda(1-\theta)/(1-\lambda\theta)$.

Also, note that, for Geo(θ), the reliability characteristics with mission time k are given by

(i) survival function $= S(k) = P[X \ge k] = \theta^k$; $k = 0, 1, 2, \ldots$

(ii) mean life time $= \mu = \theta/(1-\theta)$

(iii) failure rate $= h(k) = (1-\theta)$; $k = 0, 1, 2, \ldots$

3. Maximum Likelihood Estimation

The likelihood function of the randomly censored sample data (y_i, d_i), $i = 1, 2, \ldots, n$, from geometric distributions as discussed in Section 2, is given by

$$L = L(\underset{\sim}{y}, \underset{\sim}{d}) = \theta^{w_1}(1-\theta)^{w_2}\lambda^{w_1+w_2}(1-\lambda)^{n-w_2}$$
$$\text{where, } w_1 = \sum_{i=1}^{n} y_i, \ w_2 = \sum_{i=1}^{n} d_i. \tag{6}$$

Taking log, differentiating with respect to θ and λ, and equating to zero, we get the MLEs of θ and λ as $\hat{\theta} = w_1/(w_1 + w_2)$ and $\hat{\lambda} = (w_1 + w_2)/(n + w_1)$.

Table 1: Probability of failure.

SN	θ	λ								
		0.1	0.2	0.3	0.4	0.5	0.6	0.7	0.8	0.9
1	0.1	0.0909	0.1837	0.2784	0.3750	0.4737	0.5745	0.6774	0.7826	0.8901
2	0.2	0.0816	0.1667	0.2553	0.3478	0.4444	0.5455	0.6512	0.7619	0.8780
3	0.3	0.0722	0.1489	0.2308	0.3182	0.4118	0.5122	0.6203	0.7368	0.8630
4	0.4	0.0625	0.1304	0.2045	0.2857	0.3750	0.4737	0.5833	0.7059	0.8438
5	0.5	0.0526	0.1111	0.1765	0.2500	0.3333	0.4286	0.5385	0.6667	0.8182
6	0.6	0.0426	0.0909	0.1463	0.2105	0.2857	0.3750	0.4828	0.6154	0.7826
7	0.7	0.0323	0.0698	0.1139	0.1667	0.2308	0.3103	0.4118	0.5455	0.7297
8	0.8	0.0217	0.0476	0.0789	0.1176	0.1667	0.2308	0.3182	0.4444	0.6429
9	0.9	0.0110	0.0244	0.0411	0.0625	0.0909	0.1304	0.1892	0.2857	0.4737

By the invariance property of MLEs, we have

$$\begin{aligned}\hat{p} &= \frac{w_2}{n},\\ \hat{S}(k) &= \hat{\theta}^k,\\ \hat{\mu} &= \frac{\hat{\theta}}{(1-\hat{\theta})},\\ \hat{h}(k) &= 1-\hat{\theta}.\end{aligned} \tag{7}$$

The MLEs can be viewed as the Bayes estimators under the 0-1 loss function and the uniform prior. Also, the estimates derived by the method of moments coincide with the above MLEs in this case.

The Fisher information matrix is given by

$$\begin{aligned}I(\theta,\lambda) &= -E\begin{pmatrix}\dfrac{\partial^2 \log L}{\partial\theta^2} & \dfrac{\partial^2 \log L}{\partial\theta\partial\lambda}\\ \dfrac{\partial^2 \log L}{\partial\lambda\partial\theta} & \dfrac{\partial^2 \log L}{\partial\lambda^2}\end{pmatrix}\\ &= \begin{pmatrix}\dfrac{n\lambda}{\theta(1-\theta)(1-\theta\lambda)} & 0\\ 0 & \dfrac{n}{\lambda(1-\lambda)(1-\lambda\theta)}\end{pmatrix}\end{aligned} \tag{8}$$

so that the variances of the estimates are

$$\begin{aligned}V(\hat{\theta}) &= \frac{\theta(1-\theta)(1-\theta\lambda)}{n\lambda},\\ V(\hat{\lambda}) &= \frac{\lambda(1-\lambda)(1-\lambda\theta)}{n}.\end{aligned} \tag{9}$$

The estimates of the above variances can be obtained by replacing (θ,λ) by $(\hat{\theta},\hat{\lambda})$. The geometric distribution belongs to the exponential family of distributions; therefore, most of the properties of MLEs are valid in this case. The asymptotic, sampling distribution of $\hat{\theta}$ is normal $N(\hat{\theta},\mathrm{Var}(\hat{\theta}))$. Thus, a two-sided $100\times(1-\alpha)\%$ confidence interval for θ becomes $\hat{\theta}\pm z_{\alpha/2}\sqrt{\mathrm{var}(\hat{\theta})}$. Similarly, for λ the confidence interval is given by $\hat{\lambda}\pm z_{\alpha/2}\sqrt{\mathrm{var}(\hat{\lambda})}$. Here, $z_{\alpha/2}$ is the $(1-\alpha/2)$th percentile of standard normal distribution for $0<\alpha<1$.

4. Expected Time on Test

In life testing experiments, expected time on test (ETT) is beneficial to have an idea about the expected duration of the experiment. Since the time required for completing an experiment has a direct impact on the cost, this information is important for an experimenter to choose an appropriate sampling plan. Krishna et al. [13] developed ETT for Maxwell distribution under random censoring for the first time. In this section, we develop the mathematical formulation of ETT for randomly censored geometric distribution.

Let $Y_{(n)}=\max(Y_1,Y_2,\ldots,Y_n)$ be the nth order statistic in a randomly censored sample of size n, denoting the time to observe the nth failure or censoring time. Then, the c.d.f. of $Y_{(n)}$ is given by

$$F_{Y_{(n)}}(y) = P\left[Y_{(n)}\le y\right] = \left[1-(\theta\lambda)^{y+1}\right]^n; \quad y=0,1,2,\ldots \tag{10}$$

For randomly censored data, ETT is given by

$$\begin{aligned}\mathrm{ETT}(\mathrm{RCS}) &= E\left(Y_{(n)}\right) = \sum_{y=0}^{\infty}\left(1-F_{Y_{(n)}}(y)\right)\\ &= \sum_{j=1}^{n}\binom{n}{j}(-1)^{j+1}\frac{(\theta\lambda)^j}{1-(\theta\lambda)^j}.\end{aligned} \tag{11}$$

Similarly, in complete case, $X_{(n)}$ denotes the time to observe the nth failure in a sample of size n. Then, the ETT for

TABLE 2: Expected time on test (ETT) and its estimation.

SN	n	θ	λ = 0.2 ETT	0.2 AVOBTT	0.5 ETT	0.5 AVOBTT	0.8 ETT	0.8 AVOBTT
1	10	0.2	0.3517 (1.3248)	{0.37} [0.0014]	0.75800 (1.3248)	{0.752} [0.0003]	1.1015 (1.3248)	{1.1} [0.0005]
2	10	0.5	0.75800 (3.7255)	{0.763} [0.0004]	1.6162 (3.7255)	{1.612} [0.0001]	2.6965 (3.7255)	{2.722} [0.017]
3	10	0.8	1.1015 (12.6259)	{1.082} [0.0006]	2.6965 (12.6259)	{2.709} [0.0032]	6.0629 (12.6259)	{6.082} [0.0002]
4	20	0.2	0.5908 (1.7343)	{0.612} [0.0004]	1.0825 (1.7343)	{1.053} [0.0013]	1.4684 (1.7343)	{1.487} [0.0009]
5	20	0.5	1.0825 (4.6904)	{1.079} [0.0003]	2.0930 (4.6904)	{2.099} [0.0126]	3.4265 (4.6904)	{3.405} [0.002]
6	20	0.8	1.4684 (15.6229)	{1.508} [0.0005]	3.4265 (15.6229)	{3.44} [0.0016]	7.5614 (15.6229)	{7.519} [0.0002]
7	30	0.2	0.7550 (1.9778)	{0.741} [0.0018]	1.2508 (1.9778)	{1.25} [0.0002]	1.6743 (1.9778)	{1.672} [0.0004]
8	30	0.5	1.2508 (5.2635)	{1.267} [0.0004]	2.3815 (5.2635)	{2.399} [0.0003]	3.8598 (5.2635)	{3.862} [0.0003]
9	30	0.8	1.6743 (17.4032)	{1.674} [0.0026]	3.8598 (17.4032)	{3.891} [0.0006]	8.4516 (17.4032)	{8.46} [0.001]
10	50	0.2	0.9503 (2.2976)	{0.95} [0.0001]	1.4441 (2.2976)	{1.471} [0.0002]	1.9504 (2.2976)	{1.951} [0.0007]
11	50	0.5	1.4441 (5.9909)	{1.434} [0.0001]	2.7469 (5.9909)	{2.75} [0.0001]	4.4102 (5.9909)	{4.383} [0.0031]
12	50	0.8	1.9504 (19.6628)	{1.933} [0.0007]	4.4102 (19.6628)	{4.4} [0.0003]	9.5814 (19.6628)	{9.51} [0.0001]

This table shows the values of ETT for randomly censored sample and ( ) contains the ETT values of complete experiment. In the next column { } and [ ] show the average values & MSEs of OBTT in randomly censored sample.

complete sample is given by

$$\text{ETT (CS)} = E\left(X_{(n)}\right) = \sum_{y=0}^{\infty}\left(1 - F_{X_{(n)}}(x)\right) = \sum_{j=1}^{n}\binom{n}{j}(-1)^{j+1}\frac{\theta^j}{1-\theta^j}. \tag{12}$$

By invariance property of MLEs, the MLE of ETT for randomly censored data is obtained as

$$\widehat{\text{ETT}}\text{ (RCS)} = \sum_{j=1}^{n}\binom{n}{j}(-1)^{j+1}\frac{\left(\widehat{\theta\lambda}\right)^j}{1-\left(\widehat{\theta\lambda}\right)^j}. \tag{13}$$

Also, we can obtain the observed time on test OBTT $= Y_{(n)}$, which is the actual observed experimental time. Sometimes, OBTT is also proposed to estimate ETT.

One can compute ratio of expected experiment time (REET) for comparison purpose, as under

$$\text{REET} = \frac{\text{ETT (RCS)}}{\text{ETT (CS)}}. \tag{14}$$

For various values of θ, λ, and n, ETT (RCS) and ETT (CS) are computed and simulated in Table 2. From the table, it is observed that, for both randomly censored sample and complete sample, ETT increases with an increase in θ, λ, and n. The OBTT estimates ETT quite satisfactorily.

5. Bayesian Estimation

In Bayes estimation, the prior knowledge is updated by conducting an experiment and estimators are constructed to make inferences about the characteristics of interest. This technique provides valid alternatives to traditional estimation methods. Bayesian analysis is an important technique carried out with various types of loss functions and a variety of prior distributions. In this paper, we use the natural conjugate beta priors for the unknown parameters under generalized entropy and linear exponential loss functions.

Let the unknown parameters θ and λ follow the beta distributions of first kind with parameters (a_1, b_1) and (a_2, b_2), respectively. Here, θ and λ are regarded as random variables having the marginal prior distributions as

$$g_1(\theta) = \frac{1}{B(a_1, b_1)}\theta^{a_1-1}(1-\theta)^{b_1-1}; \quad a_1, b_1 > 0;\ 0 < \theta < 1$$
$$g_2(\lambda) = \frac{1}{B(a_2, b_2)}\lambda^{a_2-1}(1-\lambda)^{b_2-1}; \quad a_2, b_2 > 0;\ 0 < \lambda < 1. \tag{15}$$

By assuming that the prior distributions of θ and λ are independent, we have a joint prior that is incorporated with the likelihood to yield the following joint posterior distribution:

$$\pi\left(\theta,\lambda \mid \underset{\sim}{y},\underset{\sim}{d}\right) = \frac{L\left(\underset{\sim}{y},\underset{\sim}{d} \mid \theta,\lambda\right) g_1(\theta)\, g_2(\lambda)}{\int_\theta \int_\lambda L\left(\underset{\sim}{y},\underset{\sim}{d} \mid \theta,\lambda\right) g_1(\theta)\, g_2(\lambda)\, d\theta\, d\lambda} = \frac{1}{B(\alpha_1,\beta_1)} \frac{1}{B(\alpha_2,\beta_2)} \theta^{\alpha_1-1}(1-\theta)^{\beta_1-1} \cdot \lambda^{\alpha_2-1}(1-\lambda)^{\beta_2-1}; \quad 0<\theta,\lambda<1, \tag{16}$$

where $\alpha_1 = w_1 + a_1$; $\beta_1 = w_2 + b_1$, $\alpha_2 = w_1 + w_2 + a_2$; $\beta_2 = n - w_2 + b_2$

The posterior distributions of θ and λ are again independent beta distributions $B(\alpha_1,\beta_1)$ and $B(\alpha_2,\beta_2)$, respectively.

5.1. Bayes Estimates under Generalized Entropy Loss Function. The generalized entropy loss function (GELF) was proposed by Calabria and Pulcini [20] as

$$L(\theta,\theta^*) = \gamma\left[\left(\frac{\theta^*}{\theta}\right)^\delta - \delta\log\left(\frac{\theta^*}{\theta}\right) - 1\right], \quad \delta \neq 0,\ \gamma > 0, \tag{17}$$

where θ^* is the estimate of θ. γ is a constant which is cancelled out on dividing the numerator by denominator in the procedure to obtain the Bayes estimate. Thus, without loss of generality, we assume ($\gamma = 1$). For $\delta > 0$, a positive error has a more serious effect than a negative error and for $\delta < 0$, a negative error has a more serious effect than a positive error. In Bayes estimation, we choose such value of θ^* which minimizes the risk function $r(\theta,\theta^*) = E[L(\theta,\theta^*) \mid \text{data}]$. We get the Bayes estimator of θ by differentiating $r(\theta,\theta^*)$ with respect to θ^* and equating to zero. The Bayes estimator and the corresponding risk function of θ are given by

$$\theta^* = \left\{E\left(\theta^{-\delta} \mid \text{data}\right)\right\}^{-1/\delta}, \tag{18}$$

$$\text{Risk} = E\left[\log\left(\theta^\delta\right) \mid \text{data}\right] + \log\left\{E\left(\theta^{-\delta} \mid \text{data}\right)\right\} = E\left[\log\left(\theta^\delta\right) \mid \text{data}\right] - \log(\theta^*)^\delta. \tag{19}$$

From (16), the marginal posterior distribution of θ is given by

$$\pi_1\left(\theta \mid \underset{\sim}{y},\underset{\sim}{d}\right) = \frac{1}{B(\alpha_1,\beta_1)}\theta^{\alpha_1-1}(1-\theta)^{\beta_1-1}; \quad 0<\theta<1. \tag{20}$$

Thus, using (18) and (20) the Bayes estimator of θ under GELF is obtained as

$$\theta^* = \left[\frac{\Gamma(w_1+a_1-\delta)}{\Gamma(w_1+w_2+a_1+b_1-\delta)} \cdot \frac{\Gamma(w_1+w_2+a_1+b_1)}{\Gamma(w_1+a_1)}\right]^{-1/\delta}. \tag{21}$$

Now, using (19) the posterior risk function under GELF can be derived as

$$\text{Risk}(\theta^*_{\text{GELF}}) = \delta\left[\psi(w_1+a_1) - \psi(w_1+w_2+a_1+b_1)\right] + \log\left[\frac{\Gamma(w_1+a_1-\delta)}{\Gamma(w_1+a_1)}\frac{\Gamma(w_1+w_2+a_1+b_1)}{\Gamma(w_1+w_2+a_1+b_1-\delta)}\right], \tag{22}$$

where $\psi(\cdot)$ is the digamma function defined as $\psi(\theta) = (\partial/\partial\theta)\log\Gamma(\theta)$. Similarly, we can obtain the Bayes estimate and the risk function for the parameter λ.

Particular Cases of GELF. The above estimates reduce as particular cases to the Bayes estimates under other loss functions, such as the following:

(a) For $\delta = -1$, they give the estimates under popular squared error loss function (SELF).

(b) For $\delta = 1$, they coincide with the estimates under entropy loss function (ELF).

(c) For $\delta = -2$, they reduce to the estimates under precautionary loss function (PLF).

5.2. LINEX Loss Function. Varian [21] and various other authors have used the linear exponential loss function (LINEX) in different estimation problems. Under the assumption that the minimal loss occurs at $\theta^* = \theta$, the LINEX loss function can be expressed as

$$L_{\text{LINEX}}(\theta^*,\theta) = \gamma\left[\exp c(\theta^*-\theta) - c(\theta^*-\theta) - 1\right]; \quad c \neq 0,\ b > 0. \tag{23}$$

Without loss of generality, we assume that $\gamma = 1$. Under the LINEX loss function the Bayes estimator and posterior risk are define as

$$\theta^*_{\text{LINEX}} = -\frac{1}{c}\log\left[E\left\{\exp(-c\theta) \mid \text{data}\right\}\right] = -\frac{1}{c}\log\left[\frac{1}{B(\alpha_1,\beta_1)}\int_0^1 e^{-c\theta}\theta^{\alpha_1-1}(1-\theta)^{\beta_1-1}\,d\theta\right] \tag{24}$$

$$\text{Risk}(\theta^*_{\text{LINEX}}) = \log E(c\theta \mid \text{data}) + cE(\theta \mid \text{data}) = \log\left[\frac{c(w_1+a_1)}{w_1+a_1+w_2+b_1}\right] + \frac{c(w_1+a_1)}{w_1+a_1+w_2+b_1}.$$

Sometimes, a situation occurs where no formal prior information is available. Then, an improper joint prior is incorporated with the likelihood to yield a joint posterior distribution. By setting the values of $a_1 = a_2 = b_1 = b_2 = 0$ in beta priors, the Bayes estimates of θ and λ are obtained in case of noninformative prior.

5.3. Bayesian Credible and HPD Credible Intervals. If $\pi_1(\theta \mid y_i, d_i)$ is the marginal posterior distribution of the parameter θ, the credible interval for θ is obtained by

$$\int_{\theta_1}^{\theta_2} \frac{1}{B(\alpha_1, \beta_1)} \theta^{\alpha_1 - 1} (1-\theta)^{\beta_1 - 1} d\theta = 1 - \alpha \tag{25}$$

and for both sided equal tail credible intervals, we take

$$\begin{aligned} \int_{0}^{\theta_1} \frac{1}{B(\alpha_1, \beta_1)} \theta^{\alpha_1 - 1} (1-\theta)^{\beta_1 - 1} d\theta &= \frac{\alpha}{2}, \\ \int_{\theta_2}^{1} \frac{1}{B(\alpha_1, \beta_1)} \theta^{\alpha_1 - 1} (1-\theta)^{\beta_1 - 1} d\theta &= 1 - \frac{\alpha}{2}. \end{aligned} \tag{26}$$

Similarly, the credible interval for λ is obtained.

Chen and Shao [22] proposed a procedure for calculating a highest posterior density (HPD) credible interval for θ when the posterior distribution of θ is unimodal. In the present case, the posterior distribution of θ is beta distribution which is a unimodal distribution; therefore we can apply the following Chen and Shao algorithm.

Step 1. Obtain an MCMC sample θ_j, $j = 1, 2, \ldots, M$, from $\pi_1(\theta \mid y_i, d_i)$.

Step 2. Sort $\{\theta_j; j = 1, 2, \ldots, M\}$ to obtain the ordered values as $\theta_{(1)} \le \theta_{(2)} \le \cdots \le \theta_{(M)}$.

Step 3. For $0 < \alpha < 1$, compute the $100 \times (1-\alpha)\%$ credible intervals

$$\left(\theta_{(j)}, \theta_{(j+[(1-\alpha)M])}\right); \quad \text{for } j = 1, 2, \ldots, M - [(1-\alpha)M]. \tag{27}$$

Here, $[x]$ denotes the integer part of x.

Step 4. The $100 \times (1-\alpha)\%$ HPD credible interval is the one with the smallest interval width among all credible intervals obtained in Step 3.

Similarly, we can obtain the HPD credible interval for the parameter λ. For computing the above algorithm, Boa package of R can be implemented.

6. Simulation Study

Since the performance of different estimation methods cannot be compared theoretically in the present case, we therefore perform a Monte Carlo simulation study to compare the estimates, obtained from maximum likelihood and Bayes estimation techniques under various loss functions. The simulation procedure step by step is given below:

(i) Choose the values of hyperparameters (a_1, b_1) and (a_2, b_2) of informative prior distributions with fixed sample size $n = 50$. For noninformative priors take $a_1 = a_2 = b_1 = b_2 = 0$.

(ii) Take the different values of the parameters equal to the means of prior distributions as $\theta = a_1/(a_1 + b_1)$ and $\lambda = a_2/(a_2 + b_2)$, respectively.

(iii) Calculate the actual value of $\text{MTSF} = \theta/(1-\theta)$. Taking the mission time k = integer value of [MTSF/2], calculate values of the reliability characteristics $S(k)$ and $h(k)$.

(iv) Generate a randomly censored sample $(\underset{\sim}{y}, \underset{\sim}{d})$ of size $n = 50$ from the models given in (3) and (4).

(v) Calculate the maximum likelihood estimates, their variances, confidence intervals, and coverage probabilities for the parameters and MLEs of reliability characteristics.

(vi) Obtain the Bayes estimates under GELF and LINEX loss functions for the parameters and the reliability characteristics. Derive the associated credible and HPD credible intervals for the parameters along with their average length (AL) and coverage probability (CP).

(vii) Repeat steps (iv–vi), $N = 1000$ times, for different combinations of the parametric values. Compute the average values (AV) and mean square error (MSE) of the estimates obtained in steps (v-vi).

All computations were performed using the software R. The main results of the simulation study are shown in Tables 4–10. From the tables, we conclude the following:

(i) The maximum likelihood estimation method for parameters and reliability characteristics gives very good results in terms of both average values and MSEs.

(ii) The Bayes estimates are also very good in respect of bias and MSE under SELF and LINEX loss functions but Bayes estimates under LINEX are better than under SELF in respect of bias. In ELF, estimates show underestimation and in PLF, they give overestimation.

(iii) The coverage probabilities of the parameters attain their nominal levels in the cases of confidence intervals based on MLEs, credible, and HPD credible intervals. But sometimes HPD credible intervals give better coverage than the others.

7. Real Data Example

In this section, we present a real data example to illustrate the utility of our model. The data set given in Lee and Wang [23, p-231] consists of the remission times in months of 137 cancer patients. These remission times are a subset of the data from a bladder cancer study and are used here only for illustration purposes. The data set is given below in complete months.

Remission Times (Months) of 137 Cancer Patients. 4, 32, 3, 13, 19, 4, 3+, 5, 14, 5, 19+, 7, 7, 7, 20, 5, 5, 3, 46, 4+, 2, 4, 5, 2, 9, 1, 0,

TABLE 3: Estimates for real data example.

MLE	SELF	ELF	PLF	LINEX	Confidence intervals
		Bayes estimates with informative priors			
$\widehat{\theta} = 0.9048$ $\widehat{\lambda} = 0.9933$ $\widehat{p} = 0.9343$ $\widehat{ETT} = 79.73$ OBTT = 79	$\theta^* = 0.9047$ $\lambda^* = 0.9933$	$\theta^* = 0.9046$ $\lambda^* = 0.9933$	$\theta^* = 0.9047$ $\lambda^* = 0.9933$	$\theta^* = 0.9041$ $\lambda^* = 0.9924$	*Asymptotic confidence intervals* $\theta = (0.8914, 0.9174)$ $\lambda = (0.9899, 0.9962)$
		Bayes estimates with noninformative priors			
$\widehat{\theta} = 0.9048$ $\widehat{\lambda} = 0.9933$ $\widehat{p} = 0.9343$ $\widehat{ETT} = 79.73$ OBTT = 79	$\theta^* = 0.9048$, $\lambda^* = 0.9933$	$\theta^* = 0.9047$, $\lambda^* = 0.9934$	$\theta^* = 0.9048$ $\lambda^* = 0.9933$	$\theta^* = 0.9040$ $\lambda^* = 0.9932$	*HPD credible intervals* $\theta = (0.8891, 0.9205)$ $\lambda = (0.9908, 0.9976)$ *Credible intervals* $\theta = (0.8906, 0.9257)$ $\lambda = (0.9890, 0.9977)$

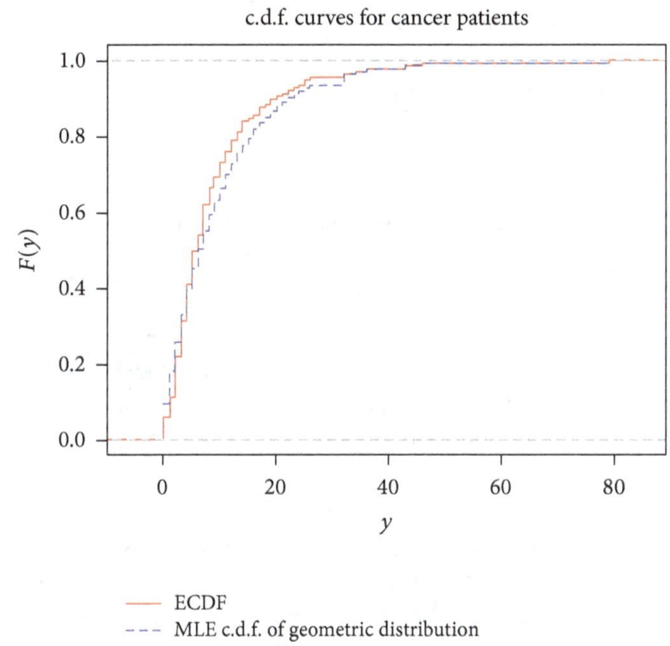

FIGURE 1

8, 3, 9, 36, 14, 26, 79, 10, 8, 4+, 2, 4, 11, 2, 4, 5, 1, 0, 11, 16, 4, 21, 10, 12, 34, 0+, 10, 6, 2, 0, 12, 0, 17, 3, 2, 1, 12, 43, 14, 2, 7, 0, 1, 7, 4, 3, 8, 1, 13, 1, 14, 6, 23, 24+, 5, 8+, 3, 10+, 18, 7, 7, 17, 25, 3, 2, 15, 17, 9, 3, 7, 1, 0, 2, 4, 9, 5, 2, 22, 6, 2, 11, 2, 7, 2, 5, 3, 4+, 8, 6, 4, 3, 0, 10, 6, 3, 5, 13, 8, 3, 5, 7, 5, 11, 2, 2, 6, 7, 4, 25, 12.

Note. Censored observations are indicated by plus sign (+).

To test the goodness of fit, we consider chi-square test and derive the following results:

Chi-square observed value = 19.34912, Chi-square tabulated value = 37.65248, and p value = 0.2199.

Here, observed value is less than the tabulated value at 5% level of significance and p value is also quite large in this case. Thus, the data set is fitted well with our model.

Also, we obtain the empirical c.d.f. and maximum likelihood estimate of c.d.f. curves for comparing the behavior of the data through the graphs. In the graph, in Figure 1, both the curves are closer to each other. Hence, we conclude that this data set is fitted very well for this model.

First of all, we obtain the estimates of the unknown parameters by maximum likelihood estimation, 95% confidence intervals, and Bayes estimation under various types of loss functions with informative priors as beta distributions having parameters (a_1, b_1) and (a_2, b_2) taking $b_1 = b_2 = 3$, $a_1 = \widehat{\theta} b_1/(1 - \widehat{\theta})$, $a_2 = \widehat{\lambda} b_2/(1 - \widehat{\lambda})$. We also estimate the parameters for noninformative priors by taking $a_1 = a_2 = b_1 = b_2 = 0$, assuming no prior information is available with us. The expected time on test and the observed time on test are also calculated for this data set. The estimates of the unknown parameters are listed in Table 3.

By using all the above criteria, we see that the maximum likelihood estimates are quite close to Bayes estimates for

TABLE 4: MLE of parameters and reliability characteristics when k = MTSF/2.

Parameters		$\hat{\theta}$		$\hat{\lambda}$			$\hat{p}$		True values			$\widehat{MTSF}$		$\hat{h}(k)$		$\hat{S}(k)$	
θ	λ	AV	MSE	AV	MSE	p	AV	MSE	MTSF	$h(k)$	$S(k)$	AV	MSE	AV	MSE	AV	MSE
0.1	0.3	0.0985	0.0062	0.2984	0.0041	0.2784	0.2771	0.0041	0.1111	0.9	0.8799	0.1185	0.0111	0.9013	0.006	0.7016	0.1608
0.2	0.4	0.1986	0.0077	0.4039	0.0045	0.3478	0.3517	0.0046	0.25	0.8	0.8178	0.263	0.0213	0.8016	0.0074	0.8011	0.009
0.4	0.2	0.3997	0.0246	0.2002	0.0029	0.1304	0.1306	0.0023	0.6667	0.6	0.7368	0.8004	0.3849	0.6026	0.0248	0.7127	0.0207
0.4	0.3	0.3978	0.0148	0.2987	0.0037	0.2045	0.2041	0.0033	0.6667	0.6	0.7368	0.7134	0.1617	0.6104	0.0151	0.7195	0.0093
0.2	0.5	0.1991	0.0058	0.4949	0.0047	0.4444	0.4398	0.0052	0.25	0.8	0.8178	0.2512	0.0143	0.8062	0.0054	0.8032	0.0059
0.2	0.7	0.1936	0.0040	0.7000	0.0038	0.6512	0.6531	0.0046	0.25	0.8	0.8178	0.2535	0.0106	0.8029	0.0039	0.8114	0.0013
0.4	0.5	0.3988	0.0080	0.4938	0.0043	0.3750	0.3697	0.0049	0.6667	0.6	0.7368	0.699	0.0711	0.6022	0.008	0.731	0.0034
0.5	0.8	0.4952	0.0041	0.7962	0.0020	0.6667	0.6635	0.0046	1	0.5	0.7071	1.0051	0.0631	0.5064	0.0039	0.7011	0.7011
0.6	0.8	0.5955	0.0032	0.7964	0.0017	0.6154	0.6128	0.0048	1.5	0.4	0.6817	1.5141	0.1335	0.4059	0.0033	0.6761	0.0024
0.8	0.5	0.7966	0.0040	0.4967	0.0031	0.1667	0.1671	0.0027	4	0.2	0.64	4.4762	1.2626	0.2053	0.0044	0.6359	0.0107
0.8	0.6	0.7978	0.0028	0.5980	0.0025	0.2308	0.2312	0.0034	4	0.2	0.64	4.3257	3.0318	0.2032	0.003	0.6379	0.0073
0.8	0.8	0.7942	0.0015	0.7977	0.0012	0.4444	0.4481	0.0047	4	0.2	0.64	4.0963	1.1591	0.2043	0.0016	0.6347	0.6347
0.5	0.7	0.4946	0.0049	0.6966	0.0025	0.5385	0.5368	0.0047	1	0.5	0.7071	1.022	0.0811	0.5038	0.0046	0.7027	0.0024
0.7	0.8	0.6963	0.0023	0.7977	0.0015	0.5455	0.5451	0.0048	2.3333	0.3	0.6596	2.3682	0.3392	0.3055	0.0027	0.6539	0.0032
0.8	0.9	0.7955	0.0012	0.8995	0.0005	0.6429	0.6463	0.0048	4	0.2	0.64	4.0182	0.643	0.2042	0.001	0.6343	0.0026
0.9	0.5	0.8967	0.0022	0.4937	0.0029	0.0909	0.0915	0.0017	9	0.1	0.6224	0.9012	0.723	0.1018	0.0021	0.6295	0.019

TABLE 5: Bayes estimates under SELF.

Parameters		Hyperparameters $b_1 = b_2 = b$			θ^*		λ^*		MTSF*		$h^*(k)$		$S^*(k)$	
θ	λ	b	a_1	a_2	AV	MSE	AV	MSE	AV	MSE	AV	MSE	AV	MSE
0.1	0.3	0.4	0.0444	0.1714	0.1016	0.0062	0.2959	0.004	0.1281	0.0129	0.9014	0.0058	0.7583	0.0590
0.2	0.4	0.5	0.125	0.3333	0.1927	0.0074	0.4022	0.0047	0.2798	0.0251	0.8012	0.0072	0.7932	0.0061
0.4	0.2	0.5	0.3333	0.125	0.3893	0.019	0.2007	0.0029	1.0339	1.9648	0.5998	0.0206	0.7110	0.0119
0.4	0.3	0.5	0.3333	0.2143	0.3972	0.0126	0.2984	0.0035	0.8221	0.2257	0.6018	0.0133	0.7196	0.0069
0.2	0.5	0.6	0.15	0.6	0.202	0.0055	0.502	0.0046	0.2648	0.017	0.8057	0.0058	0.7966	0.0041
0.2	0.7	0.6	0.15	1.4	0.1953	0.0039	0.6993	0.0032	0.2569	0.0106	0.8056	0.0039	0.8043	0.0017
0.4	0.5	0.8	0.5333	0.8	0.3927	0.007	0.4977	0.0038	0.7355	0.0796	0.6027	0.0076	0.7266	0.0035
0.5	0.8	0.8	0.8	3.2	0.4954	0.0037	0.7987	0.0019	1.0337	0.0647	0.5067	0.0038	0.6996	0.0021
0.6	0.8	0.8	1.2	3.2	0.5941	0.0033	0.7984	0.0015	1.568	0.1344	0.4046	0.003	0.6767	0.0022
0.8	0.5	0.9	3.6	0.9	0.794	0.0035	0.4963	0.0028	5.1585	7.0132	0.1997	0.0032	0.6471	0.0078
0.8	0.6	0.9	3.6	1.35	0.7975	0.0026	0.5961	0.0025	4.7292	3.8836	0.2029	0.0026	0.6407	0.0064
0.8	0.8	0.9	3.6	3.6	0.7975	0.0014	0.7977	0.0012	4.2991	1.1876	0.203	0.0014	0.6361	0.0039
0.5	0.7	1	1	2.3333	0.4962	0.0044	0.7014	0.0026	1.0742	0.0855	0.5	0.0041	0.704	0.0022
0.7	0.8	5	11.6667	20	0.699	0.0016	0.7977	0.001	2.4675	0.2653	0.3008	0.0018	0.6554	0.0028
0.8	0.9	7	28	63	0.7976	0.0007	0.9	0.0003	4.1208	0.4713	0.2028	0.0007	0.6325	0.0028
0.9	0.5	9	81	9	0.899	0.0002	0.4964	0.0019	9.9492	4.0338	0.1004	0.0002	0.6363	0.0185

informative as well as noninformative priors under various loss functions. Confidence limits also show a good coverage of the estimates.

Concluding Remarks. The present paper deals with the estimation of parameters and reliability characteristics with a randomly censored sample from geometric distribution. Maximum likelihood estimates along with their variances and confidence intervals for the parameters are derived. Bayes estimates under generalized entropy and LINEX loss functions are obtained. The estimates are shown to be almost unbiased and efficient through simulation study. The

TABLE 6: Bayes estimates under ELF.

Parameters		Hyperparameters $b_1 = b_2 = b$			θ^*		λ^*		MTSF*		$h^*(k)$		$S^*(k)$	
θ	λ	b	a_1	a_2	AV	MSE	AV	MSE	AV	MSE	AV	MSE	AV	MSE
0.1	0.3	0.4	0.0444	0.1714	0.0365	0.0117	0.2821	0.0044	0.0438	0.0141	0.9660	0.0116	0.6233	0.2716
0.2	0.4	0.5	0.125	0.3333	0.1534	0.0107	0.3911	0.0049	0.2059	0.0218	0.8398	0.0100	0.7932	0.0061
0.4	0.2	0.5	0.3333	0.125	0.3257	0.0302	0.1858	0.0032	0.6295	0.3121	0.6628	0.0317	0.7110	0.0119
0.4	0.3	0.5	0.3333	0.2143	0.3591	0.0163	0.286	0.0039	0.6288	0.1263	0.6400	0.0173	0.7054	0.0092
0.2	0.5	0.6	0.15	0.6	0.1724	0.0068	0.4931	0.0048	0.2079	0.0163	0.8358	0.0078	0.7923	0.0062
0.2	0.7	0.6	0.15	1.4	0.1751	0.0049	0.6942	0.0034	0.2185	0.0107	0.8259	0.0048	0.8026	0.0018
0.4	0.5	0.8	0.5333	0.8	0.3727	0.0084	0.4896	0.0041	0.6419	0.0621	0.6225	0.0088	0.7222	0.0038
0.5	0.8	0.8	0.8	3.2	0.4877	0.0041	0.7963	0.0019	0.9733	0.0593	0.5144	0.0042	0.6968	0.0022
0.6	0.8	0.8	1.2	3.2	0.5888	0.0035	0.7963	0.0016	1.4852	0.1185	0.4099	0.0032	0.6733	0.0023
0.8	0.5	0.9	3.6	0.9	0.7892	0.0038	0.4903	0.003	4.3333	3.0244	0.2044	0.0034	0.6327	0.0085
0.8	0.6	0.9	3.6	1.35	0.7941	0.0027	0.5919	0.0027	4.1953	2.2476	0.2063	0.0028	0.6299	0.007
0.8	0.8	0.9	3.6	3.6	0.7957	0.0014	0.7963	0.0012	4.0564	0.9404	0.2048	0.0015	0.7889	0.0076
0.5	0.7	1	1	2.3333	0.4868	0.0048	0.6976	0.0027	0.9977	0.0713	0.5093	0.0045	0.7006	0.0023
0.7	0.8	5	11.6667	20	0.6961	0.0017	0.7962	0.001	2.3577	0.2253	0.3037	0.0018	0.6552	0.0029
0.8	0.9	7	28	63	0.7965	0.0007	0.8996	0.0003	3.9877	0.4249	0.2039	0.0007	0.6345	0.0027
0.9	0.5	9	81	9	0.8983	0.0002	0.4917	0.002	9.1083	2.3059	0.1011	0.0002	0.6275	0.0192

TABLE 7: Bayes estimates under PLF.

Parameters		Hyperparameters $b_1 = b_2 = b$			θ^*		λ^*		MTSF*		$h^*(k)$		$S^*(k)$	
θ	λ	b	a_1	a_2	AV	MSE	AV	MSE	AV	MSE	AV	MSE	AV	MSE
0.1	0.3	0.4	0.0444	0.1714	0.1228	0.0072	0.3025	0.0039	0.1646	0.0202	0.8815	0.0066	0.7764	0.0427
0.2	0.4	0.5	0.125	0.3333	0.2092	0.0072	0.4075	0.0046	0.3097	0.029	0.7894	0.0068	0.0068	0.0068
0.4	0.2	0.5	0.3333	0.125	0.4137	0.0177	0.2077	0.0029	0.8368	0.7049	0.5735	0.0193	0.7195	0.0107
0.4	0.3	0.5	0.3333	0.2143	0.4135	0.012	0.3044	0.0034	0.9176	0.3777	0.5916	0.0124	0.7196	0.0067
0.2	0.5	0.6	0.15	0.6	0.2151	0.0055	0.5063	0.0045	0.3011	0.021	0.7885	0.0056	0.8011	0.0035
0.2	0.7	0.6	0.15	1.4	0.2047	0.0038	0.7018	0.0032	0.2811	0.0122	0.7935	0.0038	0.8064	0.0019
0.4	0.5	0.8	0.5333	0.8	0.4019	0.0067	0.5015	0.0038	0.7717	0.1008	0.5988	0.0074	0.7254	0.0034
0.5	0.8	0.8	0.8	3.2	0.4991	0.0036	0.7999	0.0018	1.0759	0.0747	0.5005	0.0036	0.7028	0.0019
0.6	0.8	0.8	1.2	3.2	0.5967	0.0032	0.7994	0.0015	1.5987	0.1519	0.4045	0.0032	0.6762	0.0024
0.8	0.5	0.9	3.6	0.9	0.7963	0.0034	0.4993	0.0028	5.7368	12.5795	0.1977	0.0032	0.6534	0.0078
0.8	0.6	0.9	3.6	1.35	0.7992	0.0025	0.5981	0.0025	5.0512	5.3916	0.2011	0.0025	0.6459	0.0062
0.8	0.8	0.9	3.6	3.6	0.7984	0.0013	0.7984	0.0011	4.4549	1.4551	0.2015	0.0015	0.6374	0.0041
0.5	0.7	1	1	2.3333	0.5006	0.0042	0.7032	0.0026	1.0838	0.093	0.5029	0.0043	0.7002	0.0024
0.7	0.8	5	11.6667	20	0.7004	0.0016	0.7984	0.001	2.4955	0.2796	0.3018	0.0017	0.6548	0.0029
0.8	0.9	7	28	63	0.7981	0.0007	0.9002	0.0003	4.2083	0.5192	0.2016	0.0007	0.6355	0.0026
0.9	0.5	9	81	9	0.8994	0.0002	0.4988	0.0018	10.4371	5.8746	0.1	0.0002	0.6285	0.0201

Table 8: Bayes estimates under LINEX loss function.

Parameters		Hyperparameters $b_1 = b_2 = b$			θ^*		λ^*		MTSF*		$h^*(k)$		$S^*(k)$	
θ	λ	b	a_1	a_2	AV	MSE	AV	MSE	AV	MSE	AV	MSE	AV	MSE
0.1	0.3	0.4	0.0444	0.1714	0.1021	0.0062	0.2963	0.004	0.1255	0.0127	0.9017	0.0058	0.7563	0.0607
0.2	0.4	0.5	0.125	0.3333	0.1934	0.0074	0.4026	0.0047	0.2686	0.0221	0.0221	0.007	0.789	0.0072
0.4	0.2	0.5	0.3333	0.125	0.3911	0.0191	0.2009	0.0029	0.9272	0.7253	0.7253	0.0203	0.0203	0.0203
0.4	0.3	0.5	0.3333	0.2143	0.3984	0.0126	0.2988	0.0035	0.7762	0.1916	0.6069	0.0133	0.7146	0.0074
0.2	0.5	0.6	0.15	0.6	0.2025	0.0055	0.5024	0.0046	0.2694	0.0167	0.8012	0.0056	0.7993	0.7993
0.2	0.7	0.6	0.15	1.4	0.1957	0.0039	0.6996	0.0032	0.2603	0.0107	0.8025	0.0039	0.8053	0.002
0.4	0.5	0.8	0.5333	0.8	0.3934	0.007	0.498	0.0038	0.7127	0.0778	0.6074	0.0077	0.7229	0.0036
0.5	0.8	0.8	0.8	3.2	0.4958	0.0037	0.7989	0.0019	1.0371	0.0659	0.5039	0.0037	0.7012	0.002
0.6	0.8	0.8	1.2	3.2	0.5944	0.0033	0.7986	0.0015	1.541	0.1319	0.4068	0.0033	0.6743	0.0025
0.8	0.5	0.9	3.6	0.9	0.7943	0.0035	0.4966	0.0028	4.6209	3.2105	0.1996	0.0032	0.646	0.0079
0.8	0.6	0.9	3.6	1.35	0.7978	0.0026	0.5963	0.0025	4.4168	2.36	0.2025	0.0026	0.6402	0.0064
0.8	0.8	0.9	3.6	3.6	0.7976	0.0014	0.7977	0.0012	4.2072	1.0542	0.2023	0.0015	0.6388	0.0037
0.5	0.7	1	1	2.3333	0.4966	0.0044	0.7016	0.0026	1.0353	0.0796	0.507	0.0045	0.6982	0.0024
0.7	0.8	5	11.6667	20	0.6992	0.0016	0.7977	0.001	2.4091	0.2343	0.3031	0.0017	0.6563	0.0021
0.8	0.9	7	28	63	0.7969	0.0008	0.8986	0.0006	4.0813	0.439	0.2032	0.0008	0.6386	0.0018
0.9	0.5	9	81	9	0.8989	0.0002	0.4966	0.0019	9.129	2.0268	0.1006	0.0002	0.626	0.0021

Table 9: Bayes estimates under SELF & noninformative prior.

Parameters		θ^*		λ^*		MTSF*		$h^*(k)$		$S^*(k)$	
θ	λ	AV	MSE	AV	MSE	AV	MSE	AV	MSE	AV	MSE
0.1	0.3	0.1009	0.0062	0.2998	0.0037	0.1323	0.0144	0.8991	0.0062	0.6741	0.1729
0.2	0.4	0.1992	0.0071	0.4026	0.0045	0.2808	0.0239	0.8008	0.0071	0.7827	0.0147
0.4	0.2	0.3892	0.0251	0.1950	0.0028	1.0339	1.9648	0.6108	0.0251	0.6970	0.0207
0.4	0.3	0.3954	0.0156	0.3006	0.0037	0.8367	0.2940	0.6046	0.0156	0.7141	0.0090
0.2	0.5	0.1966	0.0060	0.5002	0.0045	0.2700	0.0197	0.8034	0.0060	0.7961	0.0055
0.2	0.7	0.1944	0.0037	0.6990	0.0038	0.2566	0.0099	0.8056	0.0037	0.8053	0.0017
0.4	0.5	0.3968	0.0077	0.4976	0.0039	0.7370	0.0829	0.6032	0.0077	0.7262	0.0035
0.5	0.8	0.4916	0.0040	0.7967	0.0019	1.0286	0.0659	0.5084	0.0040	0.6982	0.0022
0.6	0.8	0.5919	0.0035	0.7976	0.0018	1.5524	0.1427	0.4081	0.0035	0.6736	0.0026
0.8	0.5	0.7942	0.0042	0.4948	0.0033	5.4377	15.0463	0.2058	0.004	0.6388	0.0100
0.8	0.6	0.7953	0.0030	0.5962	0.0026	4.7787	4.8446	0.2047	0.0030	0.6384	0.0073
0.8	0.8	0.7972	0.0016	0.7987	0.0012	4.3301	1.3504	0.2028	0.0016	0.6391	0.0039
0.5	0.7	0.4896	0.0046	0.6977	0.0029	1.0341	0.0827	0.5104	0.0046	0.6963	0.0025
0.7	0.8	0.6976	0.0024	0.7991	0.0015	2.4897	0.3811	0.3024	0.0024	0.6559	0.0029
0.8	0.9	0.7964	0.0011	0.8973	0.0005	4.1749	0.7743	0.2036	0.0011	0.6345	0.0026
0.9	0.5	0.8977	0.0022	0.4953	0.0028	9.9492	4.0338	0.1030	0.0022	0.6373	0.0192

TABLE 10: Confidence, credible, and HPD credible intervals for the parameters.

					Confidence interval				HPD credible interval				Credible interval			
θ	λ	B	a_1	a_2	θ		λ		θ		λ		θ		λ	
					AL	CP	AL	CP	AL	CP	AL	CP	AL	CP	AL	CP
0.1	0.3	0.4	0.0444	0.1714	0.2168	0.875	0.2468	0.934	0.1991	0.8781	0.2047	0.902	0.2142	0.893	0.2316	0.932
0.2	0.4	0.5	0.125	0.3333	0.3133	0.903	0.2581	0.943	0.2600	0.904	0.2134	0.913	0.2877	0.917	0.2432	0.936
0.4	0.2	0.5	0.3333	0.125	0.5606	0.908	0.2093	0.923	0.4363	0.917	0.1720	0.907	0.4826	0.920	0.1946	0.925
0.4	0.3	0.5	0.3333	0.2143	0.4565	0.924	0.2353	0.939	0.3657	0.90	0.1943	0.902	0.4081	0.912	0.2204	0.922
0.2	0.5	0.6	0.15	0.6	0.2903	0.918	0.2606	0.937	0.2328	0.905	0.2150	0.906	0.2599	0.904	0.2444	0.919
0.2	0.7	0.6	0.15	1.4	0.2379	0.912	0.2336	0.938	0.1960	0.901	0.1914	0.914	0.2203	0.924	0.2182	0.931
0.4	0.5	0.8	0.5333	0.8	0.3420	0.934	0.2461	0.935	0.2761	0.930	0.2026	0.906	0.3120	0.917	0.2315	0.938
0.5	0.8	0.8	0.8	3.2	0.2399	0.933	0.1722	0.936	0.1978	0.915	0.1400	0.910	0.2255	0.935	0.1591	0.931
0.6	0.8	0.8	1.2	3.2	0.2198	0.94	0.1606	0.94	0.1810	0.901	0.1309	0.901	0.2061	0.921	0.1492	0.928
0.8	0.5	0.9	3.6	0.9	0.2435	0.937	0.2139	0.939	0.1924	0.910	0.1770	0.904	0.2166	0.934	0.2014	0.915
0.8	0.6	0.9	3.6	1.35	0.2066	0.935	0.1954	0.943	0.1652	0.907	0.1616	0.916	0.1872	0.927	0.1844	0.939
0.8	0.8	0.9	3.6	3.6	0.1513	0.953	0.1342	0.948	0.1234	0.914	0.1101	0.905	0.1408	0.934	0.1255	0.921
0.5	0.7	1	1	2.3333	0.2666	0.938	0.2049	0.950	0.2182	0.904	0.1671	0.908	0.2484	0.935	0.1903	0.925
0.7	0.8	5	11.6667	20	0.1895	0.950	0.1476	0.947	0.1456	0.908	0.1115	0.943	0.1659	0.931	0.1276	0.960
0.8	0.9	7	28	63	0.1253	0.944	0.0883	0.942	0.0942	0.913	0.0625	0.946	0.1081	0.942	0.0712	0.963
0.9	0.5	9	81	9	0.1738	0.911	0.2055	0.956	0.0841	0.996	0.1565	0.930	0.0952	0.996	0.1784	0.957

Note. Here, AL: average length and CP: coverage probability.

concept of expected time on test in random censoring is also considered. A real data example is given for explaining the methods developed in the paper.

Competing Interests

The authors declare that there is no conflict of interests regarding the publication of this paper.

References

[1] J. P. Gilbert, *Random censorship [Ph.D. thesis]*, University of Chicago, 1962.

[2] N. Breslow and J. Crowley, "A large sample study of the life table and product limit estimates under random censorship," *The Annals of Statistics*, vol. 2, pp. 437–453, 1974.

[3] J. A. Koziol and S. B. Green, "A Cramer-von Mises statistic for randomly censored data," *Biometrika*, vol. 63, no. 3, pp. 465–474, 1976.

[4] S. Csörgó and L. Horváth, "On the Koziol-Green model for random censorship," *Biometrika*, vol. 68, no. 2, pp. 391–401, 1981.

[5] J. H. Kim, "Chi-square goodness-of-fit tests for randomly censored data," *The Annals of Statistics*, vol. 21, no. 3, pp. 1621–1639, 1993.

[6] M. Friesl and J. Hurt, "On Bayesian estimation in an exponential distribution under random censorship," *Kybernetika*, vol. 43, no. 1, pp. 45–60, 2007.

[7] M. Saleem and A. Raza, "On Bayesian analysis of the exponential survival time assuming the exponential censor time," *Pakistan Journal of Science*, vol. 63, no. 1, pp. 44–48, 2011.

[8] M. E. Ghitany, "A compound Rayleigh survival model and its application to randomly censored data," *Statistical Papers*, vol. 42, no. 4, pp. 437–450, 2001.

[9] M. Saleem and M. Aslam, "On Bayesian analysis of the Rayleigh survival time assuming the random censor time," *Pakistan Journal of Statistics*, vol. 25, no. 2, pp. 71–82, 2009.

[10] M. E. Ghitany and S. Al-Awadhi, "Maximum likelihood estimation of Burr XII distribution parameters under random censoring," *Journal of Applied Statistics*, vol. 29, no. 7, pp. 955–965, 2002.

[11] M. Y. Danish and M. Aslam, "Bayesian estimation for randomly censored generalized exponential distribution under asymmetric loss functions," *Journal of Applied Statistics*, vol. 40, no. 5, pp. 1106–1119, 2013.

[12] M. Y. Danish and M. Aslam, "Bayesian inference for the randomly censored Weibull distribution," *Journal of Statistical Computation and Simulation*, vol. 84, no. 1, pp. 215–230, 2014.

[13] H. Krishna, Vivekanand, and K. Kumar, "Estimation in Maxwell distribution with randomly censored data," *Journal of Statistical Computation and Simulation*, vol. 85, no. 17, pp. 3560–3578, 2015.

[14] M. Yaqub and A. H. Khan, "Geometric failure law in life testing," *Pure and Applied Mathematika Sciences*, vol. 14, no. 1-2, pp. 69–76, 1981.

[15] S. K. Bhattacharya and S. Kumar, "Discrete life testing," *IAPQR Transactions*, vol. 13, no. 2, pp. 71–76, 1988.

[16] S. S. Maiti, "Estimation of $PX \leq Y$ in the geometric case," *Journal of the Indian Statistical Association*, vol. 33, no. 2, pp. 87–91, 1995.

[17] H. Krishna and N. Jain, "Classical and Bayes estimation of reliability characteristics of some basic system configurations with geometric lifetimes of components," *IAPQR Transactions*, vol. 27, no. 1, pp. 35–49, 2002.

[18] A. M. Sarhan and D. Kundu, "Bayes estimators for reliability measures in geometric distribution model using masked system life test data," *Computational Statistics & Data Analysis*, vol. 52, no. 4, pp. 1821–1836, 2008.

[19] R Development Core Team, *A Language and Environment for Statistical Computing*, R Found for Statistical Computation, Vienna, Austria, 2012, https://www.R-project.org/.

[20] R. Calabria and G. Pulcini, "Point estimation under asymmetric loss functions for left-truncated exponential samples," *Communications in Statistics. Theory and Methods*, vol. 25, no. 3, pp. 585–600, 1996.

[21] H. Varian, "A Bayesian approach to real estate assessment," in *Studies in Bayesian Econometrics and Statistics in Honor of L.J. Savage*, S. E. Feinberg and A. Zellner, Eds., pp. 195–208, North Holland Publishing, Amsterdam, The Netherlands, 1974.

[22] M.-H. Chen and Q.-M. Shao, "Monte Carlo estimation of Bayesian credible and HPD intervals," *Journal of Computational and Graphical Statistics*, vol. 8, no. 1, pp. 69–92, 1999.

[23] E. T. Lee and J. W. Wang, *Statistical Methods for Survival Data Analysis*, John Wiley & Sons, Hoboken, NJ, USA, 2003.

5

Hybrid Clayton-Frank Convolution-Based Bivariate Archimedean Copula

Maxwell Akwasi Boateng,[1] **Akoto Yaw Omari-Sasu**,[2] **Richard Kodzo Avuglah**,[2] **and Nana Kena Frempong**[2]

[1] *Faculty of Engineering, Ghana Technology University College, Accra, Ghana*
[2] *Department of Mathematics, Kwame Nkrumah University of Science and Technology, Kumasi, Ghana*

Correspondence should be addressed to Maxwell Akwasi Boateng; mboateng@gtuc.edu.gh

Academic Editor: Ramón M. Rodríguez-Dagnino

This study exploits the closure property of the converse convolution operator to come up with a hybrid Clayton-Frank Archimedean copula for two random variables. Pairs of random variables were generated and the upper tail observation of the cumulative distribution function (CDF) was used to assess the right skew behavior of the proposed model. Various values of the converse convolution operator were used to see their effect on the proposed model. The simulation covered lengths $n = 10^i$, $i = 2, 3, 4, 5$, and 6. The proposed model was compared with about 40 other bivariate copulas (both Archimedean and elliptical). The proposed model had parameters that spanned the entire real line, thus removing restrictions on the parameters. The parameters theta and omega were varied for a selected interval and the hybrid Clayton-Frank model was, in most cases, found to outperform the other copulas under consideration.

1. Introduction

It is quite apparent that identifying and quantifying dependencies is the core of econometric modeling, especially when it comes to risk management. Pearson's correlation factor for dependence has over time been used in this regard although it has a deficiency when it comes to issues of nonlinear dependence. Several arguments have been put forward in this regard [1] and Embrechts et al. [2]. In order to get a measure that gives more information about the dependence structure, copulas were preferred to Pearson's correlation coefficient.

Copulas give a way of isolating the marginal behavior from the dependence structure McNeil et al. [3]. Alcock and Hatherley [4] suggested that, through copulas, the nonnormal dependence structure could be modeled by using only uniforms of the marginal distributions which allowed violation of the assumptions of normality and linear dependence. This meant that the marginals could be modeled using each type of distributions without influencing the dependence structure between them.

The use of copulas has increased greatly in all fields of study since several phenomena tend to have certain dependencies amongst or between them, no matter how small the dependencies are. Boateng et al. [5] focused on the likelihood of a pair of random variables having either an Archimedean copula or an elliptical copula. Their work involved simulating several pairs of random variables and selecting an appropriate bivariate copula family. The corresponding parameter estimates were obtained by maximum likelihood estimation. They compared AICs of the various bivariate copulas under consideration, using about forty (40) bivariate copulas, for sample sizes 30, 300, 1000, 10000, 100000, and 1000000. Their result showed that, between the Archimedean and elliptical copulas, the Archimedean copulas were the most likely to fit the simulated pairs of random variables.

There have been several applications of copulas. Some of the applications being simulation of multivariate sea storms [6]; dependence structure between the stock and foreign exchange markets [7]; operational risk management [8]; portfolio optimization in the presence of dependent financial returns with long memory [9]; risk evaluation of droughts across the Pearl River basin, China [10]; probabilistic assessment of flood risks [11]; estimation of distribution algorithms for coverage problem of wireless sensor network [12]; risk assessment of hydroclimatic variability on groundwater levels in the Manjara basin aquifer in India [13]; models of tourists' time use and expenditure behavior with self-selection [14]; modeling wind speed dependence in system reliability assessment using copulas [15]; dependence between crude oil spot and futures markets [16]; stochastic modeling of power demand [17].

This study explores the possibility of a combination of existing bivariate Archimedean copulas performing better or just as well as the individual copulas.

2. Method

2.1. Copula. A copula is a function C: $[0,1]^2 \rightarrow [0,1]$ satisfying the following requirements:

(1) Grounded: $C(0,v) = C(u,0) = 0$

(2) Uniform marginals: $C(u,1) = u$, $C(1,v) = v$

(3) 2-increasing: $C(u_1,v_1) - C(u_1,v_2) - C(u_2,v_1) + C(u_2,v_2) \geq 0$, for $u_1 > u_2$ and $v_1 > v_2$.

Generator Function (Clayton)

$$\varphi(t) = \frac{1}{\theta}\left(t^{-\theta} - 1\right), \quad \theta > 0. \tag{1}$$

Inverse Function (Clayton)

$$\varphi^{-1}(t) = \left[\left(t^{-\theta} - 1\right) + 1\right]^{-1/\theta}. \tag{2}$$

Copula (Clayton)

$$C_C(u,v) = \left(u^{-\theta} + v^{-\theta} - 1\right)^{-1/\theta}. \tag{3}$$

Kendal's Tau (Clayton)

$$\tau = \frac{\theta}{\theta + 2}. \tag{4}$$

Generator Function (Frank)

$$\varphi(t) = -\log\left(\frac{\exp(-\omega t) - 1}{\exp(-\omega) - 1}\right), \quad \omega \in (-\infty, \infty). \tag{5}$$

Inverse Function (Frank)

$$\varphi^{-1}(t) = -\frac{1}{\omega} = \log\left[\exp(-t)(\exp(-\omega) - 1) + 1\right]. \tag{6}$$

Copula (Frank)

$$C_F(u,v) = -\frac{1}{\omega}\log\left[1 + \frac{(\exp(-\omega u) - 1)(\exp(-\omega v) - 1)}{\exp(-\omega) - 1}\right]. \tag{7}$$

Kendal's Tau (Frank)

$$\tau = 1 - \frac{4}{\theta} + \frac{4D_1(\theta)}{\theta}, \tag{8}$$

where $D_1(\theta) = \int_0^\theta ((x/\theta)/(\exp(x) - 1))dx$ (Debye function).

2.2. C-Convolution. Let F and H be continuous cumulative distribution functions and C be a copula function. The C-Convolution of F and H is defined by the cumulative distribution function (CDF)

$$H *^C F(t) = \int_0^1 D_1 C\left(w, F\left(t - H^{-1}(w)\right)\right) dw, \tag{9}$$

where $D_1C(w, F(t - H^{-1}(w))) = (\partial/\partial w)C(w, F(t - H^{-1}(w)))$.

2.3. Convolution-Based Copulas

2.3.1. Proposition. Let X and Y be two real-valued random variables on the same probability space $(\Omega, \mathfrak{F}, \mathbb{P})$ with a dependence structure represented by the copula function $C_{X,Y}$ and continuous marginal distributions F_X and F_Y. Then,

$$\begin{aligned} C_{X,X+Y}(u,v) &= \int_0^u D_1 C_{X,Y}\left(w, F_Y\left(F_{X+Y}^{-1}(v) - F_X^{-1}(w)\right)\right) dw, \\ F_{X+Y}(t) &= \int_0^1 D_1 C_{X,Y}\left(w, F_Y\left(t - F_X^{-1}(w)\right)\right) dw \\ &= F_X *^C F_Y(t). \end{aligned} \tag{10}$$

2.3.2. Proposition. Let F, G, and H be three continuous CDFs, $C(w,v)$ be a copula function, and

$$\widehat{C}(u,v) = \int_0^u D_1 C\left(w, F\left(G^{-1}(v) - H^{-1}(w)\right)\right) dw. \tag{11}$$

$\widehat{C}(u,v)$ is a copula iff $G = H *^C F$.

The C-Convolution operator is closed with respect to mixtures of copula functions.

Let A and B be bivariate copula functions.

$$C(u,v) = \lambda A(u,v) + (1 - \lambda) B(u,v). \tag{12}$$

For all $\lambda \in [0,1]$ and for all CDFs H and F,

$$H *^C F = H *^{\lambda A + (1-\lambda)B} F = \lambda H *^A F + (1 - \lambda) H *^B F. \tag{13}$$

2.4. Proposed Convolution-Based Hybrid Clayton-Frank Copula. Using the fact that the C-Convolution operator is closed with respect to mixtures of copula functions, we propose a hybrid Clayton-Frank copula as follows:

$$\begin{aligned} C_{CF}(u,v) = {} & \lambda\left(u^{-\theta} + v^{-\theta} - 1\right)^{-1/\theta} + (1 - \lambda)\left(-\frac{1}{\omega}\right. \\ & \left.\cdot \log\left[1 + \frac{(\exp(-\omega u) - 1)(\exp(-\omega v) - 1)}{\exp(-\omega) - 1}\right]\right). \end{aligned} \tag{14}$$

2.4.1. Check for $C_{CF}(u,v)$ Being a Copula

(1) For It Being Grounded, That Is, $C(0,v)=C(u,0)=0$. From $C_{CF}(u,v)$ above,

$$\begin{aligned} C_{CF}(0,v) &= \lambda\left(0^{-\theta}+v^{-\theta}-1\right)^{-1/\theta}+(1-\lambda)\left(-\frac{1}{\omega}\right. \\ &\quad \left.\cdot\log\left[1+\frac{(\exp(-\omega(0))-1)(\exp(-\omega v)-1)}{\exp(-\omega)-1}\right]\right) \\ &= \lambda\left(v^{-\theta}-1\right)^{-1/\theta}+(1-\lambda)\left(-\frac{1}{\omega}\right. \\ &\quad \left.\cdot\log\left[1+\frac{(0)(\exp(-\omega v)-1)}{\exp(-\omega)-1}\right]\right)=\lambda\left(v^{-\theta}\right. \\ &\quad \left.-1\right)^{-1/\theta}+(1-\lambda)\left(-\frac{1}{\omega}\log[1]\right)=\lambda\left(v^{-\theta}-1\right)^{-1/\theta} \\ &\quad +(1-\lambda)\left(-\frac{1}{\omega}(0)\right)=\lambda\left(v^{-\theta}-1\right)^{-1/\theta}\longrightarrow 0, \end{aligned} \tag{15}$$

since the largest value of u will be one (1),

$$\begin{aligned} C_{CF}(u,0) &= \lambda\left(u^{-\theta}+0^{-\theta}-1\right)^{-1/\theta}+(1-\lambda)\left(-\frac{1}{\omega}\right. \\ &\quad \left.\cdot\log\left[1+\frac{(\exp(-\omega u)-1)(\exp(-\omega(0))-1)}{\exp(-\omega)-1}\right]\right). \end{aligned}$$

Following the steps in $C_{CF}(0,v)$, $C_{CF}(u,0)$ also gives zero (0).

(2) Uniform Marginals: $C(u,1)=u$, $C(1,v)=v$.

$$\begin{aligned} C_{CF}(u,1) &= \lambda\left(u^{-\theta}+1^{-\theta}-1\right)^{-1/\theta}+(1-\lambda)\left(-\frac{1}{\omega}\right. \\ &\quad \left.\cdot\log\left[1+\frac{(\exp(-\omega u)-1)(\exp(-\omega(1))-1)}{\exp(-\omega)-1}\right]\right) \\ &= \lambda\left(u^{-\theta}\right)^{-1/\theta}+(1-\lambda)\left(-\frac{1}{\omega}\right. \\ &\quad \left.\cdot\log[1+(\exp(-\omega u)-1)]\right), \end{aligned} \tag{16}$$

which is the marginal of u.

And

$$\begin{aligned} C_{CF}(1,v) &= \lambda\left(1^{-\theta}+v^{-\theta}-1\right)^{-1/\theta}+(1-\lambda)\left(-\frac{1}{\omega}\right. \\ &\quad \left.\cdot\log\left[1+\frac{(\exp(-\omega(1))-1)(\exp(-\omega v)-1)}{\exp(-\omega)-1}\right]\right) \\ &= \lambda\left(v^{-\theta}\right)^{-1/\theta}+(1-\lambda)\left(-\frac{1}{\omega}\right. \\ &\quad \left.\cdot\log[1+(\exp(-\omega v)-1)]\right), \end{aligned} \tag{17}$$

which is the marginal of v.

(3) 2-Increasing: $C(u_1,v_1)-C(u_1,v_2)-C(u_2,v_1)+C(u_2,v_2)\geq 0$, for $u_1>u_2$ and $v_1>v_2$.

$$\begin{aligned} C_{CF}(u_1,v_1) &= \lambda\left[u_1^{-\theta}+v_1^{-\theta}-1\right]^{-1/\theta}+(1-\lambda)\left[-\frac{1}{\omega}\right. \\ &\quad \left.\cdot\log\left\{1+\frac{(\exp(-\omega u_1)-1)(\exp(-\omega v_1)-1)}{\exp(-\omega)-1}\right\}\right], \\ C_{CF}(u_1,v_2) &= \lambda\left[u_1^{-\theta}+v_2^{-\theta}-1\right]^{-1/\theta}+(1-\lambda)\left[-\frac{1}{\omega}\right. \\ &\quad \left.\cdot\log\left\{1+\frac{(\exp(-\omega u_1)-1)(\exp(-\omega v_2)-1)}{\exp(-\omega)-1}\right\}\right], \\ C_{CF}(u_2,v_1) &= \lambda\left[u_2^{-\theta}+v_1^{-\theta}-1\right]^{-1/\theta}+(1-\lambda)\left[-\frac{1}{\omega}\right. \\ &\quad \left.\cdot\log\left\{1+\frac{(\exp(-\omega u_2)-1)(\exp(-\omega v_1)-1)}{\exp(-\omega)-1}\right\}\right], \\ C_{CF}(u_2,v_2) &= \lambda\left[u_2^{-\theta}+v_2^{-\theta}-1\right]^{-1/\theta}+(1-\lambda)\left[-\frac{1}{\omega}\right. \\ &\quad \left.\cdot\log\left\{1+\frac{(\exp(-\omega u_2)-1)(\exp(-\omega v_2)-1)}{\exp(-\omega)-1}\right\}\right]. \end{aligned} \tag{18}$$

The above ensures that the 2-increasing property is satisfied.

2.5. Conditional Distribution Function of v Given u. The conditional distribution function of v given u is obtained by finding the derivative of $C_{CF}(u,v)$ with respect to u; that is, $(\partial/\partial u)C_{CF}(u,v)$.

$$\begin{aligned} \frac{\partial}{\partial u}C_{CF}(u,v) &= \frac{\partial}{\partial u}\left[\lambda\left(u^{-\theta}+v^{-\theta}-1\right)^{-1/\theta}+(1-\lambda)\left(-\frac{1}{\omega}\log\left[1+\frac{(\exp(-\omega u)-1)(\exp(-\omega v)-1)}{\exp(-\omega)-1}\right]\right)\right], \\ \frac{\partial}{\partial u}C_{CF}(u,v) &= \lambda y^{-\theta-1}\left(u^{-\theta}+v^{-\theta}-1\right)^{-(\theta+1)/\theta}-\left[\frac{(\lambda-1)\exp(\omega)(\exp(u\omega)-1)}{-\exp(\omega(x+y))+\exp(u\omega+\omega)+\exp(v\omega+\omega)-\exp(\omega)}\right]. \end{aligned} \tag{19}$$

TABLE 1: Summary of results for lambda = 0.1.

Interval description	NLM ML estimate (mean)	AIC
Omega constant, theta varied from −4 to 1 by 0.5	$1.797693e + 308$	−1415.565
Omega constant, theta varied from 4 to 1 by 0.5	$1.797693e + 308$	−1415.565
Theta constant, omega varied from −4 to 1 by 0.5	62.07445	−4.256669
Theta constant, omega varied from 4 to 1 by 0.5	$1.797693e + 308$	−1415.565

TABLE 2: Summary of results for lambda = 0.5.

Interval description	NLM ML estimate (mean)	AIC
Omega constant, theta varied from −4 to 1 by 0.5	$1.797693e + 308$	−1415.565
Omega constant, theta varied from 4 to 1 by 0.5	$1.797693e + 308$	−1415.565
Theta constant, omega varied from −4 to 1 by 0.5	17.21489	−1.691549
Theta constant, omega varied from 4 to 1 by 0.5	$1.797693e + 308$	−1415.565

2.6. Joint Distribution Function of u and v. The joint distribution function is obtained from $(\partial^2/\partial u \partial v)C_{CF}(u, v)$.

$$\frac{\partial^2}{\partial u \partial v} C_{CF}(u, v) = \frac{(\theta+1)\lambda u^{\theta-1} v^{\theta-1} \left(u^{-\theta} + v^{-\theta} - 1\right)^{-1/\theta}}{\left(v^{\theta} - u^{\theta}\left(v^{\theta} - 1\right)\right)^2} \left[- \frac{(\lambda - 1)\omega\left(\exp(\omega) - \exp(u\omega)\right)\left(\exp(u\omega) - 1\right)\exp(v\omega + \omega)}{\left(-\exp(\omega(u+v)) + \exp(u\omega + \omega) + \exp(v\omega + \omega) - \exp(\omega)\right)^2} \right]. \tag{20}$$

2.7. Parameter Estimation. In order to obtain theta (θ), Kendal's tau relationship with the traditional Clayton copula parameter (see (4)) is employed. Similarly, to obtain omega (ω), Kendal's tau relationship with the traditional Frank copula (see (8)) is used.

3. Results and Analysis

3.1. Comparison of Clayton-Frank Copula with Other Bivariate Copulas for Two Simulated Independent Random Variables. Assuming Independence of two randomly generated variables, various starting values for the maximum likelihood estimate of the hybrid Clayton-Frank model are considered. The simulation covers lengths = 10^i, $i = 2, 3, 4, 5$.

3.2. Results for $n = 10^i$, $i = 2$. Standard normal bivariate random variables of length $n = 100$ are generated and their CDF which lies in $(0, 1)$ is used for the copula analysis. The upper tail of the CDF is analysed to check whether the hybrid model works better in right skewed observations.

The interval −4 to 4 is considered for both parameters theta and omega from the hybrid model and pairs from this interval are considered as starting values for the optimization process in the maximum likelihood estimation.

Values of lambda (λ) $(0.1, 0.5, 0.8)$ are considered to see their individual effect on the hybrid Clayton-Frank copula model.

Using the same simulated pairs from the standard normal as used in the hybrid Clayton-Frank copula, the selected bivariate copula from a number of 40 bivariate copulas was the Rotated Tawn Type 2 (180 degrees) copula with an AIC value of −4.54. This means that, compared to the other bivariate copulas, the Rotated Tawn Type 2 (180 degrees) copula had the smallest AIC value. Comparing this value with the AIC values in Table 1, it can be observed that the hybrid Clayton-Frank copula outperforms all the 40 bivariate copulas under study when omega is kept constant and theta varied from −4 to 1, when omega is kept constant and theta varied from 4 to 1, and when theta is kept constant and omega varied from 4 to 1.

Using the same simulated pairs from the standard normal as used in the hybrid Clayton-Frank copula, the selected bivariate copula from a number of 40 bivariate copulas was the Rotated Tawn Type 2 (180 degrees) copula with an AIC value of −4.54. This means that, compared to the other bivariate copulas, the Rotated Tawn Type 2 (180 degrees) copula had the smallest AIC value. Comparing this value with the AIC values in Table 2, it can be observed that the hybrid Clayton-Frank copula outperforms all the 40 bivariate copulas under study when omega is kept constant and theta varied from −4 to 1, when omega is kept constant and theta varied from 4 to 1, and when theta is kept constant and omega varied from 4 to 1.

TABLE 3: Summary of results for lambda = 0.8.

Interval description	NLM ML estimate (mean)	AIC
Omega constant, theta varied from −4 to 1 by 0.5	$1.797693e + 308$	−1415.565
Omega constant, theta varied from 4 to 1 by 0.5	$1.797693e + 308$	−1415.565
Theta constant, omega varied from −4 to 1 by 0.5	3.554999	1.46329
Theta constant, omega varied from 4 to 1 by 0.5	$1.797693e + 308$	−1415.565

TABLE 4: Summary of results for lambda = 0.1, $n = 10^i$, $i = 3$.

Interval description	NLM ML estimate (mean)	AIC
Omega constant, theta varied from −4 to 1 by 0.5	$1.797693e + 308$	−1415.565
Omega constant, theta varied from 4 to 1 by 0.5	$1.797693e + 308$	−1415.565
Theta constant, omega varied from −4 to 1 by 0.5	601.341	−8.798324
Theta constant, omega varied from 4 to 1 by 0.5	$1.797693e + 308$	−1415.565

TABLE 5: Summary of results for lambda = 0.5, $n = 10^i$, $i = 3$.

Interval description	NLM ML estimate (mean)	AIC
Omega constant, theta varied from −4 to 1 by 0.5	$1.797693e + 308$	−1415.565
Omega constant, theta varied from 4 to 1 by 0.5	$1.797693e + 308$	−1415.565
Theta constant, omega varied from −4 to 1 by 0.5	194.1866	−6.537639
Theta constant, omega varied from 4 to 1 by 0.5	$1.797693e + 308$	−1415.565

Using the same simulated pairs from the standard normal as used in the hybrid Clayton-Frank copula, the selected bivariate copula from a number of 40 bivariate copulas was the Rotated Tawn Type 2 (180 degrees) copula with an AIC value of −4.54. This means that, compared to the other bivariate copulas, the Rotated Tawn Type 2 (180 degrees) copula had the smallest AIC value. Comparing this value with the AIC values in Table 3, it can be observed that the hybrid Clayton-Frank copula outperforms all the 40 bivariate copulas under study when omega is kept constant and theta varied from −4 to 1, when omega is kept constant and theta varied from 4 to 1, and when theta is kept constant and omega varied from 4 to 1.

3.3. Results for $n = 10^i$, $i = 3$. Standard normal bivariate random variables generated of length $n = 1000$ are generated and their CDF which lies in $(0, 1)$ is used for the copula analysis. The upper tail of the CDF is analysed to check whether the hybrid model works better in right skewed observations.

The interval −4 to 4 is considered for both of the parameters theta and omega from the hybrid model too and pairs from this interval are considered as starting values for the optimization process in the maximum likelihood estimation.

Values of lambda $(0.1, 0.5, 0.8)$ are considered to see their individual effect on the overall hybrid model.

Using the same simulated pairs from the standard normal as used in the hybrid Clayton-Frank copula, the selected bivariate copula from a number of 40 bivariate copulas was the Frank Copula with an AIC value of −6.41. This means that, compared to the other bivariate copulas, the Frank copula had the smallest AIC value. Comparing this value with the AIC values in Table 4, it can be observed that the hybrid Clayton-Frank copula outperforms all the 40 bivariate copulas under study for combinations of theta and omega earlier mentioned.

Using the same simulated pairs from the standard normal as used in the hybrid Clayton-Frank copula, the selected bivariate copula from a number of 40 bivariate copulas was the Frank Copula with an AIC value of −6.41. This means that, compared to the other bivariate copulas, the Frank copula had the smallest AIC value. Comparing this value with the AIC values in Table 5, it can be observed that the hybrid Clayton-Frank copula outperforms all the 40 bivariate copulas under study for combinations of theta and omega earlier mentioned.

Using the same simulated pairs from the standard normal as used in the hybrid Clayton-Frank copula, the selected bivariate copula from a number of 40 bivariate copulas was the Frank Copula with an AIC value of −6.41. This means that, compared to the other bivariate copulas, the Frank copula had the smallest AIC value. Comparing this value with the AIC values in Table 6, it can be observed that the hybrid Clayton-Frank copula outperforms all the 40 bivariate copulas under study for combinations of theta and omega earlier mentioned except when omega is kept constant and theta varied from −4 to 1.

3.4. Results for $n = 10^i$, $i = 4$. Standard normal bivariate random variables of length $n = 10000$ are generated and their CDF which lies in $(0, 1)$ is used for the copula analysis. The upper tail of the CDF is analysed to check whether the hybrid model works better in right skewed observations.

TABLE 6: Summary of results for lambda = 0.8, $n = 10^i$, $i = 3$.

Interval description	NLM ML estimate (mean)	AIC
Omega constant, theta varied from −4 to 1 by 0.5	$1.797693e + 308$	−1415.565
Omega constant, theta varied from 4 to 1 by 0.5	$1.797693e + 308$	−1415.565
Theta constant, omega varied from −4 to 1 by 0.5	48.874	−3.778491
Theta constant, omega varied from 4 to 1 by 0.5	$1.797693e + 308$	−1415.565

TABLE 7: Summary of results for lambda = 0.1, $n = 10^i$, $i = 4$.

Interval description	NLM ML estimate (mean)	AIC
Omega constant, theta varied from −4 to 1 by 0.5	$1.797693e + 308$	−1415.565
Omega constant, theta varied from 4 to 1 by 0.5	$1.797693e + 308$	−1415.565
Theta constant, omega varied from −4 to 1 by 0.5	6360.052	−15.51558
Theta constant, omega varied from 4 to 1 by 0.5	$1.797693e + 308$	−1415.565

TABLE 8: Summary of results for lambda = 0.5, $n = 10^i$, $i = 4$.

Interval description	NLM ML estimate (mean)	AIC
Omega constant, theta varied from −4 to 1 by 0.5	$1.797693e + 308$	−1415.565
Omega constant, theta varied from 4 to 1 by 0.5	$1.797693e + 308$	−1415.565
Theta constant, omega varied from −4 to 1 by 0.5	2058.903	−13.25986
Theta constant, omega varied from 4 to 1 by 0.5	$1.797693e + 308$	−1415.565

TABLE 9: Summary of results for lambda = 0.8, $n = 10^i$, $i = 4$.

Interval description	NLM ML estimate (mean)	AIC
Omega constant, theta varied from −4 to 1 by 0.5	$1.797693e + 308$	−1415.565
Omega constant, theta varied from 4 to 1 by 0.5	$1.797693e + 308$	−1415.565
Theta constant, omega varied from −4 to 1 by 0.5	544.7788	−10.60076
Theta constant, omega varied from 4 to 1 by 0.5	$1.797693e + 308$	−1415.565

The interval −4 to 4 is considered for both of the parameters theta and omega from the hybrid model and pairs from this interval are considered as starting values for the optimization process in the maximum likelihood estimation.

Values of lambda (0.1, 0.5, 0.8) are considered to see their individual effect on the overall hybrid model.

Using the same simulated pairs from the standard normal as used in the hybrid Clayton-Frank copula, the selected bivariate copula from a number of 40 bivariate copulas was the Independence Copula with an AIC value of 0. This means that, compared to the other bivariate copulas, the Independence Copula had the smallest AIC value. Comparing this value with the AIC values in Table 7, it can be observed that the hybrid Clayton-Frank copula outperforms all the 40 bivariate copulas under study for combinations of theta and omega earlier mentioned.

Using the same simulated pairs from the standard normal as used in the hybrid Clayton-Frank copula, the selected bivariate copula from a number of 40 bivariate copulas was the Independence Copula with an AIC value of 0. This means that, compared to the other bivariate copulas, the Independence copula had the smallest AIC value. Comparing this value with the AIC values in Table 8, it can be observed that the hybrid Clayton-Frank copula outperforms all the 40 bivariate copulas under study for combinations of theta and omega earlier mentioned.

Using the same simulated pairs from the standard normal as used in the hybrid Clayton-Frank copula, the selected bivariate copula from a number of 40 bivariate copulas was the Independence Copula with an AIC value of 0. This means that, compared to the other bivariate copulas, the Independence copula had the smallest AIC value. Comparing this value with the AIC values in Table 9, it can be observed that the hybrid Clayton-Frank copula outperforms all the 40 bivariate copulas under study for combinations of theta and omega earlier mentioned.

3.5. *Results for* $n = 10^i$, $i = 5$.

Standard normal bivariate random variables of length $n = 100000$ are generated and their CDF which lies in $(0, 1)$ is used for the copula analysis. The upper tail of the CDF is analysed to check whether the hybrid model works better in right skewed observations.

TABLE 10: Summary of results for lambda = 0.1, $n = 10^i$, $i = 5$.

Interval description	NLM ML estimate (mean)	AIC
Omega constant, theta varied from −4 to 1 by 0.5	1.797693e + 308	−1415.565
Omega constant, theta varied from 4 to 1 by 0.5	1.797693e + 308	−1415.565
Theta constant, omega varied from −4 to 1 by 0.5	63509.4	−18.11789
Theta constant, omega varied from 4 to 1 by 0.5	1.797693e + 308	−1415.565

TABLE 11: Summary of results for lambda = 0.5, $n = 10^i$, $i = 5$.

Interval description	NLM ML estimate (mean)	AIC
Omega constant, theta varied from −4 to 1 by 0.5	1.797693e + 308	−1415.565
Omega constant, theta varied from 4 to 1 by 0.5	1.797693e + 308	−1415.565
Theta constant, omega varied from −4 to 1 by 0.5	20204.94	−15.82736
Theta constant, omega varied from 4 to 1 by 0.5	1.797693e + 308	−1415.565

TABLE 12: Summary of results for lambda = 0.8, $n = 10^i$, $i = 5$.

Interval description	NLM ML estimate (mean)	AIC
Omega constant, theta varied from −4 to 1 by 0.5	1.797693e + 308	−1415.565
Omega constant, theta varied from 4 to 1 by 0.5	1.797693e + 308	−1415.565
Theta constant, omega varied from −4 to 1 by 0.5	5175.958	−13.10356
Theta constant, omega varied from 4 to 1 by 0.5	1.797693e + 308	−1415.565

The interval −4 to 4 is considered for both of the parameters theta and omega from the hybrid model and pairs from this interval are considered as starting values for the optimization process in the maximum likelihood estimation.

Values of lambda (0.1, 0.5, 0.8) are considered to see their individual effect on the overall hybrid model.

Using the same simulated pairs from the standard normal as used in the hybrid Clayton-Frank copula, the selected bivariate copula from a number of 40 bivariate copulas was the Independence Copula with an AIC value of 0. This means that, compared to the other bivariate copulas, the Independence Copula had the smallest AIC value. Comparing this value with the AIC values in Table 10, it can be observed that the hybrid Clayton-Frank copula outperforms all the 40 bivariate copulas under study for combinations of theta and omega earlier mentioned.

Using the same simulated pairs from the standard normal as used in the hybrid Clayton-Frank copula, the selected bivariate copula from a number of 40 bivariate copulas was the Independence Copula with an AIC value of 0. This means that, compared to the other bivariate copulas, the Independence copula had the smallest AIC value. Comparing this value with the AIC values in Table 11, it can be observed that the hybrid Clayton-Frank copula outperforms all the 40 bivariate copulas under study for combinations of theta and omega earlier mentioned.

Using the same simulated pairs from the standard normal as used in the hybrid Clayton-Frank copula, the selected bivariate copula from a number of 40 bivariate copulas was the Independence Copula with an AIC value of 0. This means that, compared to the other bivariate copulas, the Independence copula had the smallest AIC value. Comparing this value with the AIC values in Table 12, it can be observed that the hybrid Clayton-Frank copula outperforms all the 40 bivariate copulas under study for combinations of theta and omega earlier mentioned.

3.6. Results for $n = 10^i$, $i = 6$. Standard normal bivariate random variables of length $n = 1000000$ are generated and their CDF which lies in (0, 1) is used for the copula analysis. The upper tail of the CDF is analysed to check whether the hybrid model works better in right skewed observations.

The interval −4 to 4 is considered for both of the parameters theta and omega from the hybrid model and pairs from this interval are considered as starting values for the optimization process in the maximum likelihood estimation.

Values of lambda (0.1, 0.5, 0.8) are considered to see their individual effect on the overall hybrid model.

Using the same simulated pairs from the standard normal as used in the hybrid Clayton-Frank copula, the selected bivariate copula from a number of 40 bivariate copulas was the Independence Copula with an AIC value of 0. This means that, compared to the other bivariate copulas, the Independence Copula had the smallest AIC value. Comparing this value with the AIC values in Table 13, it can be observed that the hybrid Clayton-Frank copula outperforms all the 40 bivariate copulas under study for combinations of theta and omega earlier mentioned.

Using the same simulated pairs from the standard normal as used in the hybrid Clayton-Frank copula, the selected bivariate copula from a number of 40 bivariate copulas was the Independence Copula with an AIC value of 0. This means that, compared to the other bivariate copulas, the Independence Copula had the smallest

TABLE 13: Summary of results for lambda = 0.1, $n = 10^i$, $i = 6$.

Interval description	NLM ML estimate (mean)	AIC
Omega constant, theta varied from −4 to 1 by 0.5	1.797693e + 308	−1415.565
Omega constant, theta varied from 4 to 1 by 0.5	1.797693e + 308	−1415.565
Theta constant, omega varied from −4 to 1 by 0.5	676524.3	−22.84945
Theta constant, omega varied from 4 to 1 by 0.5	1.797693e + 308	−1415.565

TABLE 14: Summary of results for lambda = 0.5, $n = 10^i$, $i = 6$.

Interval description	NLM ML estimate (mean)	AIC
Omega constant, theta varied from −4 to 1 by 0.5	1.797693e + 308	−1415.565
Omega constant, theta varied from 4 to 1 by 0.5	1.797693e + 308	−1415.565
Theta constant, omega varied from −4 to 1 by 0.5	201250.4	−20.42461
Theta constant, omega varied from 4 to 1 by 0.5	1.797693e + 308	−1415.565

TABLE 15: Summary of results for lambda = 0.8, $n = 10^i$, $i = 6$.

Interval description	NLM ML estimate (mean)	AIC
Omega constant, theta varied from −4 to 1 by 0.5	1.797693e + 308	−1415.565
Omega constant, theta varied from 4 to 1 by 0.5	1.797693e + 308	−1415.565
Theta constant, omega varied from −4 to 1 by 0.5	51307.49	−17.68948
Theta constant, omega varied from 4 to 1 by 0.5	1.797693e + 308	−1415.565

TABLE 16: Bivariate copula families with which the Hybrid Clayton-Frank copula is compared.

Copula name	Number	Parameter 1	Parameter 2
Gaussian	1	$(-1, 1)$	-
Student t	2	$(-1, 1)$	$(2, \infty)$
Clayton (survival)	3, 13	$(0, \infty)$	-
Rotated Clayton (90 and 270 degrees)	23, 33	$(-\infty, 0)$	-
Gumbel (survival)	4, 14	$[1, \infty)$	-
Rotated Gumbel (90 and 270 degrees)	24, 34	$(-\infty, -1]$	-
Frank	5	$R \setminus \{0\}$	-
Joe (survival)	6, 16	$(1, \infty)$	-
Rotated Joe (90 and 270 degrees)	26, 36	$(-\infty, -1)$	-
Clayton-Gumbel (Survival)	7, 17	$(0, \infty)$	$[1, \infty)$
Rotated Clayton-Gumbel (90 and 270 degrees)	28, 38	$(-\infty, -1]$	$(-\infty, -1]$
Joe-Clayton (Survival)	9, 19	$[1, \infty)$	$(0, \infty)$
Rotated Joe-Clayton (90 and 270 degrees)	29, 39	$(-\infty, -1)$	$(-\infty, 0)$
Joe-Frank (Survival)	10, 20	$[1, \infty)$	$(0, 1]$
Rotated Joe-Frank (90 and 270 degrees)	30, 40	$(-\infty, -1)$	$[-1, 0)$
Tawn Type 1 (Survival)	104, 114	$[1, \infty)$	$[0, 1]$
Rotated Tawn Type 1 (90 and 270 degrees)	124, 134	$(-\infty, -1)$	$[0, 1]$
Tawn Type 2 (Survival)	204, 214	$[1, \infty)$	$[0, 1]$
Rotated Tawn Type 2 (90 and 270 degrees)	224, 234	$(-\infty, -1]$	$[0, 1]$

AIC value. Comparing this value with the AIC values in Table 14, it can be observed that the hybrid Clayton-Frank copula outperforms all the 40 bivariate copulas under study for combinations of theta and omega earlier mentioned.

Using the same simulated pairs from the standard normal as used in the hybrid Clayton-Frank copula, the selected bivariate copula from a number of 40 bivariate copulas was the Independence Copula with an AIC value of 0. This means that, compared to the other bivariate copulas, the Independence Copula had the smallest AIC value. Comparing this value with the AIC values in Table 15, it can be observed that the hybrid Clayton-Frank copula outperforms all the 40 bivariate copulas under study for combinations of theta and omega earlier mentioned.

Table 16 gives the various bivariate copulas (Archimedean and elliptical) with which the proposed hybrid Clayton-Frank copula is compared.

4. Conclusion

In summary, it can be observed that applying the closure property of the C-Convolution operator yielded a hybrid Clayton-Frank copula model. The joint distribution function and conditional distribution function were also obtained for the proposed hybrid Clayton-Frank copula model for a bivariate set of random variables. Using the multivariate normal distribution, simulations were performed testing the strength of the new model against about 40 existing bivariate copulas using the upper tail cumulative distribution. The proposed model outperformed most bivariate copulas (Archimedean and elliptical) for simulations that were considered. Finally, the new model eliminates restriction on copula parameters involved in the model, that is, $-\infty \leq \theta \leq \infty$ and $-\infty \leq \omega \leq \infty$, allowing for parameter values spanning the entire real line.

Conflicts of Interest

The authors declare that there are no conflicts of interest regarding the publication of this paper.

References

[1] K. Aas, *Modeling the dependence structure of financial assets: A survey of four copulas, paper*, Norwegian Computing Center, 2004.

[2] P. Embrechts, A. J. McNeil, and D. Straumann, "Correlation and dependence in risk management: properties and pitfalls," in *Risk Management: Value at Risk and Beyond*, M. A. H. Dempster, Ed., pp. 176–223, Cambridge University Press, Cambridge, UK, 2002.

[3] A. J. McNeil, R. Frey, and P. Embrechts, *Quantitative Risk Management: Concepts,Techniques, and Tools*, Princeton University Press, Princeton, NJ, USA, 2005.

[4] J. Alcock and A. Hatherley, "Portfolio construction incorporating asymmetric dependence structures: a users guide," *Accounting and Finance*, vol. 47, pp. 447–472, 2007.

[5] M. A. Boateng, A. Y. Omari-Sasu, Richard Kodzo Avuglah, and N. K. Frempong, "Akoto Yaw Omari-Sasu, Richard Kodzo Avuglah, Nana Kena Frempong, On Two Random Variables and Archimedean Copulas," in *4*, vol. 7, pp. 228–237, 2017.

[6] S. Corbella and D. D. Stretch, "Simulating a multivariate sea storm using Archimedean copulas," *Coastal Engineering Journal*, vol. 76, pp. 68–78, 2013.

[7] Y.-C. Wang, J.-L. Wu, and Y.-H. Lai, "A revisit to the dependence structure between the stock and foreign exchange markets: A dependence-switching copula approach," *Journal of Banking & Finance*, vol. 37, no. 5, pp. 1706–1719, 2013.

[8] P. Arbenz, "Bayesian copulae distributions, with application to operational risk management-some comments," *Methodology and Computing in Applied Probability*, vol. 15, no. 1, pp. 105–108, 2013.

[9] H. Boubaker and N. Sghaier, "Portfolio optimization in the presence of dependent financial returns with long memory: A copula based approach," *Journal of Banking & Finance*, vol. 37, no. 2, pp. 361–377, 2013.

[10] Q. Zhang, M. Xiao, V. P. Singh, and X. Chen, "Copula-based risk evaluation of droughts across the Pearl River basin, China," *Theoretical and Applied Climatology*, vol. 111, no. 1-2, pp. 119–131, 2013.

[11] P. Ganguli and M. J. Reddy, "Probabilistic assessment of flood risks using trivariate copulas," *Theoretical and Applied Climatology*, vol. 111, no. 1-2, pp. 341–360, 2013.

[12] X. Wang, H. Gao, and J. Zeng, "A copula-based estimation of distribution algorithms for coverage problem of wireless sensor network," *Sensor Letters*, vol. 10, no. 8, pp. 1892–1896, 2012.

[13] M. Janga Reddy and P. Ganguli, "Risk assessment of hydroclimatic variability on groundwater levels in the Manjara basin aquifer in India using Archimedean copulas," *Journal of Hydrologic Engineering*, vol. 17, no. 12, pp. 1345–1357, 2012.

[14] H. Zhang, J. Zhang, and M. Kuwano, "An integrated model of tourists' time use and expenditure behaviour with self-selection based on a fully nested Archimedean copula function," *Tourism Management*, vol. 33, no. 6, pp. 1562–1573, 2012.

[15] K. Xie, Y. Li, and W. Li, "Modelling wind speed dependence in system reliability assessment using copulas," *IET Renewable Power Generation*, vol. 6, no. 6, pp. 392–399, 2012.

[16] K. L. Chang, "The time-varying and asymmetric dependence between crude oil spot and futures markets: Evidence from the Mixture copula-based ARJI-GARCH model," *Economic Modelling*, vol. 29, no. 6, pp. 2298–2309, 2012.

[17] A. Lojowska, D. Kurowicka, G. Papaefthymiou, and L. Van Der Sluis, "Stochastic modeling of power demand due to EVs using copula," *IEEE Transactions on Power Systems*, vol. 27, no. 4, pp. 1960–1968, 2012.

6

New Generalizations of Exponential Distribution with Applications

N. A. Rather and T. A. Rather

Department of Mathematics, University of Kashmir, Srinagar 190006, India

Correspondence should be addressed to T. A. Rather; tariqrather@rediffmail.com

Academic Editor: Chin-Shang Li

The main purpose of this paper is to present k-Generalized Exponential Distribution which among other things includes Generalized Exponential and Weibull Distributions as special cases. Besides, we also obtain three-parameter extension of Generalized Exponential Distribution. We shall also discuss moment generating functions (MGFs) of these newly introduced distributions.

1. Introduction

The gamma family distributions was discussed by Karl Pearson in 1895 as pointed out in Balakrishnan and Basu [1]. However, after a period of 35 years the Exponential Distribution is a special case of the gamma distribution to appear on its own. It is also related to Poisson process as it has been observed that the time between two successive Poisson events follows the Exponential Distribution. While discussing the sampling of standard deviation (SD), the Exponential Distribution was referred to by Kondo [2] as Pearson's Type X distribution. Steffensen [3], Teissier [4], and Weibull [5] proposed the applications of Exponential Distribution in actuarial, biological, and engineering problems, respectively.

An extension of Exponential Distribution was proposed by Weibull (1951). The Exponential Distribution is a special case wherein the shape parameter equals one. The Weibull distribution has many applications in survival analysis and reliability engineering; for reference see Lai et al. (2006). Some other applications in industrial quality control are discussed in Berrettoni (1964).

2. Some Basic Definitions and K-Generalized Exponential Distributions

We begin with some definitions which provide a base for the definition of K-Generalized Exponential Distributions.

The Euler gamma function $\Gamma(\alpha)$ is defined by the integral

$$\Gamma(\alpha) = \int_0^\infty t^{\alpha-1} e^{-t} dt, \quad R(\alpha) > 0. \tag{1}$$

A random variable X is said to have gamma distribution with parameter $\alpha > 0$, if its p.d.f. is given by

$$\begin{aligned} f(x) &= \frac{1}{\Gamma(\alpha)} x^{\alpha-1} e^{-x}, \quad 0 < x < \infty \\ f(x) &= 0, \quad \text{elsewhere.} \end{aligned} \tag{2}$$

Replacing x by x/λ, we get the following form of gamma distribution with parameters α, λ with $\alpha > 0$ and $\lambda > 0$:

$$f(x) = \frac{1}{\lambda^\alpha \Gamma(\alpha)} x^{\alpha-1} e^{-x/\lambda}, \quad 0 < x < \infty, \ \alpha > 0, \ \lambda > 0. \tag{3}$$

The gamma distribution with parameters α, λ often arises in practices, as the distribution of time one has to wait until a fixed number of events have occurred.

The beta function of two variables m and n is defined by

$$B(m, n) = \int_0^1 x^{m-1} (1-x)^{n-1} dx, \quad \operatorname{Re}(m) > 0, \ \operatorname{Re}(n) > 0. \tag{4}$$

If we take $x = 1 - y$, then, $0 \le x \le 1$ implies $0 \le y \le 1$, and we get

$$\begin{aligned} B(m,n) &= \int_0^1 (1-y)^{m-1} y^{n-1} dy \\ &= \int_0^1 y^{n-1} (1-y)^{m-1} dy \\ &= \int_0^1 x^{n-1} (1-x)^{m-1} dx = B(n,m). \end{aligned} \tag{5}$$

In the literature (for reference see [6–8]), it is known that

$$B(n,m) = \frac{\Gamma(m)\Gamma(n)}{\Gamma(m+n)}. \tag{6}$$

In (5), if we take

$$x = \frac{1}{z+1} \tag{7}$$

then $0 \le x \le 1$ implies $0 \le z < \infty$, and we get

$$\begin{aligned} &B(n,m) \\ &= \int_\infty^0 \left(\frac{1}{z+1}\right)^{n-1} \left(\frac{z}{z+1}\right)^{m-1} \left(-\frac{1}{(z+1)^2}\right) dz \\ &= \int_0^\infty \frac{z^{m-1}}{(z+1)^{m+n}} dz. \end{aligned} \tag{8}$$

A continuous random variable X is said to have beta distribution with parameters m and n, if its p.d.f. is given by

$$\begin{aligned} f(x) &= \frac{1}{B(m,n)} x^{m-1} (1-x)^{n-1}, \quad 0 \le x \le 1 \\ f(x) &= 0, \quad \text{elsewhere.} \end{aligned} \tag{9}$$

This distribution is known as beta distribution of 1st kind (for reference see [7]).

The beta distribution has an application to model a random phenomenon whose set of possible values is a finite interval $[a, b]$, which by letting a denote the origin and taking $(b - a)$ as a unit measurement can be transformed into the interval $[0, 1]$.

A continuous random variable X is said to have beta distribution of 2nd kind with parameters m and n, if its p.d.f. is given by

$$\begin{aligned} f(x) &= \frac{1}{B(m,n)} \frac{z^{m-1}}{(1+z)^{m+n}}, \quad 0 \le z < \infty, \ m, n > 0, \\ f(x) &= 0, \quad \text{elsewhere.} \end{aligned} \tag{10}$$

More recently, Rahman et al. [7] (for more details see [8–16]) have defined k-gamma and k-beta distributions and their MGFs as follows.

For $k > 0$ and $z \in \mathbb{C}$, the k-gamma function is defined by the integral

$$\Gamma_k(z) = \int_0^\infty t^{z-1} e^{-t^k/k} dt. \tag{11}$$

For $R(m) > 0$, $R(n) > 0$, the k-beta function of two variables m and n is defined by

$$B_k(m,n) = \frac{1}{k} \int_0^1 t^{m/k-1} (1-t)^{n/k-1} dt. \tag{12}$$

It is implicit in the literature (for reference see [7]) that

$$B_k(m,n) = \frac{\Gamma_k(m)\Gamma_k(n)}{\Gamma_k(m+n)}. \tag{13}$$

For the sake of completeness, we present a very simple proof of the relation (13).

We have

$$B_k(m,n) = \frac{1}{k} \int_0^1 x^{m/k-1} (1-x)^{n/k-1} dx. \tag{14}$$

Put

$$x = \cos^2\theta; \tag{15}$$

then

$$dx = -2\cos\theta \sin\theta d\theta, \tag{16}$$

and we get

$$B_k(m,n) = \frac{2}{k} \int_0^{\pi/2} (\cos\theta)^{2m/k-1} (\sin\theta)^{2n/k-1} d\theta. \tag{17}$$

Now

$$\Gamma_k(m) = \int_0^\infty t^{m-1} e^{-t^k/k} dt. \tag{18}$$

Put

$$\frac{t^k}{k} = x^2 \tag{19}$$

or

$$t = \left(kx^2\right)^{1/k}, \tag{20}$$

so that

$$dt = \frac{2}{k} k^{1/k} x^{2/k-1} dx = 2k^{1/k-1} x^{2/k-1} dx; \tag{21}$$

we get

$$\begin{aligned} \Gamma_k(m) &= 2k^{1/k-1} \int_0^\infty k^{(m-1)/k} x^{(2/k)(m-1)} e^{-x^2} x^{2/k-1} dx \\ &= 2k^{(m/k)-1} \int_0^\infty x^{2m/k-1} e^{-x^2} dx. \end{aligned} \tag{22}$$

Since the integrals involved are convergent, we have

$$\begin{aligned} \Gamma_k(m)\Gamma_k(n) &= \left(2k^{m/k-1} \int_0^\infty x^{2m/k-1} e^{-x^2} dx\right) \\ &\cdot \left(2k^{n/k-1} \int_0^\infty y^{2n/k-1} e^{-y^2} dy\right) \\ &= 4k^{(m+n)/k-2} \int_0^\infty \int_0^\infty x^{2m/k-1} y^{2n/k-1} e^{-(x^2+y^2)} dx\, dy. \end{aligned} \tag{23}$$

Put

$$x = r\cos\theta,\qquad y = r\sin\theta, \tag{24}$$

so that

$$dx\,dy = r\,dr\,d\theta; \tag{25}$$

we get

$$\begin{aligned}\Gamma_k(m)\Gamma_k(n) &= 4k^{(m+n)/k-2}\int_0^{\infty}\int_0^{\pi/2}(\cos\theta)^{2m/k-1} \\ &\cdot(\sin\theta)^{2n/k-1}e^{-r^2}r^{2m/k+2n/k-1}dr\,d\theta = \left\{\frac{2}{k}\right. \\ &\cdot\left.\int_0^{\pi/2}(\cos\theta)^{2m/k-1}(\sin\theta)^{2n/k-1}d\theta\right\} \\ &\cdot\left\{2k^{(m+n)/k-1}\int_0^{\infty}r^{2(m+n)/k-1}e^{-r^2}dr\right\}.\end{aligned} \tag{26}$$

Using (17) and (22), we get

$$\Gamma_k(m)\Gamma_k(n) = B_k(m,n)\Gamma_k(m+n), \tag{27}$$

which gives

$$B_k(m,n) = \frac{\Gamma_k(m)\Gamma_k(n)}{\Gamma_k(m+n)}. \tag{28}$$

This completes the proof of the relation (13).

Corollary 1. *One has the following:*

(i) $\Gamma_k(k) = 1$.

(ii) $B_k(k/2, k/2) = \pi/k$.

(iii) $\Gamma_k(k/2) = \sqrt{\pi/k}$.

Proof. Put $m = k$ in (22); we get

$$\Gamma_k(k) = 2\int_0^{\infty}xe^{-x^2}dx = -\left[e^{-x^2}\right]_0^{\infty} = 1. \tag{29}$$

From (13), we have

$$B_k(m,n) = \frac{\Gamma_k(m)\Gamma_k(n)}{\Gamma_k(m+n)}. \tag{30}$$

We take $m = n = k/2$; we get

$$\begin{aligned}B_k\left(\frac{k}{2},\frac{k}{2}\right) &= \frac{\Gamma_k(k/2)\Gamma_k(k/2)}{\Gamma_k(k)} = \Gamma_k\left(\frac{k}{2}\right)\Gamma_k\left(\frac{k}{2}\right) \\ &= \left[\Gamma_k\left(\frac{k}{2}\right)\right]^2.\end{aligned} \tag{31}$$

This gives

$$\left[\Gamma_k\left(\frac{k}{2}\right)\right] = \sqrt{B_k\left(\frac{k}{2},\frac{k}{2}\right)}. \tag{32}$$

Putting $m = k/2$ and $n = k/2$ in (17), we get

$$\begin{aligned}B_k\left(\frac{k}{2},\frac{k}{2}\right) &= \frac{2}{k}\int_0^{\pi/2}d\theta = \frac{2}{k}\left[\theta\right]_0^{\pi/2} = \left(\frac{2}{k}\right)\left(\frac{\pi}{2}\right) \\ &= \frac{\pi}{k},\end{aligned} \tag{33}$$

which establishes (ii).

Putting the value of $B_k(k/2, k/2)$ in (32), we get

$$\Gamma_k\left(\frac{k}{2}\right) = \sqrt{\frac{\pi}{k}}, \tag{34}$$

which proves (iii).

A random variable X of continuous type is said to have Weibull distribution if its probability density function is given by

$$\begin{aligned}f(x,\alpha,\beta,\nu) &= \frac{\beta}{\alpha}\left(\frac{x-\nu}{\alpha}\right)^{\beta-1}e^{-((x-\nu)/\alpha)^{\beta}}, \quad x > \nu \\ f(x,\alpha,\beta,\nu) &= 0, \quad \text{elsewhere, } x \le \nu.\end{aligned} \tag{35}$$

Weibull distribution is widely used in engineering practice due to its versatility. It was originally proposed for the interpretation of fatigue data but now it is also used for many other problems in engineering. In particular, in the field of life phenomenon, it is used as the distribution of lifetime of some object, particularly when the "weakest link" model is appropriate for the model; that is, consider an object consisting of many parts and suppose that the object experiences death (failure) when any of its parts fail; it has been shown [6] (both oretically and empirically) under these conditions that Weibull distribution provides a close approximation to the distribution of the lifetime of the item.

The Gamma and Weibull distributions are commonly used for analyzing any lifetime data or skewed data. Both distributions have nice physical interpretation and several desirable properties. Unfortunately both distributions have drawbacks, one major disadvantage of the gamma distribution is that the distribution function or the survival function can not be computed easily if the shape parameter is not an integer. By using mathematical tables or computer software one obtains the distribution function, the survival function, or hazard function. This makes the gamma distribution unpopular as compared to Weibull distribution whose distribution function, hazard function, or survival function is easy to compute. It is well known that even though the Weibull distribution has convenient representation of distribution function, the distribution of the sum of independent and identically distributed (i.i.d) Weibull random variables is not simple to obtain. Therefore, the distribution of the mean of random sample from Weibull distribution is not easy to compute whereas the distribution of sum of independent and identically distributed (i.i.d) gamma random variables is well known. For more details see Mudholkar, Srivastava and Freimer (1995), Mudholkar and Srivastava [17], Gupta and Kundu [18], and Gupta et al. [19].

Recently R. D. Gupta and D. Kundu have introduced three-parameter Exponential Distribution (location, scale, and shape) and studied the theoretical properties of this family and compared them with respective good studies properties of Gamma and Weibull distributions. The increasing and decreasing hazard rate of the Generalized Exponential Distribution (GED) depends on the shape parameter. Generalized Exponential Distribution (GED) has several properties that are quite similar to gamma distribution but it has distribution function similar to that of the Weibull distribution which can be computed easily. Since the Generalized Exponential family has the likelihood ratio ordering on the shape parameter, one can construct a uniformly most powerful test for testing one sided hypothesis on the shape parameter when the scale and location parameters are known.

3. Generalized Exponential Distribution (GED)

A continuous random variable X whose p.d.f. is given by

$$f(x)=\begin{cases}\beta e^{-\beta x} & \text{if } x\geq 0\\ 0 & \text{if } x<0,\end{cases}\qquad \beta>0, \tag{36}$$

is said to have an Exponential Distribution.

The Generalized Exponential Distribution (GED) introduced by Mubeen et al. [15] has p.d.f.

$$f(x,\alpha,\beta)=\alpha\beta\left(1-e^{-\beta x}\right)^{\alpha-1}e^{-\beta x},\qquad x>0 \text{ for } \alpha,\beta>0 \tag{37}$$

$$f(x,\alpha,\beta)=0,\quad \text{otherwise.}$$

The MGF of (37) is given by

$$\begin{aligned}M_X(t)=E\left(e^{tX}\right)&=\int_0^{\infty}e^{tx}\alpha\beta\left(1-e^{-\beta x}\right)^{\alpha-1}e^{-\beta x}dx\\&=\alpha B\left(\alpha,1-\frac{t}{\beta}\right),\end{aligned} \tag{38}$$

where $B(m,n)=\int_0^1 y^{n-1}(1-y)^{m-1}dy$.

Replacing β by $1/\lambda$ and x by $(x-\mu)$ in (37), we get the following form of Generalized Exponential Distribution (GED):

$$f(x,\alpha,\lambda,\mu)=\frac{\alpha}{\lambda}\left(1-e^{-(x-\mu)/\lambda}\right)^{\alpha-1}e^{-(x-\mu)/\lambda},\qquad x>\mu,\ \alpha>0,\ \lambda>0. \tag{39}$$

4. Extensions of Generalized Exponential Distribution (GED)

The main aim of this paper is to present interesting extensions of Generalized Exponential Distribution (GED) in various ways and to study their moment generating functions (MGFs). We shall first define Generalized Exponential Distribution (GED) in terms of a new parameter $k>0$ and call it k-Generalized Exponential Distribution (k-GED). In fact, we prove the following result, which included Generalized Exponential Distribution as a special case.

Theorem 2. *Let X be a random variable of continuous type and let $\alpha>0$, $\beta>0$, and $k>0$ be the parameters; then the function*

$$f(x,\alpha,\beta,k)=\alpha\beta\left(1-e^{-\beta x^k/k}\right)^{\alpha-1}x^{k-1}e^{-\beta x^k/k},\qquad x>0 \tag{40}$$

$$f(x,\alpha,\beta,k)=0,\quad \textit{elsewhere}$$

is the p.d.f. of random variable X of continuous type.

Remark 3. If we take $k=1$, it reduces to Generalized Exponential Distribution.

Proof of Theorem 2. Clearly

$$f(x,\alpha,\beta,k)\geq 0\quad \forall x>0,\ \alpha>0,\ \beta>0,\ k>0. \tag{41}$$

Now

$$\begin{aligned}&\int_0^{\infty}f(x,\alpha,\beta,k)\,dx\\&=\int_0^{\infty}\alpha\beta\left(1-e^{-\beta x^k/k}\right)^{\alpha-1}x^{k-1}e^{-\beta x^k/k}dx\\&=\alpha\int_0^{\infty}\left(1-e^{-\beta x^k/k}\right)^{\alpha-1}\left(\beta x^{k-1}e^{-\beta x^k/k}\right)dx\\&=\alpha\left[\frac{\left(1-e^{-\beta x^k/k}\right)^{\alpha}}{\alpha}\right]_0^{\infty}=\left[\left(1-e^{-\beta x^k/k}\right)^{\alpha}\right]_0^{\infty}\\&=1-0=1.\end{aligned} \tag{42}$$

Hence $f(x,\alpha,\beta,k)$ is a p.d.f. of random variable X of continuous type.

5. The Moment Generating Function (MGF) of Theorem 2

In this section, we derive MGF of the random variable X having k-Generalized Exponential Distribution in terms of new parameter $k>0$; we have

$$\begin{aligned}M_k(t)=E\left(e^{tX^k}\right)&=\int_0^{\infty}e^{tx^k}f(x,\alpha,\beta,k)\,dx\\&=k\alpha\beta\int_0^{\infty}e^{tx^k}\left(1-e^{-\beta^{x^k/k}}\right)^{\alpha-1}\left(x^{k-1}e^{-\beta x^k/k}\right)dx.\end{aligned} \tag{43}$$

Put

$$e^{-\beta x^k/k}=y; \tag{44}$$

then

$$e^{-\beta x^k/k}\left(-\beta x^{k-1}\right)dx = dy,$$
$$e^{x^k} = y^{-k/\beta} \quad (45)$$

so that

$$e^{tx^k} = y^{-tk/\beta}. \quad (46)$$

Therefore,

$$\begin{aligned} M_k(t) &= \alpha \int_0^1 y^{-tk/\beta}(1-y)^{\alpha-1}dy \\ &= \alpha \int_0^1 y^{(1-tk/\beta)-1}(1-y)^{\alpha-1}dy \\ &= \alpha \int_0^1 (1-y)^{\alpha-1} y^{(1-tk/\beta)-1}dy \quad (47) \\ &= \alpha B\left(\alpha, 1-\left(\frac{tk}{\beta}\right)\right) = \frac{\alpha\Gamma(\alpha)\Gamma(1-(tk/\beta))}{\Gamma(\alpha+1-(tk/\beta))} \\ &= \frac{\Gamma(\alpha+1)\Gamma(1-(tk/\beta))}{\Gamma(\alpha+1-(tk/\beta))}. \end{aligned}$$

Our next theorem is also a generalization of Exponential Distribution in terms of new variable k, which includes Weibull distribution as a special case.

Theorem 4. *Let X be a random variable of continuous type and let $\alpha > 0$, $\beta > 0$, and $k > 0$ be the parameters; then the function*

$$f(x,\alpha,\beta,k) = k\alpha\beta\left(1-e^{-\beta x^k}\right)^{\alpha-1} x^{k-1}e^{-\beta x^k}, \quad 0 < x < \infty \quad (48)$$

$$f(x,\alpha,\beta,k) = 0, \quad \textit{elsewhere},$$

is the p.d.f. of random variable X of continuous type.

Remark 5. For $k = 1$, K-Generalized Exponential Theorem 4 reduces to Classical Exponential Distribution.

Proof of Theorem 4. Clearly

$$f(x,\alpha,\beta,k) \geq 0 \quad \forall x > 0,\ \alpha > 0,\ \beta > 0,\ k > 0. \quad (49)$$

Now

$$\begin{aligned} &\int_0^\infty f(x,\alpha,\beta,k)\,dx \\ &= k\alpha\beta \int_0^\infty \left(1-e^{-\beta x^k}\right)^{\alpha-1} x^{k-1}e^{-\beta x^k}dx \\ &= \alpha \int_0^\infty \left(1-e^{-\beta x^k}\right)^{\alpha-1}\left(k\beta x^{k-1}\right)e^{-\beta x^k}dx \\ &= \alpha\left[\frac{\left(1-e^{-\beta x^k}\right)^\alpha}{\alpha}\right]_0^\infty = \left[\left(1-e^{-\beta x^k}\right)^\alpha\right]_0^\infty \\ &= 1-0 = 1. \end{aligned} \quad (50)$$

Hence $f(x,\alpha,\beta,k)$ is a p.d.f. of random variable X of continuous type.

Remark 6 (Weibull distribution is a special case of Theorem 4). To see this, we take $\alpha = 1$, $\beta = 1$ in Theorem 4; it follows that

$$f(x,k) = kx^{k-1}e^{-x^k} = e^{-x^k}\frac{d}{dx}\left(x^k\right), \quad x > 0$$
$$f(x,k) = 0, \quad x \leq 0. \quad (51)$$

Replacing x by $(x-\mu)/\alpha$, we get

$$f(x,\alpha,k,\nu) = e^{-((x-\nu)/\alpha)^k}\frac{d}{dx}\left(\frac{x-\nu}{\alpha}\right)^k, \quad \left(\frac{x-\nu}{\alpha}\right) > 0 \quad (52)$$

$$f(x,\alpha,k,\nu) = 0, \quad \left(\frac{x-\nu}{\alpha}\right) \leq 0$$

or

$$f(x,\alpha,k,\nu) = \frac{k}{\alpha}\left(\frac{x-\nu}{\alpha}\right)^{k-1}e^{-((x-\nu)/\alpha)^k}, \quad x > \nu,$$
$$f(x,\alpha,k,\nu) = 0, \quad x \leq \nu, \quad (53)$$

is a p.d.f. of X, which is clearly the density of Weibull Distribution.

6. The Moment Generating Function (MGF) of Theorem 4

In this section, we derive MGF of the random variable X having k-Generalized Exponential Distribution in terms of a new parameter $k > 0$; we have

$$\begin{aligned} M_k(t) &= E\left(e^{tX^k}\right) = \int_0^\infty e^{tx^k} f(x,\alpha,\beta,k)\,dx \\ &= k\alpha\beta \int_0^\infty e^{txk}\left(1-e^{-\beta x^k}\right)^{\alpha-1} x^{k-1}e^{-\beta x^k}dx. \end{aligned} \quad (54)$$

Put $e^{-\beta x^k} = y$; then

$$e^{x^k} = y^{-1/\beta},$$
$$e^{-\beta x^k}\left(-k\beta x^{k-1}\right)dx = dy. \quad (55)$$

Therefore,

$$M_k(t) = \alpha \int_0^1 y^{-t/\beta} (1-y)^{\alpha-1} dy$$
$$= \alpha \int_0^1 y^{(1-t/\beta)-1} (1-y)^{\alpha-1} dy$$
$$= \alpha \int_0^1 (1-y)^{\alpha-1} y^{(1-t/\beta)-1} dy \quad (56)$$
$$= \alpha B\left(\alpha, 1-\left(\frac{t}{\beta}\right)\right) = \frac{\alpha\Gamma(\alpha)\Gamma(1-(t/\beta))}{\Gamma(\alpha+1-(t/\beta))}$$
$$= \frac{\Gamma(\alpha+1)\Gamma(1-(t/\beta))}{\Gamma(\alpha+1-(t/\beta))}.$$

We also present the following three-parameter extension of Generalized Exponential Distribution (GED). In fact, we prove the following.

Theorem 7. *Let X be a continuous random variable; then the function*

$$f(x,\alpha,\beta,\delta) = \frac{\alpha\beta\delta}{1-(1-\delta)^\alpha}\left(1-\delta e^{-\beta x}\right)^{\alpha-1} e^{-\beta x},$$
$$x>0,\ \alpha>0,\ \beta>0,\ 0<\delta\le 1, \quad (57)$$
$$f(x,\alpha,\beta,\delta)=0, \quad \textit{otherwise}$$

is the p.d.f. of random variable X of continuous type.

Proof of Theorem 7. Clearly $f(x,\alpha,\beta,\delta)\ge 0$ for all $x>0,\ \alpha>0,\ \beta>0,\ \delta>0$.

Now

$$\int_0^\infty f(x,\alpha,\beta,\delta)\,dx$$
$$= \frac{\alpha\beta\delta}{1-(1-\delta)^\alpha}\int_0^\infty \left(1-\delta e^{-\beta x}\right)^{\alpha-1} e^{-\beta x}dx$$
$$= \frac{\alpha}{1-(1-\delta)^\alpha}\int_0^\infty \left(1-\delta e^{-\beta x}\right)^{\alpha-1}\left(\beta\delta e^{-\beta x}\right)dx \quad (58)$$
$$= \left(\frac{\alpha}{1-(1-\delta)^\alpha}\right)\left[\frac{\left(1-\delta e^{-\beta x}\right)^\alpha}{\alpha}\right]_0^\infty$$
$$= \frac{(1-(1-\delta)^\alpha)}{(1-(1-\delta)^\alpha)} = 1.$$

This shows that $f(x)$ is a p.d.f. of the random variable X. This proves Theorem 7.

If we replace β by $1/\lambda$ and x by $(x-\mu)$ in Theorem 7, we get the following form of Generalized Exponential Distribution (GED):

$$f(x,\alpha,\lambda,\mu,\delta)$$
$$= \frac{\alpha\delta}{\lambda(1-(1-\delta)^\alpha)}\left(1-\delta e^{-(x-\mu)/\lambda}\right)^{\alpha-1} e^{-(x-\mu)/\lambda}, \quad (59)$$
$$x>\mu,\ \alpha>0,\ \lambda>0,\ \mu>0,\ 0<\delta\le 1.$$
$$f(x,\alpha,\lambda,\mu,\delta)=0, \quad \text{otherwise}.$$

Remark 8. For $\delta = 1$, Theorem 7 reduced to Generalized Exponential Distribution (GED).

Remark 9. Taking $\delta=1$ in (59), we get relation (39).

7. The Moment Generating Function (MGF) of Theorem 7

$$M_X(t) = E\left(e^{tX}\right) = \int_0^\infty e^{tx} f(x,\alpha,\beta,\delta)\,dx$$
$$= \frac{\alpha\beta\delta}{(1-(1-\delta)^\alpha)}\int_0^\infty e^{tx}\left(1-\delta e^{-\beta x}\right)^{\alpha-1} e^{-\beta x}dx \quad (60)$$
$$= \frac{\alpha\delta}{(1-(1-\delta)^\alpha)}\int_0^\infty \left(1-\delta e^{-\beta x}\right)^{\alpha-1} e^{tx}\left(\beta e^{-\beta x}\right)dx,$$

Put

$$e^{-\beta x} = y, \quad (61)$$

so that

$$-\beta e^{-\beta x}dx = dy,$$
$$e^{tx} = y^{-t/\beta}; \quad (62)$$

we get

$$M(t) = \frac{\alpha\delta}{(1-(1-\delta)^\alpha)}\int_0^1 (1-\delta y)^{\alpha-1} y^{-t/\beta}dy,$$
$$0<\delta\le 1. \quad (63)$$
$$M(t) = \frac{\alpha}{(1-(1-\delta)^\alpha)} B_\delta\left(\alpha, 1-\frac{t}{\beta}\right),$$

where

$$B_\delta = B_\delta(m,n) = \int_0^1 (1-\delta y)^{m-1} y^{n-1}dy,$$
$$m>0,\ n>0,\ 0<\delta\le 1. \quad (64)$$

Remark 10. For $\delta=1$, we have

$$B_1(m,n) = \int_0^1 (1-y)^{m-1} y^{n-1}dy = B(m,n). \quad (65)$$

And (63) reduces to Remark 8.

Remark 11. If in $B_\delta(m,n) = \int_0^1 (1-\delta y)^{m-1} y^{n-1} dy$, we put $x = \delta y$;
then

$$dy = \frac{1}{\delta} dx \tag{66}$$

and we get

$$\begin{aligned} B_\delta(m,n) &= \int_0^\delta (1-x)^{m-1} \left(\frac{x}{\delta}\right)^{n-1} \frac{1}{\delta} dx \\ &= \frac{1}{\delta^n} \int_0^\delta (1-x)^{m-1} x^{n-1} dx, \quad 0 \le \delta \le 1. \end{aligned} \tag{67}$$

Remark 12. Letting $\delta \to 0$ in (57) and noting that

$$\underset{\delta \to 0}{Lt} \cdot \frac{\alpha\delta}{(1-(1-\delta)^\alpha)} = 1, \tag{68}$$

we get

$$f(x,\beta) = \beta e^{-\beta x}, \quad \text{if } x \ge 0,\ \beta > 0, \tag{69}$$

which is the p.d.f. of the Exponential Distribution.

Finally we present the following more general interesting result which among other things includes Weibull distribution as a limiting case.

Theorem 13. *Let X be a random variable of continuous type. If $\delta > 0$, $\beta > 0$, and $k > 0$ are the parameters, then the function*

$$f(x,\delta,\beta,k) = \frac{k\delta\beta}{1-(1-\delta)^k} \left(1-\delta e^{-x^\beta}\right)^{k-1} x^{\beta-1} e^{-x^\beta}, \quad x > 0,\ 0 < \delta < 1, \tag{70}$$

$$f(x,\delta,\beta,k) = 0, \quad elsewhere,\ x \le 0,$$

is the p.d.f. of the random variable X.

Proof of Theorem 13. Clearly

$$f(x,\delta,\beta,k) \ge 0 \quad \forall x > 0,\ \delta > 0,\ \beta > 0,\ k > 0. \tag{71}$$

Now

$$\begin{aligned} &\int_0^\infty f(x,\delta,\beta,k)\, dx \\ &= k\delta\beta \int_0^\infty \left(1-\delta e^{-x^\beta}\right)^{k-1} x^{\beta-1} e^{-x^\beta} dx \\ &= \frac{k}{1-(1-\delta)^k} \int_0^\infty \left(1-\delta e^{-x^\beta}\right)^{k-1} \left(\beta\delta e^{-x^\beta} x^{\beta-1}\right) dx \\ &= \frac{k}{1-(1-\delta)^k} \left[\frac{\left(1-\delta e^{-x^\beta}\right)^k}{k}\right]_0^\infty \\ &= \frac{k}{1-(1-\delta)^k} \left(\frac{\left(1-(1-\delta)^k\right)}{k}\right) = 1. \end{aligned} \tag{72}$$

This shows that $f(x,\alpha,\beta,k)$ is the p.d.f. of random variable X of continuous type. This proves Theorem 13.

Remark 14. Weibull Distribution is the limiting case of Theorem 13. To see this, we let $\delta \to 0$ in Theorem 13 and note that

$$\lim_{\delta \to 0} \frac{k\delta}{1-(1-\delta)^k} = 1, \tag{73}$$

so that

$$\begin{aligned} f(x,\beta) &= \beta e^{-x^\beta} x^{\beta-1} = e^{-x^\beta} \frac{d}{dx}\left(x^\beta\right), \quad x > 0, \\ f(x,\beta) &= 0, \quad x \le 0, \end{aligned} \tag{74}$$

is the p.d.f. of random variable X of continuous type.

Replacing x by $(x-\nu)/\alpha$, it follows that

$$f(x,\alpha,\beta,\nu) = \beta e^{-((x-\nu)/\alpha)^\beta} \frac{d}{dx}\left(\frac{x-\nu}{\alpha}\right)^\beta, \quad \frac{x-\nu}{\alpha} > 0 \tag{75}$$

$$f(x,\alpha,\beta,\nu) = 0, \quad \frac{x-\nu}{\alpha} \le 0.$$

Equivalently

$$\begin{aligned} f(x,\alpha,\beta,\nu) &= \frac{\beta}{\alpha}\left(\frac{x-\nu}{\alpha}\right)^{\beta-1} e^{-((x-\nu)/\alpha)^\beta}, \quad x > \nu, \\ f(x,\alpha,\beta,\nu) &= 0, \quad x \le \nu, \end{aligned} \tag{76}$$

is a p.d.f. of X, which is clearly density of Weibull Distribution.

8. Applications

The applications of Exponential Distribution have been widespread, which include models to determine bout criteria for analysis of animal behaviour [20]; design rainfall estimation in the Coast of Chiapas [21]; analysis of Los Angeles rainfall data [22]; software reliability growth models for vital quality metrics [23]; models for episode peak and duration for ecohydroclimatic applications [24]; estimating mean life of power system equipment with limited end-of-life failure data [25]; and cure rate modeling (Kannan et al. (2010). In related work, the closeness of the exponentiated Exponential Distribution with the Weibull, gamma, and log-normal distributions is studied in Gupta and Kundu (2003a), (2003b), (2004), (2006)) and [26]. Some generalizations of the exponentiated Exponential Distribution are discussed in Nadarajah and Kotz [27].

The Exponential Distribution is often used to model the reliability of electronic systems, which do not typically experience wear-out type failures. The distribution is called "memoryless," meaning that the calculated reliability for, say, a 10-hour mission is the same for a subsequent 10-hour mission, given that the system is working properly at the start of each mission. Given a hazard (failure) rate, λ, or mean time between failure (MTBF $= 1/\lambda$), the reliability can be determined at a specific point in time (t). The distribution function, reliability function, and hazard rate of Theorem 2 are given by

$$F(x,\alpha,\beta,k) = P(X \le x) = \int_0^x f(x,\alpha,\beta,x)\,dx = \left(1 - e^{-\beta x^k/k}\right)^{\alpha}, \quad x > 0,\ \alpha > 0,\ \beta > 0. \tag{77}$$

The survival function $S(x,\alpha,\beta,k)$ is given by

$$S(x,\alpha,\beta,k) = 1 - F(x,\alpha,\beta,k) = 1 - \left(1 - e^{-\beta x^k/k}\right)^{\alpha}, \quad x > 0. \tag{78}$$

The hazard rate $h(x,\alpha,\beta,k)$ is given by

$$S(x,\alpha,\beta,k) = \frac{F(x,\alpha,\beta,k)}{S(x,\alpha,\beta,k)} = \frac{\left(1 - e^{-\beta x^k/k}\right)^{\alpha-1} x^{k-1} e^{-\beta x^k/k}}{1 - \left(1 - e^{-\beta x^k/k}\right)^{\alpha}}, \quad x > 0. \tag{79}$$

Similarly we can also obtain the distribution function, reliability function, and hazard rate of other theorems too.

9. Conclusions

In this paper the authors conclude the following.

(i) Weibull distribution is a special case of k-Generalized Exponential Distribution. Further if δ tends to 0, then our lastly proved more general result leads to Weibull distribution.

(ii) Also if δ tends to 1, then our newly introduced 3-parameter extension of Generalized Exponential Distribution (GED) reduced to classical Exponential Distribution.

(iii) The moment generating functions (MGFs) obtained in this paper generalize the classical moment generating functions (MGFs) of the given distributions.

Conflicts of Interest

The authors declare that they have no conflicts of interest.

References

[1] N. Balakrishnan and A. P. Basu, *Exponential Distribution Theory, Methods and Applications*, John Wiley, New York, NY, USA, 1995.

[2] T. Kondo, "A theory of the sampling distribution of standard deviation," *Journal of Biometrika*, vol. 22, 1/2, pp. 34–64, 1930.

[3] J. F. Steffensen, "Some recent research in the theory of statistics and actuarial science," *Institute of Actuaries*, 1930.

[4] G. Teissier, *Dysharmonies et Discontinuités dans la Croissance*, Act. Sci. et Industr., No. 95. (Exposés de Biometrie, 1), Hermann, Paris, 1934.

[5] W. Weibull, *A statistical theory of the strength of ma-terials*, vol. 151, Ingeniors Vetenskaps Akademiens, Stockholm, Sweden, 1939.

[6] S. C. Malik, *Mathematical analysis*, Wiley Eastern Limited, New Delhi, India, 1984.

[7] G. Rahman, S. Mubeen, A. Rehman, and M. Naz, "k-Gamma and k-Beta distributions and moment generating functions," *Journal of Probability and Statistics*, vol. 2014, Article ID 982013, 6 pages, 2014.

[8] M. G. Kendall and A. Stuart, *The Advanced Theory of Statistics*, vol. 2, Charles Griffin and Company, London, UK, 1961.

[9] S. Ross, *A First Course in Probability*, University of Southern California, 8th edition.

[10] C. G. Kokologiannaki, "Properties and inequalities of generalized k-gamma, beta and zeta functions," *International Journal of Contemporary Mathematical Sciences*, vol. 5, no. 13–16, pp. 653–660, 2010.

[11] S. Mubeen, M. Naz, and G. Rahman, "A note on k-hypergemetric differential equations," *Journal of Inequalities and Special Functions*, vol. 4, no. 3, pp. 38–43, 2013.

[12] S. Mubeen, G. Rahman, A. Rehman, and M. Naz, "Contiguous function relations for k-hypergeometric functions," *ISRN Mathematical Analysis*, vol. 2014, Article ID 410801, 6 pages, 2014.

[13] S. Mubeen, M. Naz, A. Rehman, and G. Rahman, "Solutions of k-hypergeometric differential equations," *Journal of Applied Mathematics*, vol. 2014, Article ID 128787, 13 pages, 2014.

[14] F. Merovci, "Power product inequalities for the Γ_k function," *International Journal of Mathematical Analysis*, vol. 4, no. 21, pp. 1007–1012, 2010.

[15] S. Mubeen, A. Rehman, and F. Shaheen, "Properties of k-gamma, k-beta and k-functions," *Bothalia Journal*, vol. 4, pp. 371–379, 2014.

[16] A. Rehman, S. Mubeen, N. Sadiq, and F. Shaheen, "Some inequalities involving k-beta functions with applications," *Journal of Inequalities and Applications*, 2014:224, 16 pages, 2014.

[17] G. S. Mudholkar and D. K. Srivastava, "Exponentiated Weibull family for analyzing bathtub failure-rate data," *IEEE Transactions on Reliability*, vol. 42, no. 2, pp. 299–302, 1993.

[18] R. D. Gupta and D. Kundu, "Exponentiated exponential family: an alternative to gamma and Weibull distributions," Dept. of Math., Stat. & Comp. Sci, University of New Brusnwick, Saint-John, Canada, 1997.

[19] R. C. Gupta, P. L. Gupta, and R. D. Gupta, "Modeling failure time data by Lehman alternatives," *Communications in Statistics—Theory and Methods*, vol. 27, no. 4, pp. 887–904, 1998.

[20] M. P. Yeates, B. J. Tolkamp, D. J. Allcroft, and I. Kyriazakis, "The use of mixed distribution models to determine bout criteria for analysis of animal behaviour," *Journal of Theoretical Biology*, vol. 213, no. 3, pp. 413–425, 2001.

[21] C. Escalante-Sandoval, "Design rainfall estimation using exponentiated and mixed exponentiated distributions in theCoast of Chiapas," *Ingenieria Hidraulica en Mexico*, vol. 22, pp. 103–113, 2007.

[22] M. T. Madi and M. Z. Raqab, "Bayesian prediction of rainfall records using the generalized exponential distribution," *Environmetrics*, vol. 18, no. 5, pp. 541–549, 2007.

[23] R. Subburaj, G. Gopal, and P. K. Kapur, "A software reliability growth model for vital quality metrics," *The South African Journal of Industrial Engineering*, vol. 18, pp. 93–108, 2007.

[24] F. Biondi, T. J. Kozubowski, A. K. Panorska, and L. Saito, "A new stochastic model of episode peak and duration for eco-hydro-climatic applications," *Ecological Modelling*, vol. 211, no. 3-4, pp. 383–395, 2008.

[25] J. E. Cota-Felix, F. Rivas-Davalos, and S. Maximov, "An alternative method for estimating mean life of power system equipment with limited end-oflife failure data," in *Proceedings of the 2009 IEEE Bucharest PowerTech: Innovative Ideas Toward the Electrical Grid of the Future*, July 2009.

[26] D. Kundu and B. Pradhan, "Bayesian inference and life testing plans for generalized exponential distribution," *Science in China. Series A. Mathematics*, vol. 52, no. 6, pp. 1373–1388, 2009.

[27] S. Nadarajah and S. Kotz, "The exponentiated type distributions," *Acta Applicandae Mathematicae*, vol. 92, no. 2, pp. 97–111, 2006.

7

Modified Slash Lindley Distribution

Jimmy Reyes,[1] Osvaldo Venegas,[2] and Héctor W. Gómez[1]

[1]*Departamento de Matemáticas, Facultad de Ciencias Básicas, Universidad de Antofagasta, Antofagasta, Chile*
[2]*Departamento de Ciencias Matemáticas y Físicas, Facultad de Ingeniería, Universidad Católica de Temuco, Temuco, Chile*

Correspondence should be addressed to Osvaldo Venegas; ovenegas@uctemuco.cl

Academic Editor: Ramón M. Rodríguez-Dagnino

In this paper we introduce a new distribution, called the modified slash Lindley distribution, which can be seen as an extension of the Lindley distribution. We show that this new distribution provides more flexibility in terms of kurtosis and skewness than the Lindley distribution. We derive moments and some basic properties for the new distribution. Moment estimators and maximum likelihood estimators are calculated using numerical procedures. We carry out a simulation study for the maximum likelihood estimators. A fit of the proposed model indicates good performance when compared with other less flexible models.

1. Introduction

The Lindley distribution was introduced by Lindley [1] with the density function given by

$$f_X(x) = \frac{\lambda^2}{1+\lambda}(1+x)e^{-\lambda x}, \quad \lambda > 0,\ x > 0. \tag{1}$$

We denote this by writing $X \sim L(\lambda)$, where λ is the shape parameter. The corresponding cumulative distribution function (c.d.f.) is

$$F_X(x) = 1 - \frac{1+\lambda+\lambda x}{1+\lambda}e^{-\lambda x}. \tag{2}$$

The properties of the Lindley distribution are studied in detail by Ghitany et al. [2]. Jodrá [3] uses the Lambert W function for the generation of random variables with Lindley or Poisson-Lindley distributions. Ghitany et al. [4] propose the power Lindley (PL) distribution generated from a random variable $X \sim L(\lambda)$ raised to the power $1/\alpha$; that is, if $Y = X^{1/\alpha}$ then Y has PL distribution of parameters α and λ, with density function given by

$$f_Y(y) = \frac{\alpha\lambda^2}{1+\lambda}(1+y^\alpha)e^{-\lambda y^\alpha}, \quad \alpha > 0,\ \lambda > 0,\ y > 0. \tag{3}$$

We denote this by writing $Y \sim \mathrm{PL}(\alpha, \lambda)$. The half-normal distribution is suitable for fitting positive data. We say that a random variable X follows a half-normal distribution with scale parameter σ if its density function is given by

$$f_X(x) = \frac{2}{\sigma}\phi\left(\frac{x}{\sigma}\right), \quad \sigma > 0,\ x > 0, \tag{4}$$

where $\phi(\cdot)$ represents the density function of the standard normal distribution; we denote this by writing $X \sim \mathrm{HN}(\sigma)$.

Olmos et al. [5] introduce a new distribution suitable for fitting positive data called the slash half-normal distribution, which is a distribution with right-tails heavier than the HN distribution. It is obtained as a particular case when the shape parameter q tends to infinity. We say that a random variable X follows a slash half-normal (SHN) distribution with scale parameter σ and kurtosis parameter q if its density function is given by

$$f_X(x) = \frac{2q}{\sigma}\int_0^1 u^q\phi\left(\frac{xu}{\sigma}\right)du, \quad \sigma > 0,\ q > 0,\ x > 0. \tag{5}$$

We denote this by writing $X \sim \mathrm{SHN}(\sigma, q)$.

In the study of symmetric distributions with heavy tails, Reyes et al. [6] introduce a modification of the class of standard slash distributions, which will be called modified slash (MS) distribution and is described as follows: We will say that X has MS distribution with parameter $q > 0$ if it can be expressed as

$$X = \frac{Z}{U^{1/q}}, \tag{6}$$

where $Z \sim N(0,1)$ and $U \sim \text{Exp}(2)$ for $q > 0$. Here Z and U are independent random variables, denoted by $X \sim \text{MS}(0,1,q)$; the density function of X has heavier tails than the standard slash distribution and in consequence has higher kurtosis. When $q \to \infty$ we obtain the standard normal distribution. The density function of the variable $X \sim \text{MS}(0,1,q)$ is given by

$$f_X(x) = \frac{2}{\sqrt{2\pi}} \int_0^\infty v^{1/q} e^{-(1/2)x^2 v^{2/q} - 2v} dv, \quad q > 0,\ x \in \mathbb{R}, \tag{7}$$

where q is kurtosis parameter; see Reyes et al. [6] for more details. Using the same idea, Reyes et al. [7] extend the skew-normal model and Reyes et al. [8] extend the Birnbaum-Saunders model. Gui [9] introduces the slash Lindley (SL) distribution and applies it to data on precipitation and plasma ferritin concentration. Gui et al. [10] introduce the Lindley-Poisson (LP) distribution and apply it to lifetime data.

The focus of this paper is the introduction of a new distribution called modified slash Lindley (MSL) distribution. Because of its mixed approach, the newly constructed distribution will have heavier tails than its parent Lindley distribution and hence will be more suitable for modeling positive data sets that may have heavy tails and/or outliers. This new distribution is quite appropriate for modelling positive data with very atypical observations (outliers); as we can see in the application, the MSL distribution MSL better models data on the survival cancer patients with atypical remission times.

The paper is organized as follows. Section 2 is devoted to the development of a stochastic representation for the MSL distribution and its use for density function derivation and also the derivation of its moments, asymmetry and kurtosis coefficients. In Section 3, the inference is discussed for the MSL distribution using the method of moments estimators and maximum likelihood estimation. We also present an illustrative example with real data on survival times. This example shows that the proposed distribution is a very appropriate model for this data set.

2. MSL Distribution

In this section we consider a stochastic representation, the density function (with some graphical representations), and properties of the modified slash Lindley distribution.

2.1. Stochastic Representation. The stochastic representation of the new distribution is given as

$$Y = \frac{X}{U^{1/q}}, \tag{8}$$

where $X \sim L(\lambda)$ and $U \sim \text{Exp}(2)$ are independent random variables with $\lambda > 0$, $q > 0$. We called the distribution of Y the MSL distribution, and we use the notation $Y \sim \text{MSL}(\lambda, q)$.

2.2. Density Function. The following result shows that the density function of the random variable MSL can be generated using the stochastic representation in (8).

Proposition 1. *Let $Y \sim MSL(\lambda, q)$. Then, the density function of Y is given by*

$$f_Y(y) = \frac{2\lambda^2}{1+\lambda} \int_0^\infty w^{1/q} \left(1 + y w^{1/q}\right) e^{-(\lambda y w^{1/q} + 2w)} dw, \quad y > 0 \tag{9}$$

with $\lambda > 0$ and $q > 0$.

Proof. Using the stochastic representation in (8) and from the Jacobian transformation approach it follows that

$$\begin{aligned} Y &= \frac{X}{U^{1/q}} \\ W &= U \\ &\Downarrow \\ X &= Y W^{1/q} \\ U &= W \\ &\Downarrow \\ J &= \begin{vmatrix} \frac{\partial X}{\partial Y} & \frac{\partial X}{\partial W} \\ \frac{\partial U}{\partial Y} & \frac{\partial U}{\partial W} \end{vmatrix} = \begin{vmatrix} w^{1/q} & \left(\frac{1}{q}\right) y w^{(1/q)-1} \\ 0 & 1 \end{vmatrix} = w^{1/q}. \end{aligned} \tag{10}$$

Hence,

$$\begin{aligned} f_{Y,W}(y,w) &= |J|\, f_{X,U}\left(y w^{1/q}, w\right) \\ &= w^{1/q} f_X\left(y w^{1/q}\right) f_U(w) \\ &= \frac{\lambda^2}{1+\lambda} w^{1/q} \left(1 + y w^{1/q}\right) e^{-\lambda y w^{1/q}} 2 e^{-2w}, \quad y > 0,\ w > 0. \end{aligned} \tag{11}$$

By marginalizing, the result follows immediately.

The following proposition shows that the MSL distribution results from a mixture of the LI distribution on the scale parameter $w^{-1/q}$ and shape parameter λ, and the exponential distribution with parameter equals two.

Proposition 2. *Let $Y \mid W = w \sim L(w^{-1/q}, \lambda)$ and $W \sim Exp(2)$; then $Y \sim MSL(\lambda, q)$.*

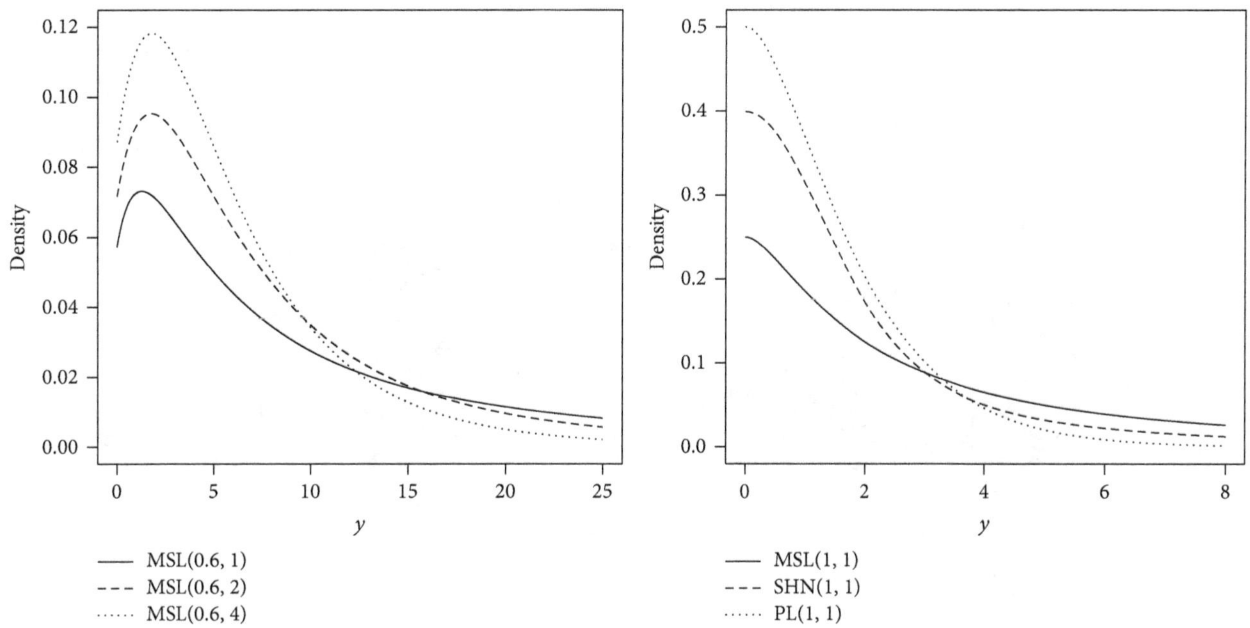

(a) (b)

FIGURE 1: Density functions for the MSL distributions with $\lambda = 0.6$ and several values of q (a), compared with SHN and PL distribution (b).

Proof. We can write

$$\begin{aligned} f_Y(y) &= \int_0^\infty \frac{\lambda^2}{(1+\lambda)\,w^{-1/q}}\left(1+\frac{y}{w^{-1/q}}\right)e^{-\lambda y/w^{-1/q}}2e^{-2w}dw \\ &= \frac{2\lambda^2}{1+\lambda}\int_0^\infty w^{1/q}\left(1+w^{1/q}y\right)e^{-(\lambda w^{1/q}y+2w)}dw. \end{aligned} \tag{12}$$

In Figure 1, we illustrate the behavior of the density function of the MSL distribution.

2.3. Moments

Proposition 3. *If $Y \sim MSL(\lambda, q)$, the rth moment of Y is given by*

$$\mu_r = E\left[Y^r\right] = \frac{r!\,(\lambda+r+1)}{\lambda^r(\lambda+1)}\left[2^{r/q}\Gamma\left(\frac{q-r}{q}\right)\right], \quad q > r. \tag{13}$$

Proof. From the stochastic representation given in (8), in which $X \sim L(\lambda)$ and $U \sim \text{Exp}(2)$ are independent random variables, we have

$$E\left[Y^r\right] = E\left[\left(\frac{X}{U^{1/q}}\right)^r\right] = E\left[X^r\right]E\left[U^{-r/q}\right], \tag{14}$$

from which both expectations are known.

Remark 4. In Figure 1 it is observed that when parameter λ is less than 1, the distribution is unimodal; see Figure 1(a).

Corollary 5. *Let $Y \sim MSL(\lambda, q)$. Then it follows that*

$$\mu_1 = E(Y) = \frac{(\lambda+2)}{\lambda(\lambda+1)}\left[2^{1/q}\Gamma\left(\frac{q-1}{q}\right)\right], \quad q > 1, \tag{15}$$

$$\mu_2 = E\left(Y^2\right) = \frac{2(\lambda+3)}{\lambda^2(\lambda+1)}\left[2^{2/q}\Gamma\left(\frac{q-2}{q}\right)\right], \quad q > 2, \tag{16}$$

$$\mu_3 = E\left(Y^3\right) = \frac{6(\lambda+4)}{\lambda^3(\lambda+1)}\left[2^{3/q}\Gamma\left(\frac{q-3}{q}\right)\right], \quad q > 3, \tag{17}$$

$$\mu_4 = E\left(Y^4\right) = \frac{24(\lambda+5)}{\lambda^4(\lambda+1)}\left[2^{4/q}\Gamma\left(\frac{q-4}{q}\right)\right], \quad q > 4. \tag{18}$$

Proof. It is a direct consequence of Proposition 3.

Corollary 6. *Let $Y \sim MSL(\lambda, q)$. Then the skewness coefficient is*

$$\gamma_1 = \frac{6(\lambda+4)(\lambda+1)^2\kappa_3 - 6(\lambda+1)(\lambda+2)(\lambda+3)\kappa_1\kappa_2 + 2(\lambda+2)^3\kappa_1^3}{\left[2(\lambda+1)(\lambda+3)\kappa_2 - (\lambda+2)^2\kappa_1^2\right]^{3/2}}, \tag{19}$$

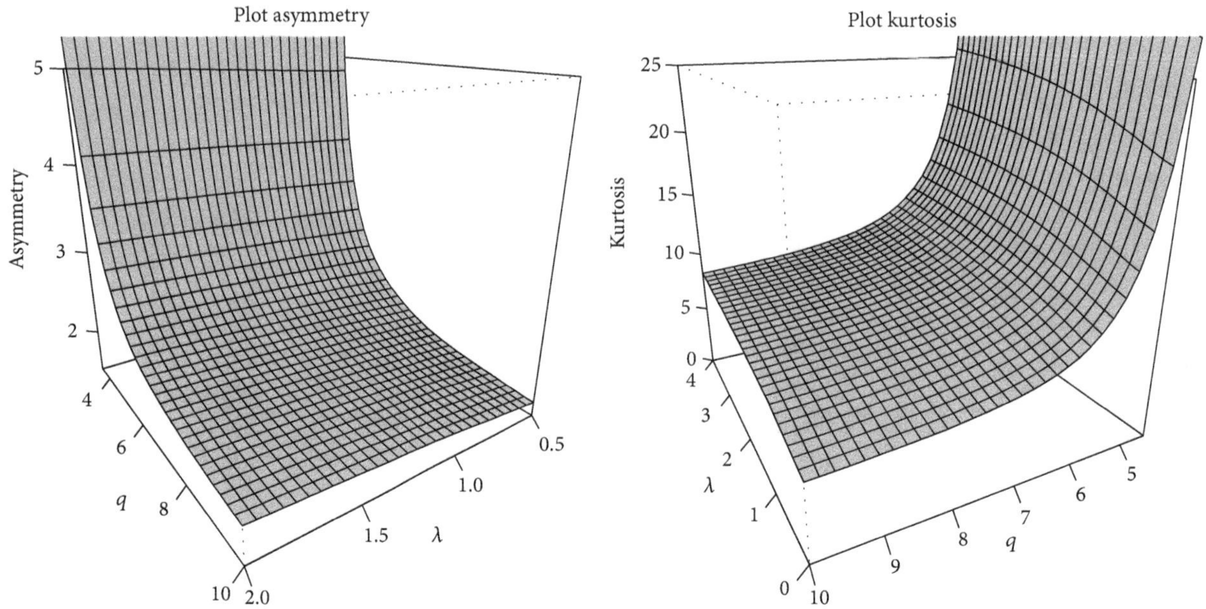

(a) (b)

Figure 2: Asymmetry (a) and kurtosis (b) for several values of λ and q for the MSL distributions.

and the kurtosis coefficient is

$$\gamma_2 = \frac{24(\lambda+1)^3(\lambda+5)\kappa_4 + 12(\lambda+1)(\lambda+2)\kappa_1\left[(\lambda+2)(\lambda+3)\kappa_1\kappa_2 - 2(\lambda+4)(\lambda+1)\kappa_3\right]}{\left[2(\lambda+1)(\lambda+3)\kappa_2 - (\lambda+2)^2\kappa_1^2\right]^2} - \frac{3(\lambda+2)^4\kappa_1^4}{\left[2(\lambda+1)(\lambda+3)\kappa_2 - (\lambda+2)^2\kappa_1^2\right]^2}, \tag{20}$$

where $\kappa_i = \Gamma((q-i)/q)$, *with* $i = 1, 2, 3, 4$.

Proof. The result is obtained by using the following skewness and kurtosis coefficients

$$\gamma_1 = \frac{\mathbb{E}[Y-\mathbb{E}(Y)]^3}{[\mathrm{Var}(Y)]^{3/2}} = \frac{\mu_3 - 3\mu_2\mu_1 + 2\mu_1^3}{(\mu_2-\mu_1^2)^{3/2}},$$
$$\gamma_2 = \frac{\mathbb{E}[Y-\mathbb{E}(Y)]^4}{[\mathrm{Var}(Y)]^2} = \frac{\mu_4 - 4\mu_1\mu_3 + 6\mu_1^2\mu_2 - 3\mu_1^4}{(\mu_2-\mu_1^2)^2}. \tag{21}$$

In Figure 2, we illustrate the behavior of the asymmetry and kurtosis of the MSL distribution as a function of λ and q.

Remark 7. It can be seen that for small values of the parameter q the asymmetry and kurtosis coefficients are high. Also, for high values of the parameter q the asymmetry and kurtosis coefficients tend to the coefficients of the L distribution.

3. Inference

In this section, we study the parameter estimation of the new model using the maximum likelihood and moments approach.

3.1. Method of Moments Estimation. Let $Y_1, Y_2, \ldots, Y_n$ be a random sample from the MSL(λ, q) distribution that has density function given by (9). Then using (15) and (16) and replacing $\overline{y}$ by $E(Y)$ and $\overline{y^2}$ by $E(Y^2)$ we have the following system:

$$\overline{y} = \frac{(\lambda+2)}{\lambda(\lambda+1)}\left[2^{1/q}\Gamma\left(\frac{q-1}{q}\right)\right],$$
$$\overline{y^2} = \frac{2(\lambda+3)}{\lambda^2(\lambda+1)}\left[2^{2/q}\Gamma\left(\frac{q-2}{q}\right)\right]. \tag{22}$$

Using an appropriate numerical method we solve the equation system for λ and q. Hence we obtain the moment estimators $\widehat{\boldsymbol{\theta}}_M = (\widehat{\lambda}_M, \widehat{q}_M)$ of $\boldsymbol{\theta} = (\lambda, q)$.

In the following proposition the asymptotic convergence of these estimators is presented.

Proposition 8. *Let $Y_1, Y_2, \ldots, Y_n$ be a random sample from the MSL(λ, q) distribution. Let $\boldsymbol{\theta} = (\lambda, q)$ and for $k = 1, 2, 3, 4$ denote $\mu_k(\boldsymbol{\theta}) = E(Y^k)$. If $\mu_4 < \infty$ and $\widehat{\boldsymbol{\theta}}_M(n)$ are corresponding moments estimators, we have*

$$\sqrt{n}\left(\widehat{\boldsymbol{\theta}}_M(n) - \boldsymbol{\theta}\right) \xrightarrow{\mathscr{D}} N_2(0, \Gamma(\boldsymbol{\theta})), \quad n \longrightarrow +\infty, \tag{23}$$

where

$$\begin{gathered}\Gamma(\boldsymbol{\theta}) = H^{-1}(\boldsymbol{\theta}) \sum \left[H^{-1}(\boldsymbol{\theta})\right]^T, \\ \sum = \left\{\left(\mu_{i+j} - \mu_i\mu_j\right)_{ij}\right\}, \\ H(\boldsymbol{\theta}) = \begin{pmatrix} -\left(\dfrac{\lambda^2 + 4\lambda + 2}{\lambda^2(\lambda+1)^2}\right)\tau_1 & \dfrac{(\lambda+2)\tau_1}{\lambda(\lambda+1)q^2}\left[\psi\left(\dfrac{q-1}{q}\right) - \log(2)\right] \\ -\left(\dfrac{4\lambda^2 + 20\lambda + 12}{\lambda^3(\lambda+1)^2}\right)\tau_2 & \dfrac{4(\lambda+3)\tau_2}{\lambda^2(\lambda+1)q^2}\left[\psi\left(\dfrac{q-2}{q}\right) - \log(2)\right]\end{pmatrix},\end{gathered} \tag{24}$$

where $A(\boldsymbol{\theta}) = (\mu_1(\boldsymbol{\theta}), \mu_2(\boldsymbol{\theta}))$, $H(\boldsymbol{\theta}) = \partial A(\boldsymbol{\theta})/\partial(\boldsymbol{\theta})$, and $\psi(\cdot)$ is the function digamma and $\tau_i = 2^{i/q}\kappa_i$, with $i = 1, 2$.

Proof. It follows directly by using standard large sample theory results for moment estimators, as discussed for example, in Sen and Singer [11].

3.2. Maximum Likelihood Estimation. We will now discuss maximum likelihood estimation. Given a random sample $Y_1, \ldots, Y_n$ from the distribution of MSL(λ, q), the log likelihood function can be written as

$$\begin{aligned}\ell(\lambda, q) &= n\log(2) + 2n\log(\lambda) - n\log(\lambda+1) \\ &\quad + \sum_{i=1}^{n}\log\left[G(y_i)\right],\end{aligned} \tag{25}$$

where $G(y_i) = \int_0^\infty u^{1/q}(1 + y_i u^{1/q})e^{-(\lambda y_i u^{1/q} + 2u)}du$ and hence the maximum likelihood equation is given by

$$\begin{aligned}\sum_{i=1}^{n}\frac{G_1(y_i)}{G(y_i)} &= -\frac{2n}{\lambda} + \frac{n}{\lambda+1}, \\ \sum_{i=1}^{n}\frac{G_2(y_i)}{G(y_i)} &= 0,\end{aligned} \tag{26}$$

where the expressions for $G_1(y_i)$ and $G_2(y_i)$ should be given

$$\begin{aligned}G_1(y_i) &= \frac{\partial}{\partial\lambda}G(y_i) = -\int_0^\infty u^{1/q}w_i(1 + w_i) \\ &\quad \cdot e^{-(\lambda w_i + 2u)}du, \\ G_2(y_i) &= \frac{\partial}{\partial q}G(y_i) = \frac{1}{q^2}\int_0^\infty u^{1/q}\log(u) \\ &\quad \cdot e^{-(\lambda w_i + 2u)}\left[(\lambda - 2)w_i + \lambda w_i^2 - 1\right]du,\end{aligned} \tag{27}$$

where $w_i = y_i u^{1/q}$.

Solutions for (26) can be obtained using numerical procedures such as the Newton-Raphson procedure.

3.2.1. Simulation Study. By using the representation considered by Ghitany et al. [2] to generate random numbers of a random variable L and the representation given in (8), it is possible to generate random numbers for the MSL(λ, q) distribution, leading to the following algorithm:

(1) Simulate $S \sim$ Exp(λ).

(2) Simulate $T \sim$ Gamma$(2, \lambda)$.

(3) Simulate $V \sim$ Bernulli$(\lambda/(1 + \lambda))$.

(4) Compute $X = VS + (1 - V)T$.

(5) Simulate $U \sim$ Exp(2).

(6) Compute $Y = X/U^{1/q}$.

It then follows that $Y \sim$ MSL(λ, q).

Table 1 shows the results of simulations studies illustrating the behavior of the MLEs for 1000 generated samples of sizes 50, 100, 150, and 200 from population distributed as MSL(λ, q). For each sample generated, MLEs were computed numerically using a Newton-Raphson procedure. Means and standard deviations (SD) are reported. It is observed that the bias becomes smaller as the sample size n increases, as one would expect.

3.3. An Illustrative Example with a Real Data Set. We consider here a data set representing the remission times (in months) of a random sample of 128 bladder cancer patients as reported in Lee and Wang [12]. The data are as follows: 0.08, 2.09, 3.48, 4.87, 6.94, 8.66, 13.11, 23.63, 0.20, 2.23, 3.52, 4.98, 6.97, 9.02, 13.29, 0.40, 2.26, 3.57, 5.06, 7.09, 9.22, 13.80, 25.74, 0.50, 2.46, 3.64, 5.09, 7.26, 9.47, 14.24, 25.82, 0.51, 2.54, 3.70, 5.17, 7.28, 9.74, 14.76, 26.31, 0.81, 2.62, 3.82, 5.32, 7.32, 10.06, 14.77, 32.15, 2.64, 3.88, 5.32, 7.39, 10.34, 14.83, 34.26, 0.90, 2.69, 4.18, 5.34, 7.59, 10.66, 15.96, 36.66, 1.05, 2.69, 4.23, 5.41, 7.62, 10.75,

TABLE 1: Empirical means and SD for the MLE estimators of λ and q of MSL distribution.

λ	q	$n = 50$		$n = 100$		$n = 150$		$n = 200$	
		$\hat{\lambda}$ (SD)	$\hat{q}$ (SD)	$\hat{\lambda}$ (SD)	$\hat{q}$ (SD)	$\hat{\lambda}$ (SD)	$\hat{q}$ (SD)	$\hat{\lambda}$ (SD)	$\hat{q}$ (SD)
	1.0	0.9851 (0.2747)	1.1886 (0.1853)	1.1223 (0.2278)	0.9640 (0.0898)	1.0552 (0.1605)	1.0387 (0.0808)	0.9750 (0.1027)	1.0379 (0.0774)
1.0	2.0	1.0954 (0.2110)	2.2455 (0.7985)	0.9489 (0.1667)	2.1419 (0.5861)	1.0248 (0.1516)	2.0774 (0.4849)	1.0227 (0.1460)	1.8972 (0.3098)
	3.0	1.0226 (0.1422)	3.1068 (0.6515)	1.0112 (0.1059)	3.0265 (0.6214)	1.0258 (0.0926)	3.0183 (0.6211)	0.9985 (0.0860)	2.9957 (0.6083)
	1.0	1.9635 (0.4886)	1.0926 (0.1390)	2.0276 (0.1597)	0.9809 (0.1084)	2.0243 (0.1293)	1.0164 (0.0456)	1.9755 (0.0347)	0.9842 (0.0110)
2.0	2.0	1.9682 (0.5096)	2.1367 (0.7198)	2.0142 (0.2867)	2.1358 (0.6465)	2.0142 (0.2867)	2.1358 (0.2661)	2.0301 (0.2264)	1.9619 (0.1602)
	3.0	1.9751 (0.2893)	2.9374 (0.8254)	2.0568 (0.1853)	3.1459 (0.6483)	2.0381 (0.1450)	3.1482 (0.5828)	2.0167 (0.1401)	3.1034 (0.5469)
	1.0	3.1778 (0.9120)	0.9904 (0.2537)	2.9683 (0.4836)	1.0102 (0.0915)	3.0395 (0.2637)	1.0109 (0.0865)	2.9991 (0.1343)	0.9842 (0.0554)
3.0	2.0	2.9809 (0.9879)	2.1940 (0.6991)	2.9259 (0.5831)	2.1793 (0.5714)	2.9457 (0.5517)	2.1249 (0.5237)	3.0159 (0.3946)	2.1202 (0.4817)
	3.0	2.9091 (0.5866)	3.1248 (0.7443)	3.0304 (0.5791)	2.9697 (0.7089)	3.0304 (0.5791)	2.9697 (0.7089)	2.9849 (0.2684)	3.0498 (0.6866)
	1.0	4.0870 (0.9069)	1.0462 (0.1096)	3.8912 (0.6294)	0.9819 (0.0913)	4.0325 (0.5819)	0.9854 (0.0862)	3.9763 (0.5440)	1.0341 (0.0645)
4.0	2.0	4.5154 (0.5064)	1.9870 (0.2169)	3.9653 (0.4912)	2.0438 (0.1484)	3.9788 (0.2978)	2.0708 (0.1426)	3.9912 (0.0041)	2.0024 (0.1401)
	3.0	4.1487 (0.8627)	3.2438 (0.7974)	4.0190 (0.4633)	3.1469 (0.4665)	4.0825 (0.3607)	3.1308 (0.3849)	4.0508 (0.2894)	3.0714 (0.2110)

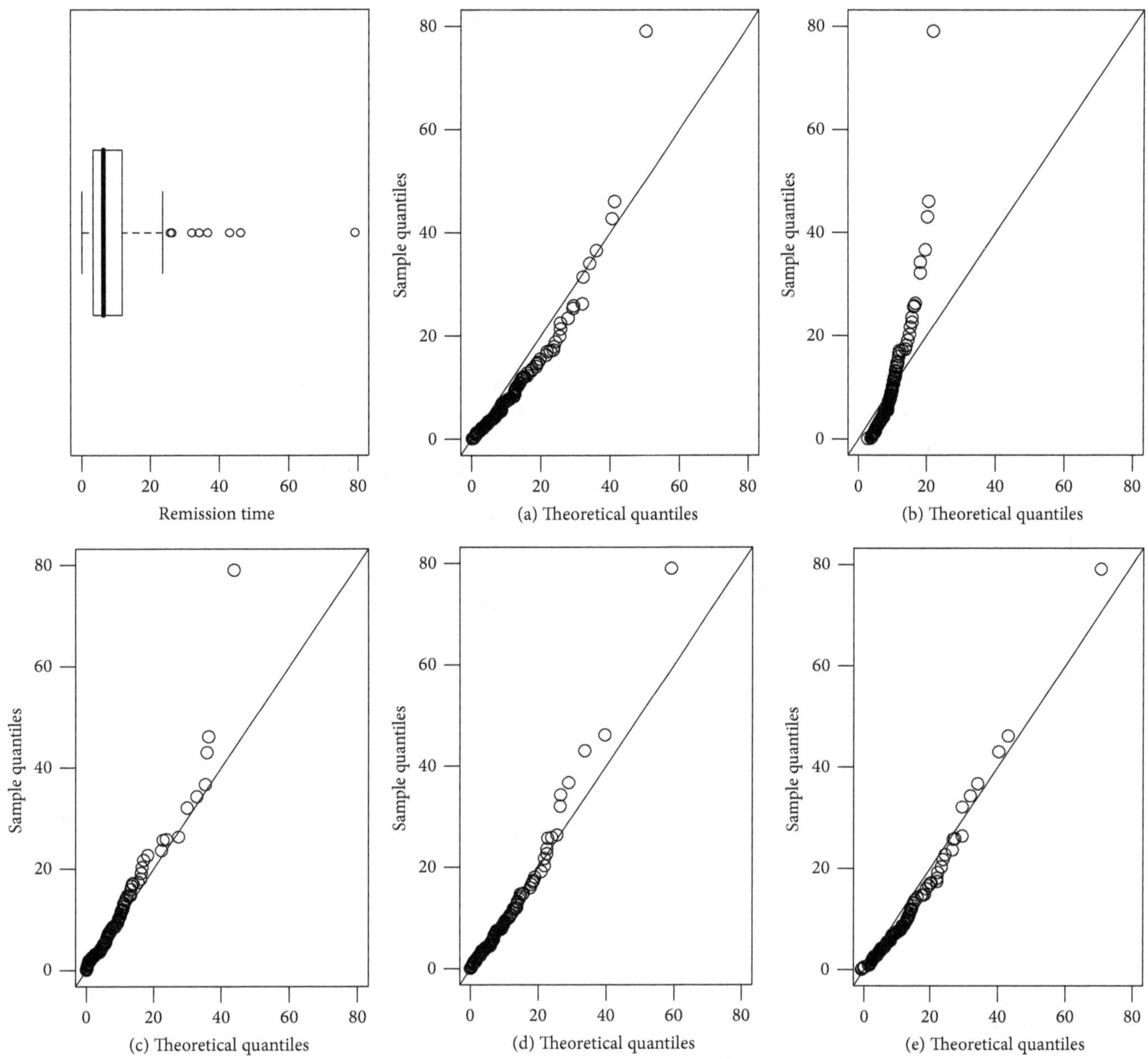

FIGURE 3: Boxplot and qqplots for the remission times data set: PL (a), LP (b), SHN (c), SL (d), and MSL (e).

TABLE 2: Descriptive statistics for data set.

n	$\overline{y}$	S	Asymmetry	Kurtosis
128	9.366	10.508	3.248	18.195

16.62, 43.01, 1.19, 2.75, 4.26, 5.41, 7.63, 17.12, 46.12, 1.26, 2.83, 4.33, 5.49, 7.66, 11.25, 17.14, 79.05, 1.35, 2.87, 5.62, 7.87, 11.64, 17.36, 1.40, 3.02, 4.34, 5.71, 7.93, 11.79, 18.10, 1.46, 4.40, 5.85, 8.26, 11.98, 19.13, 1.76, 3.25, 4.50, 6.25, 8.37, 12.02, 2.02, 3.31, 4.51, 6.54, 8.53, 12.03, 20.28, 2.02, 3.36, 6.76, 12.07, 21.73, 2.07, 3.36, 6.93, 8.65, 12.63, 22.69. For this data set we present basic descriptive statistics and the maximum likelihood parameter estimates for the MSL model; see Tables 2 and 3. We compare these results with those of the SL, SHN, LP, and PL models, indicating the corresponding standard errors in parentheses.

Using the results from Section 3.1, moment estimators were computed leading to the following values (standard errors in parentheses): $\widehat{\lambda}_M = 0.446$ (0.063) and $\widehat{q}_M = 2.024$ (0.007); these were used as initial estimates for the maximum likelihood approach.

We calculated the Akaike information criterion AIC (see Akaike [13]) and the Bayesian information criterion BIC (see Schwarz [14]). These criteria reveal that the MSL model presents the best fit for the data set studied.

The left side of Figure 3 shows the boxplot of these data. The graphs of the QQ plots for the remission times data calculated with the MSL model fitted with the maximum

TABLE 3: Summary statistics for a data set of remission times: model, ML estimates, and AIC and BIC values.

Model	ML estimates	AIC	BIC
MSL(λ, q)	$\widehat{\lambda} = 0.374$ (0.020), $\widehat{q} = 2.803$ (0.032)	**823.835**	**829.559**
SL(θ, σ, q)	$\widehat{\theta} = 0.194$ (0.169), $\widehat{\sigma} = 0.535$ (0.534), $\widehat{q} = 2.100$ (0.512)	825.445	834.002
SHN(σ, q)	$\widehat{\sigma} = 6.178$ (0.863), $\widehat{q} = 2.038$ (0.447)	826.287	831.991
LP(λ, θ)	$\widehat{\lambda} = 3.174$ (0.997), $\widehat{\theta} = 0.113$ (0.020)	826.769	832.423
PL(α, λ)	$\widehat{\alpha} = 0.830$ (0.047), $\widehat{\lambda} = 0.294$ (0.037)	830.708	836.411

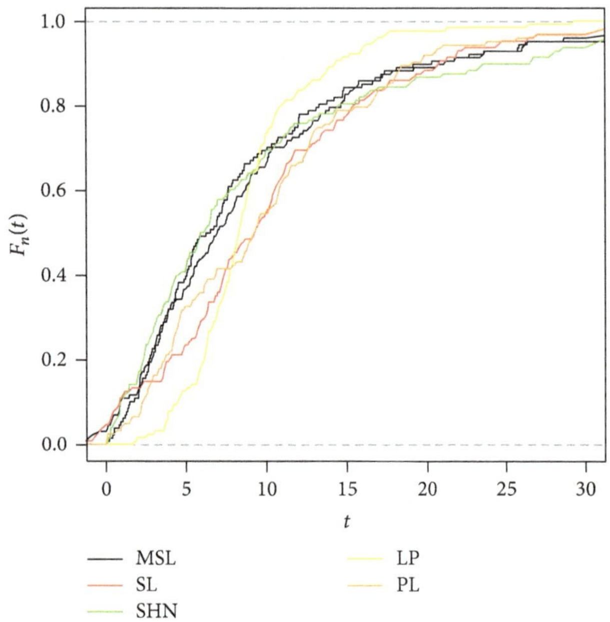

FIGURE 4: Models fitted by the maximum likelihood approach for the remission times data set: MSL (blue), SL (red), SHN (green), LP (yellow), and PL (orange).

likelihood estimates of the parameters; these are shown in Figure 3 compared with the SL, SHN, LP, and PL models. Figure 4 shows the empirical c.d.f. with the c.d.f. estimated by MSL, SL, SHN, LP, and PL; these also show the good agreement of the MSL model for the remission times data set.

4. Discussion

We introduce a new distribution with positive support based on the L distribution, called Modified Slash Lindley distribution. It is generated as the quotient of two independent random variables, one being a Lindley distribution and the other a power of the exponential distribution. The Lindley distribution is a special case. Moments estimators are computed which can be used for initializing maximum likelihood estimation using Newton-Raphson. By computing the asymmetry and kurtosis coefficients we have illustrated that MSL is able to accommodate data with higher kurtosis. We present an illustration with a real data set where we show that the MSL distribution provides a better fit to data than the SL, SHN, LP, and PL distributions.

Disclosure

Preliminary results of this manuscript were presented as an abstract in a conference proceedings "XXV COMCA Congreso de Matemática Capricornio" 2016.

Competing Interests

The authors declare that there is no conflict of interests regarding the publication of this paper.

Acknowledgments

The research of J. Reyes and H. W. Gómez was supported by SEMILLERO UA-2016 (Chile).

References

[1] D. V. Lindley, "Fiducial distributions and Bayes' theorem," *Journal of the Royal Statistical Society. Series B. Methodological*, vol. 20, pp. 102–107, 1958.

[2] M. E. Ghitany, B. Atieh, and S. Nadarajah, "Lindley distribution and its application," *Mathematics and Computers in Simulation*, vol. 78, no. 4, pp. 493–506, 2008.

[3] P. Jodrá, "Computer generation of random variables with Lindley or Poisson–Lindley distribution via the Lambert *W* function," *Mathematics and Computers in Simulation*, vol. 81, no. 4, pp. 851–859, 2010.

[4] M. E. Ghitany, D. K. Al-Mutairi, N. Balakrishnan, and L. J. Al-Enezi, "Power Lindley distribution and associated inference," *Computational Statistics and Data Analysis*, vol. 64, pp. 20–33, 2013.

[5] N. M. Olmos, H. Varela, H. W. Gómez, and H. Bolfarine, "An extension of the half-normal distribution," *Statistical Papers*, vol. 53, no. 4, pp. 875–886, 2012.

[6] J. Reyes, H. W. Gómez, and H. Bolfarine, "Modified slash distribution," *Statistics*, vol. 47, no. 5, pp. 929–941, 2013.

[7] J. Reyes, H. W. Gómez, and I. Vidal, "Modified skew-slash distribution," *Communications in Statistics. Theory and Methods*, vol. 45, no. 4, pp. 1070–1080, 2016.

[8] J. Reyes, F. Vilca, D. I. Gallardo, and H. W. Gomez, "Modified slash Birnbaum-Saunders distribution," *Hacettepe Journal of Mathematics and Statistics*, vol. 46, no. 112, pp. 1–1, 2016.

[9] W. Gui, "Statistical properties and applications of the Lindley slash distribution," *Journal of Applied Statistical Science*, vol. 20, no. 3, pp. 283–298, 2012.

[10] W. Gui, S. Zhang, and X. Lu, "The Lindley-Poisson distribution in lifetime analysis and its properties," *Hacettepe Journal of Mathematics and Statistics*, vol. 43, no. 6, pp. 1063–1077, 2014.

[11] P. K. Sen and J. M. Singer, *Large Sample Methods in Statistics*, Chapman & Hall, New York, NY, USA, 1993.

[12] E. T. Lee and J. W. Wang, *Statistical methods for survival data analysis*, Wiley Series in Probability and Statistics, Wiley-Interscience [John Wiley & Sons], Hoboken, NJ, USA, 3rd edition, 2003.

[13] H. Akaike, "A new look at the statistical model identification," *IEEE Transactions on Automatic Control*, vol. 19, no. 6, pp. 716–723, 1974.

[14] G. Schwarz, "Estimating the dimension of a model," *The Annals of Statistics*, vol. 6, no. 2, pp. 461–464, 1978.

8

Parameter Estimation in Mean Reversion Processes with Deterministic Long-Term Trend

Freddy H. Marín Sánchez and Verónica M. Gallego

Department of Mathematical Sciences, Universidad Eafit, Carrera 49 No. 7 Sur 50, Medellin, Colombia

Correspondence should be addressed to Freddy H. Marín Sánchez; fmarinsa@eafit.edu.co

Academic Editor: Chin-Shang Li

This paper describes a procedure based on maximum likelihood technique in two phases for estimating the parameters in mean reversion processes when the long-term trend is defined by a continued deterministic function. Closed formulas for the estimators that depend on observations of discrete paths and an estimation of the expected value of the process are obtained in the first phase. In the second phase, a reestimation scheme is proposed when *a priori* knowledge exists of the long-term trend. Some experimental results using simulated data sets are graphically illustrated.

1. Introduction

In the literature it is common to find studies and applications of mean reversion models, at both practical and theoretical levels, especially in fields such as energy markets, commodities, and interest rates. In these models there is a long-term trend which acts as an attractor making the process to oscillate around it and there is a random component that adds volatility to the movement. The specific characteristics of the models will depend on the structure of the trend and the volatility of the process.

Some common models of mean reversion are those in which all the parameters in the process are constant, such as Ornstein-Uhlenbeck model, the CIR model proposed by Cox et al. [1], and in general the models obtained from the CKLS structure proposed by Chan et al. [2]. In addition to these models there are other methods not so common in the literature in which the reversion is determined not by a parameter but a deterministic function or another stochastic process [3, 4]. In general, the fit of the models to specific situations is done by assigning values to the parameters and functions, either from *a priori* knowledge of the problem or from parameter estimation techniques based on historical data series that help to detect unobservable information from the process.

One of the first difficulties in the application of parameter estimation techniques is that it is necessary to have information with the same time resolution that is specified in the model. Since models of stochastic differential equations are formulated in continuous time and that the observed paths of the process can be obtained only in discrete time, it is necessary to discretize the EDE model. An approach frequently used for this purpose is given by Euler-Maruyama. With this methodology a new discrete-time process is obtained, with which it is possible to make inferences that are valid for continuous time model, as in the case of estimating the parameters.

There are different procedures that can be implemented such as methods based on distributional moments [5], Kernel methods [6, 7], least squares, and maximum likelihood [8–10]. Parameter estimation becomes fundamental too for modeling control systems. In the literature it is possible to find different methodologies for those issues such as those developed for Meng and Ding [11], where the parameter estimation in an equation-error autoregressive (EEAR) system is done through a transformation of the model to an equivalent one removing the autoregressive term and to obtain the parameters of the original model the principle of equivalence is used. Another methodology is proposed for Cao and Liu [12] where the parameter estimation in a power system is carried out from a recursive procedure to estimate simultaneously all the parameters from techniques based on the hierarchical identification principle, using gradient algorithms and least squared algorithms. Complementarily, Ding et al. [13] used methods based on the gradient algorithms and least squares iterative algorithms for system identification in output error (OE) and output error moving average (OEMA) systems, where the parameters that depend on unknown variables are computed using estimates of these unknown variables from previously estimated parameters.

In the area of multirate system identification, Ding et al. [14, 15] used the polynomial transformation technique to deal with the identification problem for dual-rate stochastic systems, while Sahebsara et al. [16] discussed the parameter estimation of multirate multi-input multioutput systems. Also, Ding and Chen [17] presented the combined parameter and state estimation algorithms of the lifted state-space models for general dual-rate systems based on the hierarchical identification method. Shi et al. [18] gave a crosstalk identification algorithm for multirate xDSL FIR systems.

In general, the auxiliary model identification idea has been used to solve the identification problem of dual-rate single-input single-output systems as shown in [19].

For more detailed information about multi-innovation stochastic gradient algorithm for multi-input multioutput systems and for multirate multi-input systems as well as auxiliary models based multi-innovation extended stochastic identification theory (see [20] and references therein). We will concentrate on one-factor stochastic model in which the parameters are estimated using maximum likelihood based on the discretized model when the long-term trend is given by a deterministic function. In this case we must estimate both the trend function and the parameters. To estimate the long-term trend function some convolution techniques and numerical differentiation are used, whereas the normality properties of residuals resulting from discretization are used for estimating parameters by maximum likelihood and in this way it is possible to obtain closed formulas for estimators based on the observations of a path of the process and the estimation of the long-term trend.

In the estimation process there may be some bias in the estimates with respect to the actual values despite the fact that they are obtained by the method of maximum likelihood. When this situation occurs it is necessary to develop alternative methodologies to have a greater adjustment to the estimates based on the initial estimates [21].

This paper is organized as follows. In Section 2 a description of mean reversion processes is presented for cases when the long-term trend is given by a continued deterministic function. The first phase of the estimation technique is shown in Section 3. Section 4 presents some examples and preliminary results for the estimation technique and in Section 5 (second phase), a procedure for reestimating the parameters from additional *a priori* knowledge is described.

2. Mean Reversion Processes with Deterministic Long-Term Trend

Mean reversion processes of one factor with constant parameters and mean given by a deterministic function can be written as

$$dX_t = \alpha\left(\mu(t) - X_t\right)dt + \sigma X_t^{\gamma}\, dB_t; \quad t \in [0, T] \tag{1}$$

with initial condition $X_0 = x$, where $\alpha > 0$, $\sigma > 0$, and $\gamma \in [0, 3/2]$ are constants, $\mu(t)$ is a deterministic function, and $\{B_t\}_{t\geq 0}$ is a Unidimensional Standard Brownian Motion defined on a complete probability space $(\Omega, \mathcal{F}, \mathbb{P})$.

α parameter is called reversion rate, $\mu(t)$ is the mean reversion level or trend of long-run equilibrium, σ is the parameter associated with the volatility, and γ determines the sensitivity of the variance to the level of X_t.

Model (1) is a generalization of the models CKLS, Chan et al. [2], where the mean reversion level is a deterministic function that captures the trend of the process. In a general way, $\mu(t)$ plays the role of an attractor at each point t in the sense that, when $X_t > \mu(t)$ the trend term $\alpha(\mu(t) - X_t) < 0$ and therefore X_t decreases and when $X_t < \mu(t)$ a similar argument establishes that X_t grows.

To establish the relation between the mean reversion function $\mu(t)$ and the expected value of the process $E[X_t]$, we can consider (1) written in integral form as

$$X_t - X_0 = \alpha \int_0^t \left(\mu(s) - X_s\right) ds + \sigma \int_0^t X_s^{\gamma} dB_s. \tag{2}$$

Taking the expected value, $E[X_t] - E[X_0] = \alpha \int_0^t (\mu(s) - E[X_s])ds$ and thus obtain the ordinary differential equation:

$$\dot{m}(t) = \alpha\left(\mu(t) - m(t)\right); \quad m(t) = E\left[X_t\right]. \tag{3}$$

The solution of this equation is $m(t) = m(0)e^{-\alpha t} + e^{-\alpha t}\int_0^t e^{\alpha s}\mu(s)ds$.

To illustrate graphically the relation between $\mu(t)$ and $m(t)$, we use $\mu(t) = a + b\sin(2\pi t/P)$ as an example. In this case the solution to (3) is given by

$$\begin{aligned} m(t) &= a + \left[m_0 - a + \frac{\alpha\beta b}{\alpha^2 + \beta^2}\right] e^{-\alpha t} \\ &\quad + \frac{\alpha b}{\alpha^2 + \beta^2}\left[\alpha \sin(\beta t) - \beta\cos(\beta t)\right], \end{aligned} \tag{4}$$

where $\beta = 2\pi/P$.

Figure 1 shows the dynamic behavior of $\mu(t)$, $m(t)$ and one path of X_t for some specific values in the parameters.

From the Figure it can be seen that the expected value of the process mimics the long-term mean: that is just the way mean reversion works.

3. Phase 1: Parameter Estimation

Similar to the procedure established in Marín Sánchez and Palacio [10] and Huzurbazar [22] the main objective is to obtain closed formulas for the parameter estimators from

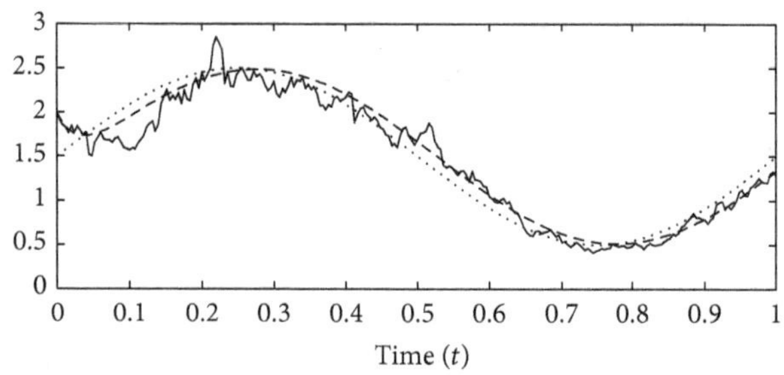

FIGURE 1: Dynamic behavior for $m(t)$, $\mu(t)$, and X_t, where $X_0 = 2$, $\alpha = 37$, $\sigma = 0.7$, $\gamma = 1$, $a = 1.5$, $b = 1$, $P = 1$, and $\beta = 2\pi$ with $\Delta t = 1/250$ for a total of 250 observations.

discrete observations of a sample path of the process. This can be achieved from the discretized maximum likelihood function, which is obtained from the process that results from the Euler-Maruyama discretization scheme applied to (1) and using the normal and independent increments properties of the Brownian Motion over the residuals of the discrete process.

Consider the differential equation (1) with the initial value $X(0) = x$. $\alpha(\mu(t) - X_t)$ and σX_t^γ are, respectively, the trend and diffusion functions that depend on t, X_t, and a vector of unknown parameters $\theta = [\alpha, \sigma, \gamma]$. Now suppose that the conditions that guarantee the existence and uniqueness of the solution Mao [23] are satisfied, so the expected value of X_t exists and $\mu(t)$ can be defined using $m(t)$ and its derivative $\dot{m}(t)$ (see (3)). In this way,

$$\mu(t) = m(t) + \frac{\dot{m}(t)}{\alpha}. \tag{5}$$

Replacing $\mu(t)$ in (1) results in

$$dX_t = \alpha\left(m(t) + \frac{\dot{m}(t)}{\alpha} - X_t\right)dt + \sigma X_t^\gamma dB_t; \quad t \in [0, T]. \tag{6}$$

Now we can assume that we have the discrete observations $X_t = X(\tau_t)$ for the times $\{\tau_0, \tau_1, \ldots, \tau_T\}$, where $\Delta_t = \tau_t - \tau_{t-1} \geq 0$, $t = 1, 2, \ldots, T$. Using the numerical scheme of Euler-Maruyama on (6) it follows that

$$\begin{aligned} X(\tau_t) &= X(\tau_{t-1}) \\ &\quad + [\alpha(m(\tau_{t-1}) - X(\tau_{t-1})) + \dot{m}(\tau_{t-1})]\Delta_t \\ &\quad + X^\gamma(\tau_{t-1})\varepsilon_{\tau_t}, \end{aligned} \tag{7}$$

where $\varepsilon_{\tau_t} \approx N(0, \sigma^2\Delta_t)$.

Define the new variable Y_t:

$$Y_t = \frac{X_t - X_{t-1} - [\alpha(m_{t-1} - X_{t-1}) + \dot{m}_{t-1}]\Delta}{X_{t-1}^\gamma} \tag{8}$$

with Δ constant.

Note that Y_t, $i = 1, 2, 3, \ldots, T$, are independent random variables distributed $N(0, \sigma^2\Delta)$.

The variable Y_t depends on the observations $X = (X_0, X_1, \ldots, X_T)'$ of the sample path of the process and on the unobservable quantities $m(t)$ and its derivative $\dot{m}(t)$. Taking into account that the sample path resumes all the information that is known from the process, it is necessary to estimate the expected value $m(t)$ from those observations. As in Tifenbach [4], p. 31-32, we can assume that the expected value of the process can be approximated using a convolution[1] over the sample path, so the variable $m(t)$ will depend only on the observations $X = (X_0, X_1, \ldots, X_T)'$.

Thus, we assume that the expected value of the process X_t can be approximated by taking a convolution of the sample path $X = (X_0, X_1, \ldots, X_T)$ for some particular convolution function $c(t)$ given by

$$m_i = \sum_{k=-N}^{N} c_k X_{i-k}. \tag{9}$$

The most common convolution is the moving average, where c_k constants are given by $c_k = 1/(2N+1)$ for all N, which is denoted by (c_t^1). Other convolutions can be defined varying the relative weight of each observation: for example, they can be defined from the coefficients of odd rows of Pascal's triangle which is denoted by (c_t^2). In this case the central observations have a greater weight than more distant observations and for this reason the convolution is close to the path. Those specific cases of convolutions are presented in Figure 2.

The convolution is used for the purpose of smoothing the sample path, eliminating the noise and capturing the trend. This objective can also be achieved using other techniques like filtering. In Figure 2 the Hodrick-Prescott [24] filter denoted by (c_t^{hp}) was used to obtain an estimation of $m(t)$ that are smoother than the convolutions (c_t^1) and (c_t^2).

At this point we have that the expected value $m(t)$ can be approximated from the sample path $X = (X_0, X_1, \ldots, X_T)$. Now, to get the derivative of $m(t)$ we can apply numerical derivatives techniques like Taylor's theorem: for example, the derivative at the observation i can be defined using a three-point rule as

$$\dot{m}_i = \frac{2m_{i+1} - 3m_i + m_{i-1}}{\Delta}, \tag{10}$$

for $i = 1, 2, \ldots, T-1$. For $i = 0$ the derivate is given by $\dot{m}_0 = (m_1 - m_0)/\Delta$ and for $i = T$ by $\dot{m}_T = (m_T - m_{T-1})/\Delta$. This procedure can be done using a different number of points to calculate the derivate at each point.

With the estimation of $m(t)$ and $\dot{m}(t)$ we have one realization of Y_t and now we can define its normal density function:

$$\begin{aligned} f(Y_i : \theta) &= \left(\frac{1}{2\pi\sigma^2\Delta}\right)^{1/2} \\ &\quad \cdot \exp\left[\frac{-1}{2\sigma^2\Delta}\left(\frac{X_i - X_{i-1} - [\alpha(m_{i-1} - X_{i-1}) + \dot{m}_{i-1}]\Delta}{X_{i-1}^\gamma}\right)^2\right]. \end{aligned} \tag{11}$$

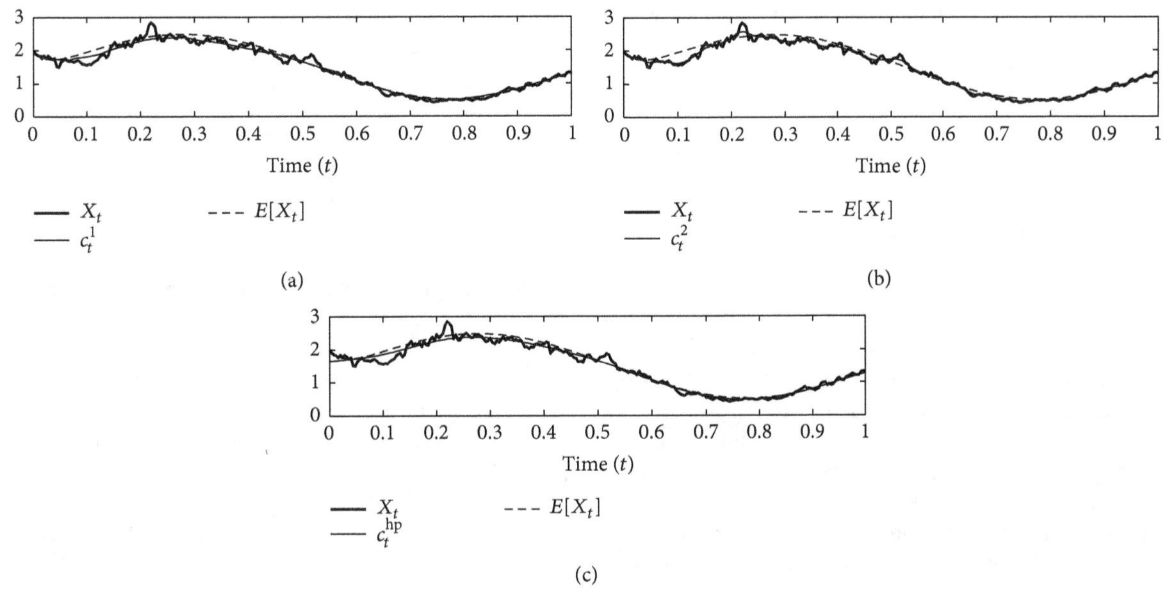

FIGURE 2: Dynamic behavior for process X_t, the expected value $m(t)$, and three types of convolution. (a) Moving average (c_t^1) with $N = 20$. (b) Pascal's weights (c_t^2) with $N = 20$. (c) Hodrick-Prescott filter (c_t^{hp}) with $\lambda_{\mathrm{hp}} = 40000$. For all the cases $\Delta t = 1/250$ for a total of 250 observations and the parameters of X_t are the same as in Figure 1.

Since each Y_i is independent the joint density function can be expressed as the product of their marginal densities:

$$f(Y_1, Y_2, \dots, Y_T) = \prod_{k=1}^{T} f(Y_k : \theta). \tag{12}$$

Consequently, the likelihood function is given by

$$L(\theta \mid \{Y_t\}) = \left(\frac{1}{2\pi\sigma^2\Delta}\right)^{T/2} \exp\left[\frac{-1}{2\sigma^2\Delta} \cdot \sum_{i=1}^{T}\left(\frac{X_i - X_{i-1} - [\alpha(m_{i-1} - X_{i-1}) + \dot{m}_{i-1}]\Delta}{X_{i-1}^{\gamma}}\right)^2\right]. \tag{13}$$

Taking the natural logarithm on both sides it follows that

$$\log L = \frac{-T}{2}\log\left(2\pi\sigma^2\Delta\right) - \frac{1}{2\sigma^2\Delta}\left[\sum_{i=1}^{T}\left(\frac{X_i - X_{i-1} - [\alpha(m_{i-1} - X_{i-1}) + \dot{m}_{i-1}]\Delta}{X_{i-1}^{\gamma}}\right)^2\right]. \tag{14}$$

Now consider the problem of maximizing the likelihood function given by

$$\frac{\partial \log(L)}{\partial\theta} = \vec{0}, \quad \text{ie}, \hat{\theta} = \arg\max_{\theta}(\log(L)). \tag{15}$$

If we assumed that the value of γ is known *a priori*, the estimation process leads to the estimated parameter α given by

$$\hat{\alpha} = \frac{\sum_{i=1}^{T}\left((X_i - X_{i-1} - \dot{m}_{i-1}\Delta)(m_{i-1} - X_{i-1})/X_{i-1}^{2\gamma}\right)}{\sum_{i=1}^{T}\left[(m_{i-1} - X_{i-1})/X_{i-1}^{\gamma}\right]^2\Delta}. \tag{16}$$

The estimator $\hat{\sigma}$ is determined by the equation:

$$\hat{\sigma} = \sqrt{\frac{1}{T\Delta}\sum_{i=1}^{T}\left(\frac{X_i - X_{i-1} - [\hat{\alpha}(m_{i-1} - X_{i-1}) + \dot{m}_{i-1}]\Delta}{X_{i-1}^{\gamma}}\right)^2}. \tag{17}$$

Finally, with the estimation of $\hat{\alpha}$ we can now approximate $\hat{\mu}(t)$ with (5).

4. Preliminary Results

The aim of this section is to show the performance of the estimators defined in the previous section. To make this we will simulate some paths of a process satisfying (1) with a set of parameters α and σ and giving a deterministic form to the $\mu(t)$ function and then find the estimators with the established procedure in order to determine—using statistical properties—if the estimates are approximate to actual parameters.

We analyze the performance of the estimators for sinusoidal and parabolic $\mu(t)$ trend functions and for a set of values of α and σ when $\gamma = 1$. Each generated path consists of 250 observations with $\Delta t = 1/250$. The statistics are calculated for 1000 different simulations. Considering that for the parameters estimation it is necessary to find the expected value of the process from each path, we compare two different convolution functions and the Hodrick-Prescott filter. For numerical derivation a five-point derivation rule is used.

The values of the parameters and $\mu(t)$ functions used in the simulations are presented in Table 1.

In Table 2 we present some basic statistics (mean $\overline{\theta}$ and standard deviation s_θ) for the estimators obtained from

TABLE 1: Simulation parameters for sinusoidal and parabolic trend functions.

	Case 1	Case 2
α	30	30
σ	0.4	0.13
X_0	2	1.5
$\mu(t)$	$1.5 + \sin(2\pi t)$	$-0.5t^2 + t + 1.5$

the simulations. The convolution functions used in the simulations are the ones defined for the weights (c_k^1) and (c_k^3). c_k^1 weights are the same defined in Section 3 for the moving average and c_k^3 weights are given by $c_{k,N}^3 = (2N - |k|)/N(3N + 1)$ for k from $-N$ to N. In this convolution the central observation has a relative weight twice the most distant observation.

From Table 2 we can conclude that the estimation technique is working well, especially in the first example (sinusoidal trend). The relative error of the estimators is less than 2% in the case of the parameter σ but there is a bias in the parameter α that depends on the convolution function and at the same time on the parameter N used in such convolution.

In Figure 3 we can see the performance of the estimation of the deterministic function $\mu(t)$. In that figure an estimation from a random path is shown instead of the average. The estimates are close to the theoretical function in both cases but having some differences at the beginning and the final of the observations. One form to measure the fit between the theoretical function and its estimation is using the Root Mean Square error (RMS). Table 3 presents that measure for the proposed estimation in cases 1 and 2.

5. Phase 2: Re-Estimation

From the simulations we can see that the estimate of $\mu(t)$ is not very accurate and that differences come from $\widehat{m}(t)$ and $\widehat{\dot{m}}(t)$, and in turn this affects the estimation of other parameters.

If it were possible to have all the paths of the process it would not be necessary for numerical approximation of the expected value or its derivative to capture information from $\mu(t)$, and it could be enough to perform an average of all paths at each time t_i and that could generate pretty accurate estimations. As this condition is not the usual but only has a path, the procedure detailed above is necessary and this involves the errors mentioned in the estimates of $m(t)$ and $\dot{m}(t)$.

To correct this deficiency it is proposed to carry out a reestimate procedure trying to adjust a function to the initial estimate from *a priori* knowledge of the process. This recalculation allows a correction in the other estimators while providing a functional form that allows simulations of future paths of the process.

The following procedure is proposed to perform the reestimate of $\mu(t)$:

(1) Get an estimate of $\widehat{\mu}(t)$ according to the procedure described in Section 4.

(2) Define a functional structure $f(\Theta, t)$ from *a priori* knowledge of the process.

(3) Estimate the vector parameters Θ minimizing the error function defined by

$$\text{Error}(\Theta) = \sum_{t=1}^{T} \left[f(\Theta, t) - \widehat{\mu}(t) \right]^2. \quad (18)$$

As an example, suppose the functional form of the long-term mean is defined as

$$f(\Theta, t) = at^2 + bt + c, \quad (19)$$

where $\Theta = [a, b, c]'$. In this case the error function would be given by

$$\text{Error}(\Theta) = \sum_{t=0}^{T} \left[\widehat{\mu}(t) - \left(at^2 + bt + c \right) \right]^2. \quad (20)$$

To estimate Θ it is necessary to solve the system of equations defined by $\nabla\,\text{Error}(\Theta) = 0$, which is equivalent to the system of linear equations $M\Theta = d$, where

$$M = \begin{bmatrix} \sum t^4 & \sum t^3 & \sum t^2 \\ \sum t^3 & \sum t^2 & \sum t \\ \sum t^2 & \sum t & (T+1) \end{bmatrix}$$
$$d = \begin{bmatrix} \sum t^2 \widehat{\mu}(t) \\ \sum t \widehat{\mu}(t) \\ \sum \widehat{\mu}(t) \end{bmatrix}. \quad (21)$$

A similar procedure can be implemented under the assumption that $f(\Theta, t) = a + b\sin((2\pi/P)t)$, to find $\Theta = [a, b]'$ if P is known. A comparison between $\mu(t)$, $\mu_{\text{est}}(t)$, and $\mu_{\text{rest}}(t) = f(\widehat{\Theta}, t)$ is shown in Figure 4 for both cases and for a random path of the process defined with the same set of parameters of Table 1.

Having a best estimate of $\mu(t)$ allows a reestimation of the parameters α and σ. To achieve this we must perform a similar procedure to that proposed in Section 4, defining

$$Y_t = \frac{X_t - X_{t-1} - \alpha\left(\mu_{t-1} - X_{t-1}\right)\Delta}{X_{t-1}^{\gamma}} \qquad \Delta \text{ constant}. \quad (22)$$

So,

$$\widehat{\widehat{\alpha}} = \frac{\sum_{i=1}^{T} \left(\left(X_i - X_{i-1}\right) \left(\widehat{\widehat{\mu}}_{i-1} - X_{i-1}\right) \Delta / X_{i-1}^{2\gamma} \right)}{\sum_{i=1}^{T} \left[\left(\widehat{\widehat{\mu}}_{i-1} - X_{i-1}\right) / X_{i-1}^{\gamma} \right]^2 \Delta},$$
$$\widehat{\widehat{\sigma}} = \sqrt{\frac{1}{T\Delta} \sum_{i=1}^{T} \left(\frac{X_i - X_{i-1} - \widehat{\widehat{\alpha}}\left(\widehat{\widehat{\mu}}_{i-1} - X_{i-1}\right)\Delta}{X_{i-1}^{\gamma}} \right)^2}. \quad (23)$$

TABLE 2: Results obtained with 1000 simulations with 250 observations each. The real values of the parameters are shown in Table 1. For case 1 in the convolutions we use $N = 22$ and for case 2 $N = 60$. For both cases we set $\lambda = 62500$ in c_k^{hp}.

		Case 1					Case 2		
$\mu_1(t)$		c_k^1	c_k^3	c_k^{hp}	$\mu_2(t)$		c_k^1	c_k^3	c_k^{hp}
α	$\overline{\alpha}$	24.8540	31.3347	32.8771	α	$\overline{\alpha}$	31.1160	35.8249	51.1093
	s_α	3.2218	3.9947	5.1406		s_α	5.8254	6.6984	10.1414
σ	$\overline{\sigma}$	0.4002	0.3942	0.4023	σ	$\overline{\sigma}$	0.1293	0.1281	0.1267
	s_σ	0.0177	0.0174	0.0174		s_σ	0.0060	0.0059	0.0060

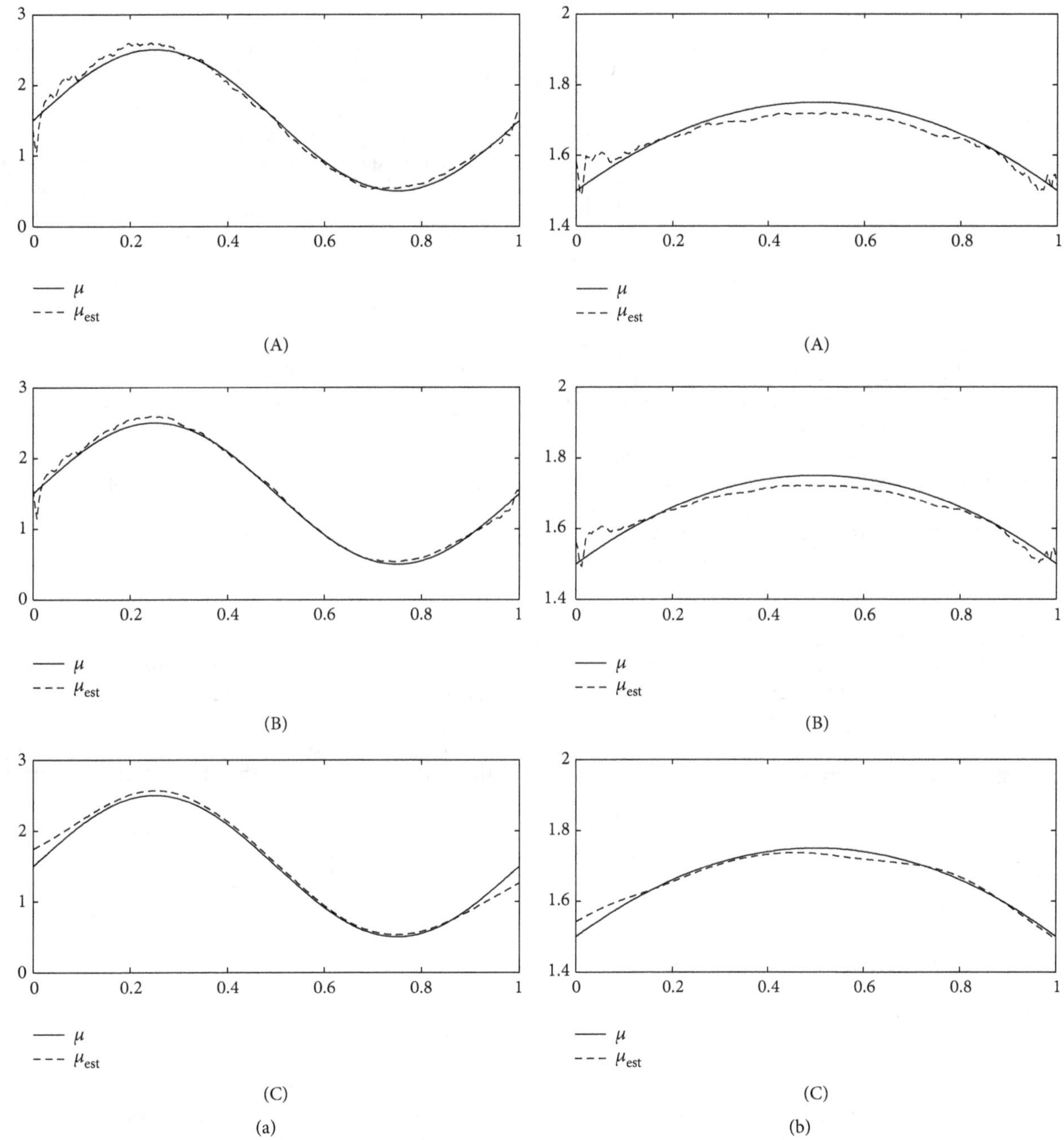

FIGURE 3: Estimated $\mu_1(t)$ and $\mu_2(t)$. The left side correspond to Case 1 (sinusoidal) and the right side to Case 2 (parabolic). The (A) plots are obtained with the convolution (c_k^1), (B) plots with (c_k^3) and (C) plots with (c_k^{hp}). The parameters involved in this figure are presented in Tables 1 and 2.

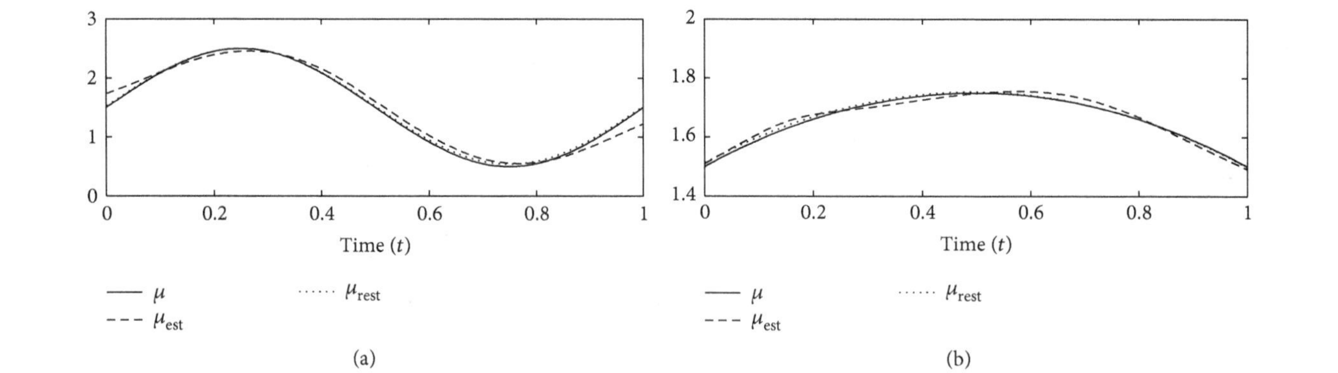

FIGURE 4: Comparison of dynamic behavior of $\mu(t)$, $\mu_{\text{est}}(t)$, and $\mu_{\text{rest}}(t)$. In case (a) $\mu_1(t) = 1.5 + \sin(2\pi t)$ and in case (b) $\mu_2(t) = -0.5t^2 + t + 1.5$.

TABLE 3: Results of RMS error for each estimation $\mu_1(t)$ and $\mu_2(t)$.

RMS	Case 1	Case 2
c_k^1	0.0782	0.0266
c_k^3	0.0578	0.0238
c_k^{hp}	0.0767	0.0140

Table 4 summarizes the results obtained from the first estimation and the reestimation procedure for the cases where $\mu(t)$ has a sinusoidal form and a parabolic form and the other parameters are defined in Table 1.

From the results of the above example we can observe how the reestimation procedure reduces the bias in the initial estimation of α for both sinusoidal and parabolic cases. The final bias in α is approximately 1.5% of the real value of the parameter, reducing almost 90% of the error in the first estimation in the case of sinusoidal function and near 60% for the parabolic function. The reestimations of σ do not significantly change from the original estimate. For the case of $\mu(t)$ the RMS error is present in Table 5 for a random path of the process. In both cases the error in the reestimation is less than the 50% of the error present in the initial estimation.

TABLE 4: Estimated and reestimated parameters. The estimations are calculated with the convolution (c_k^{hp}). Results obtained with 1000 simulations with 250 observations each for μ_1 and μ_2 as defined in Table 1: case 1 $\alpha = 30$ and $\sigma = 0.4$, case 2 $\alpha = 30$ and $\sigma = 0.13$.

$\mu_1(t)$		$\widehat{\alpha}$	$\widehat{\widehat{\alpha}}$	$\mu_2(t)$		$\widehat{\alpha}$	$\widehat{\widehat{\alpha}}$
α	$\overline{\alpha}$	24.8540	29.6620	α	$\overline{\alpha}$	31.1160	30.4616
	s_α	3.2218	3.4518		s_α	5.8254	5.6828
$\mu_1(t)$		$\widehat{\sigma}$	$\widehat{\widehat{\sigma}}$	$\mu_2(t)$		$\widehat{\sigma}$	$\widehat{\widehat{\sigma}}$
σ	$\overline{\sigma}$	0.4002	0.4021	σ	$\overline{\sigma}$	0.1293	0.1300
	s_σ	0.0177	0.0187		s_σ	0.0060	0.0060

TABLE 5: Results of RMS error for the estimation and reestimation of $\mu(t)$: case 1 sinusoidal trend and case 2 parabolic trend.

RMS	Case 1	Case 2
$\mu_{\text{est}}(t)$	0.0961	0.0137
$\mu_{\text{rest}}(t)$	0.0261	0.0058

6. Conclusions and Comments

This paper shows how starting from a path of a stochastic process is possible to find an approximation of the parameters that define the underlying dynamics It is important to highlight that, for modeling the dynamic behavior of prices of some commodities or interest rates, only one realization of the process in discrete time is available and it represents the only source of information to capture the nature of the process. Therefore, it becomes very important to obtain and interpret the invisible and visible information found on the path to make projections more adjusted to the future process behavior being studied.

It is essential that a model that seeks to reproduce the behavior of a process that is generated in the real world incorporates *a priori* information provided by an expert in the topic and that this additional information allows the model to better fit the process. This prior information is key in the development of the methodology proposed in this paper, because it allows us to adjust the initial estimates of the parameters making more reliable the results obtained with the adjusted models.

On the other hand, in the development of the proposed methodology it can be seen that it allows us to obtain closed formulas for the maximum likelihood estimators in mean reversion process when the long-term trend is defined by a deterministic function, decreasing computational effort and allowing greater understanding of procedures.

Competing Interests

The authors declare that they have no competing interests.

Endnotes

1. A convolution function $c(t)$ is an even, continuous, and real valued function satisfying the fact that $c(t) \geqslant 0$ and $\int_{-\infty}^{\infty} c(t)dt = 1$. A convolution of a curve $s(t)$ is another function $l(t)$ defined by

$$l(t) = \int_{-\infty}^{\infty} c(w)\, s(t-w)\, dw, \qquad (*)$$

where $c(t)$ is a convolution function.

In the discrete case the convolution of two sequences can be written as

$$l_i = \sum_{k=-\infty}^{\infty} c_k s_{i-k}. \qquad (**)$$

References

[1] J. C. Cox, J. Ingersoll, and S. Ross, "A theory of the term structure of interest rates," *Econometrica*, vol. 53, no. 2, pp. 385–407, 1985.

[2] K. C. Chan, G. A. Karolyi, F. A. Longstaff, and A. B. Sanders, "An empirical comparison of alternative models of the short-term interest rate," *The Journal of Finance*, vol. 47, no. 3, pp. 1209–1227, 1992.

[3] D. Pilipovic, *Energy Risk: Valuing and Managing Energy Derivatives*, McGraw-Hill, New York, NY, USA, 2007.

[4] B. Tifenbach, *Numerical Methods for Modeling Energy Spot Prices*, University of Calgary, 2000.

[5] T. Koulis and A. Thavaneswaran, "Inference for interest rate models using Milstein's approximation," *Journal of Mathematical Finance*, vol. 3, no. 1, pp. 110–118, 2013.

[6] D. Florens-Zmirou, "On estimating the diffusion coefficient from discrete observations," *Journal of Applied Probability*, vol. 30, no. 4, pp. 790–804, 1993.

[7] G. J. Jiang and J. L. Knight, "A nonparametric approach to the estimation of diffusion processes, with an application to a short-term interest rate model," *Econometric Theory*, vol. 13, no. 5, pp. 615–645, 1997.

[8] K. B. Nowman, "Gaussian estimation of single-factor continuous time models of the term structure of interest rates," *The Journal of Finance*, vol. 52, no. 4, pp. 1695–1706, 1997.

[9] J. Yu and P. Phillips, "Corrigendum to 'a Gaussian approach for continuous time models of the shortterm interest rate,'" *Econometrics Journal*, vol. 14, pp. 126–129, 2011.

[10] F. H. Marín Sánchez and J. S. Palacio, "Gaussian estimation of one-factor mean reversion processes," *Journal of Probability and Statistics*, vol. 2013, Article ID 239384, 10 pages, 2013.

[11] D. Meng and F. Ding, "Model equivalence-based identification algorithm for equation-error systems with colored noise," *Algorithms*, vol. 8, no. 2, pp. 280–291, 2015.

[12] Y. Cao and Z. Liu, "Signal frequency and parameter estimation for power systems using the hierarchical identification principle," *Mathematical and Computer Modelling*, vol. 52, no. 5-6, pp. 854–861, 2010.

[13] F. Ding, P. X. Liu, and G. Liu, "Gradient based and least-squares based iterative identification methods for OE and OEMA systems," *Digital Signal Processing*, vol. 20, no. 3, pp. 664–677, 2010.

[14] F. Ding, P. X. Liu, and Y. Shi, "Convergence analysis of estimation algorithms for dual-rate stochastic systems," *Applied Mathematics and Computation*, vol. 176, no. 1, pp. 245–261, 2006.

[15] F. Ding, P. X. Liu, and H. Z. Yang, "Parameter identification and intersample output estimation for dual-rate systems," *IEEE Transactions on Systems, Man, and Cybernetics, Part A: Systems and Humans*, vol. 38, no. 4, pp. 966–975, 2008.

[16] M. Sahebsara, T. Chen, and S. L. Shah, "Frequency-domain parameter estimation of general multi-rate systems," *Computers and Chemical Engineering*, vol. 30, no. 5, pp. 838–849, 2006.

[17] F. Ding and T. Chen, "Hierarchical identification of lifted state-space models for general dual-rate systems," *IEEE Transactions on Circuits and Systems. I. Regular Papers*, vol. 52, no. 6, pp. 1179–1187, 2005.

[18] Y. Shi, F. Ding, and T. Chen, "Multirate crosstalk identification in xDSL systems," *IEEE Transactions on Communications*, vol. 54, no. 10, pp. 1878–1886, 2006.

[19] F. Ding and T. Chen, "Combined parameter and output estimation of dual-rate systems using an auxiliary model," *Automatica*, vol. 40, no. 10, pp. 1739–1748, 2004.

[20] L. Han, J. Sheng, F. Ding, and Y. Shi, "Auxiliary model identification method for multirate multi-input systems based on least squares," *Mathematical and Computer Modelling*, vol. 50, no. 7-8, pp. 1100–1106, 2009.

[21] J. Yu, "Bias in the estimation of the mean reversion parameter in continuous time models," *Journal of Econometrics*, vol. 169, no. 1, pp. 114–122, 2012.

[22] V. S. Huzurbazar, "The likelihood equation, consistency and the maxima of the likelihood function," *Annals of Eugenics*, vol. 14, pp. 185–200, 1948.

[23] X. Mao, *Stochastic Differential Equation and Applications*, Horwood Publishing Limited, England, UK, 1997.

[24] R. J. Hodrick and E. C. Prescott, "Postwar U.S business cycles: an empirical investigation," *Journal of Money, Credit and Banking*, vol. 29, no. 1, pp. 1–16, 1997.

9

Probability of Dying in Each of the Competing Risks under Bimorbid Condition

Kalpana Singh[1,2] and Suddhendu Biswas[1]

[1]*Department of Statistics, Amity Institute of Applied Science (AIAS), Amity University, Noida 201301, India*
[2]*Public Health Foundation of India, Sector 44, Plot 47, Gurgaon, Haryana 122002, India*

Correspondence should be addressed to Kalpana Singh; nkalpanasingh@gmail.com

Academic Editor: Shesh N. Rai

The motivation of this study is to explore the respective probability of dying in either of the risks while a person has dual or more competing risks. The problem requires evolution of a model describing the probability of dying in either of risks while a person is exposed to competing risk of two or more diseases. This is obtained by choice of a model which is bivariate negative binomial distribution for obtaining the number of deaths (assuming the deaths occur in the last trial). A Bayesian method has also been used for estimating probability of dying in heart disease and chronic kidney disease in the presence of diabetes, hypertension, and stroke or heart disease with chronic kidney disease. For the analysis of the same we have used CARRS (cardiometabolic risk reduction in south Asia) surveillance data, a multistage cluster random sampling technique.

1. Introduction

Gumbell (1960) described the earliest and the simplest bivariate exponential distribution. This method has received applications in many areas, including competing risks, extreme values, failure times, regional analyses of precipitation, and reliability.

Downton [1] described the bivariate exponential distribution arising from "shocks" causing various types of failure to components which have geometric distributions. It has received applications in queueing systems and hydrology and has been used as a model for Wold's Markov dependent processes, intensity and duration of rainfall, and height of water waves.

Marshall and Olkin [2] estimated the bivariate exponential distribution in which X and Y are the lifetimes of two components subject to three kinds of shocks; these shocks are assumed to be governed by independent Poisson processes with parameters $\theta 1$, $\theta 2$, and $\theta 3$, according as whether the shock applies to component 1 only, component 2 only, or both the components. The distribution has received wide applicability in nuclear reactor safety, competing risks, reliability, and quantal response contexts.

Biswas and Noor [3] investigated a method for classifying unknown causes of death due to two competing risks into assignable causes.

The real cause of death for a person, otherwise suffering from two major risks (heart disease with hypertension, heart disease with diabetes, heart disease with chronic kidney disease, diabetes with hypertension, stoke with hypertension, stroke with diabetes, chronic kidney disease with diabetes, and chronic kidney disease with hypertension), is very difficult to estimate and remains unclassified. The purpose of the present paper is to devise a suitable probability model describing the effect of two such competing risks as $R1$ or $R2$ and obtain the expected proportion of such unclassified cases suffering from both the conditions $R1$ and $R2$ and dying either *in* $R1$ or $R2$. It has been tacitly assumed that while both the risks due to $R1$ and $R2$ may be operating at the time of death of a person, really the death in actuality occurs, principally due to either $R1$ or $R2$. Further, the development of the model is in the line of a risk model and we assume that death is caused ultimately on account of a single risk of either $R1$ *or* $R2$. Biswas and Noor [3] have investigated the same type of problem with hypothetical data with several parameters in the model. However, in this study some parameters were found redundant because of availability of data.

The approach followed by Biswas and Noor [3] was based on number of shocks corresponding to two competing risks ($R1$ and $R2$) and the number of shocks causing death by either of the two shocks taken as parameters. The shock due to risk $R1$ occurs with Poisson intensity and the effect of shock lasts for a fixed period (dead time) during which no further shock due to $R1$ or $R2$ may occur. But, in our model, we have not considered the effect of shock, which lasts for a fixed period, during which no further shocks occur due to $R1$ or $R2$.

The number of shocks for each of the competing risks that may cause death varies from individual to individual, information of which was readily available from the CARRS Cohort dataset. Therefore, we could avoid redundant parameter estimation. Hence our model may be considered as an extension of earlier Biswas and Noor paper.

Recently McCormick et al. [4] investigated the same type of problem "probabilistic cause-of-death assignment using verbal autopsies." The present application using the Bayesian method is more or less based on the same approach.

2. Study Design

As stated earlier, the present study is based on the baseline data from the cross-sectional survey of the three CARRS Study sites conducted between October 2010 and 2011 with follow-up for the investigation of mortality through December 2014. The duration of the follow-up was 3.5 years. The CARRS (cardiometabolic risk reduction in south Asia) Study design was adapted from the WHO STEPS model [5] to assess the prevalence and incidence of cardiometabolic diseases (CMDs) and their risk factors. Details of the CARRS Study design have been published [6]. In brief, participants were selected in each of the three cities (Chennai, New Delhi, and Karachi) using a multistage cluster sampling technique [7]. Municipal subdivisions, wards in Chennai and Delhi and census enumeration blocks in Karachi, were treated as the primary sampling units. Households in selected subdivisions were then manually listed and mapped. The manual household listing was used to randomly select households with equal opportunity for selection. From each randomly selected household, one eligible man and one eligible woman (aged 20+, nonpregnant, and not bedridden) were selected. For households with more than two eligible adults of the same sex, the "Kish method" [5] (all eligible participants from the household were ranked according to age in decreasing orders (males followed by females); participants were selected using KISH table identifying the last digit of household and number of eligible participants) was applied to randomly select one individual to be included in the study.

3. Data Collection

Data were collected from the households through interviews in local languages and standardized clinical examinations as well as blood sample collection either at local camps (Chennai, Delhi) or at home (Karachi). In the three cities, collective baseline response rate was 94.7% for questionnaire completion and 84.3% for specimen collection. Data were gathered by trained field teams using standardized techniques [6].

3.1. Definition of Chronic Conditions. In the present study, chronic conditions including hypertension, diabetes, chronic kidney disease, heart disease, and stroke were considered. These medical conditions were defined as follows: hypertension (self-report of prior diagnosis or measured blood pressure of 140/90 mmHg or above); diabetes (self-report of prior diagnosis or fasting plasma glucose of ≥126 mg/dL or HbA1c of ≥6.5); heart disease/stroke (self-report of prior diagnosis); CKD (self-report of prior diagnosis). All participant had already had two or more risks at the beginning of the study.

We further classified participants into two-way joint distribution of the specified chronic conditions, that is, heart disease with hypertension, heart disease with diabetes, heart disease with chronic kidney disease, diabetes with hypertension, stroke with hypertension, stroke with diabetes, chronic kidney disease with diabetes, and chronic kidney disease with hypertension.

3.2. Development of the Model. Let R_1 and R_2 be two competing risks in order.

(I) p = probability of occurrence of R_1.

(II) q = probability of occurrence of R_2, where $p + q = 1$.

(III) M_1 and M_2 represent the number of trials where the individuals were exposed to the risks R_1 and R_2, respectively.

(IV) θ = probability that a particular risk of first or second type is fatal leading to death.

(V) $1 - \theta$ = probability of surviving in R_1 and R_2.

(VI) χ is the probability of dying in either R_1 or R_2 without considering that the risks are competing.

(VII) N_1 = number of death cases exposed to the risk of R_1 or R_2 while the death might have occurred either due to R_1 or R_2, the real cause being unknown.

(VIII) N_2 = number of death cases exposed to risk of R_1 or R_2 while the real cause is unknown.

$N = N_1 + N_2$ = size of the population.

(IX) $\chi = N_1/(N_1 + N_2)$.

Then we have

$$N_1 = N\sum_{x=0}^{M'}\sum_{r=0}^{M}\binom{x+r-1}{r-1}\left[(1-\theta)\,p\right]^{r-1}\left[(1-\theta)\,q\right]^{x}\left[p\,(\theta)\right]$$

$$N_2 = N\sum_{x=1}^{M'}\sum_{r=0}^{M}\binom{x+r-1}{r-1}\left[(1-\theta)p\right]^r\left[(1-\theta)q\right]^{x-1}\left[q(\theta)\right]$$

$$\chi = \frac{N\sum_{x=0}^{M'}\sum_{r=0}^{M}\binom{x+r-1}{r-1}\left[(1-\theta)p\right]^{r-1}\left[(1-\theta)q\right]^{x}\left[p(\theta)\right]}{N\sum_{x=0}^{M'}\sum_{r=0}^{M}\binom{x+r-1}{r-1}\left[(1-\theta)p\right]^{r-1}\left[(1-\theta)q\right]^{x}\left[p(\theta)\right]+N\sum_{x=1}^{M'}\sum_{r=0}^{M}\binom{x+r-1}{r-1}\left[(1-\theta)p\right]^{r}\left[(1-\theta)q\right]^{x-1}\left[q(\theta)\right]} \quad (1)$$

χ being the conditional probability of dying in R_1 under the dual risk of both R_1 and R_2. Hence the probability of unclassified case of death being classified and put in the category of R_1 (the exact cause of death) is given by

$$z_1 = \left[\chi * \frac{Q_{t-1}}{Q_{t-1}+Q_{t-2}}\right] \quad (2)$$

and that of being classified in R_2 is given by

$$z_2 = \left[\chi * \frac{Q_{t-2}}{Q_{t-1}+Q_{t-2}}\right], \quad (3)$$

where Q_{t-1} is the crude probability of dying in risk R_1 in presence of risk R_2 in a duration $(0,t)$. It is given by

$$Q_{t-1} = \int_0^t e^{-\int_0^t \lambda d\lambda}\lambda_1 d\lambda, \quad (4)$$

where

$$\begin{aligned}\lambda &= \lambda_1+\lambda_2\\ \lambda &= \frac{\lambda_1}{\lambda}\left[1-e^{-\lambda t}\right];\\ &\cong \frac{\lambda_1}{\lambda}\left[1-(1-\lambda t)\right];\\ &\cong \frac{\lambda_1}{\lambda} * \lambda t.\end{aligned} \quad (5)$$

Similarly,

$$\begin{aligned}Q_{t-2} &= \int_0^t e^{-\int_0^t \lambda d\lambda}\lambda_2 d\lambda,\\ \lambda &= \frac{\lambda_2}{\lambda}\left[1-e^{-\lambda t}\right];\\ &\cong \frac{\lambda_2}{\lambda}\left[1-(1-\lambda t)\right];\\ &\cong \frac{\lambda_2}{\lambda} * \lambda t.\end{aligned} \quad (6)$$

Following Freund [7], we make further generalization of the above parameters given that while the risk R_1 occurred with intensity λ_1, then the risk R_2 following R_1 occurred with greater intensity λ_2' (instead of λ_2) ($\lambda_2' > \lambda_2$): because of greater probability of the risk R_2. Similarly, with a reverse reasoning it can be shown that risk R_2 occurring earlier to R_1 will be associated with greater intensity λ_1' (instead of λ_1).

We illustrate this concept by taking Freund bivariate exponential distribution given by

$$f(x,y) = \begin{cases}\lambda_1\lambda_2' e^{-\lambda_2' y-(\lambda_1+\lambda_2-\lambda_2')x} & \text{for } 0 < x < y\\ \lambda_2\lambda_1' e^{-\lambda_1' x-(\lambda_1'+\lambda_2-\lambda_1')y} & \text{for } 0 < y < x.\end{cases} \quad (7)$$

We present in Table 1 the relative comparison of level of joint risks in several competing chronic conditions as follows: namely, diabetes versus heart disease, hypertension versus heart disease, chronic kidney disease versus heart disease, diabetes versus stroke, hypertension versus stroke, diabetes versus chronic kidney disease, hypertension versus chronic kidney disease, and diabetes versus hypertension.

Diabetes, hypertension, heart disease, chronic kidney disease, and stroke are not recurrent. Survival data basically are of two types, namely, (i) parallel type and (ii) series or competing type.

Parallel type of data occurs in the analysis of multicomponent parallel systems, particularly parallel paired components like two kidneys, two eyes, two lungs, and so forth. Since survival of the system depends on the longevity of the last surviving component in this case, paired data are available irrespective of whether they are censored or not.

The series types of data generally arise in case of several competing risk, say, $R_1, R_2, \ldots, R_k$, which may be imagined to be competing each other for taking the life of an individual or a system. But since the death or failure may occur because of a particular risk (in presence of other risk), we do not get paired or multivariate data on the longevities due to several risk bases on the same patients. Hence there exists no sample space and as such bivariate or multivariate models to measure the longevities under competing risk are not possible. This may happen more particularly in respect of requirement of paired data under the joint competing risk like diabetes with heart disease and chronic kidney disease with diabetes. Since death may occur ultimately as one of the two risks, therefore, it is not possible to get paired data or bivariate sample space.

We have considered number of correlated shocks analogous as competing risks, so the model of Biswas and Noor [3] can approximately be taken in the present investigation.

The diabetes and hypertension are the risk factors that may cause other problems that may lead to death.

Statistically, the other diseases may occur due to interaction of hypertension and diabetes. The nature of other diseases and their magnitude are unknown; therefore such interactions have been ignored assuming they will not affect the estimation.

TABLE 1: Relative comparison of risks in several chronic conditions of mortality.

Parameters	Diabetes, heart disease	Hypertension, heart disease	Chronic kidney disease, heart disease
p	0.04	0.03	0.09
q	0.96	0.97	0.91
N_1	0.0007	0.0008	0.005
N_2	0.006	0.006	0.006
χ_1	0.104	0.118	0.45
χ_2	0.896	0.882	0.55
Q_{t-1}	0.06	0.05	0.14
Q_{t-2}	0.14	0.14	0.14
Z_1	0.03	0.03	0.23
Z_2	0.64	0.65	0.27
Parameters	Diabetes, chronic kidney disease	Hypertension, chronic kidney disease	Diabetes, hypertension
p	0.04	0.04	0.04
q	0.96	0.96	0.96
N_1	0.0007	0.0008	0.0007
N_2	0.005	0.005	0.0008
χ_1	0.141	0.16	0.429
χ_2	0.877	0.86	0.571
Q_{t-1}	0.06	0.05	0.06
Q_{t-2}	0.14	0.14	0.05
Z_1	0.04	0.04	0.23
Z_2	0.63	0.64	0.27
Parameters	Diabetes, stroke	Hypertension, stroke	
p	0.04	0.03	
q	0.96	0.97	
N_1	0.0007	0.0008	
N_2	0.0138	0.0138	
χ_1	0.048	0.055	
χ_2	0.952	0.945	
Q_{t-1}	0.06	0.05	
Q_{t-2}	0.15	0.15	
Z_1	0.01	0.01	
Z_2	0.7	0.72	

We do not have enough data to calculate the additional joint distribution (such as heart disease with diabetes and kidney disease). The present study is devoted to joint distribution of two competing risks of mortality on the basis of relevant models. However, the methodology can be generalized on more than two competing risks.

There were 14.7% (CARRS data) individuals who have 2 or more competing risks. A substitutional proportion of 56.6% had no competing risk at baseline. In the analysis, we have considered the follow-up data of chronic conditions and mortality during the 3.5 years of follow-up.

Heart disease and chronic kidney disease have 60 to 65% chance of mortality with hypertension or diabetes. Stroke has 70% chance of mortality with hypertension or diabetes.

4. Conditional Cause Specific Mortality Rate by Bayesian Approach

We consider a Bayesian method of estimating the conditional probabilities of dying in heart disease and chronic kidney disease in presence of hypertension, diabetes, and stroke or heart disease with chronic kidney disease. Data consisted of n individuals with observed set S_i of binary indicators of symptoms, $S_i = (s_{i1}; s_{i2}; s_{i3}; s_{i4})$, and an indicator y_i that denotes which one of C causes was responsible for persons i's death.

Defining $p(y_i = c_i \mid S_i)$ is the conditional probability of dying in c_i given s_i binary system including c_i. The empirical Bayes estimator has been obtained from the prior data which was available from CARRS. Usually simple Bayesian estimate without prior distribution can be calculated assuming all the risks have been same probability distribution of death. Using Bayes' rule [4],

$$p(y_i = c_i \mid S_i) = \frac{p(S_i \mid y_i = c)\, p(y_i = c)}{p(S_i \mid y_i = c)\, p(y_i = c) + p(S_i \mid y_i \neq c)\, p(y_i \neq c)}$$

$$p(\text{dying in CKD} \mid \text{the symptoms of hypertension, diabletes or stroke})$$

$$= \frac{p(\text{CKD} \mid \text{CKD, HTN, DM, Stroke}) * p(\text{dying in CKD})}{p(\text{CKD} \mid \text{CKD, HTN, DM, Stroke}) * p(\text{dying in CKD}) + p(\text{non CKD} \mid \text{CKD, HTN, DM, Stroke}) * p(\text{not dying in CKD})}$$

$$= \mathbf{0.31} \tag{8}$$

$$p(\text{dying in Heart disease} \mid \text{the symptoms of hypertension, diabletes, chronic kidney disease or stroke})$$

$$= \frac{p(\text{HD} \mid \text{HD, CKD, HTN, DM, Stroke}) * p(\text{dying in HD})}{p(\text{HD} \mid \text{HD, CKD, HTN, DM, Stroke}) * p(\text{dying in HD}) + p(\text{non HD} \mid \text{HD, CKD, HTN, DM, Stroke}) * p(\text{not dying in HD})}$$

$$= \mathbf{0.25}$$

although the Bayes method provides full information of the data and is based on sound logic of inverse probability.

5. Conclusion

Heart diseases (60%), chronic kidney disease (65%), and stroke (70%) have higher chances of dying in the presence of hypertension and diabetes. Therefore, a future research question would be whether the individual disease is alone or it is enhanced by the competing risk.

It would be worthwhile to carry out similar studies for replications as well as enhancing an understanding of epidemiology.

Disclosure

Suddhendu Biswas was formerly the dean of Faculty of Mathematical Sciences, University of Delhi.

Conflicts of Interest

The authors have no conflicts of interest to disclose.

Acknowledgments

The CARRS Study was funded in part by the National Heart, Lung, and Blood Institute of the National Institutes of Health, Department of Health and Human Services (Contract no. HHSN268200900026C), and the United Health Group (Minneapolis, MN, USA). The CARRS Study Group includes the following: steering committee: Dorairaj Prabhakaran, KM Venkat Narayan, K Srinath Reddy, Nikhil Tandon, V Mohan, Muhammed M Kadir, and Mohammed K Ali; conduct and operations: Dorairaj Prabhakaran, Nikhil Tandon, KM Venkat Narayan, Mohammed K Ali, Roopa Shivashankar, Imran Naeem, R Pradeepa, and M Deepa; coordinating centre (Delhi): Dorairaj Prabhakaran, Nikhil Tandon, Roopa Shivashankar, Vamadevan S Ajay, and Deksha Kapoor; data management and statistical team: Dimple Kondal, Naveen, Garima, and Vandana; laboratory: Lakshmy Ramakrishnan, Ruby Gupta, and Savita. Regarding reporting, Kalpana Singh and Professor Suddhendu Biswas had done the analysis and written the manuscript. The authors are very thankful to Professor Dorairaj Prabhakaran for editing and revising the manuscript.

References

[1] F. Downton, "Bivariate exponential distributions in reliability theory," *Journal of the Royal Statistical Society. Series B (Methodological)*, vol. 32, no. 3, pp. 408–417, 1970.

[2] A. W. Marshall and I. Olkin, "A multivariate exponential distribution," *Journal of the American Statistical Association*, vol. 62, no. 317, pp. 30–44, 1967.

[3] S. Biswas and H. S. Noor, "On a probability model for classifying the unknown causes of death due to two competing risks into assignable causes," *Biometrical Journal*, vol. 30, no. 7, pp. 827–833, 1988.

[4] H. T. McCormick, Z. Richard Li, C. Calvert, C. A. Crampin, K. Kahn, and J. S. Clark, "Probabilistic cause-of-death assignment using verbal autopsies: Center for Statistics and the Social Sciences: Working Paper Series," Working Paper no. 147, 2014.

[5] World Health Organization, "STEPwise approach to surveillance (STEPS)," 2014, http://www.who.int/chp/steps/en/.

[6] M. Nair, M. K. Ali, V. S. Ajay et al., "CARRS Surveillance study: design and methods to assess burdens from multiple perspectives," *BMC Public Health*, vol. 12, no. 1, article 701, 2012.

[7] J. E. Freund, "A bivariate extension of the exponential distribution," *Journal of the American Statistical Association*, vol. 56, no. 296, pp. 971–977, 1961.

10

Estimating the Proportion of True Null Hypotheses in Multiple Testing Problems

Oluyemi Oyeniran[1] and Hanfeng Chen[2]

[1]*Manufacturing, Toxicology and Applied Statistical Sciences, Janssen Research & Development, Spring House, PA 19002, USA*
[2]*Department of Mathematics and Statistics, Bowling Green State University, Bowling Green, OH 43403, USA*

Correspondence should be addressed to Hanfeng Chen; hchen@bgsu.edu

Academic Editor: Shein-chung Chow

The problem of estimating the proportion, π_0, of the true null hypotheses in a multiple testing problem is important in cases where large scale parallel hypotheses tests are performed independently. While the problem is a quantity of interest in its own right in applications, the estimate of π_0 can be used for assessing or controlling an overall false discovery rate. In this article, we develop an innovative nonparametric maximum likelihood approach to estimate π_0. The nonparametric likelihood is proposed to be restricted to multinomial models and an EM algorithm is also developed to approximate the estimate of π_0. Simulation studies show that the proposed method outperforms other existing methods. Using experimental microarray datasets, we demonstrate that the new method provides satisfactory estimate in practice.

1. Introduction

Estimating the proportion π_0 of true null hypotheses in a multiple testing setup is very crucial in wanting to assess and/or control false discovery rate, which is quite significant in genomics, disease discovery, and cancer discovery. Langaas et al. [1] remarked "An important reason for wanting to estimate π_0 is that it is a quantity of its own right. In addition, a reliable estimate of π_0 is important when we want to assess or control multiple error rates, such as the false discovery rate FDR of Benjamini and Hochberg [2]." In the case of testing for differential expression in DNA microarrays, the proportion of differentially expressed genes is $1 - \pi_0$, and it is important to know whether 5% or 35% of the genes, for example, are differentially expressed, even if we cannot identify these genes (see Langaas et al. [1]). Multiple testing refers to any instance that involves the simultaneous testing of several hypotheses. A common feature in genomes studies is the analysis of a large number of simultaneous measurements in a small number of samples. One must decide whether the findings are truly causative correlations or just the byproducts of multiple hypothesis testing (Gyorffy et al. [3]). If one does not take the multiplicity of tests into account, then the probability that some of the true null hypotheses are rejected by chance alone may be unduly large.

In a multiple hypothesis testing problem, m null hypotheses are tested simultaneously; that is, we test

$$H_{0i} \text{ versus } H_{1i}, \tag{1}$$

for $i = 1, 2, \ldots, m$, simultaneously. Assume that the m tests are constructed based on the observed p values, $p_1, \ldots, p_m$, respectively. The unknown quantity π_0 to be estimated is the proportion of the true null hypotheses among $H_{01}, \ldots, H_{0m}$. Introduce the i.i.d Bernoulli random variables $H_1, \ldots, H_m$ with $P(H_i = 0) = \pi_0$. Then H_i can be interpreted in terms of the multiple testing problems as follows:

$$H_i = \begin{cases} 0, & \text{if } H_{0i} \text{ is true,} \\ 1, & \text{otherwise} \end{cases} \tag{2}$$

for $i = 1, \ldots, m$.

Table 1: Outcomes from m null hypotheses tests.

	Not rejected	Rejected	Total
H_0 true	U	V	m_0
H_0 false	T	S	$m - m_0$
Total	W	R	m

We assume the p values, $p_1, \ldots, p_m$, are continuous and independent random variables, so that the p values are independently and identically distributed as Unif(0, 1) when the null hypotheses are all true. One chooses to reject or fail to reject each null hypothesis based on the corresponding p value. Consequences of the tests are summarized in Table 1.

In Table 1 m_0 is the number of null hypotheses; R is the observable random variable representing the number of hypotheses rejected. Note that all other random variables U, V, S, and T in Table 1 are unobservable.

The problem of estimating the proportion π_0 has naturally arisen in assessing or controlling an overall false rejection rate in a simultaneous hypotheses testing problems. A reliable estimate of π_0 is crucial when we want to control and/or assess the false discovery rate (FDR) proposed by Benjamini and Hochberg [2], defined as

$$\text{FDR} = E\{QI(R > 0)\}, \tag{3}$$

where $Q = V/R$ and

$$I(R > 0) = \begin{cases} 1 & \text{if } R > 0, \\ 0 & \text{otherwise.} \end{cases} \tag{4}$$

Benjamini and Hochberg [2] prove that Simes' procedure (Simes [4]) has the FDR controlled at level α if the underlying test statistics and the corresponding p values are continuous and identically and independently distributed. Specifically, they show

$$\text{FDR} \leq \frac{m_0}{m}\alpha = \pi_0 \alpha. \tag{5}$$

Consequently, if γ is the FDR level that one wishes to achieve and if π_0 can be estimated efficiently, say by $\widehat{\pi}_0$, then α can be chosen to be $\alpha = \gamma/\widehat{\pi}_0$ to gain additional testing powers in the multiple testing problem with FDR being under control. If $\widehat{\pi}_0$ is overestimated substantially, however, the value of α is underestimated substantially, leading to significantly narrower simultaneous confidence intervals for multiple comparisons and significant reduction of testing power for the multiple testing problems. On the other hand, if α is chosen via some other procedure, say the Bonferroni method in the multiple comparison problem, assessing FDR accurately via estimating π_0 efficiently is then of great interest.

The paper is organized as follows. Section 2 is a review of existing estimating methods; Section 3 introduces the new estimating procedure; Section 4 contains simulation results and Section 5 presents application of the new estimating procedure to real life examples.

2. Existing Estimating Methods

2.1. Mixture Model Framework. A mixture model can be used to fit the p values in order to estimate the proportion, π_0, of the true hypotheses, where large scale parallel hypotheses are performed independently. The estimate for π_0 can be based on the mixture model for the common density f of the p values described as follows:

$$f = \pi_0 + (1 - \pi_0) h, \tag{6}$$

where $h(p)$ is the conditional probability density function of the p value under an alternative (see Langaas et al. [1]). Using this mixture representation, we are able to characterize the maximum likelihood estimate for f. The estimation method is derived under the assumption of independent and identically distributed p values. The null p values are uniformly distributed on $[0, 1]$. We should note that $h(p)$ describes the configuration of true alternative populations among the m underlying populations; it seems that a nonparametric approach to the estimation of π_0 is more appealing. Without loss of generality, we define $H = 1$ as the alternative hypothesis, which will be used throughout this section. Three nonparametric estimators proposed recently by other authors are described and discussed in the sequel subsections.

2.2. Storey's Method. Consider the common marginal mixture density of the p values, for any $\lambda \in (0, 1)$:

$$P(p > \lambda) = \pi(1 - \lambda) + (1 - \pi) P(p > \lambda \mid H = 1). \tag{7}$$

On the basis that $P(p > \lambda \mid H = 1)$ is typically small, a large majority of the p values in the interval $[\lambda, 1]$ should be corresponding to the true null hypothesis and thus uniformly distributed on the interval (0, 1), of which most should be close to 1, so that $(1 - \pi)P(p > \lambda \mid H = 1)$ is approximately zero. Let $W(\lambda) = \#\{p_i : p_i > \lambda\}$. Note that $W(\lambda)$ should be approximately equal to the product of $m\pi_0$ and the length of the interval $[\lambda, 1]$; that is, $E(W(\lambda)) \approx m\pi_0(1 - \lambda)$. Hence, Storey [6] proposes that the proportion of true null hypotheses, π_0, is estimated by (with an appropriately chosen λ)

$$\widehat{\pi}_0^s(\lambda) = \frac{W(\lambda)}{(1 - \lambda) m}. \tag{8}$$

The value of λ has impacts on the behavior of $\widehat{\pi}_0^s(\lambda)$. $\widehat{\pi}_0^s(\lambda)$ has a large bias and small variance when λ is small and a small bias and large variance when λ is big, respectively. Since both extreme values of λ have a bias-variance trade-off, Storey et al. [7] propose bootstrapping, which is a resampling technique, to choose λ when estimating $\widehat{\pi}_0^s(\lambda)$ so as to minimize the mean square error of $\widehat{\pi}_0^s(\lambda)$. The resulting estimator is denoted by $\widehat{\pi}_0^s$.

Note that the idea leading to Storey's estimator $\widehat{\pi}_0^s$ is to treat the addition $(1 - \pi)P(p > \lambda \mid H = 1)$ term as zero to get rid of the complication caused by the unknown alternative distribution. As a consequence, Storey's estimator $\widehat{\pi}_0^s$ will tend to overestimate π_0 in a size of $(1 - \pi_0)P(p > \lambda \mid H = 1)/(1 - \lambda)$, at least theoretically. However, the anticipated size of overestimate becomes less significant when π_0 is closer to 1; that is, $1-\pi_0$ is closer to 0. The biasedness of $\widehat{\pi}_0^s$ as an estimator for $P(p > \lambda)/(1 - \lambda)$ can be dominating, when π_0 is close to 1, as evident in the simulation results in Section 4 and as observed by other authors as well, for example, Langaas et al. [1].

2.3. Convest Method. Let f and h be defined in (6). Langaas et al. [1] prove that if $h(p)$ is twice differentiable, convex, and decreasing, f can be expressed as

$$f(p) = \int_0^1 f_\theta(p)\,\mu(d\theta), \tag{9}$$

where the kernel density

$$f_\theta(p) = \begin{cases} \dfrac{2(\theta - p)}{\theta^2 I_{(0,\theta)}(p)} & \text{if } \theta \in (0, 1], \\ I_{[0,1]}(p) & \theta = 0, \end{cases} \tag{10}$$

with μ being any probability measure on $[0, 1]$. Thus Langaas et al. [1] are able to characterize the nonparametric maximum likelihood estimate for f for the density $f(p)$ as

$$\widehat{f}(p) = \int_0^1 f_\theta(p)\,\widehat{\mu}(d\theta), \tag{11}$$

where $\widehat{\mu}$ is the nonparametric maximum likelihood estimate of μ. Let p values $p_{(1)}, \ldots, p_{(m)}$ be the ordered statistics of the p values. The nonparametric maximum likelihood estimator of $f(\cdot)$ is given by

$$\widehat{f} = \arg\max \prod_{i=1}^{m} \widehat{f}\left(p_{(i)}\right). \tag{12}$$

Langaas et al. [1] then propose to estimate the proportion π_0 by

$$\widehat{\pi}_0^c = \widehat{f}(1). \tag{13}$$

It is noted that as the estimator $\widehat{\pi}_0^c$ is constructed via a density estimate $\widehat{f}(p)$ at or about the upper bound of support, namely, $\widehat{f}(1)$, it can be conservative and overestimate π_0 when the assumption $h(1) = 0$ is questionable or $h(\text{p}) \to 0$ slowly as $p \to 1$. The overestimating problem can be even more severe when π_0 is not large; that is, $1 - \pi_0$ is not so small. However, the remarks in the end of Section 2.2 are applicable to $\widehat{\pi}_0^c$ as well.

2.4. Average Estimate Approach. The average estimate approach was motivated by Storey's method. Jiang and Doerge [8] observe (and many other authors point out as well) that Storey's estimator has a large bias and small variance when λ is small and a small bias and large variance when λ is big. Since both extremes of λ have a bias-variance trade-off, Jiang and Doerge [8] propose to combine Storey's estimate with different values of λ that vary from a small extreme to a large extreme.

Let $0 < \lambda_1 < \lambda_2 < \cdots < \lambda_n < 1$, and suppose that, for each λ_i, we have Storey's estimate

$$\widehat{\pi}_0(\lambda_i) = \frac{W(\lambda_i)}{(1 - \lambda_i)\, m}. \tag{14}$$

Jiang and Doerge [8] propose to estimate π_0 by the average of $\widehat{\pi}_0(\lambda)$ over the values of λ_i; that is,

$$\widehat{\pi}_0^a = \frac{1}{n}\sum_{i=1}^{n} \widehat{\pi}_0(\lambda_i). \tag{15}$$

This estimate aims to minimize both the bias and the variance at the same time, if λ_i's are selected appropriately.

Define $0 = t_1 < t_2 < \cdots < t_B < t_{B+1} = 1$ as equally spaced points in the interval $[0, 1]$ such that the interval $[0, 1]$ is divided into B small intervals with equal length $1/B$; specifically, $t_i = (i - 1)/B$. Let $b_i = \#\{p_k : p_k \geq t_i\}$ and $c_i = \#\{p_k : t_i \leq p_k < t_{i+1}\}$ for $i = 1, \ldots, B$. Define

$$i^* = \min\left\{i : c_i \leq \frac{b_i}{(B - i + 1)}\right\}. \tag{16}$$

Jiang and Doerge [8] then propose to estimate π_0 by

$$\widehat{\pi}_0(B) = \frac{1}{B - i^* + 1}\sum_{j=1}^{B} \widehat{\pi}_0(t_j). \tag{17}$$

To apply the estimate, one has to choose a value of B. Jiang and Doerge [8] develop a bootstrapping algorithm to pick the optimal B. It should be noted that, as an average of Storey's estimates, this estimate is expected to inherit conservativeness from Storey's method.

3. New Method

We propose a finite mixture model of the uniform distribution and a multinomial distribution $\text{M}(k, q_1, \ldots, q_k)$ with a mixing proportion π_0 to fit the p values. Denote $\text{UM}(\pi_0; k, q_1, \ldots, q_k)$ for this finite mixture distribution. By this approach, the alternative distribution h defined in (6) is restricted to the multinomial distribution family, $\text{M}(k, q_1, \ldots, q_k)$. Alternatively, the multinomial distribution can be viewed as a parametric approximation to the nonparametric unknown density h, similarly to the idea of the empirical likelihood method (see Owen [9]). So this procedure is considered as nonparametric.

To apply the approach, we need to settle two things first: (a) convert the continuous-type observations $p_1, \ldots, p_m$ into discrete data with k categories and (b) select an integer k.

3.1. Selection of k. It is often the case in applications that the p values are highly skewed (see Storey and Tibshirani [10], Zhao et al. [11], and Markitsis and Lai [12]). In this case, we recommend Doane's modification of Sturges' rule for selection of k as follows, to count for skewness of the mixture distribution (6) (see Doane [13] and Sturges [14]):

$$\widehat{k} = \log_2(1 + \widehat{\gamma}), \tag{18}$$

where $\hat{\gamma}$ is an estimate of the skewness coefficient. In the case of symmetry, we recommend to adapt Sturges' rule to determine the selection of k:

$$\widehat{k} = 1 + \log_2 m. \tag{19}$$

3.2. Transformation of p_i's. To transform p_i's, partition the unit interval into $\widehat{k}$ subintervals with equal width $1/k$. Define

$$\xi_{ij} = \begin{cases} 1, & \text{if } \frac{(j-1)}{\widehat{k}} \le p_i < \frac{j}{\widehat{k}}, \\ 0, & \text{otherwise} \end{cases} \tag{20}$$

for $i = 1, \dots, m$, $j = 1, \dots, \widehat{k}$. Keep in mind that

$$\sum_{j=1}^{\widehat{k}} \xi_{ij} = 1, \quad i = 1, \dots, m,$$
$$\sum_{i=1}^{m} \sum_{j=1}^{\widehat{k}} \xi_{ij} = m. \tag{21}$$

From Storey's Bayesian interpretation for the multiple testing problem, given that the alternative is true with a probability of $1 - \pi_0$, the p value follows the distribution h. In the same way, the transformed data ξ_{ij}'s can be interpreted as follows. Given that the alternative is true with a probability of $1 - \pi_0$, $\xi_i = (\xi_{i1}, \dots, \xi_{ik})'$ is a multinomial random vector and is distributed as $\mathrm{M}(k, q_1, \dots, q_k)$, for $i = 1, \dots, m$. Therefore, $\xi_1, \dots, \xi_m$ are independently and identically distributed as the finite mixture distribution $\mathrm{UM}(\pi_0; k, q_1, \dots, q_k)$ and the maximum likelihood estimate $\widehat{\pi}_0$ for π_0 thus results from the transformed data ξ_{ij}'s. Explicitly, $\widehat{\pi}_0$ together with $(\widehat{q}_1, \dots, \widehat{q}_k)$ maximizes the log-likelihood function

$$l(\pi_0, q_1, \dots, q_k) = \sum_{i=1}^{m} \log \left\{ \pi_0 + (1 - \pi_0) \prod_{j=1}^{k} q_j^{\xi_{ij}} \right\}. \tag{22}$$

3.3. EM Algorithm. Note that maximizing the nonlinear log-likelihood function (22) can be complicating. The EM algorithm may be used easily to obtain an approximation to $\widehat{\pi}_0$. In order to do so, we introduce a latent Bernoulli variable w that indicates the component membership in the finite mixture distribution $\mathrm{UM}(\pi_0; k, q_1, \dots, q_k)$. That is, $P(w = 1) = \pi_0$ and given $w = 0$, ξ_1 follows $\mathrm{M}(k, q_1, \dots, q_k)$. Note that (w, ξ_1) has the distribution

$$g(w, \xi_1 \mid \pi_0, q_1, \dots, q_k) = \pi_0^w \left\{ (1 - \pi_0) q_1^{\xi_{11}} q_2^{xi_{12}} \cdots q_k^{\xi_{1k}} \right\}^{1-w}, \tag{23}$$

for $w = 0$ or 1 and $\xi_{1j} = 0$ or 1, $j = 1, \dots, k$.

Let $(w_i; \xi_{i1}, \dots, \xi_{ik})$, $i = 1, \dots, m$, be a random sample of size m from model (23). In the present problem, ξ_{ij}'s are only available data for analysis and w_i's are unobservable and so considered as missing values. This defines a missing-value model. The log-likelihood with the complete data is given by

$$l^* (\pi_0, q_1, \dots, q_k) = T \log \pi_0 + (m - T) \log (1 - \pi_0) + \sum_{j=1}^{k} (m - T) \xi_{.j} \log q_j, \tag{24}$$

where $T = \sum_{i=1}^{m} w_i$, and $\xi_{.j} = \sum_{i=1}^{m} \xi_{ij}$, for $j = 1, \dots, k$. We are ready to describe the EM algorithm.

E-Step. Let $\pi_0^{(s)}$ and $q^{(s)}$ be the current approximations to the maximum likelihood estimates under the model $\mathrm{UM}(\pi_0, k, q_1, \dots, q_k)$. For the next approximation, the E-step establishes the expected log-likelihood function as

$$\begin{aligned} Q(\pi_0, q) &= E_{\pi_0^{(s)}, q^{(s)}} \{ l^* (\pi_0, q_1, \dots, q_k \mid \eta_1, \dots, \eta_m) \} \\ &= E_{\pi_0^{(s)}, q^{(s)}} (T \mid \eta) \log \pi_0 \\ &\quad + \left(m - E_{\pi_0^{(s)}, q^{(s)}} (T \mid \eta) \right) \log (1 - \pi_0) \\ &\quad + \sum_{j=1}^{k} \left(m - E_{\pi_0^{(s)}, q^{(s)}} (T \mid \eta) \right) \xi_{.j} \log q_j \end{aligned} \tag{25}$$

M-Step. In the M-step, $Q(\pi_0, q)$ is maximized to yield the next approximation $\pi_0^{(s+1)}$ and $q^{(s+1)}$. Consider

$$\frac{\partial Q}{\partial \pi_0} = 0 \Longrightarrow$$
$$\pi_0^{(s+1)} = \frac{E_{\pi_0^{(s)}, q^{(s)}} (T \mid \eta)}{m} = \sum_{i=1}^{m} E_{\pi_0^{(s)}, q^{(s)}} (w_i \mid \eta). \tag{26}$$

where

$$\begin{aligned} E_{\pi_0^{(s)}, q^{(s)}} (w_i \mid \eta) &= P_{\pi_0^{(s)}, q^{(s)}} (w_i = 1 \mid \eta) \\ &= \frac{\pi_0^{(s)}}{\pi_0^{(s)} + (1 - \pi_0^{(s)}) \prod_{j=1}^{k} [q_j^{(s)}]^{\xi_{ij}}}. \end{aligned} \tag{27}$$

From $\partial Q / \partial q = 0$, we have

$$q_j^{(s+1)} = \frac{\xi_{.j}}{m - E_{\pi_0^{(s)}, q^{(s)}} (T \mid \eta)} = \frac{\xi_{.j}}{m (1 - \pi_0^{(s+1)})}. \tag{28}$$

To sum up, let $\pi_0^{(s)}, q_1^{(s)}, \dots, q_k^{(s)}$ be the sth approximations to the maximum likelihood estimates of $\widehat{\pi}_0, \widehat{q}_1, \dots, \widehat{q}_k$ that maximize the log-likelihood function defined in (22). Then the $(s + 1)$th approximation with the EM algorithm is given by

$$\begin{aligned} \pi_0^{(s+1)} &= \frac{1}{m} \sum_{i=1}^{m} \frac{\pi_0^{(s)}}{\pi_0^{(s)} + (1 - \pi_0^{(s)}) \prod_{j=1}^{k} [q_j^{(s)}]^{\xi_{ij}}}, \\ q_j^{(s+1)} &= \frac{\xi_{.j}}{m (1 - \pi_0^{(s+1)})}. \end{aligned} \tag{29}$$

So the EM iteration process goes as shown in Algorithm 1.

Input: $\xi_{ij}, i = 1, \ldots, k, j = 1, \ldots, m$ transformed from the observed p-values $p_1, \ldots, p_m$
Output: Estimate of $\pi_0 : \widehat{\pi}_0$
(1) **begin**
(2) Initialization: Set $\pi_0 = \pi_0^{(o)}, q_j = q_j^{(o)}, 1 \le j \le k$.
(3) **repeat**
(4) Set $\pi_0 = \pi_0^{(s)}, q_j = q_j^{(s)}$ be the sth (current) approximation.
Compute

$$\pi_0^{(s+1)} = \frac{1}{m} \sum_{i=1}^{m} \frac{\pi_0^{(s)}}{\pi_0^{(s)} + \left(1 - \pi_0^{(s)}\right) \prod_{j=1}^{k} \left[q_j^{(s)}\right]^{\xi_{ij}}},$$

$$q_j^{(s+1)} = \frac{\xi_{.j}}{m(1 - \pi_0^{(s+1)})}$$

(5) **until** $l(\pi_0^{(s+1)}, q_1^{(s+1)}, \ldots, q_k^{(s+1)}) - l(\pi_0^{(s)}, q_1^{(s)}, \ldots, q_k^{(s)}) \le \epsilon$;
(6) Then $\widehat{\pi}_0 = \pi_0^{s+1}$.

Algorithm 1: EM iteration.

It is well known that each EM iteration gets closer to the maximum of log-likelihood but only in a linear convergency rate. If the components are similar in their densities, then the convergence is extremely slow. The convergence will also be slow when the maximum likelihood solution requires some of the weight parameters to be zero, because the algorithm can never reach such a boundary point. An additional and related problem is that of deciding when to stop the algorithm. One risk to a naive user is the natural tendency to use a stopping rule for the algorithm based on the changes in the parameters or the likelihood being sufficiently small. Taking smalls steps does not indicate that we are close to the solution.

To combat this problem, Lindsay [15] exploited the regularity of the EM algorithm process to predict, via the device known as Aitken acceleration, the value of the log-likelihood at the maximum likelihood solution. The Aitken's acceleration rule is usually recommended to predict the maximum efficiently and suitably whenever one is using a linearly convergent algorithm with a slow rate of convergence. If we let l_{i-2}, l_{i-1}, and l_i be the log-likelihood values for the three consecutive iterations, then the predicted final value is

$$l_i^{\infty} = l_{i-2} + \frac{l_{i-1} - l_{i-2}}{(1 - c_i)}, \quad (30)$$

where $c_i = (l_i - l_{i-1})/(l_{i-1} - l_{i-2})$. Terminate the EM iteration when $l_i^{\infty} - l_i$ is sufficiently small.

4. Simulation Studies

In order to investigate the properties of the estimators described in Section 2 and compare the performance to that of the newly developed estimating procedure described in Section 3, we conducted simulation experiments. The generation of simulated data and the calculation of the estimates were both done in the language R. Simulation studies were conducted with the p values based on a one-sided z-test in the finite normal mixture model $z \sim \pi_0 N(0,1) + (1 - \pi_0)N(1,1)$ with various values of m and π_0 for performance comparisons.

Monte Carlo data were simulated independently from $z \sim \pi_0 N(0,1) + (1 - \pi_0)N(1,1)$ and each p value was computed by $p_i = 1 - \Phi(z)$. where Φ is the cumulative distribution function of standard normal $N(0,1)$. For the generation of simulated data, three true values of π_0 were considered, namely, 0.25, 0.5, 0.75, with the sample size m = 200, 500, and 1,000. In each case, 1,000 Monte Carlo trials were performed. In implementation of the EM algorithm to compute the proposed new estimate $\widehat{\pi}_0$, the determination rule of the EM algorithm is $l_i^{\infty} - l_i \le 10^{-4}$. The EM algorithm converged for all the simulated datasets. In each case, the performance of the proposed new estimate $\widehat{\pi}_0$ was compared with those of several existing procedures through the same set of Monte Carlo data. Specifically, in the simulations, we considered the following existing methods:

(1) Storey's bootstrap estimate denoted by $\widehat{\pi}_0^b$
(2) Langaas et al.'s convex estimate denoted by $\widehat{\pi}_0^c$
(3) Jiang and Doerge's average method estimate denoted by $\widehat{\pi}_0^a$

Storey's estimate $\widehat{\pi}_0^s$ was computed by using the R package *nFDR* and the convex estimate $\widehat{\pi}_0^c$ by the R package *limma* with the function *convest*. Table 2 summarizes the simulation results. It is clearly shown that the new estimate outperformed substantially over the existing methods with comparable standard errors, while $\widehat{\pi}_0^s(0.5)$ performs much worse than $\widehat{\pi}_0^s$ as expected.

5. Application to Real Life Microarray Data

To further evaluate the performance of the new method in comparison to the three existing methods, consider the real life data from the DNA microarray experiments reported in Golub et al. [5] that can be downloaded from the R package *multtest*. The dataset was used by many authors (see [11, 16]

TABLE 2: Monte Carlo empirical estimates with standard deviations (in parentheses) by the new innovative approach along with existing estimators. The Monte Carlo size is 1,000 in each case. Denote $\widehat{\pi}_0$ for the new estimator, $\widehat{\pi}_0^s$ for Storey's estimator, $\widehat{\pi}_0^s(0.5)$ for Storey's estimator with $\lambda = 0.5$, $\widehat{\pi}_0^c$ for the convex method estimator, and $\widehat{\pi}_0^a$ for the average method estimator.

m	π_0	$\widehat{\pi}_0$	$\widehat{\pi}_0^s$	$\widehat{\pi}_0^s(0.5)$	$\widehat{\pi}_0^c$	$\widehat{\pi}_0^a$
		Estimates of π_0				
200	0.25	0.235 (0.016)	0.30 (0.021)	0.500 (0.035)	0.314 (0.022)	0.430 (0.03)
	0.5	0.489 (0.035)	0.613 (0.043)	0.660 (0.047)	0.584 (0.041)	0.626 (0.044)
	0.75	0.78 (0.055)	0.717 (0.051)	0.79 (0.056)	0.685 (0.048)	0.773 (0.054)
500	0.25	0.249 (0.011)	0.34 (0.015)	0.528 (0.024)	0.33 (0.014)	0.491 (0.022)
	0.5	0.52 (0.023)	0.646 (0.028)	0.720 (0.032)	0.623 (0.027)	0.704 (0.031)
	0.75	0.72 (0.032)	0.806 (0.036)	0.860 (0.038)	0.785 (0.035)	0.819 (0.035)
1000	0.25	0.242 (0.005)	0.393 (0.012)	0.482 (0.015)	0.349 (0.011)	0.394 (0.013)
	0.5	0.51 (0.016)	0.61 (0.019)	0.66 (0.021)	0.61 (0.019)	0.63 (0.019)
	0.75	0.72 (0.023)	0.67 (0.021)	0.86 (0.027)	0.79 (0.025)	0.81 (0.026)

and the references therein) to illustrate the applications of proposed estimating methods for the proportion of true null hypotheses in multiple testing problems. The dataset consists of 38 bones marrow samples, 27 acute lymphoblastic leukemia (ALL) samples and 11 acute myeloid leukemia (AML) samples, obtained from acute leukemia patients at the time of diagnosis, before chemotherapy. RNA prepared from bone marrow mononuclear cells was hybridized to high-density oligonucleotide microarrays, produced by Affymetrix and containing probes for 7,129 human genes. But after pre-processing, there were only 3,051 genes accurately readable resulting in a microarray 3051×38 matrix. For comparison of gene expressions between two groups, let μ_1 and μ_2 be the true mean intensities for each gene, in groups 1 and 2, respectively, to determine whether the gene is differentially expressed, that is, to test

$$H_0 : \mu_1 = \mu_2 \text{ vs } H_1 : \mu_1 \neq \mu_2. \quad (31)$$

With each of 3,051 genes, a two-sample Welch t-statistic is computed and its two-sided p value is computed under a central t-student distribution with degree 36 of freedom. A histogram of these p values is displayed in Figure 1. It is clear that the p values are highly skewed to the right.

The four estimates for the proportion π_0 of nondifferentially expressed genes among 3,051 genes, π_0, $\widehat{\pi}_0$, $\widehat{\pi}_0^s$, $\widehat{\pi}_0^c$, and $\widehat{\pi}_0^a$, are given in Table 3. It appears that the four estimates are significantly different, varying from as low as 0.4150 to as high as 0.4913. Noting that it has been commented by

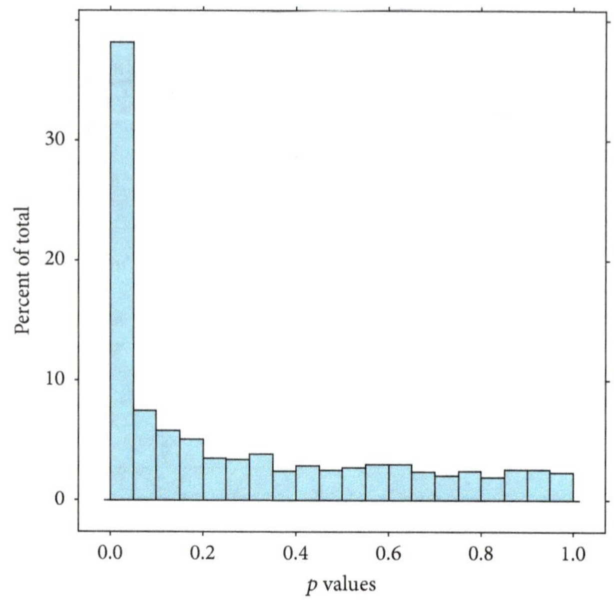

FIGURE 1: Histogram of the p values for leukemia data reported in Golub et al. [5].

TABLE 3: The estimates of the proportion of true null hypotheses with the microarray data of Golub et al. [5].

$\widehat{\pi}_0$	$\widehat{\pi}_0^s$	$\widehat{\pi}_0^c$	$\widehat{\pi}_0^a$
0.4150	0.4643	0.4701	0.4913

many authors that the estimates $\widehat{\pi}_0^s$, $\widehat{\pi}_0^c$, and $\widehat{\pi}_0^a$ are usually conservative, we conclude that the lower estimate of 0.42 by the proposed new estimating procedure appears as expected and might be closer to the true value of π_0.

Competing Interests

The authors declare that there is no conflict of interests regarding the publication of this paper.

References

[1] M. Langaas, B. H. Lindqvist, and E. Ferkingstad, "Estimating the proportion of true null hypotheses, with application to DNA microarray data," *Journal of the Royal Statistical Society, Series B: Statistical Methodology*, vol. 67, no. 4, pp. 555–572, 2005.

[2] Y. Benjamini and Y. Hochberg, "Controlling the false discovery rate: a practical and powerful approach to multiple testing," *Journal of the Royal Statistical Society, Series B: Methodological*, vol. 57, no. 1, pp. 289–300, 1995.

[3] B. Gyorffy, A. Gyorffy, and Z. Tulassay, "The problem of multiple testing and solutions for genome-wide studies," *Orvosi Hetilap*, vol. 146, pp. 559–563, 2005.

[4] R. J. Simes, "An improved Bonferroni procedure for multiple tests of significance," *Biometrika*, vol. 73, no. 3, pp. 751–754, 1986.

[5] T. R. Golub, D. K. Slonim, P. Tamayo et al., "Molecular classification of cancer: class discovery and class prediction by gene expression monitoring," *Science*, vol. 286, no. 5439, pp. 531–537, 1999.

[6] J. D. Storey, "A direct approach to false discovery rates," *Journal of the Royal Statistical Society. Series B. Statistical Methodology*, vol. 64, no. 3, pp. 479–498, 2002.

[7] J. D. Storey, J. E. Taylor, and D. Siegmund, "Strong control, conservative point estimation and simultaneous conservative consistency of false discovery rates: a unified approach," *Journal of the Royal Statistical Society, Series B: Statistical Methodology*, vol. 66, no. 1, pp. 187–205, 2004.

[8] H. Jiang and R. W. Doerge, "Estimating the proportion of true null hypotheses for multiple comparisons," *Cancer Informatics*, vol. 6, pp. 25–32, 2008.

[9] A. Owen, "Empirical likelihood for linear models," *The Annals of Statistics*, vol. 19, no. 4, pp. 1725–1747, 1991.

[10] J. D. Storey and R. Tibshirani, "Statistical significance for genomewide studies," *Proceedings of the National Academy of Sciences of the United States of America*, vol. 100, no. 16, pp. 9440–9445, 2003.

[11] H. Zhao, X. Wu, H. Zhang, and H. Chen, "Estimating the proportion of true null hypotheses in nonparametric exponential mixture model with appication to the leukemia gene expression data," *Communications in Statistics. Simulation and Computation*, vol. 41, no. 9, pp. 1580–1592, 2012.

[12] A. Markitsis and Y. Lai, "A censored beta mixture model for the estimation of the proportion of non-differentially expressed genes," *Bioinformatics*, vol. 26, no. 5, pp. 640–646, 2010.

[13] D. P. Doane, "Aesthetic frequency classifications," *The American Statistician*, vol. 30, no. 4, pp. 181–183, 1976.

[14] H. A. Sturges, "The choice of a class interval," *Journal of the American Statistical Association*, vol. 21, no. 153, pp. 65–66, 1926.

[15] B. G. Lindsay, "Mixture models: theory, geometry and applications," in *Proceedings of the NSF-CBMS Regional Conference Series in Probability and Statistics*, Institute for Mathematical Statistics, 1995.

[16] Z. Guan, B. Wu, and H. Zhao, "Application of Bernstein polynomial in the estimation of false-discovery-rate," *Statistica Sinica*, vol. 18, pp. 905–923, 2008.

11

Bayesian Estimation in Delta and Nabla Discrete Fractional Weibull Distributions

M. Ganji and F. Gharari

Department of Statistics, University of Mohaghegh Ardabili, Ardabil, Iran

Correspondence should be addressed to M. Ganji; mganji@uma.ac.ir

Academic Editor: Ramón M. Rodríguez-Dagnino

We use discrete fractional calculus for showing the existence of delta and nabla discrete distributions and then apply time scales for definitions of delta and nabla discrete fractional Weibull distributions. Also, we study the Bayesian estimation of the functions of parameters of these distributions.

1. Introduction

One of the active areas of research in statistics is to model discrete life time data by developing discretized version of suitable continuous lifetime distributions. The discretization of a continuous distribution using different methods has attracted renewed attention of researchers in the last few years; for example, see [1–9]. Recently, these different methods are classified based on different criteria of discretization in detail by Chakraborty (see [10]).

In this article, we present a new method for discretization of most of continuous distributions, where their pdfs consist of the monomial Taylor and exponential function. As an example, we do discretization for Weibull distribution. Our discretization method in comparison with prior methods for discretization of continuous distributions has two main advantages. First, for a given continuous distribution, it is possible to generate two types (delta and nabla types) of corresponding discrete distributions. Second, the main result of this paper is a unification of the continuous distributions and their corresponding discrete distributions, which is at the same time a distribution to the case of so-called time scale. We use discrete fractional calculus for showing the existence of delta and nabla discrete distributions and then apply time scales for definition of delta and nabla discrete distributions and as a unification theory under which continuous and discrete distributions are subsumed. Finally, we study the Bayesian estimation of functions of parameters of these distributions.

2. Preliminaries

In this section, we provide a collection of definitions and related results which are essential and will be used in the next discussions. As mentioned in [11, 12] the definitions and theorem are as follows.

A time scale $\mathbb{T}$ is an arbitrary nonempty closed subset of the real numbers $\mathbb{R}$. The most well-known examples are $\mathbb{T} = \mathbb{R}$ and $\mathbb{T} = \mathbb{Z}$. The forward (backward) jump operator is defined by $\sigma(t) := \inf\{s \in \mathbb{T} : s > t\}$ ($\rho(t) := \sup\{s \in \mathbb{T} : s < t\}$), where $\inf \emptyset := \sup \mathbb{T}$ and $\sup \emptyset := \inf \mathbb{T}$. A point $t \in \mathbb{T}$ is said to be right-dense if $t < \sup \mathbb{T}$ and $\sigma(t) = t$ (left-dense if $t > \inf \mathbb{T}$ and $\rho(t) = t$) and right-scattered if $\sigma(t) > t$ (left-scattered if $\rho(t) < t$). The forward (backward) graininess function $\mu : \mathbb{T} \to [0, \infty)$ ($\nu : \mathbb{T} \to [0, \infty)$) is defined by $\mu(t) := \sigma(t) - t$ ($\nu(t) := t - \rho(t)$). More generally, we denote all $\rho(t)$, $\sigma(t)$, and t with $\eta(t)$.

Definition 1. A function $f : \mathbb{T} \to \mathbb{R}$ is called regulated if its right-sided limits exist at all right-dense points in $\mathbb{T}$ and its left-sided limits exist at all left-dense points in $\mathbb{T}$.

Definition 2. A function $f : \mathbb{T} \to \mathbb{R}$ is called rd-continuous (ld-continuous) if it is continuous at right-dense (left-dense) points in $\mathbb{T}$ and its left-sided (right-sided) limits exist at left-dense (right-dense) points in $\mathbb{T}$.

The set $\mathbb{T}_k$ ($\mathbb{T}_k^*$) is derived from the time scale $\mathbb{T}$ as follows: If $\mathbb{T}$ has a left-scattered maximum (right-scattered minimum)

m, then $\mathbb{T}_k := \mathbb{T} - \{m\}$ ($\mathbb{T}_k^* := \mathbb{T} - \{m\}$). Otherwise, $\mathbb{T}_k := \mathbb{T}$ ($\mathbb{T}_k^* := \mathbb{T}$).

Definition 3. A function $f : \mathbb{T} \to \mathbb{R}$ is said to be delta (nabla) differentiable at a point $t \in \mathbb{T}_k$ ($t \in \mathbb{T}_k^*$) if there exists a number $f^{\Delta}(t)$ ($f^{\nabla}(t)$) with the property that, given any $\epsilon > 0$, there exists a neighborhood U of t such that

$$\begin{aligned} &\left|f(\sigma(t)) - f(s) - f^{\Delta}(t)(\sigma(t) - s)\right| \le \epsilon\left|\sigma(t) - s\right|, \\ &\left(\left|f(\rho(t)) - f(s) - f^{\nabla}(t)(\rho(t) - s)\right| \le \epsilon\left|\rho(t) - s\right|\right) \end{aligned} \tag{1}$$

for all $s \in U$.

For a function $f : \mathbb{T} \to \mathbb{R}$ it is possible to introduce a derivative $f^{\Delta}(t)$ ($f^{\nabla}(t)$) and an integral $\int_a^b f(t)\Delta t$ ($\int_a^b f(t)\nabla t$) in such a manner that $f^{\Delta}(t) = f'$ ($f^{\nabla}(t) = f'$) and $\int_a^b f(t)\Delta t = \int_a^b f(t)dt$ ($\int_a^b f(t)\nabla t = \int_a^b f(t)dt$) in the case $\mathbb{T} = \mathbb{R}$ and $f^{\Delta}(t) = \Delta f$ ($f^{\nabla}(t) = \nabla f$) and $\int_a^b f(t)\Delta t = \sum_a^{b-1} f(t)$ ($\int_a^b f(t)\nabla t = \sum_{a+1}^b f(t)$) in the case $\mathbb{T} = \mathbb{Z}$, where the forward and backward difference operators are defined by $\Delta f = f(t+1) - f(t)$ and $= f(t) - f(t-1)$, respectively. Also, we define the iterated operators $\Delta^n = \Delta(\Delta^{n-1})$ and $\nabla^n = \nabla(\nabla^{n-1})$ for $n \in \mathbb{N}$.

Definition 4. A function $p : \mathbb{T} \to \mathbb{R}$ is called μ-regressive (ν-regressive) provided that $1 + \mu(t)p(t) \neq o$ ($1 - \nu(t)p(t) \neq o$) for all $t \in \mathbb{T}_k$ ($t \in \mathbb{T}_k^*$).

The set $\mathbb{R}_\mu$ ($\mathbb{R}_\nu$) of all μ-regressive and rd-continuous (ν-regressive and ld-continuous) functions forms an Abelian group under the circle plus addition $\oplus$ defined by $(p\oplus q)(t) := p(t)+q(t)+\mu(t)p(t)q(t)$ ($(p\oplus q)(t) := p(t)+q(t)-\nu(t)p(t)q(t)$) for all $t \in \mathbb{T}_k$ ($t \in \mathbb{T}_k^*$). The additive inverse $\ominus p$ of $p \in \mathbb{R}_\mu$ ($p \in \mathbb{R}_\nu$) is defined by

$$\begin{aligned} &(\ominus p)(t) := -\frac{p(t)}{1 + \mu(t)p(t)} \\ &\left((\ominus p)(t) := -\frac{p(t)}{1 - \nu(t)p(t)}\right) \end{aligned} \tag{2}$$

for all $t \in \mathbb{T}_k$ ($t \in \mathbb{T}_k^*$).

For real numbers a and b we denote $\mathbb{N}_a = \{a, a+1, \ldots\}$ and ${}_b\mathbb{N} = \{b, b-1, \ldots\}$.

Theorem 5. *Let $p \in \mathbb{R}_\mu$ ($p \in \mathbb{R}_\nu$) and $t_0 \in \mathbb{T}$ be a fixed point. Then the delta (nabla) exponential function $e_p(\cdot, t_0)$ ($e_p^*(\cdot, t_0)$) is the unique solution of the initial value problem*

$$\begin{aligned} &y^{\Delta} = p(t)y, \quad y(t_0) = 1 \\ &\left(y^{\nabla} = p(t)y, \ y(t_0) = 1\right). \end{aligned} \tag{3}$$

If $\mathbb{T} = \mathbb{N}_a$, when $p(t) \equiv p$, where $p \in \mathbb{R}_\mu$ ($p \in \mathbb{R}_\nu = \mathbb{C}\setminus\{1\}$) and $t_0 = a$, it is easy to see that $e_p(t,a) = (1+p)^{t-a}$ ($e_p^*(t,a) = (1-p)^{a-t}$) and if $\mathbb{T} = \mathbb{R}$, $e_p(t,a) = e^{p(t-a)}$ ($e_p^*(t,a) = e^{p(t-a)}$), where "e" is ordinary exponential function. Moreover, in the special case, $e_1(t,0) = 2^t$ ($e_{1/2}^*(t,0) = 2^t$). More generally, we will denote all $e_p(t,a)$, $e_p^*(t,a)$, and $e^{p(t-a)}$ with $\widehat{e}_p(t,a)$.

Definition 6. The delta (nabla) Taylor monomials are the functions $h_n : \mathbb{T}\times\mathbb{T} \to \mathbb{R}$, $n \in \mathbb{N}_0$, and are defined recursively as follows:

$$\begin{aligned} &h_0(t,s) = 1, \\ &h_{n+1}(t,s) = \int_s^t h_n(\tau,s)\Delta\tau, \\ &\qquad \forall t,s \in \mathbb{T}, \\ &\left(h_0^*(t,s) = 1, \ h_{n+1}^*(t,s) = \int_s^t h_n^*(\tau,s)\nabla\tau, \ \forall t,s \in \mathbb{T}\right). \end{aligned} \tag{4}$$

We consider three cases for the time scale $\mathbb{T}$.

(a) If $\mathbb{T} = \mathbb{R}$, then $\sigma(t) = \rho(t) = t$ and the Taylor monomials can be written explicitly as

$$h_n(t,s) = h_n^*(t,s) = \frac{(t-s)^n}{n!}, \quad t,s \in \mathbb{R}, \ n \in \mathbb{N}_0. \tag{5}$$

For each $\alpha \in \mathbb{R}\setminus\{-\mathbb{N}\}$, define the αth Taylor monomial to be

$$h_\alpha(t,s) = \frac{(t-s)^\alpha}{\Gamma(\alpha+1)}, \tag{6}$$

and Γ denoted the special gamma function.

In this paper, we only consider the special case, $h_\alpha(t) := h_\alpha(t,0) = t^\alpha/\Gamma(\alpha+1)$ as Taylor monomial (tm).

(b) If $\mathbb{T} = \mathbb{Z}$, then $\sigma(t) = t+1$ and the Taylor monomials can be written explicitly as

$$h_n(t,s) = \frac{(t-s)^{\underline{n}}}{n!}, \quad t,s \in \mathbb{Z}, \ n \in \mathbb{N}_0, \tag{7}$$

where

$$t^{\underline{n}} = \prod_{j=0}^{n-1}(t-j) = \frac{\Gamma(t+1)}{\Gamma(t+1-j)} \tag{8}$$

and product is zero when $t+1-j = 0$ for some j. More generally, for α arbitrary define $t^{\underline{\alpha}} = \Gamma(t+1)/\Gamma(t+1-\alpha)$, where the convention of that division at pole yields zero. This generalized falling function allows us to extend (7) to define a general Taylor monomial that will serve us well in the probability distributions setting.

For each $\alpha \in \mathbb{R}\setminus\{-\mathbb{N}\}$, define the delta αth Taylor monomial to be

$$h_\alpha(t,s) = \frac{(t-s)^{\underline{\alpha}}}{\Gamma(\alpha+1)}. \tag{9}$$

In this paper, we only consider the special case $h_{\underline{\alpha}}(t) := h_\alpha(t,0) = t^{\underline{\alpha}}/\Gamma(\alpha+1)$ as delta Taylor monomial (dtm).

(c) If $\mathbb{T} = \mathbb{Z}$, then $\rho(t) = t-1$ and the Taylor monomials can be written explicitly as

$$h_n(t,s) = \frac{(t-s)^{\overline{n}}}{n!}, \quad t,s \in \mathbb{Z}, \ n \in \mathbb{N}_0, \tag{10}$$

where

$$t^{\overline{n}} = \prod_{j=0}^{n-1} (t+j) = \frac{\Gamma(t+n)}{\Gamma(t)}. \quad (11)$$

More generally, for any $\alpha \in \mathbb{R} \setminus \{-\mathbb{N}\}$, the α rising function is defined as $t^{\overline{\alpha}} = \Gamma(t+\alpha)/\Gamma(t)$ and $0^{\overline{\alpha}} = 0$. The rising $t^{\overline{\alpha}}$ and falling $t^{\underline{\alpha}}$ are related by $t^{\overline{\alpha}} = (t+\alpha-1)^{\underline{\alpha}}$.

This rising function allows us to extend (10) in order to define a general Taylor monomial that will serve us well in the probability distributions setting.

For each $\alpha \in \mathbb{R} \setminus \{-\mathbb{N}\}$ define the nabla αth Taylor monomial to be

$$h_{\alpha}^{*}(t,s) = \frac{(t-s)^{\overline{\alpha}}}{\Gamma(\alpha+1)}. \quad (12)$$

In this paper, we only consider the special case $h_{\overline{\alpha}}(t) := h_{\alpha}^{*}(t,0) = t^{\overline{\alpha}}/\Gamma(\alpha+1)$ as nabla Taylor monomial (ntm).

More generally, we will denote all $h_{\underline{\alpha}}(t)$, $h_{\overline{\alpha}}(t)$, and $h_{\alpha}(t)$ with $\widehat{h}_{\alpha}(t)$.

Let b be a real number and $f : {}_{b}\mathbb{N} \to \mathbb{R}$. The delta Riemann right fractional sum of order $\alpha > 0$ is defined by Abdeljawad [13] as

$$\Delta^{-\alpha} f(t) = \frac{1}{\Gamma(\alpha)} \sum_{s=t+\alpha}^{b} (\rho(s)-t)^{\underline{\alpha-1}} f(s), \quad t \in {}_{b-\alpha}\mathbb{N}. \quad (13)$$

We define the nabla Riemann right fractional sum of order $\alpha > 0$ as

$$\nabla^{-\alpha} f(t) = \frac{1}{\Gamma(\alpha)} \sum_{s=t}^{b-1} (\sigma(s)-t)^{\overline{\alpha-1}} f(s), \quad t \in {}_{b-1}\mathbb{N}. \quad (14)$$

The delta Riemann right fractional difference of order $\alpha > 0$ is defined by Abdeljawad [13] as

$$\Delta^{\alpha} f(t) = (-1)^{n} \nabla^{n} \Delta^{-(n-\alpha)} f(t), \quad (15)$$

for $t \in {}_{b-(n-\alpha)}\mathbb{N}$ and $n = [\alpha]+1$, where $[\alpha]$ is the greatest integer less than α. Also, the nabla Riemann right fractional difference of order $\alpha > 0$ is defined by

$$\nabla^{\alpha} f(t) = (-1)^{n} \Delta^{n} \nabla^{-(n-\alpha)} f(t), \quad (16)$$

for $t \in {}_{b-n}\mathbb{N}$.

In [14], authors have obtained the following alternative definition for delta Riemann right fractional difference

$$\Delta^{\alpha} f(t) = \frac{1}{\Gamma(-\alpha)} \sum_{s=t-\alpha}^{b} (\rho(s)-t)^{\underline{-\alpha-1}} f(s). \quad (17)$$

Similarly, we can prove the following formula for nabla Riemann right fractional difference:

$$\nabla^{\alpha} f(t) = \frac{1}{\Gamma(-\alpha)} \sum_{s=t}^{b-1} (\sigma(s)-t)^{\overline{-\alpha-1}} f(s). \quad (18)$$

For an introduction to discrete fractional calculus, the reader is referred to [15–18].

3. Generating Discrete Distributions by Discrete Fractional Calculus

The following results show the relationship between continuous and discrete fractional calculus and statistics and also allow us to define different types of discrete distributions.

Suppose that X is a positive continuous random variable. The expectation of the tm function, $h_{\alpha-1}(X)$, coincides with Riemann-Liouville right fractional integral of the pdf at the origin for $\alpha > 0$ and Marchaud fractional derivative of the pdf at the origin for $-1 < \alpha < 0$; that is, we have

$$E[h_{\alpha-1}(X)] = \begin{cases} (I_{-}^{\alpha} f)(0), & \alpha > 0 \\ (D_{-}^{\alpha} f)(0), & -1 < \alpha < 0, \end{cases} \quad (19)$$

where

$$(I_{-}^{\alpha} f)(t) = \frac{1}{\Gamma(\alpha)} \int_{0}^{\infty} x^{\alpha-1} f(x+t)\, dx \quad (20)$$

is the Riemann-Liouville right fractional integral, while

$$(D_{-}^{\alpha} f)(t) = \frac{1}{\Gamma(-\alpha)} \int_{0}^{\infty} x^{-\alpha-1} \{f(x+t) - f(t)\}\, dx \quad (21)$$

is the Marchaud fractional derivative [19]. The definitions of fractional operators can be found in [20].

It can be seen that the limits of the above integrals are equal to the support of random variable X. Considering this point, we present the following theorems for discrete random variable X.

Theorem 7. *Suppose that X is a discrete random variable. The expectation of the dtm function, $h_{\underline{\alpha-1}}(X)$, coincides with delta Riemann right fractional sum of the pmf at -1 for $\alpha > 0$ and delta Riemann right fractional difference of the pmf at -1 for $\alpha < 0$, $\alpha \notin \{-\mathbb{N}\}$; that is,*

$$E\left[h_{\underline{\alpha-1}}(X)\right] = \begin{cases} (\Delta^{-\alpha} f)(-1), & \alpha > 0 \\ (\Delta^{\alpha} f)(-1), & \alpha < 0,\ \alpha \notin \{\dots,-2,-1\}, \end{cases} \quad (22)$$

where

$$(\Delta^{-\alpha} f)(t) = \sum_{x=\alpha-1}^{b-1-t} \frac{x^{\underline{\alpha-1}}}{\Gamma(\alpha)} f(x+t+1) \quad (23)$$

is the delta Riemann right fractional sum, while

$$(\Delta^{\alpha} f)(t) = \sum_{x=-\alpha-1}^{b-t-1} \frac{x^{\underline{-\alpha-1}}}{\Gamma(-\alpha)} f(x+t+1) \quad (24)$$

is the delta Riemann right fractional difference.

Proof. For $\alpha > 0$, substitute $x = \rho(s) - t$ in expression (13) and also, for $\alpha < 0$ and $\alpha \notin \{-1,-2,\dots\}$, in expression (17).

Here, considering the limits of summation we can define the discrete distributions with the support $\mathbb{N}_{\alpha-1}$ or a finite subset of it. In this case, we will call X delta discrete random variable. In this work, we will define the delta discrete fractional Weibull distribution. Another example is the delta discrete uniform distribution, $DU\{\alpha-1, \alpha, \ldots, \alpha+\beta\}$, where $\alpha \in \mathbb{R}$ and $\beta \in \mathbb{N}_{-1}$.

Theorem 8. *We suppose that X is a discrete random variable. The expectation of the ntm function, $h_{\overline{\alpha-1}}(X)$, coincides with nabla Riemann right fractional sum of the pmf at 1 for $\alpha > 0$ and nabla Riemann right fractional difference of the pmf at 1 for $\alpha < 0$, $\alpha \notin \{-\mathbb{N}\}$; that is,*

$$E\left[h_{\overline{\alpha-1}}(X)\right] = \begin{cases} (\nabla^{-\alpha} f)(1), & \alpha > 0 \\ (\nabla^{\alpha} f)(1), & \alpha < 0,\ \alpha \notin \{\ldots,-2,-1\}, \end{cases} \tag{25}$$

where

$$(\nabla^{-\alpha} f)(t) = \sum_{x=1}^{b-t} \frac{x^{\overline{\alpha-1}}}{\Gamma(\alpha)} f(x+t-1) \tag{26}$$

is the nabla Riemann right fractional sum, while

$$(\nabla^{\alpha} f)(t) = \sum_{x=1}^{b-t} \frac{x^{\overline{-\alpha-1}}}{\Gamma(-\alpha)} f(x+t-1) \tag{27}$$

is the nabla Riemann right fractional difference.

Proof. For $\alpha > 0$, substitute $x = \sigma(s) - t$ in expression (14) and also for $\alpha < 0$ and $\alpha \notin \{-1, -2, \ldots\}$, in expression (18). □

Then, considering the limits of summation in recent theorem we can define the discrete distributions with support $\mathbb{N}_1$ or a finite subset of it. In this case, we will call X nabla discrete random variable. In this work, we will define the nabla discrete fractional Weibull distribution. Another example is the nabla discrete uniform distribution, $DU\{1, 2, \ldots, \alpha-\beta+1\}$, where $\alpha \in \mathbb{R}$ and $\beta \in {}_{\alpha}\mathbb{N}$.

4. The Delta and Nabla Discrete Fractional Weibull Distributions

In this section, we will introduce delta and nabla discrete fractional Weibull distributions, by substituting continuous Taylor monomial and exponential functions with their corresponding discrete types (on the discrete time scale) in continuous Weibull distribution.

4.1. The Nabla Discrete Fractional Weibull Distribution

Definition 9. It is said that the random variable X has a nabla discrete fractional Weibull distribution with (α, λ) parameters, denoted by $W^{\nabla}(\alpha, \lambda)$, if its pmf is given by

$$\begin{aligned} \Pr[X = x] &= \frac{\lambda \Gamma(\alpha+1)\, h_{\overline{\alpha-1}}(x)}{e_{\lambda}^{*}\left(\rho\left(\Gamma(\alpha+1)\, h_{\overline{\alpha}}(x)\right), 0\right)} \\ &= \alpha \lambda x^{\overline{\alpha-1}} (1-\lambda)^{x^{\overline{\alpha}}-1}, \quad x = \mathbb{N}_1, \end{aligned} \tag{28}$$

where $\alpha > 0$, $0 < \lambda < 1$.

Now we show that

$$\sum_{x=1}^{\infty} \alpha \lambda x^{\overline{\alpha-1}} (1-\lambda)^{x^{\overline{\alpha}}-1} = 1. \tag{29}$$

For this purpose, by using Theorems 4.3 and 3.8 (integration by substitution) from [21] and considering $(x^{\overline{\alpha}})^{\nabla} = \alpha x^{\overline{\alpha-1}}$ and under the substitution $u = x^{\overline{\alpha}}$, we have

$$\begin{aligned} \sum_{x=1}^{\infty} \alpha \lambda x^{\overline{\alpha-1}} (1-\lambda)^{x^{\overline{\alpha}}-1} &= \int_0^{\infty} \alpha \lambda x^{\overline{\alpha-1}} (1-\lambda)^{x^{\overline{\alpha}}-1} \nabla x \\ &= \int_0^{\infty} \lambda (1-\lambda)^{u-1} \nabla u = \sum_{u=1}^{\infty} \lambda (1-\lambda)^{u-1} = 1. \end{aligned} \tag{30}$$

(i) Particular Cases. (a) For $\alpha = 1$, $W^{\nabla}(\alpha, \lambda)$ in (28) reduces to a one-parameter nabla discrete fractional Weibull with pmf

$$\Pr[X = x] = \lambda (1-\lambda)^{\rho(x)}, \quad x = 1, 2, \ldots. \tag{31}$$

Obviously, this is the pmf of geometric distribution (the number of independent trials required for first success) or nabla discrete exponential distribution.

(b) For $\alpha = n$, $n \in \mathbb{N}$, $W^{\nabla}(\alpha, \lambda)$ in (28) is a new delta discrete distribution with pmf

$$\Pr[X = x] = n! \binom{x+n-2}{x-1} \lambda (1-\lambda)^{n!\binom{x+n-1}{x-1}-1}, \quad x = 1, 2, \ldots. \tag{32}$$

If we substitute $\rho(x) = x$, (31) and (32) are given by

$$\Pr[X = x] = \lambda (1-\lambda)^{x}, \quad x = 0, 1, 2, \ldots, \tag{33}$$

$$\Pr[X = x] = n! \binom{x+n-1}{x} \lambda (1-\lambda)^{n!\binom{x+n}{x}-1}, \quad x = 0, 1, 2, \ldots, \tag{34}$$

respectively. It can be seen that (33) is the same geometric distribution (the number of failures for first success). Then (34), as a general case of (33), is a type of negative binomial distribution.

(c) For $\alpha = 2$, $W^{\nabla}(\alpha, \lambda)$ in (28) is a nabla discrete distribution with pmf

$$\Pr[X = x] = 2x\lambda (1-\lambda)^{x(x+1)-1}, \quad x = 1, 2, \ldots, \tag{35}$$

where we will call it nabla discrete Rayleigh distribution.

(ii) Statistical Properties. If $X \sim W^{\nabla}(\alpha, \lambda)$, then the survival function, the hazard function, and the mean of random variable X are given by

$$s(x) = (1-\lambda)^{(\rho(x))^{\overline{\alpha}}}, \tag{36}$$

$$h(x) = \alpha \lambda x^{\overline{\alpha-1}} (1-\lambda)^{2(\rho(x))^{\overline{\alpha}} (\rho(x))^{\overline{\alpha/2}-1}}, \tag{37}$$

$$E[X] = \sum_{x=1}^{\infty} (1-\lambda)^{(\rho(x))^{\overline{\alpha}}}, \quad (38)$$

respectively.

To continue, we use the beta type I, beta type II, and Kummer-beta distributions. These distributions can be found in [22–25].

(iii) Bayesian Estimation in Nabla Discrete Fractional Weibull Distribution. In this section, we study the Bayesian estimation of functions of parameter λ of nabla discrete fractional Weibull distribution. The likelihood function of λ, in this case, is given by

$$L(\lambda) \propto \lambda^n (1-\lambda)^{\sum x_i^{\overline{\alpha}} - n}. \quad (39)$$

We take a prior distribution given below:

$$\pi(\lambda) = \frac{\lambda^{p-1}(1-\lambda)^{q-1} e^{\mu(1-\lambda)}}{\mathrm{B}(p,q)\, {}_1F_1(q; p+q; \mu)}, \quad 0 < \lambda < 1,\ p, q > 0,\ \mu \in \mathbb{R}, \quad (40)$$

where $\mathrm{B}(p,q) = \Gamma(p)\Gamma(q)/\Gamma(p+q)$ and

$${}_1F_1(a;b;z) = \sum_{k=0}^{\infty} \left(a^{\overline{k}} z^k / b^{\overline{k}} k! \right). \quad (41)$$

This prior density is known as the Kummer-beta density and denoted by $\mathrm{KB}(p,q,\mu)$. The posterior probability density function of λ, corresponding to $\pi(\lambda)$, is given by

$$\begin{aligned} \pi(\lambda \mid x) &\propto L(\lambda)\pi(\lambda) \\ &\propto \lambda^{n+p-1}(1-\lambda)^{\sum x_i^{\overline{\alpha}} - n + q - 1} e^{\mu(1-\lambda)} \Longrightarrow \\ \lambda \mid x &\sim \mathrm{KB}\left(n+p, \sum x_i^{\overline{\alpha}} - n + q, \mu\right). \end{aligned} \quad (42)$$

$\pi(\lambda)$ is a natural conjugate prior density. Note that, for $\mu = 0$, the above prior density simplifies to a beta type I density with parameters p and q. Under the squared error loss function (SELF) given by $\mathfrak{L}(\gamma(\lambda), \delta) = (\gamma(\lambda) - \delta)^2$, where $\gamma(\lambda)$ is a function of λ and δ is a decision, the Bayesian estimate $\widehat{\gamma}_{\mathrm{B}}$ of $\gamma(\lambda) = \lambda^r$, corresponding to posterior density $\pi(\lambda \mid x)$, is given by

$$\begin{aligned} \widehat{\lambda}_{\mathrm{B}}^r &= \frac{\int_0^1 \lambda^{n+r+p-1}(1-\lambda)^{\sum x_i^{\overline{\alpha}} - n + q - 1} e^{\mu(1-\lambda)} d\lambda}{\mathrm{B}(n+p, \sum x_i^{\overline{\alpha}} - n + q)\, {}_1F_1(\sum x_i^{\overline{\alpha}} - n + q; \sum x_i^{\overline{\alpha}} + p + q; \mu)} \\ &= \frac{\mathrm{B}\left(n+p+r, \sum x_i^{\overline{\alpha}} - n + q\right) {}_1F_1\left(\sum x_i^{\overline{\alpha}} - n + q; \sum x_i^{\overline{\alpha}} + p + q + r; \mu\right)}{\mathrm{B}(n+p, \sum x_i^{\overline{\alpha}} - n + q)\, {}_1F_1(\sum x_i^{\overline{\alpha}} - n + q; \sum x_i^{\overline{\alpha}} + p + q; \mu)}. \end{aligned} \quad (43)$$

Similarly, under the weighted squared error loss function (WSELF) given by $\mathfrak{L}(\gamma(\lambda), \delta) = w(\lambda)(\gamma(\lambda) - \delta)^2$, where $w(\lambda)$ is a function of λ, the Bayesian estimate $\widehat{\gamma}_w$ of $\gamma(\lambda)$, for two different forms of $w(\lambda)$, is given below.

(a) Let $w(\lambda) = \lambda^{-2}$. The Bayesian estimate $\widehat{\gamma}_m$ of $\gamma(\lambda)$, known as the minimum expected loss (MEL) estimate, for $\gamma(\lambda) = \lambda^r$, corresponding to posterior density $\pi(\lambda \mid x)$, is given by

$$\begin{aligned} \widehat{\lambda}_m^r &= \frac{\int_0^1 \lambda^{(n+r+p-2)-1}(1-\lambda)^{\sum x_i^{\overline{\alpha}} - n + q - 1} e^{\mu(1-\lambda)} d\lambda}{\int_0^1 \lambda^{(n+p-2)-1}(1-\lambda)^{\sum x_i^{\overline{\alpha}} - n + q - 1} e^{\mu(1-\lambda)} d\lambda} \\ &= \frac{\mathrm{B}\left(n+r+p-2, \sum x_i^{\overline{\alpha}} - n + q\right) {}_1F_1\left(\sum x_i^{\overline{\alpha}} - n + q; \sum x_i^{\overline{\alpha}} + p + q + r - 2; \mu\right)}{\mathrm{B}(n+p-2, \sum x_i^{\overline{\alpha}} - n + q)\, {}_1F_1(\sum x_i^{\overline{\alpha}} - n + q; \sum x_i^{\overline{\alpha}} + p + q - 2; \mu)}. \end{aligned} \quad (44)$$

This loss function was used by Tummala and Sathe [26] for estimating the reliability of certain life time distributions and by Zellner and Park [27] for estimating functions of parameters of some econometric models.

(b) Let $w(\lambda) = \lambda^{-2} e^{\nu(1-\lambda)}$, $\nu < 0$. The Bayesian estimate $\widehat{\gamma}_e$ of $\gamma(\lambda)$, known as the exponentially weighted minimum expected loss (EWMEL) estimate, for $\gamma(\lambda) = \lambda^r$, corresponding to posterior density $\pi(\lambda \mid x)$, is given by

$$\begin{aligned} \widehat{\lambda}_e^r &= \frac{\int_0^1 \lambda^{(n+r+p-2)-1}(1-\lambda)^{\sum x_i^{\overline{\alpha}} - n + q - 1} e^{(\mu+\nu)(1-\lambda)} d\lambda}{\int_0^1 \lambda^{(n+p-2)-1}(1-\lambda)^{\sum x_i^{\overline{\alpha}} - n + q - 1} e^{(\mu+\nu)(1-\lambda)} d\lambda} \\ &= \frac{\mathrm{B}\left(n+r+p-2, \sum x_i^{\overline{\alpha}} - n + q\right) {}_1F_1\left(\sum x_i^{\overline{\alpha}} - n + q; \sum x_i^{\overline{\alpha}} + p + q + r - 2; \mu+\nu\right)}{\mathrm{B}(n+p-2, \sum x_i^{\overline{\alpha}} - n + q)\, {}_1F_1(\sum x_i^{\overline{\alpha}} - n + q; \sum x_i^{\overline{\alpha}} + p + q - 2; \mu+\nu)}. \end{aligned} \quad (45)$$

4.2. The Delta Discrete Fractional Weibull Distribution

Definition 10. It is said that the random variable X has a delta discrete fractional Weibull distribution with (α, λ) parameters, denoted by $W^{\Delta}(\alpha, \lambda)$, if its pmf is given by

$$\begin{aligned} \Pr[X = x] &= \frac{\lambda \Gamma(\alpha+1) h_{\underline{\alpha-1}}(x)}{e_{\lambda}\left(\sigma\left(\Gamma(\alpha+1) h_{\underline{\alpha}}(x)\right), 0\right)} \\ &= \frac{\alpha \lambda x^{\underline{\alpha-1}}}{(1+\lambda)^{x^{\underline{\alpha}}+1}}, \quad x = \mathbb{N}_{\alpha-1}, \end{aligned} \tag{46}$$

where $\alpha > 0,\ \lambda > 0$.

Now we show that

$$\sum_{x=\alpha-1}^{\infty} \left(\frac{\alpha \lambda x^{\underline{\alpha-1}}}{(1+\lambda)^{x^{\underline{\alpha}}+1}} \right) = 1. \tag{47}$$

For this purpose, we apply Theorem 5.40 (change of variable) from [12]. Considering $(x^{\underline{\alpha}})^{\Delta} = \alpha x^{\underline{\alpha-1}}$ and then under the substitution $u = x^{\underline{\alpha}}$, we have

$$\begin{aligned} \sum_{x=\alpha-1}^{\infty} \frac{\alpha \lambda x^{\underline{\alpha-1}}}{(1+\lambda)^{x^{\underline{\alpha}}+1}} &= \int_{\alpha-1}^{\infty} \frac{\alpha \lambda x^{\underline{\alpha-1}}}{(1+\lambda)^{x^{\underline{\alpha}}+1}} \Delta x \\ &= \int_{0}^{\infty} \frac{\lambda \Delta u}{(1+\lambda)^{u+1}} = \sum_{u=0}^{\infty} \frac{\lambda}{(1+\lambda)^{1+u}} \\ &= 1. \end{aligned} \tag{48}$$

(i) Particular Cases. (a) For $\alpha = 1$, $W^{\Delta}(\alpha, \lambda)$ in (46) reduces to a one-parameter delta discrete fractional Weibull with pmf

$$\Pr[X = x] = \lambda (1+\lambda)^{-\sigma(x)} = \left(\frac{\lambda}{1+\lambda}\right)\left(\frac{1}{1+\lambda}\right)^{x}, \quad x = 0, 1, \ldots. \tag{49}$$

Obviously, this is the pmf of geometric distribution (the number of failures for first success) or delta discrete exponential distribution.

(b) For $\alpha = n$, $n \in \mathbb{N}$, $W^{\Delta}(\alpha, \lambda)$ in (46) is a new delta discrete distribution with pmf

$$\begin{aligned} &\Pr[X = x] \\ &\quad = n! \binom{x}{x-n+1} \left(\frac{\lambda}{1+\lambda}\right)\left(\frac{1}{1+\lambda}\right)^{n!\binom{x}{x-n}}, \quad x = \mathbb{N}_{n-1}. \end{aligned} \tag{50}$$

If we substitute $\sigma(x) = x$, (49) and (50) are given by

$$\Pr[X = x] = \left(\frac{\lambda}{1+\lambda}\right)\left(\frac{1}{1+\lambda}\right)^{x-1}, \quad x = 1, 2, \ldots, \tag{51}$$

$$\Pr[X = x] = n! \binom{x-1}{x-n} \left(\frac{\lambda}{1+\lambda}\right)\left(\frac{1}{1+\lambda}\right)^{n!\binom{x-1}{x-n-1}}, \quad x = n, n+1, \ldots, \tag{52}$$

respectively. It can be seen that (51) is the same geometric distribution (the number of independent trials required for first success). Then (52), as a general case of (51), is a type of negative binomial distribution.

(c) For $\alpha = 2$, $W^{\Delta}(\alpha, \lambda)$ in (46) is a delta discrete distribution with pmf

$$\Pr[X = x] = 2x \left(\frac{\lambda}{1+\lambda}\right)\left(\frac{1}{1+\lambda}\right)^{x(x-1)}, \quad x = 1, 2, \ldots, \tag{53}$$

where we will call it delta discrete Rayleigh distribution.

(ii) Statistical Properties. If $X \sim W^{\Delta}(\alpha, \lambda)$, then the survival function, the hazard function, and the mean of random variable X are given by

$$s(x) = \frac{1}{(1+\lambda)^{x^{\underline{\alpha}}}}, \tag{54}$$

$$h(x) = \alpha \left(\frac{\lambda}{1+\lambda}\right) x^{\underline{\alpha-1}}, \tag{55}$$

$$E[X] = \sum_{x=\alpha-1}^{\infty} \frac{1}{(1+\lambda)^{x^{\underline{\alpha}}}}, \quad \alpha \geq 1, \tag{56}$$

respectively.

(iii) Bayesian Estimation in Delta Discrete Fractional Weibull Distribution. In this section, we study the Bayesian estimation of functions of parameter λ of delta discrete fractional Weibull distribution. The likelihood function of λ, in this case, is given by

$$L(\lambda) \propto \frac{\lambda^{n}}{(1+\lambda)^{\sum x_i^{\underline{\alpha}}+n}}. \tag{57}$$

We take a prior distribution given below:

$$\pi(\lambda) = \frac{\lambda^{p-1}}{\mathrm{B}(p,q)(1+\lambda)^{p+q}}, \quad \lambda > 0,\ p, q > 0. \tag{58}$$

This prior density is known as the beta type II density and denoted by $\mathrm{B}^{\mathrm{II}}(p, q)$.

The posterior probability density function of λ, corresponding to $\pi(\lambda)$, is given by

$$\begin{aligned} &\pi(\lambda \mid x) \propto L(\lambda)\pi(\lambda) \propto \frac{\lambda^{n+p-1}}{(1+\lambda)^{\sum x_i^{\underline{\alpha}}+n+p+q}} \Longrightarrow \\ &\lambda \mid x \sim \mathrm{B}^{\mathrm{II}}\left(n+p, \sum x_i^{\underline{\alpha}} + q\right). \end{aligned} \tag{59}$$

$\pi(\lambda)$ is a natural conjugate prior density. Note that $\pi(\lambda)$ is a special case of inverted hypergeometric function type I density, which is given by Gupta and Nagar [28] and Nagar and Alvarez [29]. Under SELF, the Bayesian estimate of $\gamma(\lambda) = \lambda^r$, corresponding to posterior density $\pi(\lambda \mid x)$, is given by

$$\widehat{\lambda}_{\mathrm{B}}^{r} = \int_0^{\infty} \frac{\lambda^{n+r+p-1}}{\mathrm{B}\left(n+p,\sum x_i^{\underline{\alpha}}+q\right)(1+\lambda)^{\sum x_i^{\underline{\alpha}}+n+p+q}} d\lambda = \frac{\mathrm{B}\left(n+p+r,\sum x_i^{\underline{\alpha}}+q-r\right)}{\mathrm{B}\left(n+p,\sum x_i^{\underline{\alpha}}+q\right)}. \quad (60)$$

Similarly, under WSELF, when $w(\lambda) = \lambda^{-2}$, the MEL estimate of $\gamma(\lambda) = \lambda^r$, corresponding to posterior density $\pi(\lambda \mid x)$, is given by

$$\widehat{\lambda}_{m}^{r} = \frac{\int_0^{\infty} \lambda^{(n+r+p-2)-1}(1+\lambda)^{-(\sum x_i^{\underline{\alpha}}+n+p+q)} d\lambda}{\int_0^{\infty} \lambda^{(n+p-2)-1}(1+\lambda)^{-(\sum x_i^{\underline{\alpha}}+n+p+q)} d\lambda} = \frac{\mathrm{B}\left(n+p+r-2,\sum x_i^{\underline{\alpha}}+q-r+2\right)}{\mathrm{B}\left(n+p-2,\sum x_i^{\underline{\alpha}}+q+2\right)}. \quad (61)$$

5. Unification of the Continuous and Discrete Weibull Distributions

For a given time scale $\mathbb{T}$, we present the construction of pdf of Weibull distribution, such that the density function on time scales is

$$f_X(x) = \frac{\lambda\Gamma(\alpha+1)\widehat{h}_{\alpha-1}(x)}{\widehat{e}_{\lambda}\left(\eta\left(\Gamma(\alpha+1)\widehat{h}_{\alpha-1}(x)\right),0\right)}, \quad x \in \mathbb{T}. \quad (62)$$

In order that the reader sees how continuous Weibull distribution and delta and nabla discrete fractional Weibull distributions follow from (62), it is only at this point necessary to know that

$$\widehat{h}_{\alpha-1}(x) = h_{\alpha-1}(x) = \frac{x^{\alpha-1}}{\Gamma(\alpha)}, \quad \eta(x) = x, \quad \widehat{e}_{\lambda}(\eta(X),0) = e^{\lambda x} \quad (63)$$

if $\mathbb{T} = \mathbb{R}^+$,

$$\widehat{h}_{\alpha-1}(x) = h_{\underline{\alpha-1}}(x) = \frac{x^{\underline{\alpha-1}}}{\Gamma(\alpha)}, \quad \eta(x) = \sigma(x), \quad \widehat{e}_{\lambda}(\eta(X),0) = (1+\lambda)^{\sigma(x)} \quad (64)$$

if $\mathbb{T} = \mathbb{N}_{\alpha-1}$, $\alpha > 0$, and

$$\widehat{h}_{\alpha-1}(x) = h_{\overline{\alpha-1}}(x) = \frac{x^{\overline{\alpha-1}}}{\Gamma(\alpha)}, \quad \eta(x) = \rho(x), \quad \widehat{e}_{\lambda}(\eta(X),0) = (1-\lambda)^{-\rho(x)}, \quad (65)$$

if $\mathbb{T} = \mathbb{N}_1$.

Competing Interests

The authors declare that they have no competing interests.

References

[1] A. M. Abouammoh and N. S. Alhazzani, "On discrete gamma distribution," *Communications in Statistics—Theory and Methods*, vol. 44, no. 14, pp. 3087–3098, 2015.

[2] S. Chakraborty and D. Chakravarty, "Discrete gamma distributions: properties and parameter estimations," *Communications in Statistics—Theory and Methods*, vol. 41, no. 18, pp. 3301–3324, 2012.

[3] M. A. Jazi, C.-D. Lai, and M. H. Alamatsaz, "A discrete inverse Weibull distribution and estimation of its parameters," *Statistical Methodology*, vol. 7, no. 2, pp. 121–132, 2010.

[4] N. L. Johnson, A. W. Kemp, and S. Kots, *Univariate Discrete Distributions*, John Wiley & Sons, New York, NY, USA, 2005.

[5] A. W. Kemp, "The discrete half-normal distribution," in *Proceedings of the International Conference on Mathematical and Statistical Modeling in Honor of Enrique Castillo*, June 2008.

[6] M. S. A. Khan, A. Khalique, and A. M. Abouammoh, "On estimating parameters in a discrete Weibull distribution," *IEEE Transactions on Reliability*, vol. 38, no. 3, pp. 348–350, 1989.

[7] H. Krishna and P. S. Pundir, "Discrete Burr and discrete Pareto distributions," *Statistical Methodology*, vol. 6, no. 2, pp. 177–188, 2009.

[8] T. Nakagawa and S. Osaki, "The discrete weibull distribution," *IEEE Transactions on Reliability*, vol. 24, no. 5, pp. 300–301, 1975.

[9] D. Roy, "The discrete normal distribution," *Communications in Statistics—Theory and Methods*, vol. 32, no. 10, pp. 1871–1883, 2003.

[10] S. Chakraborty, "Generating discrete analogues of continuous probability distributions—a survey of methods and constructions," *Journal of Statistical Distributions and Applications*, vol. 2, no. 6, pp. 1–30, 2015.

[11] M. Bohner and A. Peterson, *Dynamic Equations on Time Scales*, Birkhäuser, Boston, Mass, USA, 2001.

[12] M. Bohner and A. Peterson, *Advances in Dynamic Equations on Time Scales*, Birkhäuser, Boston, Mass, USA, 2003.

[13] T. Abdeljawad, "Dual identities in fractional difference calculus within Riemann," *Advances in Difference Equations*, vol. 2013, article 36, pp. 47–55, 2013.

[14] T. Abdeljawad, D. Baleanu, F. Jarad, and R. P. Agarwal, "Fractional sums and differences with binomial coefficients," *Discrete Dynamics in Nature and Society*, vol. 2013, Article ID 104173, 6 pages, 2013.

[15] T. Abdeljawad, "On delta and nabla caputo fractional differences and dual identities," *Discrete Dynamics in Nature and Society*, vol. 2013, Article ID 406910, 12 pages, 2013.

[16] T. Abdeljawad and D. F. M. Torres, "Symmetric duality for left and right Riemann-Liouville and Caputo fractional differences," *Arab Journal of Mathematical Sciences*, 2016.

[17] C. Goodrich and A. C. Peterson, *Discrete Fractional Calculus*, Springer, Berline, Germany, 2015.

[18] M. Holm, *The theory of discrete fractional calculus development and application [M.S. dissertation]*, University of Nebraska, Lincoln, NE, USA, 2011.

[19] M. Ganji and F. Gharari, "The generalized random variable appears the trace of fractional calculus in statistics," *Applied*

Mathematics Information Sciences Letters, vol. 3, no. 2, pp. 61–67, 2015.

[20] K. S. Miller and B. Ross, *An Introduction to the Fractional Calculus And Fractional Differential Equations*, John Wiley & Sons, New York, NY, USA, 1993.

[21] D. Mozyrska, E. Pawluszewicz, and D. M. Torres, "The Riemann-Stieltjes integral on time scales," *The Australian Journal of Mathematical Analysis and Applications*, vol. 7, no. 1, pp. 1–14, 2010.

[22] M. B. Gordy, "Computationally convenient distributional assumptions for common-value auctions," *Computational Economics*, vol. 12, no. 1, pp. 61–78, 1998.

[23] N. L. Johnson, S. Kots, and N. Balakrishnan, *Continuous Univariate Distributions, Volume 2*, John Wiley & Sons, New York, NY, USA, 1994.

[24] D. K. Nagar and E. Zarrazola, "Distributions of the product and the quotient of independent Kummer-beta variables," *Scientiae Mathematicae Japonicae*, vol. 61, no. 1, pp. 109–117, 2005.

[25] L. E. Sanchez and D. K. Nagar, "Distributions of the product and quotient of independent beta type 3 variables," *Far East Journal of Theoretical Statistics*, vol. 17, no. 2, pp. 239–251, 2005.

[26] V. M. R. Tummala and P. T. Sathe, "Minimum expected loss estimators of reliability and parameters of certain lifetime distributions," *IEEE Transactions on Reliability*, vol. 27, no. 4, pp. 283–285, 1978.

[27] A. Zellner and S.-B. Park, "Minimum Expected Loss (MELO) estimators for functions of parameters and structural coefficients of econometric models," *Journal of American Statistical Association*, vol. 74, no. 365, pp. 185–193, 1979.

[28] A. K. Gupta and D. K. Nagar, *Matrix Variate Distributions*, Chapman & Hall/CRC, Boca Raton, Fla, USA, 2000.

[29] D. K. Nagar and J. A. Alvarez, "Properties of the hypergeometric function type I distribution," *Advances and Applications in Statistics*, vol. 5, no. 3, pp. 341–351, 2005.

12

Multivariate Macdonald Distribution and Its Properties

Daya K. Nagar, Edwin Zarrazola, and Luz Estela Sánchez

Instituto de Matemáticas, Universidad de Antioquia, Calle 67, No. 53–108, Medellín, Colombia

Correspondence should be addressed to Daya K. Nagar; dayaknagar@yahoo.com

Academic Editor: Ramón M. Rodríguez-Dagnino

We give multivariate generalization of Macdonald distribution and study several of its properties. We also define the multivariate Macdonald-gamma distribution and derive a number of results pertaining to it.

1. Introduction

The three parameter Macdonald distribution (Nagar et al. [1, 2]) is defined by the probability density function (p.d.f.):

$$f_{\mathrm{M}}(y;\alpha,\beta,\sigma)=\frac{\sigma^{-\beta}y^{\beta-1}\Gamma\left(\alpha;\sigma^{-1}y\right)}{\Gamma(\beta)\Gamma(\alpha+\beta)},\qquad y>0,\ \sigma>0,\ \beta>0,\ \alpha+\beta>0, \tag{1}$$

where the extended gamma function, $\Gamma(\nu;\sigma)$, is defined as

$$\Gamma(\nu;\sigma)=\int_0^\infty t^{\nu-1}\exp\left(-t-\frac{\sigma}{t}\right)dt,\quad \sigma>0. \tag{2}$$

For $\mathrm{Re}(\nu) > 0$ and by taking $\sigma = 0$, it is clear that the extension of the gamma function reduces to the classical gamma function, $\Gamma(\nu;0) = \Gamma(\nu)$. The generalized gamma function (extended) has been proved very useful in various problems in engineering and physics; see, for example, Aslam Chaudhry and Zubair [3–8].

We will denote by $Y \sim \mathrm{M}(\alpha,\beta,\sigma)$ if the random variable Y follows the Macdonald distribution. If $\sigma = 1$ in the density above, then we will simply write $Y \sim \mathrm{M}(\alpha,\beta)$. It has been shown in Nagar et al. [9] that the product of two independent gamma variables follows a Macdonald distribution. A random variable W is said to have a two parameter gamma distribution, denoted by $W \sim \mathrm{Ga}(\kappa,\theta)$, if its p.d.f. is given by

$$\frac{w^{\kappa-1}\exp(-w/\theta)}{\theta^\kappa\Gamma(\kappa)},\quad w>0. \tag{3}$$

Note that for $\theta = 1$, the above distribution reduces to a standard gamma distribution and in this case we write $W \sim \mathrm{Ga}(\kappa)$. There are several univariate, multivariate, and matrix variate generalizations of gamma distribution; for example, see standard text such as Johnson et al. [10], Kotz et al. [11], and Gupta and Nagar [12]. Kalla et al. [13], by using the generalized gamma function defined and studied by Al-Musallam and Kalla [14, 15], have defined a generalization of the gamma distribution which includes a number of distributions as special cases. In a recent article, Gupta et al. [16] have generalized the matrix variate gamma distribution. If W_1 and W_2 are independent, $W_1 \sim \mathrm{Ga}(\kappa_1,\theta_1)$ and $W_2 \sim \mathrm{Ga}(\kappa_2,\theta_2)$, then $W_1W_2 \sim \mathrm{M}(\kappa_2-\kappa_1,\kappa_1,\theta_1\theta_2)$.

Recently, Nagar et al. [17] have used the Macdonald distribution to construct a new bivariate distribution which they call *Macdonald-gamma* distribution. The random variables X and Y are said to have a Macdonald-gamma distribution, denoted by $(X,Y) \sim \mathrm{MG}(\alpha,\beta,\sigma)$, if their joint p.d.f. is given by

$$f(x,y;\alpha,\beta,\sigma)=\frac{x^{\alpha-1}y^{\beta-1}\exp(-x-y/\sigma x)}{\sigma^\beta\Gamma(\beta)\Gamma(\alpha+\beta)},\qquad x>0,\ y>0, \tag{4}$$

where $\beta > 0$, $\alpha+\beta > 0$, and $\sigma > 0$. The bivariate distribution defined by the above density has many interesting features. For example, the marginal and the conditional distributions of Y are Macdonald and gamma, the marginal distribution of X is gamma, and the conditional distribution of X given Y is extended gamma. The Macdonald-gamma distribution is positively likelihood ratio dependent (PLRD). Several results pertaining to this distribution such as marginal and conditional distributions, moments, entropies, information matrix, and distributions of sum are derived in Nagar et al. [17]. The distributions of the product and the ratio of independent or correlated Macdonald random variables are derived in Nagar et al. [9].

In this article, we study multivariate generalization of the Madcdonal distribution defined by the density (1) and derive properties such as marginal and conditional distributions, moments, distribution of sums, and factorization. We also give a multivariate generalization of Macdonald-gamma and study its properties.

2. Some Definitions and Preliminary Results

In this section, we give some definitions and results which are used in subsequent sections.

The gamma function was first introduced by Leonard Euler in 1729 as the limit of a discrete expression and later as an absolutely convergent improper integral:

$$\Gamma(\nu) = \int_0^\infty t^{\nu-1} \exp(-t)\, dt, \quad \operatorname{Re}(\nu) > 0. \tag{5}$$

The gamma function has many beautiful properties and has been used in almost all the branches of science and engineering. Replacing t by z/σ, $\sigma > 0$, in (5), the gamma function with an additional parameter $\sigma > 0$ can be given as

$$\Gamma(\nu) = \sigma^{-\nu} \int_0^\infty z^{\nu-1} \exp\left(-\frac{z}{\sigma}\right) dz, \quad \operatorname{Re}(\nu) > 0. \tag{6}$$

The extended gamma function is very similar to the modified Bessel function of type 2. An integral representation of the modified Bessel function type 2 (Gradshteyn and Ryzhik [18, Eq. 3.471.9]) is given by

$$K_\nu\left(2\sqrt{ab}\right) = \frac{1}{2}\left(\frac{a}{b}\right)^{\nu/2} \int_0^\infty t^{\nu-1}\left[-\left(at + \frac{b}{t}\right)\right] dt, \tag{7}$$

where $\operatorname{Re}(a) > 0$ and $\operatorname{Re}(b) > 0$. Comparing (2) and (7), it can easily be seen that

$$\Gamma(\nu;\sigma) = 2\sigma^{\nu/2} K_\nu\left(2\sqrt{\sigma}\right). \tag{8}$$

Also, substituting $x = \sigma/t$ in (2), it can be checked that

$$\Gamma(\nu;\sigma) = \sigma^\nu \Gamma(-\nu;\sigma). \tag{9}$$

Finally, we define beta type 1 and beta type 2 distributions. These definitions can be found in Johnson et al. [19].

Definition 1. The random variable X is said to have a beta type 1 distribution with parameters (a,b), $a > 0, b > 0$, denoted as $X \sim B1(a,b)$, if its p.d.f. is given by

$$\frac{x^{a-1}(1-x)^{b-1}}{B(a,b)}, \quad 0 < x < 1, \tag{10}$$

where $B(a,b)$ is the beta function defined by

$$B(a,b) = \frac{\Gamma(a)\Gamma(b)}{\Gamma(a+b)}, \quad \operatorname{Re}(a) > 0,\ \operatorname{Re}(b) > 0. \tag{11}$$

Definition 2. The random variable X is said to have a beta type 2 distribution with parameters (a,b), denoted as $X \sim B2(a,b)$, $a > 0, b > 0$, if its p.d.f. is given by

$$\frac{x^{a-1}(1+x)^{-(a+b)}}{B(a,b)}, \quad x > 0. \tag{12}$$

Definition 3. The random variables $X_1,\dots,X_n$ are said to have a Dirichlet type 1 distribution with parameters $\alpha_i > 0$, $i = 1,\dots,n$, and $\beta > 0$, denoted by $(X_1,\dots,X_n) \sim D1(\alpha_1,\dots,\alpha_n;\beta)$ if their joint p.d.f. is

$$\frac{\Gamma\left(\sum_{i=1}^n \alpha_i + \beta\right)}{\prod_{i=1}^n \Gamma(\alpha_i)\Gamma(\beta)} \prod_{i=1}^n x_i^{\alpha_i-1}\left(1 - \sum_{i=1}^n x_i\right)^{\beta-1}, \quad x_i > 0,\ i = 1,\dots,n,\ \sum_{i=1}^n x_i < 1. \tag{13}$$

The matrix variate generalizations of beta type 1 and beta type 2 distributions have been defined and studied extensively. For example, see Gupta and Nagar [12].

If $Y \sim M(\alpha,\beta,\sigma)$, then the rth moment of Y is given as

$$E(Y^r) = \frac{\sigma^r \Gamma(\beta+r)\Gamma(\alpha+\beta+r)}{\Gamma(\beta)\Gamma(\alpha+\beta)}. \tag{14}$$

Theorem 4. *If $(X,Y) \sim MG(\alpha,\beta,\sigma)$, then the p.d.f. of the sum $S = X + Y$ is given by*

$$f_S(s;\alpha,\beta,\sigma) = \frac{\Gamma(\alpha+1)s^{\alpha+\beta-1}\exp(-s)}{\sigma^\beta\Gamma(\beta)\Gamma(\alpha+\beta)} \cdot \sum_{m=0}^\infty L_m\left(\frac{1}{\sigma}\right)\frac{\Gamma(\beta+m)}{\Gamma(\alpha+\beta+m+1)} \times {}_1F_1(\beta+m;\alpha+\beta+m+1;s), \quad s > 0, \tag{15}$$

where $\alpha+1 > 0$, $\beta > 0$, $\sigma > 0$, $L_m(x)$ is the Laguerre polynomial of degree m, and ${}_1F_1$ represents the confluent hypergeometric function.

Proof. See Nagar et al. [17].

3. Density Function

We propose a multivariate generalization of the Macdonald distribution as follows.

Definition 5. The random variables $Y_1,\dots,Y_n$ are said to have a multivariate Macdonald distribution with parameters $\alpha, \beta_1,\dots,\beta_n$, and σ, denoted as $(Y_1,\dots,Y_n) \sim MM(\alpha,\beta_1,\dots,\beta_n,\sigma)$, if their joint p.d.f. is given by

$$C(\alpha,\beta_1,\dots,\beta_n,\sigma)\prod_{i=1}^n y_i^{\beta_i-1}\Gamma\left(\alpha;\frac{\sum_{i=1}^n y_i}{\sigma}\right), \quad y_i > 0,\ i = 1,\dots,n, \tag{16}$$

where $\beta_i > 0$, $i = 1,\dots,n$, $\alpha + \sum_{i=1}^n \beta_i > 0$, $\sigma > 0$, and $C(\alpha,\beta_1,\dots,\beta_n,\sigma)$ is the normalizing constant.

Since, the integration of the p.d.f. (16) over its support set is one, we have

$$C(\alpha,\beta_1,\ldots,\beta_n,\sigma) \cdot \int_0^\infty \cdots \int_0^\infty \prod_{i=1}^n y_i^{\beta_i-1}\Gamma\left(\alpha;\frac{\sum_{i=1}^n y_i}{\sigma}\right)dy_1\cdots dy_n = 1. \tag{17}$$

Replacing $\Gamma(\alpha;\sum_{i=1}^n y_i/\sigma)$ by its equivalent integral, namely,

$$\Gamma\left(\alpha;\frac{\sum_{i=1}^n y_i}{\sigma}\right) = \int_0^\infty x^{\alpha-1}\exp\left(-x-\frac{\sum_{i=1}^n y_i}{\sigma x}\right)dx \tag{18}$$

in (17), the normalizing constant is derived as

$$\begin{aligned}[C(\alpha,\beta_1,\ldots,\beta_n,\sigma)]^{-1} &= \int_0^\infty x^{\alpha-1}\exp(-x) \cdot \left[\prod_{i=1}^n \int_0^\infty y_i^{\beta_i-1}\exp\left(-\frac{y_i}{\sigma x}\right)dy_i\right]dx \\ &= \sigma^{\sum_{i=1}^n \beta_i}\prod_{i=1}^n \Gamma(\beta_i)\int_0^\infty x^{\alpha+\sum_{i=1}^n\beta_i-1}\exp(-x)\,dx \\ &= \sigma^{\sum_{i=1}^n \beta_i}\prod_{i=1}^n \Gamma(\beta_i)\,\Gamma\left(\alpha+\sum_{i=1}^n\beta_i\right).\end{aligned} \tag{19}$$

Theorem 6. *If $(Y_1,\ldots,Y_n) \sim \mathrm{MM}(\alpha,\beta_1,\ldots,\beta_n,\sigma)$, then for $1 \le s \le n$, $(Y_1,\ldots,Y_s) \sim \mathrm{MM}(\alpha+\sum_{i=s+1}^n\beta_i,\beta_1,\ldots,\beta_s,\sigma)$.*

Proof. By using (18), the joint density of $Y_1,\ldots,Y_n$ can be written as

$$C(\alpha,\beta_1,\ldots,\beta_n,\sigma) \cdot \prod_{i=1}^n y_i^{\beta_i-1}\int_0^\infty x^{\alpha-1}\exp\left(-x-\frac{\sum_{i=1}^n y_i}{\sigma x}\right)dx, \tag{20}$$

where $y_i > 0, i = 1,\ldots,n$. Now, integrating $y_{s+1},\ldots,y_n$ in the above density, we get the marginal density of $Y_1,\ldots,Y_s$ as

$$C(\alpha,\beta_1,\ldots,\beta_n,\sigma)\,\sigma^{\sum_{i=s+1}^n\beta_i}\prod_{i=s+1}^n\Gamma(\beta_i)\prod_{i=1}^s y_i^{\beta_i-1} \times \int_0^\infty x^{\alpha+\sum_{i=s+1}^n\beta_i-1}\exp\left(-x-\frac{\sum_{i=1}^s y_i}{\sigma x}\right)dx. \tag{21}$$

Now, writing the above integral in terms of extended gamma function and simplifying, we get the desired result.

Corollary 7. *If $(Y_1,\ldots,Y_n) \sim \mathrm{MM}(\alpha,\beta_1,\ldots,\beta_n,\sigma)$, then for $s = 1,\ldots,n$ the marginal distribution of Y_s is Macdonald, $Y_s \sim \mathrm{M}(\alpha+\sum_{i(\neq s)=1}^n\beta_i,\beta_s,\sigma)$.*

Theorem 8. *Let $(Y_1,\ldots,Y_n) \sim \mathrm{MM}(\alpha,\beta_1,\ldots,\beta_n,\sigma)$ and define $Z_i = Y_i/Z$ for $i = 1,\ldots,n-1$ and $Z = \sum_{j=1}^n Y_j$. Then, $(Z_1,\ldots,Z_{n-1})$ and Z are independent, $(Z_1,\ldots,Z_{n-1}) \sim \mathrm{D1}(\beta_1,\ldots,\beta_{n-1};\beta_n)$, and $Z \sim \mathrm{M}(\alpha,\sum_{i=1}^n\beta_i,\sigma)$.*

Proof. Substituting $z_i = y_i/z$ for $i = 1,\ldots,n-1$ and $z = \sum_{j=1}^n y_j$ with the Jacobian $J(y_1,\ldots,y_{n-1},y_n \to z_1,\ldots,z_{n-1},z) = z^{n-1}$ in (16), the joint density of $Z_1,\ldots,Z_{n-1}$ and Z is derived as

$$C(\alpha,\beta_1,\ldots,\beta_n,\sigma)\,z^{\sum_{i=1}^n\beta_i-1}\Gamma\left(\alpha,\frac{z}{\sigma}\right)\prod_{j=1}^{n-1}z_j^{\beta_j-1} \cdot \left(1-\sum_{j=1}^{n-1}z_j\right)^{\beta_n-1}, \tag{22}$$

where $z_i > 0$, $i = 1,\ldots,n-1$, $\sum_{i=1}^{n-1}z_i < 1$, and $z > 0$. From the above factorization, it is clear that $(Z_1,\ldots,Z_{n-1})$ and Z are independent, $(Z_1,\ldots,Z_{n-1}) \sim \mathrm{D1}(\beta_1,\ldots,\beta_{n-1};\beta_n)$, and $Z \sim \mathrm{M}(\alpha,\sum_{i=1}^n\beta_i,\sigma)$. □

Corollary 9. *If $(Y_1,\ldots,Y_n) \sim \mathrm{MM}(\alpha,\beta_1,\ldots,\beta_n,\sigma)$, then*

$$\frac{\sum_{i=1}^s Y_i}{\sum_{j=1}^n Y_j} \sim \mathrm{B1}\left(\sum_{i=1}^s\beta_i,\sum_{i=s+1}^n\beta_i\right), \quad s < n. \tag{23}$$

4. Joint Moments

We derive the joint moments of random variables jointly distributed as multivariate Macdonald. These moments will facilitate for us to compute several expected values such as mean and variance.

Using (16) and (19), the joint moments of $Y_1,\ldots,Y_n$ are obtained as

$$\begin{aligned}E(Y_1^{r_1}\cdots Y_n^{r_n}) &= C(\alpha,\beta_1,\ldots,\beta_n,\sigma) \cdot \int_0^\infty\cdots\int_0^\infty\prod_{i=1}^n y_i^{\beta_i+r_i-1}\Gamma\left(\alpha,\frac{\sum_{i=1}^n y_i}{\sigma}\right)dy_1\cdots dy_n \\ &= \frac{C(\alpha,\beta_1,\ldots,\beta_n,\sigma)}{C(\alpha,\beta_1+r_1,\ldots,\beta_n+r_n,\sigma)} \\ &= \frac{\sigma^{\sum_{i=1}^n r_i}\prod_{i=1}^n\Gamma(\beta_i+r_i)\,\Gamma[\alpha+\sum_{i=1}^n(\beta_i+r_i)]}{\prod_{i=1}^n\Gamma(\beta_i)\,\Gamma(\alpha+\sum_{i=1}^n\beta_i)}.\end{aligned} \tag{24}$$

From the above expression the (r,s)th moment of Y_i and Y_j, for $i \neq j$, is given by

$$E(Y_i^rY_j^s) = \frac{\sigma^{r+s}\Gamma(\beta_i+r)\,\Gamma(\beta_j+s)\,\Gamma(\alpha+\beta_i+\beta_j+r+s)}{\Gamma(\beta_i)\,\Gamma(\beta_j)\,\Gamma(\alpha+\beta_i+\beta_j)}. \tag{25}$$

Substituting $s = r$, $\beta_i = \beta$, and $\beta_j = \beta + 1/2$ in (25) and using the duplication formula, the rth moment of $2\sqrt{Y_iY_j}$ is obtained as

$$E\left[\left(2\sqrt{Y_iY_j}\right)^r\right] = \frac{\sigma^r\Gamma(2\beta+r)\,\Gamma(\alpha+2\beta+1/2+r)}{\Gamma(2\beta)\,\Gamma(\alpha+2\beta+1/2)}. \tag{26}$$

Now, comparing the above moment expression with (14), we can conclude that $2\sqrt{Y_iY_j} \sim \mathrm{M}(\alpha + 1/2, 2\beta, \sigma)$. For $s = -r$, the above expression reduces to

$$E\left(Y_i^r Y_j^{-r}\right) = \frac{\Gamma(\beta_i + r)\Gamma(\beta_j - r)}{\Gamma(\beta_i)\Gamma(\beta_j)}, \quad (27)$$

which shows that Y_i/Y_j has a standard beta type 2 distribution with parameters β_i and β_j. Substituting appropriately in (25), means and variances of Y_i and Y_j and the covariance between Y_i and Y_j are computed as

$$\begin{aligned}
E(Y_k) &= \sigma\beta_k(\alpha + \beta_i + \beta_j), \quad k = i, j, \\
E(Y_k^2) &= \sigma^2\beta_k(\beta_k + 1)(\alpha + \beta_i + \beta_j)(\alpha + \beta_i + \beta_j + 1), \quad k = i, j, \\
\mathrm{Var}(Y_k) &= \sigma^2\beta_k(\alpha + \beta_i + \beta_j)(\alpha + \beta_i + \beta_j + \beta_k + 1), \quad k = i, j, \\
E(Y_iY_j) &= \sigma^2\beta_i\beta_j(\alpha + \beta_i + \beta_j)(\alpha + \beta_i + \beta_j + 1), \\
\mathrm{Cov}(Y_i, Y_j) &= \sigma^2\beta_i\beta_j(\alpha + \beta_i + \beta_j).
\end{aligned} \quad (28)$$

The correlation coefficient between Y_i and Y_j is given by

$$\rho_{Y_i,Y_j} = \sqrt{\frac{\beta_i\beta_j}{(\alpha + 2\beta_i + \beta_j + 1)(\alpha + \beta_i + 2\beta_j + 1)}}. \quad (29)$$

Further, for $\beta_i = \beta_j = \beta$, the correlation coefficient between Y_i and Y_j is given by

$$\rho_{Y_i,Y_j} = \frac{\beta}{\alpha + 3\beta + 1}. \quad (30)$$

5. Factorizations

In this section, we give several factorizations of the multivariate Macdonald density.

Theorem 10. *Let $(Y_1, \ldots, Y_n) \sim \mathrm{MM}(\alpha, \beta_1, \ldots, \beta_n, \sigma)$. For $i = 1, \ldots, n-1$, define $U_i = \sum_{j=1}^i Y_j / \sum_{j=1}^{i+1} Y_j$ and $U_n = \sum_{j=1}^n Y_j$. Then, the random variables $U_1, \ldots, U_n$ are independent, $U_i \sim \mathrm{B1}(\sum_{j=1}^i \beta_j, \beta_{i+1})$, $i = 1, \ldots, n-1$, and $U_n \sim \mathrm{M}(\alpha, \sum_{i=1}^n \beta_i, \sigma)$.*

Proof. From the transformation given in the theorem, we obtain $y_1 = u_n\prod_{i=1}^{n-1} u_i$, $y_2 = u_n(1 - u_1)\prod_{i=2}^{n-1} u_i, \ldots, y_{n-1} = u_n(1 - u_{n-2})u_{n-1}$, and $y_n = u_n(1 - u_{n-1})$ with the Jacobian $J(y_1, \ldots, y_n \to u_1, \ldots, u_n) = \prod_{i=2}^n u_i^{i-1}$. Now, making appropriate substitutions in the joint p.d.f. of $Y_1, \ldots, Y_n$, we obtain

$$C(\alpha, \beta_1, \ldots, \beta_n, \sigma)\left(u_n\prod_{i=1}^{n-1} u_i\right)^{\beta_1 - 1} \cdot \prod_{j=2}^n \left[u_n(1 - u_{j-1})\prod_{i=j}^{n-1} u_i\right]^{\beta_j - 1} \Gamma\left(\alpha; \frac{u_n}{\sigma}\right)\prod_{i=2}^n u_i^{i-1}. \quad (31)$$

Further, writing

$$C(\alpha, \beta_1, \ldots, \beta_n, \sigma) = \left[\prod_{j=1}^{n-1} B\left(\sum_{i=1}^j \beta_i, \beta_{j+1}\right)\right]^{-1} \times \left[\sigma^{\sum_{i=1}^n \beta_i}\Gamma\left(\sum_{i=1}^n \beta_i\right)\Gamma\left(\alpha + \sum_{i=1}^n \beta_i\right)\right]^{-1}, \quad (32)$$

the above expression is simplified as

$$\left[\prod_{j=1}^{n-1} \frac{u_j^{\sum_{i=1}^j \beta_i - 1}(1 - u_j)^{\beta_{j+1} - 1}}{B(\sum_{i=1}^j \beta_i, \beta_{j+1})}\right] \cdot \frac{u_n^{\sum_{i=1}^n \beta_i - 1}\Gamma(\alpha; u_n/\sigma)}{\sigma^{\sum_{i=1}^n \beta_i}\Gamma(\sum_{i=1}^n \beta_i)\Gamma(\alpha + \sum_{i=1}^n \beta_i)}, \quad (33)$$

where $0 < u_1, \ldots, u_{n-1} < 1$ and $u_n > 0$. Now, from the above factorization, we get the result.

Theorem 11. *Let $(Y_1, \ldots, Y_n) \sim \mathrm{MM}(\alpha, \beta_1, \ldots, \beta_n, \sigma)$. Define $Z_n = \sum_{j=1}^n Y_j$ and $Z_i = Y_{i+1}/\sum_{j=1}^i Y_j$, for $i = 1, \ldots, n-1$. Then, $Z_1, \ldots, Z_n$ are independent, $Z_i \sim \mathrm{B2}(\beta_{i+1}, \sum_{j=1}^i \beta_j)$, $i = 1, \ldots, n-1$, and $Z_n \sim \mathrm{M}(\alpha, \sum_{i=1}^n \beta_i, \sigma)$.*

Proof. This result is obtained from Theorem 10, by observing that $Z_i = (1 - U_i)/U_i$, for $i = 1, \ldots, n-1$, $Z_n = U_n$, and $(1 - U_i)/U_i \sim \mathrm{B2}(\beta_{i+1}, \sum_{j=1}^i \beta_j)$, where $U_i \sim \mathrm{B1}(\sum_{j=1}^i \beta_j, \beta_{i+1})$. □

Theorem 12. *Let $(Y_1, \ldots, Y_n) \sim \mathrm{MM}(\alpha, \beta_1, \ldots, \beta_n, \sigma)$. Define $W_n = \sum_{j=1}^n Y_j$ and $W_i = \sum_{j=1}^i Y_j/Y_{i+1}$, for $i = 1, \ldots, n-1$. Then, $W_1, \ldots, W_n$, are independent, $W_i \sim \mathrm{B2}(\sum_{j=1}^i \beta_j, \beta_{i+1})$ for $i = 1, \ldots, n-1$ and $W_n \sim \mathrm{M}(\alpha, \sum_{i=1}^n \beta_i, \sigma)$.*

Proof. The result is obtained from Theorem 11 by taking into account that $W_i = 1/Z_i$ for $i = 1, \ldots, n-1$, $W_n = Z_n$, and $1/Z_i \sim \mathrm{B2}(\sum_{j=1}^i \beta_j, \beta_{i+1})$, where $Z_i \sim \mathrm{B2}(\beta_{i+1}, \sum_{j=1}^i \beta_j)$. □

Theorem 13. *Let $(Y_1, \ldots, Y_n) \sim \mathrm{MM}(\alpha, \beta_1, \ldots, \beta_n, \sigma)$. Define $V_n = \sum_{j=1}^n Y_j$ and for $i = 1, \ldots, n-1$ $V_i = Y_i/\sum_{j=i}^n Y_j$. Then, $V_1, \ldots, V_n$ are independent, $V_i \sim \mathrm{B1}(\beta_i, \sum_{j=i+1}^n \beta_j)$, for $i = 1, \ldots, n-1$ and $V_n \sim \mathrm{M}(\alpha, \sum_{i=1}^n \beta_i, \sigma)$.*

Proof. Making the substitutions $y_1 = v_n v_1$, $y_2 = v_n v_2(1-v_1), \ldots, y_{n-1} = v_n v_{n-1}(1-v_1)\cdots(1-v_{n-2})$, and $y_n = v_n(1-v_1)\cdots(1-v_{n-1})$ with the Jacobian $J(y_1,\ldots,y_n \to v_1,\ldots,v_n) = v_n^{n-1}\prod_{i=1}^{n-2}(1-v_i)^{n-i-1}$ in (16), we obtain the joint p.d.f. of $V_1,\ldots,V_n$ as

$$C(\alpha,\beta_1,\ldots,\beta_n,\sigma)\prod_{i=1}^{n-1}\left[v_n v_i \prod_{j=1}^{i-1}(1-v_j)\right]^{\beta_i-1} \cdot \left[v_n \prod_{j=1}^{n-1}(1-v_j)\right]^{\beta_n-1}\Gamma\left(\alpha;\frac{v_n}{\sigma}\right) \cdot v_n^{n-1}\prod_{i=1}^{n-2}(1-v_i)^{n-i-1}, \tag{34}$$

which can be rewritten as

$$\left[\prod_{i=1}^{n-1}\frac{v_i^{\beta_i-1}(1-v_i)^{\sum_{j=i+1}^n\beta_j-1}}{B\left(\beta_i,\sum_{j=i+1}^n\beta_j\right)}\right] \cdot \frac{u_n^{\sum_{i=1}^n\beta_i-1}\Gamma(\alpha;v_n/\sigma)}{\sigma^{\sum_{i=1}^n\beta_i}\Gamma\left(\sum_{i=1}^n\beta_i\right)\Gamma\left(\alpha+\sum_{i=1}^n\beta_i\right)}. \tag{35}$$

Now, the desired result follows from the above factorization. □

Theorem 14. *Let* $(Y_1,\ldots,Y_n) \sim \mathrm{MM}(\alpha,\beta_1,\ldots,\beta_n,\sigma)$. *Define* $Z_n = \sum_{j=1}^n Y_j$ *and* $Z_i = Y_i/\sum_{j=i+1}^n Y_j$, *for* $i = 1,\ldots,n-1$. *Then,* $Z_1,\ldots,Z_n$, *are independent,* $Z_i \sim \mathrm{B2}(\beta_i,\sum_{j=i+1}^n\beta_j)$, *for* $i = 1,\ldots,n-1$ *and* $Z_n \sim \mathrm{M}(\alpha,\sum_{i=1}^n\beta_i,\sigma)$.

Proof. The result is obtained from Theorem 13, by noting that $Z_i = V_i/(1-V_i)$ for $i = 1,\ldots,n-1$, $Z_n = V_n$ and $V_i/(1-V_i) \sim \mathrm{B2}(\beta_i,\sum_{j=i+1}^n\beta_j)$ for $V_i \sim \mathrm{B1}(\beta_i,\sum_{j=i+1}^n\beta_j)$.

Theorem 15. *Let* $(Y_1,\ldots,Y_n) \sim \mathrm{MM}(\alpha,\beta_1,\ldots,\beta_n,\sigma)$. *Define* $W_n = \sum_{j=1}^n Y_j$ *and* $W_i = \sum_{j=i+1}^n Y_j/Y_i$, *for* $i = 1,\ldots,n-1$. *Then,* $W_1,\ldots,W_n$ *are independent,* $W_i \sim \mathrm{B2}(\sum_{j=i+1}^n\beta_j,\beta_i)$ *for* $i = 1,\ldots,n-1$ *and* $W_n \sim \mathrm{M}(\alpha,\sum_{i=1}^n\beta_i,\sigma)$.

Proof. This result follows from Theorem 14, by observing that $W_i = 1/Z_i$ for $i = 1,\ldots,n-1$, $W_n = Z_n$ and $1/Z_i \sim \mathrm{B2}(\sum_{j=i+1}^n\beta_j,\beta_i)$, where $Z_i \sim \mathrm{B2}(\beta_i,\sum_{j=i+1}^n\beta_j)$.

6. The Multivariate Macdonald-Gamma Distribution

We propose a multivariate generalization of the Macdonald-gamma distribution as follows.

Definition 16. The random variables $X, Y_1,\ldots,Y_n$ are said to have a multivariate Macdonald-gamma distribution with parameters $\alpha,\beta_1,\ldots,\beta_n$, and σ, denoted as $(X,Y_1,\ldots,Y_n) \sim \mathrm{MMG}(\alpha,\beta_1,\ldots,\beta_n,\sigma)$, if their joint p.d.f. is

$$C(\alpha,\beta_1,\ldots,\beta_n,\sigma)\,x^{\alpha-1}\prod_{i=1}^n y_i^{\beta_i-1}\exp\left(-x-\frac{\sum_{i=1}^n y_i}{\sigma x}\right), \quad x>0,\ y_i>0,\ i=1,\ldots,n, \tag{36}$$

where $\beta_i > 0$, $i = 1,\ldots,n$, $\alpha+\sum_{i=1}^n\beta_i > 0$, $\sigma > 0$, and the normalizing constant $C(\alpha,\beta_1,\ldots,\beta_n,\sigma)$ is given by (19).

Theorem 17. *If* $(X,Y_1,\ldots,Y_n) \sim \mathrm{MMG}(\alpha,\beta_1,\ldots,\beta_n,\sigma)$, *then for* $1 \le s \le n$, $(X,Y_1,\ldots,Y_s) \sim \mathrm{MMG}(\alpha+\sum_{i=s+1}^n\beta_i,\beta_1,\ldots,\beta_s,\sigma)$.

Proof. Integrating $y_{s+1},\ldots,y_n$ in (36), we get the marginal density of $X, Y_1,\ldots,Y_s$ as

$$C\left(\alpha+\sum_{i=s+1}^n\beta_i,\beta_1,\ldots,\beta_s,\sigma\right)x^{\alpha+\sum_{i=s+1}^n\beta_i-1}\prod_{i=1}^s y_i^{\beta_i-1} \cdot \exp\left(-x-\frac{\sum_{i=1}^s y_i}{\sigma x}\right), \tag{37}$$

where $x > 0$ and $y_i > 0$, $i = 1,\ldots,s$.

Corollary 18. *If* $(X,Y_1,\ldots,Y_n) \sim \mathrm{MMG}(\alpha,\beta_1,\ldots,\beta_n,\sigma)$, *then for* $s = 1,\ldots,n$ *the joint distribution of* X *and* Y_s *is Macdonald-gamma,* $(X,Y_s) \sim \mathrm{MG}(\alpha+\sum_{i(\ne s)=1}^n\beta_i,\beta_s,\sigma)$.

Theorem 19. *If* $(X,Y_1,\ldots,Y_n) \sim \mathrm{MMG}(\alpha,\beta_1,\ldots,\beta_n,\sigma)$, *then for* $1 \le s \le n$, $(Y_1,\ldots,Y_s) \sim \mathrm{MM}(\alpha+\sum_{i=s+1}^n\beta_i,\beta_1,\ldots,\beta_s,\sigma)$.

Proof. Integrating x in (37) by using the definition of the extended gamma function, we obtain the desired result.

Corollary 20. *If* $(X,Y_1,\ldots,Y_n) \sim \mathrm{MMG}(\alpha,\beta_1,\ldots,\beta_n,\sigma)$, *then for* $s = 1,\ldots,n$ *the marginal distribution of* Y_s *is Macdonald,* $Y_s \sim \mathrm{M}(\alpha+\sum_{i(\ne s)=1}^n\beta_i,\beta_s,\sigma)$.

Theorem 21. *Let* $(X,Y_1,\ldots,Y_n) \sim \mathrm{MMG}(\alpha,\beta_1,\ldots,\beta_n,\sigma)$ *and define* $Z_i = Y_i/Z$ *for* $i = 1,\ldots,n-1$ *and* $Z = \sum_{j=1}^n Y_j$. *Then,* $(Z_1,\ldots,Z_{n-1})$ *and* (X,Z) *are independent,* $(Z_1,\ldots,Z_{n-1}) \sim \mathrm{D1}(\beta_1,\ldots,\beta_{n-1};\beta_n)$ *and* $(X,Z) \sim \mathrm{MG}(\alpha,\sum_{i=1}^n\beta_i,\sigma)$.

Proof. Substituting $z_i = y_i/z$ for $i = 1,\ldots,n-1$ and $z = \sum_{j=1}^n y_j$ with the Jacobian $J(y_1,\ldots,y_{n-1},y_n \to z_1,\ldots,z_{n-1},z) = z^{n-1}$ in (36), the joint density of $Z_1,\ldots,Z_{n-1}$, X and Z is derived as

$$C(\alpha,\beta_1,\ldots,\beta_n,\sigma)\,x^{\alpha-1}z^{\sum_{i=1}^n\beta_i-1}\exp\left(-x-\frac{z}{\sigma x}\right) \cdot \prod_{j=1}^{n-1}z_j^{\beta_j-1}\left(1-\sum_{j=1}^{n-1}z_j\right)^{\beta_n-1}, \tag{38}$$

where $x > 0$, $z_i > 0$, $i = 1,\ldots,n-1$, $\sum_{i=1}^{n-1}z_i < 1$, and $z > 0$. From the above factorization, it is clear that

$(Z_1, \ldots, Z_{n-1})$ and (X, Z) are independent, $(Z_1, \ldots, Z_{n-1}) \sim \mathrm{D1}(\beta_1, \ldots, \beta_{n-1}; \beta_n)$, and $(X, Z) \sim \mathrm{MG}(\alpha, \sum_{i=1}^n \beta_i, \sigma)$.

Corollary 22. *If $(X, Y_1, \ldots, Y_n) \sim \mathrm{MMG}(\alpha, \beta_1, \ldots, \beta_n, \sigma)$, then*

$$\frac{\sum_{i=1}^s Y_i}{\sum_{j=1}^n Y_j} \sim \mathrm{B1}\left(\sum_{i=1}^s \alpha_i, \sum_{i=s+1}^n \alpha_i\right), \quad s < n. \tag{39}$$

Corollary 23. *If $(X, Y_1, \ldots, Y_n) \sim \mathrm{MMG}(\alpha, \beta_1, \ldots, \beta_n, \sigma)$ and $Z_i = Y_i/Z$ for $i = 1, \ldots, n-1$ and $Z = \sum_{j=1}^n Y_j$. Then, $(Z_1, \ldots, Z_{n-1})$ and Z/X are independent, $(Z_1, \ldots, Z_{n-1}) \sim \mathrm{D1}(\beta_1, \ldots, \beta_{n-1}; \beta_n)$ and $Z/X \sim \mathrm{G}(\sum_{i=1}^n \beta_i, \sigma)$.*

Theorem 24. *If $(X, Y_1, \ldots, Y_n) \sim \mathrm{MMG}(\alpha, \beta_1, \ldots, \beta_n, \sigma)$, then the p.d.f. of $S = X + \sum_{j=1}^n Y_j$ is given by*

$$\begin{aligned} &\frac{\Gamma(\alpha+1)\, s^{\alpha+\sum_{i=1}^n \beta_i - 1} \exp(-s)}{\sigma^{\sum_{i=1}^n \beta_i} \Gamma(\sum_{i=1}^n \beta_i) \Gamma(\alpha + \sum_{i=1}^n \beta_i)} \\ &\cdot \sum_{m=0}^{\infty} L_m\left(\frac{1}{\sigma}\right) \frac{\Gamma(\sum_{i=1}^n \beta_i + m)}{\Gamma(\alpha + \sum_{i=1}^n \beta_i + m + 1)} \\ &\times {}_1F_1\left(\sum_{i=1}^n \beta_i + m; \alpha + \sum_{i=1}^n \beta_i + m + 1; s\right), \quad s > 0, \end{aligned} \tag{40}$$

where $\alpha + 1 > 0$ and $\sigma > 0$.

Proof. From Theorem 21, the joint distribution of X and $\sum_{j=1}^n Y_j$ is Macdonald-gamma, $(X, \sum_{j=1}^n Y_j) \sim \mathrm{MG}(\alpha, \sum_{j=1}^n \beta_j, \sigma)$. Further, from Theorem 4, we get the distribution of S.

Using (36) and (19), the joint moments of $X, Y_1, \ldots, Y_n$ are obtained as

$$\begin{aligned} E(X^s Y_1^{r_1} \cdots Y_n^{r_n}) &= C(\alpha, \beta_1, \ldots, \beta_n, \sigma) \int_0^\infty x^{\alpha+s-1} \\ &\times \int_0^\infty \cdots \int_0^\infty \prod_{i=1}^n y_i^{\beta_i + r_i - 1} \\ &\cdot \exp\left(-x - \frac{\sum_{i=1}^n y_i}{\sigma x}\right) dy_1 \cdots dy_n dx \\ &= \frac{C(\alpha + s, \beta_1 + r_1, \ldots, \beta_n + r_n, \sigma)}{C(\alpha, \beta_1, \ldots, \beta_n, \sigma)} \\ &= \frac{\sigma^{\sum_{i=1}^n r_i} \prod_{i=1}^n \Gamma(\beta_i + r_i) \Gamma[\alpha + \sum_{i=1}^n (\beta_i + r_i) + s]}{\prod_{i=1}^n \Gamma(\beta_i) \Gamma(\alpha + \sum_{i=1}^n \beta_i)}. \end{aligned} \tag{41}$$

Next, we give several factorizations of the multivariate Macdonald-gamma density.

Theorem 25. *Let $(X, Y_1, \ldots, Y_n) \sim \mathrm{MMG}(\alpha, \beta_1, \ldots, \beta_n, \sigma)$. For $i = 1, \ldots, n-1$, define $U_i = \sum_{j=1}^i Y_j / \sum_{j=1}^{i+1} Y_j$ and $U_n = \sum_{j=1}^n Y_j$. Then, $U_1, \ldots, U_{n-1}$ and (X, U_n) are independent, $U_i \sim \mathrm{B1}(\sum_{j=1}^i \beta_j, \beta_{i+1})$, $i = 1, \ldots, n-1$, and $(X, U_n) \sim \mathrm{MG}(\alpha, \sum_{i=1}^n \beta_i, \sigma)$.*

Proof. Similar to the proof of Theorem 10.

Theorem 26. *Let $(X, Y_1, \ldots, Y_n) \sim \mathrm{MM}(\alpha, \beta_1, \ldots, \beta_n, \sigma)$. Define $Z_n = \sum_{j=1}^n Y_j$ and $Z_i = Y_{i+1} / \sum_{j=1}^i Y_j$, for $i = 1, \ldots, n-1$. Then, $Z_1, \ldots, Z_{n-1}$ and (X, Z_n) are independent, $Z_i \sim \mathrm{B2}(\beta_{i+1}, \sum_{j=1}^i \beta_j)$, $i = 1, \ldots, n-1$, and $(X, Z_n) \sim \mathrm{MG}(\alpha, \sum_{i=1}^n \beta_i, \sigma)$.*

Proof. This result is obtained from Theorem 25, by observing that $Z_i = (1 - U_i)/U_i$, for $i = 1, \ldots, n-1$, $Z_n = U_n$, and $(1 - U_i)/U_i \sim \mathrm{B2}(\beta_{i+1}, \sum_{j=1}^i \beta_j)$, where $U_i \sim \mathrm{B1}(\sum_{j=1}^i \beta_j, \beta_{i+1})$.

Theorem 27. *Let $(X, Y_1, \ldots, Y_n) \sim \mathrm{MMG}(\alpha, \beta_1, \ldots, \beta_n, \sigma)$. Define $W_n = \sum_{j=1}^n Y_j$ and $W_i = \sum_{j=1}^i Y_j / Y_{i+1}$, for $i = 1, \ldots, n-1$. Then, $W_1, \ldots, W_{n-1}$, and (X, W_n) are independent, $W_i \sim \mathrm{B2}(\sum_{j=1}^i \beta_j, \beta_{i+1})$ for $i = 1, \ldots, n-1$, and $(X, W_n) \sim \mathrm{MG}(\alpha, \sum_{i=1}^n \beta_i, \sigma)$.*

Proof. The result is obtained from Theorem 26 by taking into account that $W_i = 1/Z_i$ for $i = 1, \ldots, n-1$, $W_n = Z_n$, and $1/Z_i \sim \mathrm{B2}(\sum_{j=1}^i \beta_j, \beta_{i+1})$, where $Z_i \sim \mathrm{B2}(\beta_{i+1}, \sum_{j=1}^i \beta_j)$.

Theorem 28. *Let $(X, Y_1, \ldots, Y_n) \sim \mathrm{MM}(\alpha, \beta_1, \ldots, \beta_n, \sigma)$. Define $V_n = \sum_{j=1}^n Y_j$ and for $i = 1, \ldots, n-1$ $V_i = Y_i / \sum_{j=i}^n Y_j$. Then, $V_1, \ldots, V_{n-1}$ and (X, V_n) are independent, $V_i \sim \mathrm{B1}(\beta, \sum_{j=i+1}^n \beta_j)$, for $i = 1, \ldots, n-1$, and $(X, V_n) \sim \mathrm{MG}(\alpha, \sum_{i=1}^n \beta_i, \sigma)$.*

Proof. Similar to the proof of Theorem 13.

Theorem 29. *Let $(X, Y_1, \ldots, Y_n) \sim \mathrm{MM}(\alpha, \beta_1, \ldots, \beta_n, \sigma)$. Define $Z_n = \sum_{j=1}^n Y_j$ and $Z_i = Y_i / \sum_{j=i+1}^n Y_j$, for $i = 1, \ldots, n-1$. Then, $Z_1, \ldots, Z_{n-1}$ and (X, Z_n) are independent, $Z_i \sim \mathrm{B2}(\beta_i, \sum_{j=i+1}^n \beta_j)$, for $i = 1, \ldots, n-1$, and $(X, Z_n) \sim \mathrm{MG}(\alpha, \sum_{i=1}^n \beta_i, \sigma)$.*

Proof. The result is obtained from Theorem 28, by noting that $Z_i = V_i/(1 - V_i)$ for $i = 1, \ldots, n-1$, $Z_n = V_n$, and $V_i/(1 - V_i) \sim \mathrm{B2}(\beta_i, \sum_{j=i+1}^n \beta_j)$ for $V_i \sim \mathrm{B1}(\beta_i, \sum_{j=i+1}^n \beta_j)$.

Theorem 30. *Let $(X, Y_1, \ldots, Y_n) \sim \mathrm{MM}(\alpha, \beta_1, \ldots, \beta_n, \sigma)$. Define $W_n = \sum_{j=1}^n Y_j$ and $W_i = \sum_{j=i+1}^n Y_j / Y_i$, for $i = 1, \ldots, n-1$. Then, $W_1, \ldots, W_{n-1}$ and (X, W_n) are independent, $W_i \sim \mathrm{B2}(\sum_{j=i+1}^n \beta_j, \beta_i)$ for $i = 1, \ldots, n-1$ and $(X, W_n) \sim \mathrm{MG}(\alpha, \sum_{i=1}^n \beta_i, \sigma)$.*

Proof. This result follows from Theorem 29, by observing that $W_i = 1/Z_i$ for $i = 1, \ldots, n-1$, $W_n = Z_n$, and $1/Z_i \sim \mathrm{B2}(\sum_{j=i+1}^n \beta_j, \beta_i)$, where $Z_i \sim \mathrm{B2}(\beta_i, \sum_{j=i+1}^n \beta_j)$.

Competing Interests

The authors declare that there is no conflict of interests regarding the publication of this article.

Acknowledgments

The research work of Daya K. Nagar and Luz Estela Sánchez was supported by the Sistema Universitario de Investigación, Universidad de Antioquia, under the Project no. IN10231CE.

References

[1] D. K. Nagar, A. Roldán-Correa, and A. K. Gupta, "Extended matrix variate gamma and beta functions," *Journal of Multivariate Analysis*, vol. 122, pp. 53–69, 2013.

[2] D. K. Nagar, A. Roldán-Correa, and A. K. Gupta, "Matrix variate Macdonald distribution," *Communications in Statistics: Theory and Methods*, vol. 45, no. 5, pp. 1311–1328, 2016.

[3] M. Aslam Chaudhry and S. M. Zubair, "On the decomposition of generalized incomplete gamma functions with applications to Fourier transforms," *Journal of Computational and Applied Mathematics*, vol. 55, no. 1, pp. 99–123, 1994.

[4] M. Aslam Chaudhry and S. M. Zubair, "Generalized incomplete gamma functions with applications," *Journal of Computational and Applied Mathematics*, vol. 55, no. 1, pp. 99–124, 1994.

[5] M. Aslam Chaudhry and S. M. Zubair, "Extended gamma and digamma functions," *Fractional Calculus & Applied Analysis*, vol. 4, no. 3, pp. 303–324, 2001.

[6] M. Aslam Chaudhry and S. M. Zubair, "Extended incomplete gamma functions with applications," *Journal of Mathematical Analysis and Applications*, vol. 274, no. 2, pp. 725–745, 2002.

[7] M. Aslam Chaudhry and S. M. Zubair, *On a Class of Incomplete Gamma Functions with Applications*, Chapman & Hall/CRC, Boca Raton, Fla, USA, 2002.

[8] M. Aslam Chaudhry and S. M. Zubair, "On an extension of generalized incomplete gamma functions with applications," *Journal of the Australian Mathematical Society, Series B: Applied Mathematics*, vol. 37, no. 3, pp. 392–405, 1996.

[9] D. K. Nagar, E. Zarrazola, and L. E. Sánchez, "Product and ratio of Macdonald random variables," *International Journal of Mathematical Analysis*, vol. 10, no. 13, pp. 639–649, 2016.

[10] N. L. Johnson, S. Kotz, and N. Balakrishnan, *Continuous Univariate Distributions. Vol. 1*, Wiley Series in Probability and Mathematical Statistics: Applied Probability and Statistics. A Wiley-Interscience Publication, John Wiley & Sons, New York, NY, USA, 1994.

[11] S. Kotz, N. Balakrishnan, and N. L. Johnson, *Continuous Multivariate Distributions: Models and Applications*, vol. 1 of *Wiley Series in Probability and Statistics*, John Wiley & Sons, New York, NY, USA, 2nd edition, 2000.

[12] A. K. Gupta and D. K. Nagar, *Matrix Variate Distributions*, Chapman & Hall/CRC, Boca Raton, Fla, USA, 2000.

[13] S. L. Kalla, B. N. Al-Saqabi, and H. G. Khajah, "A unified form of gamma-type distributions," *Applied Mathematics and Computation*, vol. 118, no. 2-3, pp. 175–187, 2001.

[14] F. Al-Musallam and S. L. Kalla, "Asymptotic expansions for generalized gamma and incomplete gamma functions," *Applicable Analysis*, vol. 66, no. 1-2, pp. 173–187, 1997.

[15] F. Al-Musallam and S. L. Kalla, "Further results on a generalized gamma function occurring in diffraction theory," *Integral Transforms and Special Functions*, vol. 7, no. 3-4, pp. 175–190, 1998.

[16] A. K. Gupta, D. K. Nagar, and L. Estela Sánchez, "Properties of matrix variate confluent hypergeometric function distribution," *Journal of Probability and Statistics*, vol. 2016, Article ID 2374907, 12 pages, 2016.

[17] D. K. Nagar, E. Zarrazola, and L. E. Sánchez, "A bivariate distribution whose marginal laws are gamma and Macdonald," *International Journal of Mathematical Analysis*, vol. 10, no. 10, pp. 455–467, 2016.

[18] I. S. Gradshteyn and I. M. Ryzhik, *Table of Integrals, Series, and Products*, Elsevier/Academic Press, Amsterdam, The Netherlands, 8th edition, Translation edited and with a preface by Daniel Zwillinger and Victor Moll, 2015.

[19] N. L. Johnson, S. Kotz, and N. Balakrishnan, *Continuous Univariate Distributions*, vol. 2 of *Wiley Series in Probability and Mathematical Statistics: Applied Probability and Statistics. A Wiley-Interscience Publication*, John Wiley & Sons, New York, NY, USA, 1995.

13

Upper Bound of the Generalized p Value for the Population Variances of Lognormal Distributions with Known Coefficients of Variation

Rada Somkhuean, Sa-aat Niwitpong, and Suparat Niwitpong

Department of Applied Statistics, Faculty of Applied Science, King Mongkut's University of Technology North Bangkok, Bangkok 10800, Thailand

Correspondence should be addressed to Sa-aat Niwitpong; sa-aat.n@sci.kmutnb.ac.th

Academic Editor: Shein-chung Chow

This paper presents an upper bound for each of the generalized p values for testing the one population variance, the difference between two population variances, and the ratio of population variances for lognormal distribution when coefficients of variation are known. For each of the proposed generalized p values, we derive a closed form expression of the upper bound of the generalized p value. Numerical computations illustrate the theoretical results.

1. Introduction

The problem of statistical inference for the population variances has been widely discussed by various authors; see, for example, Singh et al. [1], Agrawal and Sthapit [2], Arcos Cebrián and Rueda García [3], and Arcos et al. [4]. Kadilar and Cingi [5] proposed some ratio estimators for the population variance in simple and stratified random sampling. Cojbasic and Tomovic [6] proposed the bootstrap methods to construct the confidence intervals of the population variance for one sample and the difference of variances of two samples. Cojbasic and Loncar [7] proposed one-sided bootstrap method to construct the confidence intervals of the population variance of skewed distributions. Rajic et al. [8] proposed the new method for the testing one population variance and the difference of variances of two samples, based on t-statistics and bootstrap method. Singh and Malik [9] proposed a family of estimators for the population variance using auxiliary attributes. In this paper, we used the generalized p values, proposed by Tsui and Weerahandi [10] and Weerahandi [11], to find new generalized p values for testing: (1) one population variance, (2) the difference between two population variances, and (3) the ratio of population variances of lognormal distributions when coefficients of variation are known. This problem is analogous to the Behrens-Fisher problem; see, for example, Tang and Tsui [12] and Somkhuean et al. [13]. In Section 2, the outline for some basic steps to construct the generalized p value for testing hypothesis in this problem is presented. The process of deriving each of the upper bounds as mentioned above is presented in Section 3. The numerical results are shown in Section 4 and the conclusion is presented in Section 5.

2. Generalized p Values

The concept of the generalized p values has been introduced by Tsui and Weerahandi [10] and Weerahandi [11]. We briefly review this concept as follows.

Let X be a random variable with a density function $f(X \mid \xi)$, where $\xi = (\theta, \delta)$, θ is the parameter of interest, and δ is a nuisance parameter.

Suppose we want to test

$$H_0\colon \theta \le \theta_0 \quad \text{vs. } H_1\colon \theta > \theta_0, \tag{1}$$

where θ_0 is a specified quantity. Let x be a particular observed sample. The generalized test variable, $T(X, x, \xi)$, is required to satisfy the following conditions:

(A1) For fixed x and $\xi = (\theta, \delta)$, the distribution $T(X, x, \xi)$ is free from the nuisance parameter δ.

(A2) $t_{\text{obs}} = T(x, x, \xi)$ is free from any unknown parameter.

(A3) $T(X, x, \xi)$ is either stochastically increasing or decreasing in θ for any given t and fixed values of x and δ.

Under the above conditions, if $T(X, x, \xi)$ is a stochastically increasing test variable then the subset of space is extreme region C. For the one-sided hypothesis given above they defined a data-based extreme region C_x is of the form

$$C_x = \{X : T(X, x, \xi) \geq 0\}. \tag{2}$$

Given the observed sample x, the generalized p value is defined as

$$\begin{aligned} p(x) &= \sup_{\theta \in H_0} P(X \in C \mid \theta) \\ &= \sup_{\theta \in H_0} P(\{X : T(X, x, \xi) \geq 0\}), \end{aligned} \tag{3}$$

for further details and for several applications based on the generalized p value; we refer to the book by Weerahandi [11].

Moreover, Tsui and Weerahandi [10] used the generalized p value $p(x)$ for the Behrens-Fisher problem of testing the difference of two independent normal distribution means with possibly unequal variances. Later, Tang and Tusi [12] extended the works of Weerahandi [11], Gamage and Weerahandi [14] to derive the formula of the upper bound (r) of the generalized p value $p(x)$ which is in the form of (see also, e.g., Kabaila and Lloyd [15])

$$P(p(x) \leq r) \leq r. \tag{4}$$

In this paper, we also extend Tang and Tsui [12] work to find upper bound of each of the generalized p values, $p(x)$, for hypotheses testing of the one population variance, the difference between two population variances, and the ratio of population variances of lognormal distributions with known coefficients of variation.

3. Main Results for the Population Variance of Lognormal Distributions with Known Coefficients of Variation

Let $X_i = \{X_{ij}, X_{ij}, \ldots, X_{ij}\}$, for $(i = 1, 2,\ j = 1, 2, \ldots, n_i)$ be random samples having lognormal distribution and let $Y_{ij} = \ln(X_{ij}) \sim N(\mu_i, \sigma_i^2)$ where μ_i and σ_i^2 denote the mean and variance of Y_i, respectively. In particular, the mean, variance, and the coefficient of variation for lognormal distribution are, respectively, given by

$$\begin{aligned} E(X_i) &= \exp\left(\mu_i + \frac{\sigma_i^2}{2}\right), \\ \operatorname{var}(X_i) &= \exp\left(2\mu_i + \sigma_i^2\right)\left(\exp\left(\sigma_i^2\right) - 1\right), \\ \mathrm{CV}_i &= \sqrt{\exp\left(\sigma_i^2\right) - 1}, \end{aligned} \tag{5}$$

where CV_i denotes the coefficient of variation of X_i which is computed from $\sqrt{\operatorname{var}(X_i)}/E(X_i)$.

It is easy to see that

$$\begin{aligned} \tau_i &= \mathrm{CV} = \sqrt{\exp\left(\sigma_i^2\right) - 1}, \\ \tau_i^2 &= \exp\left(\sigma_i^2\right) - 1, \\ \tau_i^2 + 1 &= \exp\left(\sigma_i^2\right), \\ \sigma_i^2 &= \ln\left(\tau_i^2 + 1\right) \end{aligned} \tag{6}$$

and the parameters of interest are

$$\begin{aligned} \theta_1 &= \operatorname{var}(X_1) = \exp\left(2\mu_1 + \ln\left(\tau_1^2 + 1\right)\right)\tau_1^2 \\ &= k_1 \exp\left(2\mu_1 + l_1\right), \quad k_1 = \tau_1^2,\ l_1 = \ln\left(\tau_1^2 + 1\right), \\ \theta_2 &= \operatorname{var}(X_1) - \operatorname{var}(X_2) \\ &= \exp\left(2\mu_1 + \ln\left(\tau_1^2 + 1\right)\right)\tau_1^2 \\ &\quad - \exp\left(2\mu_2 + \ln\left(\tau_2^2 + 1\right)\right)\tau_2^2 \\ &= k_1 \exp\left(2\mu_1 + l_1\right) - k_2 \exp\left(2\mu_2 + l_2\right), \\ &\qquad k_i = \tau_i^2,\ l_i = \ln\left(\tau_i^2 + 1\right),\ i = 1, 2, \\ \theta_3 &= \frac{\operatorname{var}(X_1)}{\operatorname{var}(X_2)} = \frac{k_1 \exp\left(2\mu_1 + l_1\right)}{k_2 \exp\left(2\mu_2 + l_2\right)}. \end{aligned} \tag{7}$$

For testing the null hypothesis, H_{01}: $\theta_1 \leq \theta_{01}$ vs H_{a1}: $\theta_1 > \theta_{01}$, H_{02}: $\theta_2 \leq \theta_{02}$ vs H_{a2}: $\theta_2 > \theta_{02}$, and H_{03}: $\theta_3 \leq \theta_{03}$ vs H_{a3}: $\theta_3 > \theta_{03}$, the sufficient statistics involving these problems are $(\overline{Y}_i, S_i^2)$, where

$$\begin{aligned} \overline{Y}_i &= \frac{1}{n_i}\sum_{i=1}^{n_i} Y_i, \\ S_i^2 &= \frac{1}{n_i - 1}\sum_{i=1}^{n_i}\left(Y_{ij} - \overline{Y}_i\right)^2. \end{aligned} \tag{8}$$

It is known that the probability distributions of the statistics below are independent:

$$\begin{aligned} \overline{Y}_i &\sim N\left(\mu_i, \frac{\sigma_i^2}{n_i}\right), \\ U_i &= \frac{(n_i - 1) S_i^2}{\sigma_i^2} \sim \chi_{n_i - 1}^2. \end{aligned} \tag{9}$$

We denote $D = \{X_{11}, X_{12}, \ldots, X_{1n_1}\}$, $d = \{x_{11}, x_{12}, \ldots, x_{1n_1}\}$ and $Q = \{X_{11}, X_{12}, \ldots, X_{1n_1}, X_{21}, X_{22}, \ldots, X_{2n_2}\}$, $q = \{x_{11}, x_{12}, \ldots, x_{1n_1}, x_{21}, x_{22}, \ldots, x_{2n_2}\}$. Here d and q are the vector of the observed samples. Let $(\overline{y}_i, s_i^2)$ be the observed value of the sufficient statistic $(\overline{Y}_i, S_i^2)$. Following Tang and Tsui [12] and Somkhuean et al. [13], the repeated sampling, $(\overline{y}_i, s_i^2)$ follows the same probability distributions as (9).

Case 1. The hypothesis testing is

$$H_{01}: \theta_1 \le \theta_{01}, \qquad H_a: \theta_1 > \theta_{01}. \tag{10}$$

The parameter of the one population variance of lognormal distribution when the coefficient of variation is known is

$$\theta_1 = k_1 \exp(2\mu_1 + l_1). \tag{11}$$

Using the generalized test variable for θ_1 which is

$$\begin{aligned} T(X_1, x_1, \theta_1) &= k_1 \exp(2\mu_1 + l_1) \\ &= k_1 \exp\left(2\left(\overline{y}_1 - \frac{\overline{Y}_1 - \mu_1}{S_1/\sqrt{n_1}} \frac{s_1}{\sqrt{n_1}}\right) + l_1\right) \\ &= k_1 \exp\left(2\left(\overline{y}_1 - \frac{Z}{\sqrt{U_1}/\sqrt{n_1 - 1}} \frac{s_1}{\sqrt{n_1}}\right) + l_1\right), \end{aligned} \tag{12}$$

where $Z = \sqrt{n_1}(\overline{Y}_1 - \mu_1)/\sigma_1 \sim N(0,1)$, $U_1 = (n_1 - 1)S_1^2/\sigma_1^2 \sim \chi^2_{n_1 - 1}$.

It is easy to see that $T(X_1, x_1, \theta_1)$ in (12) satisfies conditions (A1)–(A3) in Section 2.

The generalized p value, $p(d)$, is defined, under the null hypothesis H_{01}, to be

$$\begin{aligned} p(d) &= \sup_{H_{01}} P(T_1(X_1, x_1, \theta_1) \ge T_1(x_1, x_1, \theta_1)) \\ &= P(T_1(X_1, x_1, \theta_1) \ge T_1(x_1, x_1, \theta_1)). \end{aligned} \tag{13}$$

Following (13) the generalized p value for (10) can be defined as

$$\begin{aligned} p(d) &= P(T_1(X_1, x_1, \theta_1) \ge T_1(x_1, x_1, \theta_1)) \\ &= P\left(k_1 \exp\left(2\left(\overline{y}_1 - \frac{Z}{\sqrt{U_1}/\sqrt{n_1 - 1}} \frac{s_1}{\sqrt{n_1}}\right) + l_1\right) \ge \theta_{01}\right) = P\left(\ln(k_1) + 2\left(\overline{y}_1 - \frac{Z}{\sqrt{U_1}/\sqrt{n_1 - 1}} \frac{s_1}{\sqrt{n_1}}\right) + l_1 \ge \ln(\theta_{01})\right) = P\left(\overline{y}_1 - \frac{Z}{\sqrt{U_1}/\sqrt{n_1 - 1}} \frac{s_1}{\sqrt{n_1}} \ge \frac{1}{2}\left(\ln\left(\frac{\theta_{01}}{k_1}\right) - l_1\right)\right) = P\left(Z \le \frac{\overline{y}_1 - (1/2)(\ln(\theta_{01}/k_1) - l_1)}{s_1/\sqrt{n_1}} \left(\frac{\sqrt{U_1}}{\sqrt{n_1 - 1}}\right)\right) \\ &= E_{U_1}\left(\Phi\left(\frac{\overline{y}_1 - (1/2)(\ln(\theta_{01}/k_1) - l_1)}{s_1/\sqrt{n_1}} \left(\frac{\sqrt{U_1}}{\sqrt{n_1 - 1}}\right)\right)\right), \end{aligned} \tag{14}$$

where $E_{U_1}(\cdot)$ is an expectation operator with respect to $U_1 = (n_1 - 1)S^2/\sigma^2 \sim \chi^2_{n_1 - 1}$ and $\Phi(\cdot)$ is cdf of the standard normal distribution.

Theorem 1. *If $f(u_1) = \Phi(T_{n_1 - 1}(\sqrt{u_1}/\sqrt{n_1 - 1}) - m)$ then $f(u_1)$ is a convex function of u_1 where $T_{n_1 - 1}$ is t- distribution with $n_1 - 1$ degrees of freedom.*

Proof. See Appendix.

Theorem 2. *The upper bound of $p(d)$ in (14) takes the form $\Psi_{n_1 - 1}(\Phi^{-1}(r) + m)$, for $0 < r < 0.5$, $m = (1/2)(\ln(\theta_{01}/k_1) - l_1)/(\sigma_1/\sqrt{n})$, where $\Psi(\cdot)$ is a cdf of t-distribution function with $(n_1 - 1)$ degrees of freedom and $\Phi(\cdot)$ is a cdf of the standard normal distribution.*

Proof. See Appendix.

Case 2. The hypothesis testing is

$$H_{02}: \theta_2 \le \theta_{02} \qquad \text{vs } H_{a2}: \theta_2 > \theta_{02}. \tag{15}$$

The parameter of the difference between two population variances for lognormal distributions is

$$\theta_2 = k_1 \exp(2\mu_1 + l_1) - k_2 \exp(2\mu_2 + l_2). \tag{16}$$

Without loss of generality, suppose $\theta_2 = \theta_{02} = 0$,

$$\begin{aligned} \theta_2 &= k_1 \exp(2\mu_1 + l_1) - k_2 \exp(2\mu_2 + l_2) \\ &= 2(\mu_1 - \mu_2) + \left(\ln\left(\frac{k_1}{k_2}\right) + (l_1 - l_2)\right) \\ &= 2(\mu_1 - \mu_2) + D_1. \end{aligned} \tag{17}$$

Using generalized test variable for θ_2 is

$$\begin{aligned} T_2(X_1, X_2, x_1, x_2, \theta_2) &= 2\left(\overline{y}_1 - \overline{y}_2 + \frac{\overline{Y}_2 - \overline{Y}_1 - (\mu_2 - \mu_1)}{\sqrt{\sigma_1^2/n_1 + \sigma_2^2/n_2}} \sqrt{\frac{\sigma_1^2 s_1^2}{n_1 S_1^2} + \frac{\sigma_2^2 s_2^2}{n_2 S_2^2}}\right) + D_1 \\ &= 2\left(\overline{y}_1 - \overline{y}_2 + \frac{Z}{\sqrt{U_1 + U_2}} \cdot \sqrt{\frac{((n_1 - 1)/n_1) s_1^2}{U_1/(U_1 + U_2)} + \frac{((n_2 - 1)/n_2) s_2^2}{U_2/(U_1 + U_2)}}\right) + D_1 \end{aligned}$$

$$= 2\left(\bar{y}_1 - \bar{y}_2 + \frac{1}{\sqrt{n_1+n_2-2}} \cdot \frac{Z}{\sqrt{(U_1+U_2)/(n_1+n_2-2)}} \cdot \sqrt{\frac{((n_1-1)/n_1)s_1^2}{U_1/(U_1+U_2)} + \frac{((n_2-1)/n_2)s_2^2}{U_2/(U_1+U_2)}}\right) + D_1$$

$$= 2\left(\bar{y}_1 - \bar{y}_2 + \frac{T_{n_1+n_2-2}}{\sqrt{n_1+n_2-2}} \cdot \sqrt{\frac{((n_1-1)/n_1)s_1^2}{B} + \frac{((n_2-1)/n_2)s_2^2}{1-B}}\right) + D_1. \quad (18)$$

It easy to see that $T_2(X_1, X_2, x_1, x_2, \theta_2)$ in (18) satisfies conditions (A1)–(A3) in Section 2.

The generalized p value, $p(q)$, is defined, under the null hypothesis H_{02}, to be

$$p(q) = \sup_{H_{02}} P(T_2(X_1, X_2, x_1, x_2, \theta_2) \geq T_2(x_1, x_2, x_1, x_2, \theta_2)) = P(T_2(X_1, X_2, x_1, x_2, \theta_2) \geq T_2(x_1, x_2, x_1, x_2, \theta_2)). \quad (19)$$

Following (19) the generalized p value for (15) can be defined as

$$p(q) = P(T_2(X_1, X_2, x_1, x_2, \theta_2) \geq T_2(x_1, x_2, x_1, x_2, \theta_2))$$
$$= P\left(2\left(\bar{y}_1 - \bar{y}_2 + \frac{T_{n_1+n_2-2}}{\sqrt{n_1+n_2-2}} \cdot \sqrt{\frac{((n_1-1)/n_1)s_1^2}{B} + \frac{((n_2-1)/n_2)s_2^2}{1-B}}\right) + D_1 \geq 0\right)$$
$$= P\left(\frac{T_{n_1+n_2-2}}{\sqrt{n_1+n_2-2}} \cdot \sqrt{\frac{((n_1-1)/n_1)s_1^2}{B} + \frac{((n_2-1)/n_2)s_2^2}{1-B}} \geq \bar{y}_2 - \bar{y}_1 - \frac{D_1}{2}\right) = P\left(T_{n_1+n_2-2} \leq \left((\bar{y}_1 - \bar{y}_2) + \frac{D_1}{2}\right) \cdot \sqrt{\frac{n_1+n_2-2}{((n_1-1)/n_1)s_1^2/B + ((n_2-1)/n_2)s_2^2/(1-B)}}\right)$$
$$= E_B\left(\Psi_{n_1+n_2-2}\left(\left((\bar{y}_1 - \bar{y}_2) + \frac{D_1}{2}\right) \cdot \sqrt{\frac{n_1+n_2-2}{((n_1-1)/n_1)s_1^2/B + ((n_2-1)/n_2)s_2^2/(1-B)}}\right)\right), \quad (20)$$

where $\Psi_{n_1+n_2-2}(\cdot)$ is a cdf of t-distribution with $n_1 + n_2 - 2$ degrees of freedom and $E_B(\cdot)$ is an expectation operator with respect to $B \cdot (B \sim \text{Beta}((n_1-1)/2, (n_2-1)/2))$.

Theorem 3. *If $f(b) = \Psi_v((z + w)\sqrt{1/(k_1/b + k_2/b)})$ then for fixed $z \leq 0$ and $w \geq 0$, $f(b)$ is convex function for b.*

Proof. See Appendix.

Theorem 4. *If $g(t) = P(\Psi_{n_1+n_2-2}((z + w)\sqrt{1/(tC_{n_1-1}/(n_1-1) + (1-t)C_{n_2-1}/(n_2-1))}) \leq r)$ and z, C_{n_1-1}, C_{n_2-1} are independent random variable such that $z \sim N(0,1)$, $C_{n_1-1} \sim \chi^2_{n_1-1}$, $C_{n_2-1} \sim \chi^2_{n_2-1}$, then $g(t)$ is convex function for t.*

Proof. See Appendix.

Theorem 5. *The upper bound of $p(q)$ is $\Psi_{\min\{n_1-1,n_2-1\}}(\Psi^{-1}_{n_1+n_2-2}(r) - w)$ for $0 < r < 0.5$, $w = (D_1/2)/\sqrt{\sigma_1^2/n_1 + \sigma_2^2/n_2}$, where $\Psi_{n_1+n_2-2}(\cdot)$ is the t-distribution function with $n_1 + n_2 - 2$ degrees of freedom and $\Psi^{-1}_{n_1+n_2-2}(\cdot)$ is the inverse function of $\Psi_{n_1+n_2-2}(\cdot)$.*

Proof. See Appendix.

Case 3. The hypothesis testing is

$$H_{03}: \theta_3 \leq \theta_{03} \quad \text{vs } H_{a3}: \theta_3 > \theta_{03}. \quad (21)$$

The parameter of the ratio of population variances for log-normal distributions is

$$\theta_3 = \frac{k_1 \exp(2\mu_1 + l_1)}{k_2 \exp(2\mu_2 + l_2)}, \quad (22)$$

where

$$\theta_3 = \frac{k_1 \exp(2\mu_1 + l_1)}{k_2 \exp(2\mu_2 + l_2)} = \theta_{03}$$
$$= \ln\left(\frac{k_1 \exp(2\mu_1 + l_1)}{k_2 \exp(2\mu_2 + l_2)}\right) = \ln(\theta_{03})$$
$$= 2(\mu_1 - \mu_2) - \left(\ln\left(\frac{k_2}{k_1}\right) + (l_2 - l_1) + \ln\theta_{03}\right) \quad (23)$$
$$= 2(\mu_1 - \mu_2) - D_2 = (\mu_1 - \mu_2) - \frac{D_2}{2},$$
$$D_2 = \ln\left(\frac{k_2}{k_1}\right) + (l_2 - l_1) + \ln\theta_{03}.$$

Using generalized test variable for θ_3 is

$$\begin{aligned} T_3(X_1, X_2, x_1, x_2, \theta_3) &= \overline{y}_1 - \overline{y}_2 \\ &+ \frac{\overline{Y}_2 - \overline{Y}_1 - (\mu_2 - \mu_1)}{\sqrt{\sigma_1^2/n_1 + \sigma_2^2/n_2}} \sqrt{\frac{\sigma_1^2 s_1^2}{n_1 S_1^2} + \frac{\sigma_2^2 s_2^2}{n_2 S_2^2} - \frac{D_2}{2}} = \overline{y}_1 \\ &- \overline{y}_2 + Z\sqrt{\frac{(n_1-1)s_1^2}{n_1 U_1} + \frac{(n_2-1)s_2^2}{n_2 U_2} - \frac{D_2}{2}} = \overline{y}_1 \\ &- \overline{y}_2 + \frac{Z}{\sqrt{U_1+U_2}} \\ &\cdot \sqrt{\frac{((n_1-1)/n_1)s_1^2}{U_1/(U_1+U_2)} + \frac{((n_2-1)/n_2)s_2^2}{U_2/(U_1+U_2)} - \frac{D_2}{2}} \\ &= \overline{y}_1 - \overline{y}_2 + \frac{1}{\sqrt{n_1+n_2-2}} \\ &\cdot \frac{Z}{\sqrt{(U_1+U_2)/(n_1+n_2-2)}} \\ &\cdot \sqrt{\frac{((n_1-1)/n_1)s_1^2}{U_1/(U_1+U_2)} + \frac{((n_2-1)/n_2)s_2^2}{U_2/(U_1+U_2)} - \frac{D_2}{2}} \\ &= \overline{y}_1 - \overline{y}_2 + \frac{T_{n_1+n_2-2}}{\sqrt{n_1+n_2-2}} \\ &\cdot \sqrt{\frac{((n_1-1)/n_1)s_1^2}{B} + \frac{((n_2-1)/n_2)s_2^2}{1-B} - \frac{D_2}{2}}. \end{aligned} \tag{24}$$

It easy to see that $T_3(X_1, X_2, x_1, x_2, \theta_3)$ in (24) satisfies conditions (A1)–(A3) in Section 2.

The generalized p value, $p(q_2)$, is defined, under the null hypothesis H_{03}, to be

$$\begin{aligned} p(q_2) &= \sup_{H_{03}} P(T_3(X_1, X_2, x_1, x_2, \theta_3) \\ &\geq T_3(x_1, x_2, x_1, x_2, \theta_3)) \\ &= P(T_3(X_1, X_2, x_1, x_2, \theta_3) \\ &\geq T_3(x_1, x_2, x_1, x_2, \theta_3)). \end{aligned} \tag{25}$$

Following (25) the generalized p value for (21) can be defined as

$$\begin{aligned} p(q_2) &= P(T_3(X_1, X_2, x_1, x_2, \theta_3) \geq T_3(x_1, x_2, x_1, x_2, \theta_3)) \\ &= P\left(\overline{y}_1 - \overline{y}_2 + \frac{T_{n_1+n_2-2}}{\sqrt{n_1+n_2-2}} \right. \\ &\left. \cdot \sqrt{\frac{((n_1-1)/n_1)s_1^2}{B} + \frac{((n_2-1)/n_2)s_2^2}{1-B} - \frac{D_2}{2}} \geq 0\right) \\ &= P\left(\frac{T_{n_1+n_2-2}}{\sqrt{n_1+n_2-2}}\right. \\ &\cdot \sqrt{\frac{((n_1-1)/n_1)s_1^2}{B} + \frac{((n_2-1)/n_2)s_2^2}{1-B}} \geq \overline{y}_2 - \overline{y}_1 \\ &\left. + \frac{D_2}{2}\right) = P\left(T_{n_1+n_2-2} \leq \left((\overline{y}_1 - \overline{y}_2) - \frac{D_2}{2}\right)\right. \\ &\left. \cdot \sqrt{\frac{n_1+n_2-2}{((n_1-1)/n_1)s_1^2/B + ((n_2-1)/n_2)s_2^2/(1-B)}}\right) \\ &= E_B\left(\Psi_{n_1+n_2-2}\left(\left((\overline{y}_1 - \overline{y}_2) - \frac{D_2}{2}\right)\right.\right. \\ &\left.\left. \cdot \sqrt{\frac{n_1+n_2-2}{((n_1-1)/n_1)s_1^2/B + ((n_2-1)/n_2)s_2^2/(1-B)}}\right)\right), \end{aligned} \tag{26}$$

where $\Psi(\cdot)$ is a cdf of t-distribution with $n_1 + n_2 - 2$ degrees of freedom and $E(\cdot)$ is an expectation operator with respect to $B \cdot (B \sim \text{Beta}(n_1 - 1/2, n_2 - 1/2))$.

Theorem 6. *The upper bound of* $p(q_2)$ *is* $\Psi_{\min\{n_1-1,n_2-1\}}(\Psi^{-1}_{n_1+n_2-2}(r)+w)$ *for* $0 < r < 0.5$, $w = (D_2/2)/\sqrt{\sigma_1^2/n_1 + \sigma_2^2/n_2}$, *where* $\Psi_{n_1+n_2-2}(\cdot)$ *is the t-distribution function with* $n_1 + n_2 - 2$ *degrees of freedom and* $\Psi^{-1}_{n_1+n_2-2}(\cdot)$ *is the inverse function of* $\Psi_{n_1+n_2-2}(\cdot)$.

Proof. It is similar to Theorem 5.

4. Numerical Results

In this section, we used functions written in the **R** program [16] to compute the values of the upper bounds of the generalized p values proposed in Theorems 2, 5, and 6. For given values of $n_1, n_2, m, W, \theta_{01}, \theta_{03}$, and r, we computed the upper bounds of $p(d)$, $p(q)$, and $p(q_2)$, by using the results from Theorems 2, 5, and 6 shown in Tables 1–3. As we can see in these tables, all results of the upper bounds of the generalized p values proposed in Theorems 2, 5, and 6 depend mainly on a variety of values of $n_1, n_2, m, W, \theta_{01}, \theta_{03}$, and r. As a result, these upper bounds confirm our proof in Theorems 2, 5, and 6.

5. Conclusion

We proposed three new generalized p values for testing the hypotheses of (1) one population variance, (2) the difference between two population variances, and (3) the ratio of population variances of lognormal distributions when the coefficients of variation are known. We also proved new upper bounds for our proposed generalized p values. We note here that the results for these results for case (1), case (2), and case (3) were analogous to the upper bound of the generalized p value for the Behrens-Fisher problem

TABLE 1: The upper bound of p value for hypothesis (10), for $\tau = 1$.

n_1	θ_{01}	m	$r = 0.01$	$r = 0.02$	$r = 0.05$	$r = 0.10$	$r = 0.20$
	2.001	0.000949333	0.02254277	0.03514731	0.06730417	0.1161747	0.2111518
10	2.002	0.001898192	0.02257782	0.03520131	0.06740329	0.1163342	0.2114042
	2.003	0.002846578	0.02261291	0.03525537	0.06750250	0.1164939	0.2116567
	2.001	0.000949333	0.01779592	0.02963687	0.06122800	0.1105716	0.2073298
15	2.002	0.001898192	0.01782805	0.02968858	0.06132703	0.1107338	0.2075869
	2.003	0.002846578	0.01786022	0.02974036	0.06142616	0.110896	0.2078441
	2.001	0.000949333	0.01563906	0.02706209	0.05832057	0.1078731	0.2054935
20	2.002	0.001898192	0.01566962	0.02711254	0.05841947	0.1080364	0.2057529
	2.003	0.002846578	0.01570022	0.02716304	0.05851845	0.1081999	0.2060124

TABLE 2: The upper bound of p value for hypothesis (15), for $\tau_1 = 1$, $\tau_2 = 1$, and $W = 0$.

n_1, n_2	$r = 0.01$	$r = 0.02$	$r = 0.05$	$r = 0.10$	$r = 0.20$
5, 5	0.02213745	0.03526116	0.06823592	0.1174918	0.2121440
10, 10	0.01553660	0.02705864	0.05846833	0.1080566	0.2055204
15, 15	0.01356674	0.02457767	0.05550640	0.1052267	0.2035670
20, 20	0.01262788	0.02338536	0.05407854	0.1038673	0.2026341
30, 30	0.01172026	0.02222509	0.05268530	0.1025436	0.2017292

TABLE 3: The upper bound of p value for hypothesis (21), for $\tau_1 = 1$, $\tau_2 = 1$, and $W = 0$.

n_1, n_2	θ_{03}	W	$r = 0.01$	$r = 0.02$	$r = 0.05$	$r = 0.10$	$r = 0.20$
	1.01	0.00944854	0.02234846	0.03562201	0.06898679	0.1188115	0.2144116
5, 5	1.02	0.01880399	0.02255971	0.03598345	0.06973902	0.1201326	0.2166765
	1.03	0.02806817	0.02277119	0.03634547	0.07049260	0.1214551	0.2189384
	1.01	0.00944854	0.01577932	0.02747895	0.05934179	0.1095594	0.2079965
10, 10	1.02	0.01880399	0.0160234	0.02790139	0.06021843	0.1110648	0.2104685
	1.03	0.02806817	0.01626883	0.02832593	0.06109821	0.1125725	0.2129362
	1.01	0.00944854	0.01381588	0.02501398	0.05641761	0.1067866	0.2061060
15, 15	1.02	0.01880399	0.01406687	0.02545308	0.05733271	0.1083493	0.2086403
	1.03	0.02806817	0.01431970	0.02589493	0.05825161	0.1099147	0.2111700
	1.01	0.009448542	0.01287934	0.02382890	0.05500805	0.1054549	0.2052033
20, 20	1.02	0.01880399	0.01313292	0.02427558	0.05594180	0.1070456	0.2077678
	1.03	0.02806817	0.01338861	0.02472537	0.05687972	0.1086391	0.2103272
	1.01	0.009448542	0.01197343	0.02267532	0.05363272	0.1041586	0.2043280
30, 30	1.02	0.01880399	0.01222900	0.02312906	0.05458475	0.1057767	0.2069217
	1.03	0.02806817	0.01248697	0.02358628	0.05554131	0.1073979	0.2095102

proposed by Tang and Tsui [12]. Numerical results shown in Tables 1–3 confirmed our results of the upper bounds of the generalized p values proposed in Theorems 2, 5, and 6; we also found that the proposed upper bounds are increasing up on the parameter values of m and W. For example, for $n_1 = 10$, $m = 0.002846578$, and $r = 0.01$, the upper bound of p value using Theorem 2 is 0.022261291 and this upper bound of p value approaches $r = 0.01$ when n_1 is increasing. Similar results are applied for other cases. For the two-tailed test, that is, H_{01}: $\theta_1 = \theta_{01}$ and H_a: $\theta_1 \neq \theta_{01}$; it is easy to apply the results of Theorem 2 of Tang and Tsui [12] to all hypotheses in this paper. So we skipped it.

Appendix

Proof of Theorem 1. Defining $h(u_1) = T_{n_1-1}(\sqrt{u_1}/\sqrt{n_1-1}) - m$, we have $f(u_1) = \Phi(h(u_1))$. Let ϕ be the probability density function of u_1.

Hence

$$f''(u_1) = (f'(u_1))' = (\phi(h(u_1))h'(u_1))' = \phi'(h(u_1))(h'(u_1))^2 + \phi(h(u_1))h''(u_1). \tag{A.1}$$

For $T_{n_1-1} \leq 0$, $h(u_1) < 0$. Hence $\phi(h(z)) > 0$ and $\phi'(h(z)) > 0$.

Moreover

$$h''(u_1) = \left(\frac{1}{2}T_{n_1-1}\frac{u_1^{-1/2}}{\sqrt{n_1-1}}\right)' = -\frac{1}{4}T_{n_1-1}\frac{u_1^{-3/2}}{\sqrt{n_1-1}} > 0. \quad \text{(A.2)}$$

Hence $f''(u_1) > 0$, and $f(u_1)$ is convex in u_1.

Proof of Theorem 2. Denote $z = \overline{y}/(\sigma_1/\sqrt{n})$ and $U_1 = (n_1 - 1)S^2/\sigma^2 \sim \chi^2_{n_1-1}$.

From (14), we have

$$\begin{aligned} p(d) &= E_{U_1}\left(\Phi\left(\frac{\overline{y}-(1/2)(\ln(\theta_{01}/k_1)-l_1)}{s_1/\sqrt{n_1}}\left(\frac{\sqrt{U_1}}{\sqrt{n_1-1}}\right)\right)\right) \\ &= E_{U_1}\left(\Phi\left(\left(\frac{\overline{y}}{\sigma_1/\sqrt{n_1}}\frac{\sqrt{U_1}}{\sqrt{n_1-1}} - \frac{(1/2)(\ln(\theta_{01}/k_1)-l_1)}{\sigma_1/\sqrt{n_1}}\frac{\sqrt{U_1}}{\sqrt{n_1-1}}\right)\left(\frac{\sigma_1/\sqrt{n_1}}{s_1/\sqrt{n_1}}\right)\right)\right) \\ &= E_{U_1}\left(\Phi\left(\left(z\frac{\sqrt{U_1}}{\sqrt{n_1-1}} - \frac{(1/2)(\ln(\theta_{01}/k_1)-l_1)}{\sigma_1/\sqrt{n_1}}\frac{\sqrt{U_1}}{\sqrt{n_1-1}}\right)\left(\frac{\sqrt{n_1-1}}{\sqrt{U_1}}\right)\right)\right) \\ &= E_U\left(\Phi\left(T_{n_1-1}\frac{\sqrt{U_1}}{\sqrt{n_1-1}} - m\right)\right), \end{aligned} \quad \text{(A.3)}$$

$$m = \frac{(1/2)(\ln(\theta_{01}/k_1)-l_1)}{\sigma_1/\sqrt{n_1}}.$$

For any $r < 0.5$ and $p(d) < r$, hence, by Theorem 1, we have $E(f(U_1)) \geq f(E(U_1))$

$$\begin{aligned} p(d) &= E_{U_1}\left(\Phi\left(T_{n_1-1}\frac{\sqrt{U_1}}{\sqrt{n_1-1}} - m\right)\right) \\ &\geq \Phi\left(T_{n_1-1}E\left(\frac{\sqrt{U_1}}{\sqrt{n_1-1}}\right) - m\right) \\ &= \Phi(T_{n-1} - m) = p_1(d). \end{aligned} \quad \text{(A.4)}$$

For $0 < r < 0.5$, we have

$$\begin{aligned} P_d(\{d : p(d) \leq r\}) &\leq P_d(\{d : p_1(d) \leq r\}) \\ &= P(p_1(d) \leq r) \\ &= P(\Phi(T_{n_1-1} - m) \leq r) \\ &= P(T_{n_1-1} \leq \Phi^{-1}(r) + m) \\ &= E(\Psi_{n_1-1}(\Phi^{-1}(r) + m)) \\ &= \Psi_{n_1-1}(\Phi^{-1}(r) + m), \end{aligned} \quad \text{(A.5)}$$

$$\text{where } m = \frac{(1/2)(\ln(\theta_{01}/k_1)-l_1)}{\sigma_1/\sqrt{n}}.$$

Proof of Theorem 3. Define $f(b) = f(h(b))$, $h(b) = (z - w)\sqrt{1/(k_1/b + k_2/b)}$, and ψ_v is the probability density function of t-distribution of v degrees of freedom. Hence

$$\begin{aligned} f''(h(b)) &= (f'(h(b)))' = (\psi_v(h(b))h'(b))' \\ &= \psi'_v(h(b))(h'(b))^2 + \phi(h(b))h''(b). \end{aligned} \quad \text{(A.6)}$$

For $z \leq 0$, $w \geq 0$ implies that $h(b) < 0$, $\psi'_v(h(b)) \geq 0$, and $\psi_v(h(b)) \geq 0$.

We have

$$\begin{aligned} h''(b) &= (z+w)\left(\left(\frac{k_1}{b} + \frac{k_2}{1-b}\right)^{-3/2}\cdot\left(-\frac{1}{2}\right)\left(-\frac{k_1}{b^2} + \frac{k_2}{(1-b)^2}\right)\right)' \\ &= -(z+w)\left(\frac{k_1^2/2b^2 + k_2^2/2(1-b)^4 + 2k_1k_2/b(1-b)^3 + 2k_1k_2/b^3(1-b) + 3k_1^2k_2^2/b^2(1-b)^2}{(k_1/b + k_2/(1-b))^{5/2}}\right) > 0. \end{aligned} \quad \text{(A.7)}$$

We have $f''(h(b)) > 0$. Hence $f(b)$ is convex function for b.

Proof of Theorem 4.

$$\begin{aligned} g(t) &= P\left(\Psi_{n_1+n_2-2}\left((z+w)\cdot\sqrt{\frac{1}{tC_{n_1-1}/(n_1-1) + (1-t)C_{n_2-1}/(n_2-1)}}\right) \leq r\right) \\ &= P\left(\left((z+w)\cdot\sqrt{\frac{1}{tC_{n_1-1}/(n_1-1) + (1-t)C_{n_2-1}/(n_2-1)}}\right)\right. \end{aligned}$$

$$\leq \Psi^{-1}_{n_1+n_2-2}(r)\Bigg) = P\Bigg(z \leq \sqrt{\frac{tC_{n_1-1}}{n_1-1}+\frac{(1-t)C_{n_2-1}}{n_2-1}}\left(\Psi^{-1}_{n_1+n_2-2}(r)\right)-w\Bigg) = E\left(\Phi\left(\sqrt{\frac{tC_{n_1-1}}{n_1-1}+\frac{(1-t)C_{n_2-1}}{n_2-1}}\left(\Psi^{-1}_{n_1+n_2-2}(r)\right)-w\right)\right), \tag{A.8}$$

where Φ is the cdf of a standard normal distribution.

Let $g_1(t) = \Phi(h_1(t))$, $h_1(t) = \sqrt{tC_{n_1-1}/(n_1-1)+(1-t)C_{n_2-1}/(n_2-1)}(\Psi^{-1}_{n_1+n_2-2}(r)) - w$, and ϕ is the probability density function of a standard normal distribution. Hence

$$\begin{aligned} g_1''(t) &= \left(g_1'(t)\right)' = \left(\phi\left(h_1(t)\right)h_1'(t)\right)' \\ &= \phi'\left(h_1(t)\right)\left(h_1'(t)\right)^2 + \phi\left(h_1(t)\right)h_1''(t). \end{aligned} \tag{A.9}$$

We have

$$h_1''(t) = \left(\frac{1}{2}\Psi^{-1}_{n_1+n_2-2}(r) \cdot \left(\frac{tC_{n_1-1}}{n_1-1}+\frac{(1-t)C_{n_2-1}}{n_2-1}\right)^{-1/2} \cdot \left(\frac{C_{n_1-1}}{n_1-1}-\frac{C_{n_2-1}}{n_2-1}\right)\right)' = \left(-\frac{1}{4}\Psi^{-1}_{n_1+n_2-2}(r) \cdot \left(\frac{tC_{n_1-1}}{n_1-1}+\frac{(1-t)C_{n_2-1}}{n_2-1}\right)^{-3/2} \cdot \left(\frac{C_{n_1-1}}{n_1-1}-\frac{C_{n_2-1}}{n_2-1}\right)^2\right)' \geq 0. \tag{A.10}$$

We have $g_1''(t) \geq 0$. Hence $g_1(t)$ is convex function for t. As a result, $g(t) = E(g_1(t))$ is convex in t.

Proof of Theorem 5. Denote

$$\begin{aligned} T &= \frac{\sigma_1^2/n_1}{\sigma_1^2/n_1+\sigma_2^2/n_2}, \\ z &= \frac{(\bar{y}_1-\bar{y}_2)}{\sqrt{\sigma_1^2/n_1+\sigma_2^2/n_2}}, \\ C_{n_1-1} &= \frac{(n_1-1)s_1^2}{\sigma_1^2}, \\ C_{n_2-1} &= \frac{(n_2-1)s_2^2}{\sigma_2^2}. \end{aligned} \tag{A.11}$$

From (24)

$$\begin{aligned} p(q) &= E_B\left(\Psi_{n_1+n_2-2}\left(\frac{(\bar{y}_1-\bar{y}_2)+D_1/2}{\sqrt{\sigma_1^2/n_1+\sigma_2^2/n_2}}\sqrt{\frac{n_1+n_2-2}{((n_1-1)/n_1)s_1^2/B+((n_2-1)/n_2)s_2^2/(1-B)}\left(\frac{1}{1/(\sigma_1^2/n_1+\sigma_2^2/n_2)}\right)}\right)\right) \\ &= E_B\left(\Psi_{n_1+n_2-2}\left((z+w)\sqrt{\frac{n_1+n_2-2}{\left(C_{n_1-1}\sigma_1^2/n_1\right)/B(\sigma_1^2/n_1+\sigma_2^2/n_2)+\left(C_{n_2-1}\sigma_2^2/n_2\right)/(1-B)}}\right)\right), \\ w &= \frac{D_1/2}{\sqrt{\sigma_1^2/n_1+\sigma_2^2/n_2}} \\ &= E_B\left(\Psi_{n_1+n_2-2}\left((z+w)\sqrt{\frac{n_1+n_2-2}{TC_{n_1-1}/B+(1-T)C_{n_2-1}/(1-B)}}\right)\right). \end{aligned} \tag{A.12}$$

For any $r < 0.5$ and $p(q) < r$, hence, by Theorem 1, such that $E_B(f(B)) \geq f(E(B))$

$$p(q) = E_B\left(\Psi_{n_1+n_2-2}\left((z+w) \cdot \sqrt{\frac{n_1+n_2-2}{TC_{n_1-1}/B+(1-T)C_{n_2-1}/(1-B)}}\right)\right) \geq \Psi_{n_1+n_2-2}\left(E_B\left((z+w)\right.\right.$$

$$\cdot \sqrt{\frac{n_1+n_2-2}{TC_{n_1-1}/B+(1-T)C_{n_2-1}/(1-B)}}\Bigg)\Bigg)$$
$$= \Psi_{n_1+n_2-2}\Bigg((z+w)$$
$$\cdot \sqrt{\frac{1}{TC_{n_1-1}/(n_1-1)+(1-T)C_{n_2-1}/(n_2-1)}}\Bigg)$$
$$= p_1(q). \tag{A.13}$$

For $0 < r < 0.5$, we have

$$P_q(\{q : p(q) \le r\}) \le P_q(\{q : q_1(q) \le r\}) = g(T),$$

by Theorem 4,

$$g(T) \le \max\{g(0), g(1)\}$$
$$= \max\left\{P\left(\Psi_{n_1+n_2-2}\left((z+w)\sqrt{\frac{1}{C_{n_1-1}/(n_1-1)}}\right)\le r\right),\right.$$
$$\left.P\left(\Psi_{n_1+n_2-2}\left((z+w)\sqrt{\frac{1}{C_{n_2-1}/(n_2-1)}}\right)\le r\right)\right\}$$
$$= \max\left\{P\left(z\sqrt{\frac{1}{C_{n_1-1}/(n_1-1)}} \le -w\sqrt{\frac{1}{C_{n_1-1}/(n_1-1)}}\right.\right.$$
$$\left.+\Psi^{-1}_{n_1+n_2-2}(r)\right), P\left(z\sqrt{\frac{1}{C_{n_2-1}/(n_2-1)}}\right.$$
$$\left.\left.\le -w\sqrt{\frac{1}{C_{n_2-1}/(n_2-1)}}+\Psi^{-1}_{n_1+n_2-2}(r)\right)\right\} \tag{A.14}$$
$$= \max\left\{E\left(\Psi_{n_1-1}\left(-w\sqrt{\frac{1}{C_{n_1-1}/(n_1-1)}}+\Psi^{-1}_{n_1+n_2-2}(r)\right)\right),\right.$$
$$\left.E\left(\Psi_{n_2-1}\left(-w\sqrt{\frac{1}{C_{n_2-1}/(n_2-1)}}+\Psi^{-1}_{n_1+n_2-2}(r)\right)\right)\right\}$$
$$\ge \max\left\{\Psi_{n_1-1}\left(E\left(-w\sqrt{\frac{1}{C_{n_1-1}/(n_1-1)}}\right)+\Psi^{-1}_{n_1+n_2-2}(r)\right),\right.$$
$$\left.\Psi_{n_2-1}\left(E\left(-w\sqrt{\frac{1}{C_{n_2-1}/(n_2-1)}}\right)+\Psi^{-1}_{n_1+n_2-2}(r)\right)\right\}$$
$$= \Psi_{\min\{n_1-1,n_2-1\}}\left(\Psi^{-1}_{n_1+n_2-2}(r)-w\right),\quad w=\frac{D_1/2}{\sqrt{\sigma_1^2/n_1+\sigma_2^2/n_2}}.$$

Competing Interests

The authors declare that there is no conflict of interests regarding the publication of this paper.

Acknowledgments

The second author is grateful to Grant no. KMUTNB-GOV-59-36 from King Mongkut's University of Technology North Bangkok.

References

[1] H. P. Singh, L. N. Upadhyaya, and U. D. Namjoshi, "Estimation of finite population variance," *Current Science*, vol. 57, no. 24, pp. 1331-1334, 1988.

[2] M. C. Agrawal and A. B. Sthapit, "Unbiased ratio-type variance estimation," *Statistics and Probability Letters*, vol. 25, no. 4, pp. 361–364, 1995.

[3] A. Arcos Cebrián and M. Rueda García, "Variance estimation using auxiliary information: an almost unbiased multivariate ratio estimator," *Metrika. International Journal for Theoretical and Applied Statistics*, vol. 45, no. 2, pp. 171–178, 1997.

[4] A. Arcos, M. Rueda, M. D. Martínez, S. González, and Y. Román, "Incorporating the auxiliary information available in variance estimation," *Applied Mathematics and Computation*, vol. 160, no. 2, pp. 387–399, 2005.

[5] C. Kadilar and H. Cingi, "Ratio estimators for the population variance in simple and stratified random sampling," *Applied Mathematics and Computation*, vol. 173, no. 2, pp. 1047–1059, 2006.

[6] V. Cojbasic and A. Tomovic, "Nonparametric confidence intervals for population variance of one sample and the difference of variances of two samples," *Computational Statistics & Data Analysis*, vol. 51, no. 12, pp. 5562–5578, 2007.

[7] V. Cojbasic and D. Loncar, "One-sided confidence intervals for population variances of skewed distributions," *Journal of Statistical Planning and Inference*, vol. 141, no. 5, pp. 1667–1672, 2011.

[8] V. C. Rajic, J. Kocovic, D. Loncar, and T. R. Antic, "Testing population variance in case of one sample and the difference of variances in case of two samples: example of wage and pension data sets in Serbia," *Economic Modelling*, vol. 29, no. 3, pp. 610–613, 2012.

[9] R. Singh and S. Malik, "Improved estimation of population variance using information on auxiliary attribute in simple random sampling," *Applied Mathematics and Computation*, vol. 235, pp. 43–49, 2014.

[10] K.-W. Tsui and S. Weerahandi, "Generalized *p*-values in significance testing of hypotheses in the presence of nuisance parameters," *Journal of the American Statistical Association*, vol. 84, no. 406, pp. 602–607, 1989.

[11] S. Weerahandi, *Exact Statistical Methods for Data Analysis*, Springer, New York, NY, USA, 1995.

[12] S. Tang and K.-W. Tsui, "Distributional properties for the generalized *p*-value for the Behrens-Fisher problem," *Statistics & Probability Letters*, vol. 77, no. 1, pp. 1–8, 2007.

[13] R. Somkhuean, S. Niwitpong, and S.-A. Niwitpong, "Upper bounds of generalized p-values for testing the coefficients of variation of lognormal distributions," *Chiang Mai Journal of Science*, vol. 43, pp. 671–681, 2016.

[14] J. Gamage and S. Weerahandi, "Size performance of some tests in one-way ANOVA," *Communications in Statistics—Simulation and Computation*, vol. 27, no. 3, pp. 625–640, 1998.

[15] P. Kabaila and C. J. Lloyd, "Tight upper confidence limits from discrete data," *The Australian Journal of Statistics*, vol. 39, no. 2, pp. 193–204, 1997.

[16] The R Development Core Team, *An Introduction to R*, R Foundation for Statistical Computing, Vienna, Austria, 2010, http://www.R-project.org.

14

The Half-Logistic Lomax Distribution for Lifetime Modeling

Masood Anwar and Jawaria Zahoor

Department of Mathematics, COMSATS Institute of Information Technology, Park Road, Chak Shahzad, Islamabad, Pakistan

Correspondence should be addressed to Masood Anwar; masoodanwar@comsats.edu.pk

Academic Editor: Chin-Shang Li

We introduce a new two-parameter lifetime distribution called the half-logistic Lomax (HLL) distribution. The proposed distribution is obtained by compounding half-logistic and Lomax distributions. We derive some mathematical properties of the proposed distribution such as the survival and hazard rate function, quantile function, mode, median, moments and moment generating functions, mean deviations from mean and median, mean residual life function, order statistics, and entropies. The estimation of parameters is performed by maximum likelihood and the formulas for the elements of the Fisher information matrix are provided. A simulation study is run to assess the performance of maximum-likelihood estimators (MLEs). The flexibility and potentiality of the proposed model are illustrated by means of real and simulated data sets.

1. Introduction

The commonly used lifetime distributions (exponential, gamma, Weibull, Lomax, lognormal, etc.) have a limited range of behavior and do not provide adequate fit to complex data sets in different sciences. Generalizations of these distributions offer more flexibility and provide reasonable parametric fits to complex data sets. Motivated by the various applications of Lomax and half-logistic distributions in areas of income and wealth inequality, firm size, size of cities, queuing problems, actuarial science, medical and biological sciences, and engineering, we propose a two-parameter continuous lifetime distribution by compounding the half-logistic and the Lomax distribution called *half-logistic Lomax (HLL) distribution.*

The Lomax [1] (or Pareto Type-II) distribution was introduced to model business failure data. For more detail about the Lomax distribution, we refer the readers to Rady et al. [2], Tahir et al. [3], and the references therein. In literature, there are several generalizations of the Lomax distribution. Abdul-Moniem [4] developed the exponentiated Lomax distribution, and Al-Awadhi and Ghitany [5] introduced the discrete Poisson–Lomax distribution by using the Lomax distribution as a mixing distribution for the Poisson parameter. Asgharzadeh et al. [6] proposed the Pareto Poisson–Lindely distribution, and Cordeiro et al. [7] investigated the gamma-Lomax distribution and studied its properties. Ghitany et al. [8] and Gupta et al. [9] considered the Marshal–Olkin approach and extended the Lomax distribution, and Lemonte and Cordeiro [10] proposed and studied the McDonald-Lomax, the beta Lomax, and the Kumaraswamy Lomax distributions. Other models constitute flexible family of distributions in terms of the variates of shapes and hazard functions; see, for example, Al-Zahrani and Sagor [11], El-Bassiouny et al. [12], Rady et al. [2], Kilany [13], and Tahir et al. [3]. These generalizations of the Lomax distribution are considered to be useful life distribution models.

The cumulative distribution function (cdf) of Lomax distribution is given by

$$H(x;\alpha,\beta) = 1 - (1+\beta x)^{-\alpha}, \quad x > 0, \tag{1}$$

where $\alpha > 0$ is a shape parameter and $\beta > 0$ is a scale parameter. The probability density function (pdf) corresponding to (1) is

$$h(x;\alpha,\beta) = \alpha\beta(1+\beta x)^{-(\alpha+1)}, \quad x > 0. \tag{2}$$

Cordeiro et al. [14] define the cdf of the new *type I half-logistic-G* (TIHL-G) family of distributions by

$$F(x;\lambda,\theta)=\int_0^{-\ln(1-G(x;\theta))}\frac{2\lambda e^{-\lambda t}}{(1+e^{-\lambda t})^2}dt = \frac{1-[1-G(x;\theta)]^{\lambda}}{1+[1-G(x;\theta)]^{\lambda}}, \tag{3}$$

where $G(x;\theta)$ is the baseline cdf depending on a parameter vector θ and an additional shape parameter $\lambda>0$. As a special case, if $\lambda=1$, then the TIHL-G is the half-logistic-G (HL-G) distribution with cdf

$$F(x;\theta)=\frac{G(x;\theta)}{1+\overline{G}(x;\theta)}, \quad \text{where } \overline{G}(x;\theta)=1-G(x;\theta). \tag{4}$$

The corresponding pdf to (4) is given by

$$f(x;\theta)=\frac{2g(x;\theta)}{\left[1+\overline{G}(x;\theta)\right]^2}, \quad \text{where } g(x)=\frac{dG(x)}{dx}. \tag{5}$$

This paper aims to provide a new lifetime model with a minimum number of parameters by compounding the half-logistic and the Lomax distribution called half-logistic Lomax (HLL) distribution. The proposed distribution is heavy-tailed and has a decreasing or upside-down bathtub (or unimodal) shaped hazard rates depending on its parameters. Upside-down bathtub shaped hazard rates are common in reliability, engineering, and survival analysis. The HLL distribution can also be applied in engineering as the Lomax [1] distribution and can be a useful alternative to other well-known densities in lifetime applications. It is interesting to note that the HLL distribution is a special case of Marshall-Olkin–Lomax distribution introduced by Ghitany et al. [8]. We obtain some mathematical properties of the proposed distribution and parameters of the model are estimated by the maximum-likelihood estimation method.

The rest of this paper is organized as follows. In Section 2, we introduce the half-logistic Lomax distribution and provide plots of its density function. In Section 3, we investigate various mathematical properties of the HLL distribution including survival and hazard rate function, quantile function, moments, mean residual life function, mean deviation from the mean and the mean deviation from the median, entropies, and order statistics. In Section 4, estimation of parameters is given by MLE method and the asymptotic distribution of the estimators is studied via Fisher's information matrix. Simulation results on the behavior of the MLEs are presented in Section 5. A real data application is conducted in Section 6. Finally, in Section 7, we conclude that the HLL distribution is the best model as compared to other competing models.

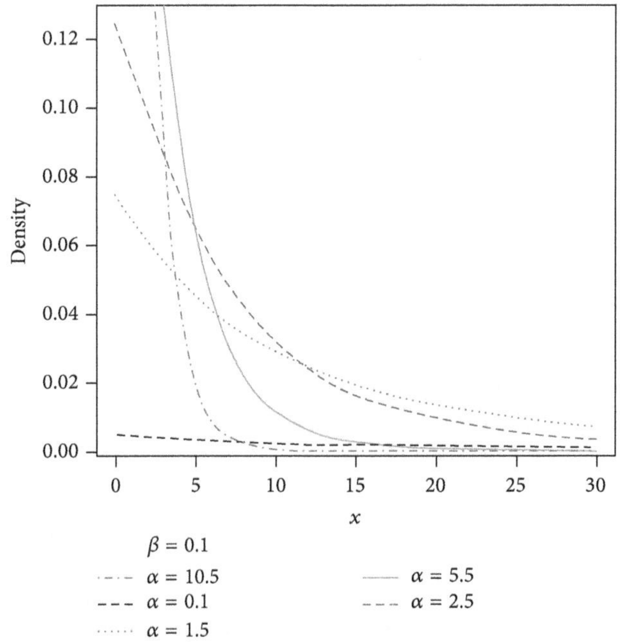

Figure 1: Plots of the HLL pdf for some parameter values.

2. The HLL Distribution

We define the *half-logistic-Lomax* (HLL) density function by inserting (1) and (2) into (5). So, we obtain

$$f(x;\alpha,\beta)=\frac{2\alpha\beta(1+\beta x)^{-(\alpha+1)}}{\left[1+(1+\beta x)^{-\alpha}\right]^2}, \quad x>0;\ \alpha,\beta>0. \tag{6}$$

The corresponding cumulative density function (cdf) follows from (1) and (4) and is given by

$$F(x)=\frac{\left[1-(1+\beta x)^{-\alpha}\right]}{\left[1+(1+\beta x)^{-\alpha}\right]}. \tag{7}$$

Hereafter, we will denote a random variable X having pdf (6) by $X \sim \mathrm{HLL}(\alpha,\beta)$. The limit of the HLL density (6) as $x\to\infty$ is 0 and the limit as $x\to 0$ is $\alpha\beta/2$. Figure 1 depicts some of the possible shapes of density (6) for selected parameter values. The mode of density (6) is obtained from solving $[d\log[f(x)]/dx]_{x=x_0}=0$, which is given by

$$\text{Mode}=x_0=\frac{1}{\beta}\left[\left(\frac{\alpha-1}{\alpha+1}\right)^{1/\alpha}-1\right]. \tag{8}$$

3. Mathematical Properties

This section describes mathematical properties of the HLL distribution.

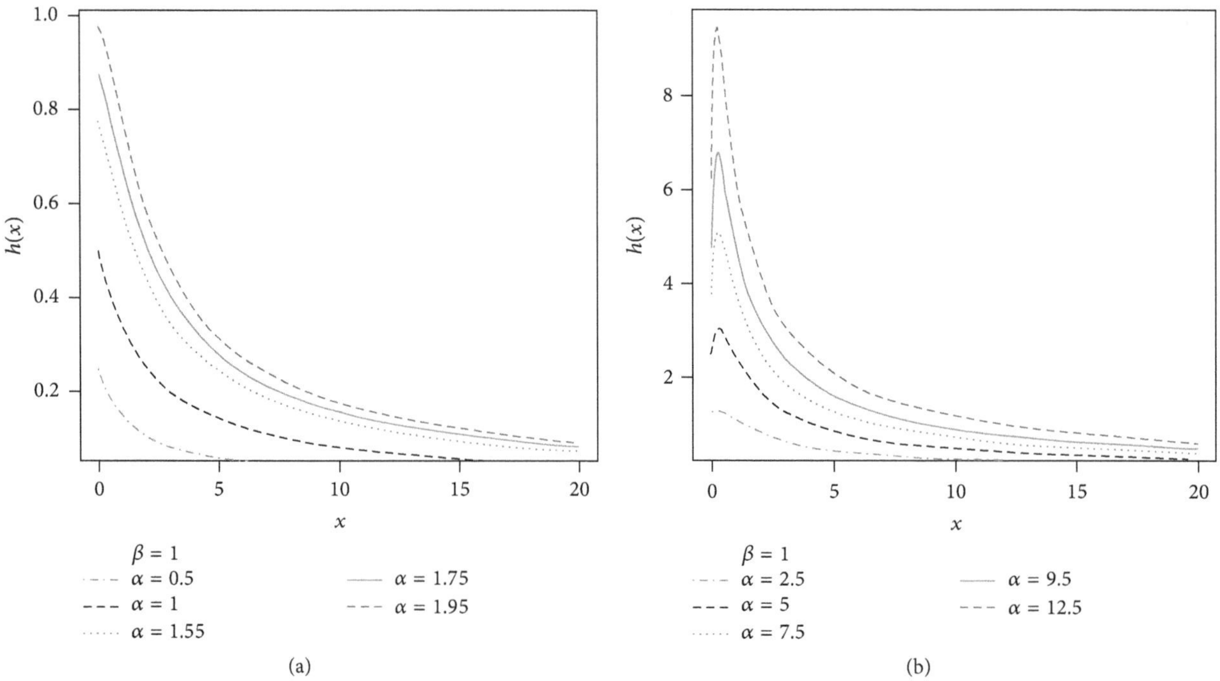

FIGURE 2: Plots of the HLL hrf for some parameter values.

3.1. The Survivor and Hazard Rate Functions. Using (6) and (7), the survival function (sf) and hazard rate function (hrf) of X are, respectively, given by

$$S(x) = \frac{2}{\left[1 + (1+\beta x)^{\alpha}\right]}, \qquad h(x) = \frac{\alpha\beta}{(1+\beta x)\left[1 + (1+\beta x)^{-\alpha}\right]}. \tag{9}$$

The limit of $h(x)$ as $x \to 0^{+}$ is $\alpha\beta/2$; that is, $h(0)$ is bounded from below and continuous in its parameters. The limit of $h(x)$ as $x \to \infty$ is 0. According to Glaser [15], we determine the parameter intervals for which the hazard rate function of the HLL distribution is decreasing or upside-down bathtub. Let

$$\eta(x) = -\frac{d}{dx}\ln f(x) = \frac{\beta\left[(\alpha+1) - (\alpha-1)(1+\beta x)^{-\alpha}\right]}{(1+\beta x)\left[1 + (1+\beta x)^{-\alpha}\right]}. \tag{10}$$

Then its first derivative is given by

$$\eta'(x) = \frac{\beta^{2}\left[2\left(\alpha^{2}-1\right)(1+\beta x)^{\alpha-2} + (\alpha-1)(1+\beta x)^{-2} - (\alpha+1)(1+\beta x)^{2(\alpha-1)}\right]}{\left[1 + (1+\beta x)^{\alpha}\right]^{2}}. \tag{11}$$

If $0 < \alpha \leq 2$, then $\eta'(x) < 0$ for all x. Then the hazard rate function is decreasing. By Glaser's theorem [15], it is sufficient to show that there exists $x_0 > 0$ such that $\eta'(x) > 0$ for all $x \in (0, x_0)$. $\eta'(x_0) = 0$ implies that $x_0 = (1/\beta)[\{(\alpha-1)/(\alpha+1)\}^{1/\alpha} - 1]$ and $\eta'(x) < 0$ for all $x > x_0$, which implies unimodal shape of the hazard rate function. For $\alpha > 2$, $\eta'(x) > 0$ for all $x < x_0$ and $\eta'(x) < 0$ for all $x > x_0$. Thus, the hazard rate function is upside-down bathtub-shaped for $\alpha > 2$. Figure 2 illustrates some of the possible shapes of hazard rate function for selected values of the parameters.

3.2. Quantile and Random Number Generation. The cumulative distribution function is given by (7). Inverting $F(x) = u$, we obtain

$$F^{-1}(x) = \frac{1}{\beta}\left[\left(\frac{1-u}{1+u}\right)^{1/\alpha} - 1\right], \quad 0 < u < 1. \tag{12}$$

Equation (12) can be used to simulate the HLL variable.

3.3. Moments. The following theorem gives moments of the HLL distribution and its mean moments and cumulants.

Theorem 1. *The rth moment about the origin of X ~ HLL(α, β) is given by*

$$\mu_r' = E\left[X^r\right] = \frac{2\alpha}{\beta^r}\sum_{k=1}^{\infty}\sum_{j=0}^{r}\binom{r}{j}\frac{k(-1)^{r+j+k-1}}{(k\alpha-j)}, \quad r = 1, 2, \ldots. \tag{13}$$

Proof. We have

$$E\left[X^r\right] = 2\alpha\beta\int_0^{\infty} x^r\frac{(1+\beta x)^{-(\alpha+1)}}{\left[1+(1+\beta x)^{-\alpha}\right]^2}dx. \tag{14}$$

Setting $y = (1+\beta x)$, $dx = dy/\beta$ yields

$$\begin{aligned} E\left[X^r\right] &= \frac{2\alpha}{\beta^r}\int_1^{\infty}(y-1)^r y^{-(\alpha+1)}\left(1+y^{-\alpha}\right)^{-2}dy \\ &= \frac{2\alpha}{\beta^r}\int_1^{\infty}\left\{\sum_{j=o}^{r}\binom{r}{j}y^{j-\alpha-1}(-1)^{r+j} \times\sum_{k=1}^{\infty}(-1)^{k-1}k\left(y^{-\alpha}\right)^{k-1}\right\}dy = \frac{2\alpha}{\beta^r} \\ &\cdot\sum_{k=1}^{\infty}\sum_{j=o}^{r}\binom{r}{j}(-1)^{r+j+k-1}k\int_1^{\infty}y^{j-k\alpha-1}dy = \frac{2\alpha}{\beta^r} \\ &\cdot\sum_{k=1}^{\infty}\sum_{j=o}^{r}\binom{r}{j}\frac{k(-1)^{r+j+k-1}}{(k\alpha-j)}. \end{aligned} \tag{15}$$

Corollary 2. *The first four moments of X ~ HLL(α, β) are given, respectively, by*

$$\begin{aligned} \mu_1' &= E[X] = \frac{2}{\beta}\sum_{k=1}^{\infty}(-1)^{k+1}\times\frac{1}{(k\alpha-1)}, \\ \mu_2' &= E\left[X^2\right] = \frac{4}{\beta^2}\sum_{k=1}^{\infty}(-1)^{k+1}\times\frac{1}{(k\alpha-1)(k\alpha-2)}, \\ \mu_3' &= E\left[X^3\right] = \frac{12}{\beta^3}\sum_{k=1}^{\infty}(-1)^{k+1} \times\frac{1}{(k\alpha-1)(k\alpha-2)(k\alpha-3)}, \\ \mu_4' &= E\left[X^4\right] = \frac{48}{\beta^4}\sum_{k=1}^{\infty}(-1)^{k+1} \times\frac{1}{(k\alpha-1)(k\alpha-2)(k\alpha-3)(k\alpha-4)}. \end{aligned} \tag{16}$$

Proof. Putting $r = 1, 2, 3, 4$ in (13) yields the desired result.

The mean moments (μ_r) and cumulants (κ_r) of X can be obtained by the following relation;

$$\begin{aligned} \mu_r &= \sum_{j=0}^{r}(-1)^j\binom{r}{j}\mu_{r-j}'\left(\mu_1'\right)^j, \\ \kappa_r &= \mu_r' - \sum_{j=1}^{r-1}\binom{r-1}{j-1}\kappa_j\mu_{r-j}'. \end{aligned} \tag{17}$$

Here $\kappa_1 = \mu_1'$. We have $\kappa_2 = \mu_2' - \mu_1'^2$, $\kappa_3 = \mu_3' - 3\mu_2'\mu_1' + 2\mu_1'^3$, $\kappa_4 = \mu_4' - 4\mu_3'\mu_1' - 3\mu_2'^2 + 12\mu_2'\mu_1'^2 - 6\mu_1'^4$, and so on.

The skewness $\gamma_1 = \kappa_3/\kappa_2^{3/2}$ and kurtosis $\gamma_2 = \kappa_4/\kappa_2^2$ follow from the second, third, and fourth cumulants.

The variance of X, denoted by $\sigma^2 = \mu_2' - \mu_1'^2$, is given by

$$\begin{aligned} \sigma^2 &= \frac{4}{\beta^2}\sum_{k=1}^{\infty}(-1)^{k+1}\frac{1}{(k\alpha-1)(k\alpha-2)} \\ &\quad - \left[\frac{2}{\beta}\sum_{k=1}^{\infty}(-1)^{k+1}\frac{1}{(k\alpha-1)}\right]^2. \end{aligned} \tag{18}$$

The rth descending factorial moment of X is

$$\begin{aligned} \mu_{(r)}' &= E\left[X^{(r)}\right] = E\left[X(X-1)\times\cdots\times(X-r+1)\right] \\ &= \sum_{j=0}^{r}s(r,j)\mu_j', \end{aligned} \tag{19}$$

where

$$s(r,j) = \frac{1}{j!}\left[\frac{d^j}{dx^j}x^{(r)}\right]_{x=0} \tag{20}$$

is the Sterling number of the first kind which counts the number of ways to permute a list of r items into j cycles.

Theorem 3. *The moment generating function of X ~ HLL(α, β) is given by*

$$M_X(t) = 2\alpha e^{-t/\beta}\sum_{k=1}^{\infty}\sum_{i=0}^{\infty}\left(\frac{t}{\beta}\right)^i\frac{k(-1)^{k-1}}{(k\alpha-i)\,i!}. \tag{21}$$

Proof. We have

$$E\left[e^{tX}\right] = 2\alpha\beta\int_0^{\infty}\frac{e^{tx}(1+\beta x)^{-(\alpha+1)}}{\left[1+(1+\beta x)^{-\alpha}\right]^2}dx. \tag{22}$$

Setting $y = (1+\beta x)$,$dx = dy/\beta$ yields

$$E\left[e^{tX}\right] = 2\alpha e^{-t/\beta}\int_1^{\infty}y^{-(\alpha+1)}e^{ty/\beta}\left[1+y^{-\alpha}\right]^{-2}dy. \tag{23}$$

After simplification, we get

$$M_X(t) = 2\alpha e^{-t/\beta}\sum_{k=1}^{\infty}\sum_{i=0}^{\infty}\left(\frac{t}{\beta}\right)^i\frac{k(-1)^{k-1}}{(k\alpha-i)\,i!}. \tag{24}$$

3.4. Mean Residual Life (MRL) Function. The MRL function is important in reliability, survival analysis, actuarial sciences, economics, and social sciences for characterizing lifetime distributions. It also plays an important role in repair and replacement strategies and summarizes the entire residual life function.

Theorem 4. *The MRL function of $X \sim HLL(\alpha, \beta)$ is given by*

$$m(x) = E(X - x \mid X > x) = \frac{1}{\beta}\left[1 + (1+\beta x)^{\alpha}\right]\sum_{k=1}^{\infty}(-1)^{k-1}\frac{(1+\beta x)^{-(k\alpha-1)}}{(k\alpha-1)}. \quad (25)$$

Proof. The MRL function is defined as

$$m(x) = E(X - x \mid X > x) = \frac{1}{1-F(x)}\int_{x}^{\infty}[1-F(t)]\,dt. \quad (26)$$

Therefore, we have

$$m(x) = \left[1 + (1+\beta x)^{\alpha}\right]\int_{x}^{\infty}\frac{(1+\beta t)^{-\alpha}}{\left[1+(1+\beta t)^{-\alpha}\right]}dt. \quad (27)$$

Setting $y = (1+\beta x)$,$dx = dy/\beta$ yields

$$m(x) = \frac{1}{\beta}\left[1 + (1+\beta x)^{\alpha}\right]\int_{(1+\beta x)}^{\infty}\frac{y^{-\alpha}}{\left[1+y^{-\alpha}\right]}dy. \quad (28)$$

After simplification, we get (25).

3.5. Mean Deviations. The mean deviations about mean and median are, respectively, defined by

$$\delta_1(\mu) = 2\mu F(\mu) - 2\mu + 2\int_{\mu}^{\infty}xf(x)\,dx, \quad (29)$$

$$\delta_2(M) = 2MF(M) - M - \mu + 2\int_{M}^{\infty}xf(x)\,dx. \quad (30)$$

Theorem 5. *The mean deviations about mean and mean deviation about median of $X \sim HLL(\alpha, \beta)$ are given by*

$$\delta_1(\mu) = \frac{-4\mu}{\left[1+(1+\beta\mu)^{\alpha}\right]} + \frac{4\alpha}{\beta}\sum_{k=1}^{\infty}k(-1)^{k-1} \cdot \left[\frac{1}{(k\alpha-1)(1+\beta\mu)^{k\alpha-1}} - \frac{1}{k\alpha(1+\beta\mu)^{k\alpha}}\right], \quad (31)$$

$$\delta_2(M) = M\frac{\left[(1+\beta M)^{\alpha}-3\right]}{\left[(1+\beta M)^{\alpha}+1\right]} - \frac{2}{\beta}\sum_{k=1}^{\infty}\frac{1}{(k\alpha-1)} + \frac{4\alpha}{\beta} \cdot \sum_{k=1}^{\infty}k(-1)^{k-1} \cdot \left[\frac{1}{(k\alpha-1)(1+\beta M)^{k\alpha-1}} - \frac{1}{k\alpha(1+\beta M)^{k\alpha}}\right]. \quad (32)$$

Proof. Let $I = \int_{\mu}^{\infty}xf(x)dx$.

By inserting pdf of the HLL into (29) and setting $y = (1+\beta x)$, $dx = dy/\beta$, we get

$$I = \frac{2\alpha}{\beta}\int_{(1+\beta\mu)}^{\infty}\left(y^{-\alpha} - y^{-(\alpha+1)}\right)\left(1+y^{-\alpha}\right)^{-2}dy. \quad (33)$$

Using the series representations $(1+w)^{-2} = \sum_{k=1}^{\infty}k(-1)^{k-1}w^{k-1}$, where k is positive integer, we get

$$I = \frac{2\alpha}{\beta}\sum_{k=1}^{\infty}k(-1)^{k-1} \cdot \left[\frac{1}{(k\alpha-1)(1+\beta\mu)^{k\alpha-1}} - \frac{1}{k\alpha(1+\beta\mu)^{k\alpha}}\right]. \quad (34)$$

Setting I into (29) and after some manipulations we get the desired result.

Similarly, the measure $\delta_2(M)$ can be obtained.

3.6. Entropies. The entropy of a random variable X is a measure of variation of uncertainty. Two popular entropies are the Rényi and Shannon entropies.

The Rényi entropy is defined as

$$I_R(\gamma) = \frac{1}{1-\gamma}\log\left\{\int_{0}^{\infty}f^{\gamma}(x)\,dx\right\}, \quad \text{where } \gamma > 0,\ \gamma \neq 1. \quad (35)$$

The Rényi entropy for the HLL distribution is given by

$$\int_{0}^{\infty}f^{\gamma}(x)\,dx = (2\alpha\beta)^{r} \cdot \int_{0}^{\infty}(1+\beta x)^{-\gamma(\alpha+1)}\left[1+(1+\beta x)^{-\alpha}\right]^{-2\gamma}dx. \quad (36)$$

Setting $y = (1+\beta x)$, $dx = dy/\beta$ yields

$$\int_{0}^{\infty}f^{\gamma}(x)\,dx = \frac{(2\alpha\beta)^{r}}{\beta}\int_{1}^{\infty}y^{-\gamma(\alpha+1)}\left[1+y^{-\alpha}\right]^{-2\gamma}dy = \frac{(2\alpha\beta)^{r}}{\beta}\sum_{j=0}^{\infty}\frac{(-1)^{j}(2\gamma)_{j}}{j!(\gamma\alpha+j\alpha+\gamma-1)}. \quad (37)$$

Thus, we have

$$I_R(\gamma) = \frac{1}{(1-\gamma)}\log\left[\frac{(2\alpha\beta)^{r}}{\beta}\sum_{j=0}^{\infty}\frac{(-1)^{j}(2\gamma)_{j}}{j!(\gamma\alpha+j\alpha+\gamma-1)}\right]. \quad (38)$$

Another entropy measure defined by $E\{-\log[f(x)]\}$ is known as the Shannon entropy and plays a similar role as the kurtosis measure in comparing the shapes of densities and measuring tail heaviness. We have

$$E\{-\log[f(x)]\} = -\log(2\alpha\beta) + (\alpha+1)E\left[\log(1+\beta x)\right] + 2E\left[\log\left\{1+(1+\beta x)^{-\alpha}\right\}\right]. \quad (39)$$

By using the binomial series expansion and (2.727) of Gradshteyn and Ryzhik [16], it follows that

$$E\left[\log\left(1+\beta x\right)\right]=\frac{\log\left(4\right)}{\alpha},$$
$$E\left[\log\left\{1+\left(1+\beta x\right)^{-\alpha}\right\}\right]=\left[1-\log\left(2\right)\right]. \tag{40}$$

Hence, the Shannon entropy can be expressed in the form

$$E\left\{-\log\left[f\left(x\right)\right]\right\}=\frac{\log\left(4\right)}{\alpha}-\log\left(2\alpha\beta\right)+2. \tag{41}$$

3.7. Order Statistics. Let $X_1, X_2, \ldots, X_n$ be a random sample of size n from a distribution with pdf $f(x)$ and cdf $F(x)$. Let $X_{1:n}, X_{2:n}, \ldots, X_{n:n}$ denote the corresponding order statistics. Then the pdf and cdf of $X_{r:n}$ are

$$\begin{aligned}f_{r:n}\left(x\right)&=\frac{1}{B\left(r,n-r+1\right)}f\left(x\right)\left[F\left(x\right)\right]^{r-1}\cdot\left[1-F\left(x\right)\right]^{n-r}\\&=\frac{1}{B\left(r,n-r+1\right)}f\left(x\right)\sum_{i=0}^{n-r}\binom{n-r}{i}\left(-1\right)^{i}\left[F\left(x\right)\right]^{i+r-1},\end{aligned} \tag{42}$$

$$F_{r:n}\left(x\right)=\sum_{k=r}^{n}\binom{n}{k}\left[F\left(x\right)\right]^{k}\left[1-F\left(x\right)\right]^{n-k},\quad 1\le r\le n, \tag{43}$$

where $B(a,b)$ is the complete beta function.

Theorem 6. *Let $f(x)$ and $F(x)$ be the pdf and cdf of $X\sim HLL(\alpha,\beta)$. Then the pdf and cdf of $X_{r:n}$ are*

$$\begin{aligned}f_{r:n}\left(x\right)&=\frac{1}{B\left(r,n-r+1\right)}\frac{2\alpha\beta\left(1+\beta x\right)^{-(\alpha+1)}}{\left[1+\left(1+\beta x\right)^{-\alpha}\right]^{2}}\cdot\sum_{i=0}^{n-r}\binom{n-r}{i}\left(-1\right)^{i}\left\{\frac{1-\left(1+\beta x\right)^{-\alpha}}{1+\left(1+\beta x\right)^{-\alpha}}\right\}^{i+r-1},\\F_{r:n}\left(x\right)&=\sum_{k=r}^{n}\binom{n}{k}\left\{\frac{1-\left(1+\beta x\right)^{-\alpha}}{1+\left(1+\beta x\right)^{-\alpha}}\right\}^{k}\cdot\left[\frac{2}{1+\left(1+\beta x\right)^{\alpha}}\right]^{n-k}.\end{aligned} \tag{44}$$

Proof. Inserting (6) and (7) into (44), we get the result.

4. Estimation and Asymptotic Distribution

We consider the maximum-likelihood estimation of the parameters α, β. The distributional properties of $\widehat{\alpha}$, $\widehat{\beta}$ are obtained using Fisher information matrix.

4.1. Maximum-Likelihood Estimation. Let $x_1, x_2, \ldots, x_n$ be a random sample of size n from HLL(α,β) distribution. Then the corresponding likelihood function is

$$L\left(\alpha,\beta\right)=\left(2\alpha\beta\right)^{n}\prod_{i=1}^{n}\frac{\left(1+\beta x_i\right)^{-(\alpha+1)}}{\left[1+\left(1+\beta x_i\right)^{-\alpha}\right]^{2}}. \tag{45}$$

The log-likelihood function is given by

$$\ell\left(x;\alpha,\beta\right)=n\log\left(2\alpha\beta\right)-\left(\alpha+1\right)\sum_{i=1}^{n}\log\left(1+\beta x_i\right)-2\sum_{i=1}^{n}\log\left[1+\left(1+\beta x_i\right)^{-\alpha}\right]. \tag{46}$$

Calculating the first-order partial derivatives of (46) with respect to α, β and equating them to zero, we get the following nonlinear equations:

$$\begin{aligned}\frac{\partial\ell}{\partial\alpha}&=\frac{n}{\alpha}-\sum_{i=1}^{n}\log\left(1+\beta x_i\right)+2\sum_{i=1}^{n}\frac{\log\left(1+\beta x_i\right)}{\left[1+\left(1+\beta x_i\right)^{\alpha}\right]}=0,\\\frac{\partial\ell}{\partial\beta}&=\frac{n}{\beta}-\left(\alpha+1\right)\sum_{i=1}^{n}\frac{x_i}{\left(1+\beta x_i\right)}+2\alpha\sum_{i=1}^{n}\frac{x_i}{\left(1+\beta x_i\right)\left[1+\left(1+\beta x_i\right)^{\alpha}\right]}=0.\end{aligned} \tag{47}$$

To find out the maximum-likelihood estimates (MLEs) of α, β, we have to solve the above nonlinear equations. Apparently, there is no closed form solution in α, β. We have to use a numerical technique method, such as Newton-Raphson method, to obtain the solution.

4.2. Asymptotic Distribution. The normal approximation for the large-sample for the MLE can be used to compute confidence intervals (CIs). For finding the observed information matrix of α, β, we compute the second-order partial derivatives of l (46) in the appendix.

The variance-covariance matrix may be approximated by $\Sigma = I^{-1}$, where I is the observed information matrix. Since Σ involves the parameters α, β, we replace the parameters by the corresponding MLEs in order to obtain an estimate of Σ, which is denoted by $\widehat{\Sigma}=\widehat{I}^{-1}$, where $\widehat{I}_{ij}=I_{ij}$, when $(\widehat{\alpha},\widehat{\beta})$is replaced by α, β. Using these results, an approximate $100(1-\vartheta)\%$ CI for α, β, respectively, is given by

$$\begin{aligned}&\widehat{\alpha}\pm z_{\vartheta/2}\sqrt{\widehat{\Sigma}_{11}},\\&\widehat{\beta}\pm z_{\vartheta/2}\sqrt{\widehat{\Sigma}_{22}},\end{aligned} \tag{48}$$

where $z_{\vartheta/2}$ is the upper $(\vartheta/2)$th percentile of the standard normal distribution.

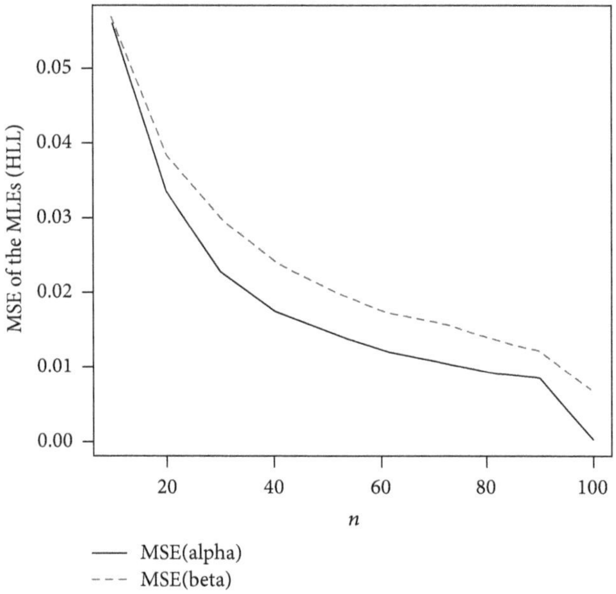

FIGURE 3: Plots of the MSE($\widehat{\alpha}$) and MSE($\widehat{\beta}$) for $\alpha = 2$, $\beta = 1$ based on 10,000 replications.

5. Simulation Study

The assessment of the maximum-likelihood estimates is based on a simulation study. The following steps were followed:

(1) Generate 5,000 samples of size n from (6). The HLL variables are generated using $X = (1/\beta)[((1-U)/(1+U))^{1/\alpha} - 1]$ given in (12), where $U \sim \text{Uniform}(0,1)$.

(2) Compute the MLEs for the 5,000 samples, say $(\widehat{\alpha}_i, \widehat{\beta}_i)$ for $i = 1, 2, \ldots, 5000$.

(3) Compute the mean square errors (MSE) for each parameter.

We repeat these steps for $n = 10, 20, \ldots, 100$ with $\alpha = 2$, $\beta = 1$. The comparison is based on MSEs. Figure 3 plots the MSEs of the MLEs of two parameters. The assessment based on this simulation study is that the MSEs for each parameter decrease to zero with increasing sample size.

6. Application

6.1. Cancer Data. We have considered a real data set, which records the remission times (in months) of a random sample of 128 bladder cancer patients reported in [17]. We have fitted the HLL distribution to the data set using MLE and compared the proposed distribution with the following distributions:

(i) The Lomax distribution, introduced by Lomax [1], with pdf is

$$f_{\text{Lomax}}(x) = \alpha\beta(1+\beta x)^{-(\alpha+1)}, \quad \text{where } x, \alpha, \beta > 0. \tag{49}$$

(ii) The McLomax distribution, introduced by Lemonte and Cordeiro [10], with pdf is

$$f_{\text{McLomax}}(x) = \frac{c\alpha\beta^{\alpha}(\beta+x)^{-(\alpha+1)}}{B(ac^{-1}, \eta+1)} \left\{1 - \left(\frac{\beta}{\beta+x}\right)^{\alpha}\right\}^{a-1} \cdot \left[1 - \left\{1 - \left(\frac{\beta}{\beta+x}\right)^{\alpha}\right\}^{c}\right]^{\eta}, \tag{50}$$

where $x, \alpha, \beta, a, c > 0$, $0 \le \eta$.

(iii) The beta Lomax (BLomax) distribution is a submodel of McLomax for $c = 1$ with pdf

$$f_{\text{BLomax}}(x) = \frac{\alpha\beta^{\alpha}(\beta+x)^{-(\alpha+1)}}{B(a, \eta+1)} \left\{1 - \left(\frac{\beta}{\beta+x}\right)^{\alpha}\right\}^{a-1} \cdot \left[1 - \left\{1 - \left(\frac{\beta}{\beta+x}\right)^{\alpha}\right\}\right]^{\eta}, \tag{51}$$

where $x, \alpha, \beta, a > 0$, $0 \le \eta$.

(iv) The Kumaraswamy Lomax (KwLomax) distribution is a submodel of McLomax for $a = c$, with pdf

$$f_{\text{KwLomax}}(x) = \frac{a\alpha\beta^{\alpha}(\beta+x)^{-(\alpha+1)}}{(\eta+1)} \left\{1 - \left(\frac{\beta}{\beta+x}\right)^{\alpha}\right\}^{a-1} \cdot \left[1 - \left\{1 - \left(\frac{\beta}{\beta+x}\right)^{\alpha}\right\}^{a}\right]^{\eta}, \tag{52}$$

where $x, \alpha, \beta, a > 0$, $0 \le \eta$.

(v) The Exponentiated Standard Lomax (ESLomax) distribution, considered by Gupta et al. [18], with pdf is

$$f_{\text{ESLomax}}(x) = \frac{\alpha}{a}(1+x)^{-(\alpha+1)}\left\{1 - (1+x)^{-\alpha}\right\}^{a-1}, \tag{53}$$

where $x, \alpha, a > 0$.

(vi) The Lomax-Logarithmic (LL) distribution, introduced by Al-Zahrani and Sagor [11], with pdf is

$$f_{\text{LL}}(x) = \frac{(\eta-1)\alpha\beta(1+\beta x)^{-(\alpha+1)}}{\left\{1 - (1-\eta)(1+\beta x)^{-\alpha}\right\}\ln(\eta)}, \tag{54}$$

where $x, \alpha, \beta, \eta > 0$.

For identification of the shape of the hazard function of the cancer data set, we use a graphical method based on the Total Time on Test (TTT) plot [19]. Let T be a random variable with nonnegative values, which represents the survival time. The empirical TTT curve is constructed by plotting $T(r/n) = [\sum_{i=1}^{r} Y_{i:n} + (n-r)Y_{r:n}]/(\sum_{i=1}^{n} Y_{i:n})$, against r/n, where $r = 1, 2, \ldots, n$ and $Y_{i:n}$ $(i = 1, 2, \ldots, n)$ are the order statistics of the sample. If the empirical TTT transform is straight

Table 1: Descriptive statistics.

Mean	Median	Mode	SD	Variance	Skewness	Kurtosis	Min.	Max.
9.37	6.40	5.0	10.51	110.43	3.287	15.48	0.08	79.05

Table 2: MLEs (standard errors in parentheses) and the measures AIC, BIC, HQIC, and CAIC for cancer data.

Model	Estimates					Measures			
	α	β	a	η	c	AIC	BIC	HQIC	CAIC
HLL	5.2814	0.0367				**827.69**	**833.40**	**830.01**	**827.79**
	(2.0847)	(0.0177)							
Lomax	13.9384	121.0225				831.67	837.37	833.98	831.76
	(15.368)	(142.714)							
McLomax	0.8085	11.2929	1.5060	4.1886	2.1046	829.82	844.09	835.62	830.31
	(3.364)	(15.818)	(0.234)	(25.029)	(3.079)				
BLomax	3.9191	23.9281	1.5853	0.1572		828.14	839.55	832.78	828.46
	(18.192)	(27.338)	(0.280)	(5.024)					
KwLomax	0.3911	12.2973	1.5162	11.0323		827.88	839.29	832.52	828.21
	(2.386)	(17.316)	(0.228)	(87.144)					
ESLomax	1.0877		4.6575			856.61	862.31	833.98	856.71
	(0.086)		(0.686)						
LL	4.8696	20.1652		7.6938		830.88	839.44	834.36	831.07
	(2.1317)	(18.3747)		(11.8362)					

diagonal, hrf is constant; if it is convex, hrf is decreasing; if it is concave, hrf is increasing; if it is convex and then concave, hrf is bathtub; and if it is concave and then convex, hrf is upside-down bathtub [19]. Figure 4 shows the empirical TTT plot for the data set. The shape of the data set is unimodal, as its TTT plot is concave and then convex. The model selection is carried out using Akaike information criterion (AIC), the Bayesian information criterion (BIC), the consistent Akaike information criterion (CAIC), and the Hannan-Quinn information criterion (HQIC).

$$
\begin{aligned}
\mathrm{AIC} &= -2l\left(\hat{\theta}\right) + 2q, \\
\mathrm{BIC} &= -2l\left(\hat{\theta}\right) + q\log(n), \\
\mathrm{HQIC} &= -2l\left(\hat{\theta}\right) + 2q\log(\log(n)), \\
\mathrm{CAIC} &= -2l\left(\hat{\theta}\right) + \frac{2qn}{n-q-1},
\end{aligned}
\tag{55}
$$

where $l(\hat{\theta})$ denotes the log-likelihood function evaluated at the MLEs, q is the number of model parameters, and n is the sample size. The model with the lowest values for these statistics could be chosen as the best model to fit the data. First, we describe the data set in Table 1. Then, we report the maximum-likelihood estimators (and the corresponding standard errors in parentheses) of the parameters and values of AIC, BIC, HQIC, and CAIC in Table 2. In Table 2, we provide some results reported in [10, 11]. We conclude that our new model HLL has the smallest AIC, BIC, HQIC, and CAIC values among all fitted models, and so it could be chosen as the best model. These results are obtained using the AdequacyModel package version 1.0.8 available for the programming language R [20].

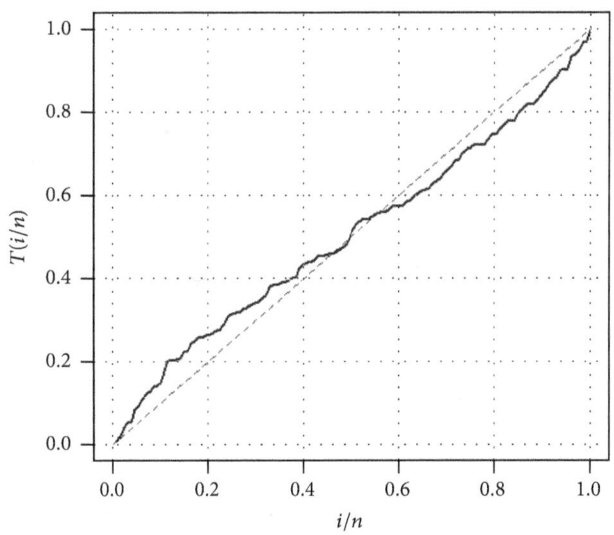

Figure 4: Empirical TTT plot for the cancer data.

6.2. Simulated Data. Here, we used a simulation study to check the flexibility of the proposed distribution. We have generated a sample of size $n = 100$ from the six distributions and compared the fit of the HLL distribution with competing models.

The details of generated samples with the specified values of the parameters from the six models are as follows:

TABLE 3: MLEs (standard errors in parentheses) and the measures AIC, BIC, HQIC, and CAIC for simulated data from Lomax distribution with $\alpha = 1.5$ and $\beta = 1$.

Model	Estimates					Measures			
	α	β	a	η	c	AIC	BIC	HQIC	CAIC
HLL	1.8032 (0.4230)	1.2386 (0.4764)				**257.03**	**262.24**	**259.14**	**257.16**
Lomax	2.2441 (0.7434)	0.5337 (0.2437)				257.39	262.60	259.50	257.52
McLomax	2.7515 (2.4395)	2.3568 (0.2818)	0.9418 (0.1849)	−0.1016 (0.8005)	2.8189 (3.079)	262.90	275.93	268.17	263.54
BLomax	1.4533 (8.1920)	2.2421 (0.9818)	0.9361 (0.1699)	0.6659 (0.0624)		260.90	271.32	265.12	261.32
KwLomax	8.5753 (0.7854)	2.0767 (7.8953)	0.9188 (0.8728)	−0.7243 (7.1544)		260.90	271.32	265.12	261.32
ESLomax	1.5909 (0.0860)		1.1135 (0.6861)			258.50	263.70	260.60	258.62
LL	2.6536 (0.3217)	0.3359 (0.23747)		0.5196 (0.08362)		258.96	266.77	262.12	259.21

TABLE 4: MLEs (standard errors in parentheses) and the measures AIC, BIC, HQIC, and CAIC for simulated data from McLomax distribution with parameters $\alpha = 5.25$, $\beta = 2.55$, $a = 1.45$, $\eta = 0.67$, and $c = 3.81$.

Model	Estimates					Measures			
	α	β	a	η	c	AIC	BIC	HQIC	CAIC
HLL	1.2207 (2.0847)	2.4477 (0.0177)				268.28	273.50	270.39	268.41
Lomax	1.4243 (0.4009)	1.0314 (0.4619)				267.49	272.70	269.60	267.61
McLomax	14.1535 (0.1364)	3.2796 (5.3498)	0.8091 (0.9234)	−0.8197 (9.1029)	2.9511 (4.0079)	266.95	279.97	272.22	267.58
BLomax	1.9585 (6.0364)	3.7068 (3.4102)	0.6586 (0.1077)	0.3007 (4.0353)		265.67	276.09	269.89	266.09
KwLomax	0.3924 (0.6603)	7.7169 (12.8494)	0.6896 (0.0966)	6.1821 (11.7374)		265.39	275.81	269.61	265.81
ESLomax	1.2764 (0.1748)		0.8217 (0.1031)			264.94	**270.15**	**267.04**	265.06
LL	8.8655 (15.7662)	0.03389 (0.0695)		0.0359 (0.0278)		**264.02**	271.83	267.18	**264.27**

(1) Lomax distribution with $\alpha = 1.5$ and $\beta = 1$ (see Table 3)

(2) McLomax distribution with parameters $\alpha = 5.25$, $\beta = 2.55$, $a = 1.45$, $\eta = 0.67$, and $c = 3.81$ (see Table 4)

(3) BLomax distribution with parameters $\alpha = 0.85$, $\beta = 12.55$, $a = 1.75$, and $\eta = 4.67$ (see Table 5)

(4) KwLomax distribution with parameters $\alpha = 0.85$, $\beta = 12.55$, $a = 1.75$, and $\eta = 4.67$ (see Table 6)

(5) ESLomax distribution with parameters $\alpha = 2.5$ and $a = 1.15$ (see Table 7)

(6) LL distribution with parameters $\alpha = 2.75$, $\beta = 0.45$, and $\eta = 0.75$ (see Table 8)

The comparison is based upon the formal goodness-of-fit tests given above. The values of best fitted model are highlighted in the tables. Based on the goodness-of-fit tests, we conclude that the HLL distribution has better fit than the other competing models.

7. Concluding Remarks

In this paper, we introduce a two-parameter HLL distribution by compounding the half-logistic and the Lomax distributions. The shape of hazard function of the new compounding distribution can be monotonically decreasing or upside-down bathtub (unimodal). Some mathematical and statistical properties of the new model including moments and moment generating functions, mean deviations from mean and median, quantile function, mean residual life function, mode, median, order statistics, and entropies are studied.

TABLE 5: MLEs (standard errors in parentheses) and the measures AIC, BIC, HQIC, and CAIC for simulated data from BLomax distribution with parameters $\alpha = 0.85$, $\beta = 12.55$, $a = 1.75$, and $\eta = 4.67$.

Model	Estimates					Measures			
	α	β	a	η	c	AIC	BIC	HQIC	CAIC
HLL	24.4487 (13.7405)	0.0112 (0.0065)				533.90	**539.11**	**536.01**	534.02
Lomax	23.6200 (11.1025)	0.0081 (0.0038)				544.04	549.25	546.15	544.17
McLomax	1.3135 (9.1629)	18.6449 (73.1176)	1.4939 (0.3102)	9.3587 (78.4737)	2.4534 (3.8777)	534.96	547.99	540.23	535.60
BLomax	2.0366 (26.5471)	93.9001 (159.3791)	1.7237 (0.2569)	13.8059 (194.682)		533.49	543.91	537.71	533.91
KwLomax	0.9258 (2.386)	35.7112 (17.316)	1.5236 (0.228)	21.0737 (139.842)		**533.05**	543.47	537.27	**533.47**
ESLomax	1.3483 (0.1185)		5.0645 (0.8554)			561.43	566.64	563.53	561.55
LL	10.6838 (5.5325)	0.0392 (0.0261)		16.8989 (15.5951)		535.87	543.69	539.04	536.12

TABLE 6: MLEs (standard errors in parentheses) and the measures AIC, BIC, HQIC, and CAIC for simulated data from KwLomax distribution with parameters $\alpha = 0.85$, $\beta = 12.55$, $a = 1.75$, and $\eta = 4.67$.

Model	Estimates					Measures			
	α	β	a	η	c	AIC	BIC	HQIC	CAIC
HLL	3.2092 (1.0615)	0.1872 (0.0847)				**454.88**	**460.09**	**456.99**	**455.02**
Lomax	5.2637 (3.0135)	0.0650 (0.0436)				455.74	460.95	457.85	455.87
McLomax	36.3053 (49.9841)	64.3913 (87.0837)	1.4745 (0.2732)	−0.6614 (0.1299)	26.4402 (21.5904)	455.68	468.70	460.95	456.32
BLomax	20.4375 (18.192)	8.2358 (1.3201)	1.6963 (0.3709)	−0.8265 (0.0270)		455.46	465.89	459.68	455.89
KwLomax	1.4823 (2.386)	3.9164 (7.0316)	1.5331 (0.3464)	0.8043 (8.0144)		455.73	466.15	459.95	456.15
ESLomax	1.3936 (0.1429)		2.6629 (0.4120)			457.01	462.22	459.12	457.13
LL	3.9058 (2.9870)	0.1236 (0.1943)		2.3957 (4.9588)		457.48	465.30	460.65	457.73

TABLE 7: MLEs (standard errors in parentheses) and the measures AIC, BIC, HQIC, and CAIC for simulated data from ESLomax distribution with parameters $\alpha = 2.5$ and $a = 1.15$.

Model	Estimates					Measures			
	α	β	a	η	c	AIC	BIC	HQIC	CAIC
HLL	2.9230 (0.8789)	1.4697 (0.6175)				70.15	75.35	72.25	70.27
Lomax	4.4945 (2.3241)	0.5392 (0.3345)				71.43	76.64	73.54	71.55
McLomax	9.2546 (18.1860)	0.8577 (0.7067)	1.9039 (0.9458)	−0.6952 (0.5267)	2.5132 (3.9407)	72.37	85.40	77.65	73.01
BLomax	0.5806 (1.7619)	0.4735 (0.3488)	1.6416 (0.4155)	3.3194 (13.6281)		70.61	81.04	74.83	71.04
KwLomax	2.4388 (1.1386)	0.5530 (0.6732)	1.6260 (0.1428)	−0.0069 (7.9604)		70.64	81.07	74.86	71.07
ESLomax	3.4021 (0.4127)		1.3948 (0.1956)			**67.33**	**72.54**	**69.44**	**67.46**
LL	3.1584 (1.3691)	1.2813 (1.4460)		3.5394 (6.8140)		72.86	80.68	76.03	73.12

TABLE 8: MLEs (standard errors in parentheses) and the measures AIC, BIC, HQIC, and CAIC for simulated data from LL distribution with parameters $\alpha = 2.75$, $\beta = 0.45$, and $\eta = 0.75$.

Model	Estimates					Measures			
	α	β	a	η	c	AIC	BIC	HQIC	CAIC
HLL	1.8358 (0.4189)	1.2344 (0.4553)				**252.97**	**258.19**	**255.08**	**253.10**
Lomax	2.1811 (0.6717)	0.5669 (0.2408)				253.22	258.43	255.33	253.34
McLomax	0.7992 (2.8691)	1.2223 (1.7168)	0.9887 (0.1874)	1.6530 (10.1760)	1.9169 (3.9962)	259.10	272.12	264.37	259.73
BLomax	1.6711 (94.5364)	1.7124 (1.0299)	1.0130 (0.1929)	0.2907 (73.1088)		257.21	267.63	261.43	257.63
KwLomax	8.6899 (7.6698)	1.7774 (0.8502)	0.9917 (0.2342)	−0.7481 (0.2298)		257.22	267.64	261.44	257.64
ESLomax	1.6418 (0.2057)		1.1489 (0.1530)			254.10	259.31	256.21	254.23
LL	1.9522 (0.5707)	1.0375 (1.7363)		3.1618 (11.1845)		255.07	262.89	258.24	255.32

We estimate the model parameters by maximum likelihood and determined the observed information matrix. We present a simulation study to illustrate the performance of MLEs. The flexibility and potentiality of the proposed model are illustrated by means of a real data set. We hope that the HLL distribution may attract wider range of applications in areas such as engineering, survival and lifetime data, economics, meteorology, hydrology, and others.

Appendix

Elements of Observed Information Matrix

The elements of 2×2 observed information matrix (I) are given by

$$\begin{aligned} I_{11} &= \frac{\partial^2 l}{\partial \alpha^2} = -\frac{n}{\alpha^2} + 2\sum_{i=1}^{n} \frac{(1+\beta x_i)^{\alpha}\left[\log(1+\beta x_i)\right]^2}{\left[1+(1+\beta x_i)^{\alpha}\right]^2}, \\ I_{12} = I_{21} &= \frac{\partial^2 l}{\partial\alpha\partial\beta} \\ &= -\sum_{i=1}^{n} \frac{x_i}{(1+\beta x_i)} \\ &\quad + 2\sum_{i=1}^{n} \frac{x_i\left[1+(1+\beta x_i)^{\alpha}\right] + \alpha x_i \log(1+\beta x_i)}{(1+\beta x_i)^{\alpha+1}\left[1+(1+\beta x_i)^{-\alpha}\right]^2}, \\ I_{22} &= \frac{\partial^2 l}{\partial \beta^2} \\ &= -\frac{n}{\beta^2} + (\alpha+1)\sum_{i=1}^{n} \frac{x_i^2}{(1+\beta x_i)^2} \\ &\quad - 2\alpha \sum_{i=1}^{n} \frac{x_i^2\left[(\alpha+1)(1+\beta x_i)^{\alpha}+1\right]}{(1+\beta x_i)\left[1+(1+\beta x_i)^{\alpha}\right]}. \end{aligned} \tag{A.1}$$

Disclosure

The authors acknowledge that the abstract of the manuscript had been presented in a conference paper under the link http://archive.stat.uconn.edu/lida17/program.pdf.

Conflicts of Interest

The authors declare that they have no conflicts of interest.

Authors' Contributions

The authors have equally contributed to this work. All authors read and approved the final manuscript.

References

[1] K. S. Lomax, "Business failures: another example of the analysis of failure data," *Journal of the American Statistical Association*, vol. 49, no. 268, pp. 847–852, 1954.

[2] E.-H. A. Rady, W. A. Hassanein, and T. A. Elhaddad, "The power Lomax distribution with an application to bladder cancer data," *SpringerPlus*, vol. 5, 1838 pages, 2016.

[3] M. H. Tahir, G. M. Cordeiro, M. Mansoor, and M. Zubair, "The Weibull-Lomax distribution: properties and applications," *Hacettepe Journal of Mathematics and Statistics*, vol. 44, no. 2, pp. 455–474, 2015.

[4] I. B. Abdul-Moniem, "Recurrence relations for moments of lower generalized order statistics from exponentiated Lomax distribution and its characterization," *International Journal of Mathematical Archive*, vol. 3, pp. 2144–2150, 2012.

[5] S. A. Al-Awadhi and M. E. Ghitany, "Statistical properties of Poisson-Lomax distribution and its application to repeated accidents data," *Journal of Applied Statistical Science*, vol. 10, no. 4, pp. 365–372, 2001.

[6] A. Asgharzadeh, H. S. Bakouch, and L. Esmaeili, "Pareto Poisson-Lindley distribution with applications," *Journal of Applied Statistics*, vol. 40, no. 8, pp. 1717–1734, 2013.

[7] G. M. Cordeiro, E. M. Ortega, and B. z. Popović, "The gamma-Lomax distribution," *Journal of Statistical Computation and Simulation*, vol. 85, no. 2, pp. 305–319, 2015.

[8] M. E. Ghitany, F. A. Al-Awadhi, and L. A. Alkhalfan, "Marshall-Olkin extended Lomax distribution and its application to censored data," *Communications in Statistics—Theory and Methods*, vol. 36, no. 9-12, pp. 1855–1866, 2007.

[9] R. C. Gupta, M. E. Ghitany, and D. K. Al-Mutairi, "Estimation of reliability from Marshall-Olkin extended Lomax distributions," *Journal of Statistical Computation and Simulation*, vol. 80, no. 7-8, pp. 937–947, 2010.

[10] A. J. Lemonte and G. M. Cordeiro, "An extended Lomax distribution," *Statistics*, vol. 47, no. 4, pp. 800–816, 2013.

[11] B. Al-Zahrani and H. Sagor, "Statistical analysis of the Lomax-logarithmic distribution," *Journal of Statistical Computation and Simulation*, vol. 85, no. 9, pp. 1883–1901, 2015.

[12] A. H. El-Bassiouny, N. F. Abdo, and H. S. Shahen, "Exponential lomax distribution," *International Journal of Computer Applications*, vol. 121, no. 13, pp. 24–29, 2015.

[13] N. M. Kilany, "Weighted Lomax distribution," *SpringerPlus*, vol. 5, no. 1, article no. 1862, 2016.

[14] G. M. Cordeiro, M. Alizadeh, and P. R. D. Marinho, "The type I half-logistic family of distributions," *Journal of Statistical Computation and Simulation*, vol. 86, no. 4, pp. 707–728, 2016.

[15] R. E. Glaser, "Bathtub and related failure rate characterizations," *Journal of the American Statistical Association*, vol. 75, no. 371, pp. 667–672, 1980.

[16] I. S. Gradshteyn and I. M. Ryzhik, *Tables of Integrals, Series, and Products*, Academic Press, New York, NY, USA, 2007.

[17] E. T. Lee and J. W. Wang, *Statistical Methods for Survival Data Analysis*, John Wiley and Sons, New York, NY, USA, 3rd edition, 2003.

[18] R. C. Gupta, P. L. Gupta, and R. D. Gupta, "Modeling failure time data by Lehman alternatives," *Communications in Statistics—Theory and Methods*, vol. 27, no. 4, pp. 887–904, 1998.

[19] M. V. Aarset, "How to identify a bathtub hazard rate," *IEEE Transactions on Reliability*, vol. 36, no. 1, pp. 106–108, 1987.

[20] R Core Team, R: A Language and Environment for Statistical Computing, R Foundation for Statistical Computing, Austria, 2015, https://www.R-project.org/.

15

Accuracy of Approximation for Discrete Distributions

Tamás F. Móri

Department of Probability Theory and Statistics, Eötvös Loránd University, Pázmány P. s. 1/C, Budapest 1117, Hungary

Correspondence should be addressed to Tamás F. Móri; mori@math.elte.hu

Academic Editor: Ramón M. Rodríguez-Dagnino

The paper is a contribution to the problem of estimating the deviation of two discrete probability distributions in terms of the supremum distance between their generating functions over the interval $[0,1]$. Deviation can be measured by the difference of the kth terms or by total variation distance. Our new bounds have better order of magnitude than those proved previously, and they are even sharp in certain cases.

1. Introduction

Dealing with random combinatorial structures needs to estimate the deviation of two discrete probability distributions in terms of the maximal distance between their generating functions over $[0,1]$. This is often the case when given a couple of not necessarily independent random events, one needs to estimate the number N of those that occur. Among many popular methods of Poisson approximation, sieve methods with generalized Bonferroni bounds, such as the graph sieve [1], are at hand. They provide estimates not only for the probability that none of the events occur, but also for the difference between the generating function of N and that of the corresponding Poisson distribution over the interval $[0,1]$ (see [2] for more details). This raises the following problem.

Let $\mathbf{p} = (p_0, p_1, \ldots)$ and $\mathbf{q} = (q_0, q_1, \ldots)$ be discrete probability distributions with generating functions $g_{\mathbf{p}}(z)$ and $g_{\mathbf{q}}(z)$, respectively. Given $\Delta = \max_{0\le z\le 1}|g_{\mathbf{p}}(z) - g_{\mathbf{q}}(z)|$, estimate the distance $|p_k - q_k|$ of kth terms, or the total variation distance between $\mathbf{p}$ and $\mathbf{q}$ defined as $\|\mathbf{p} - \mathbf{q}\| = \sum_{k=0}^{\infty} |p_k - q_k|$.

The difficulty lies in the constraint that the difference of the generating functions is only available over the real interval $[0,1]$, and not over the whole complex unit disc, which would make it possible to apply standard methods of characteristic functions.

Several positive and negative results were achieved in the last three decades, beginning with [3]; see Section 2. Lower and upper estimates got closer and closer, but the final answer is still ahead. The aim of the present note is to provide new bounds that have better order of magnitude than those proved previously. They are even sharp in certain cases.

The paper is organized as follows. In Section 2 we introduce the necessary notions and notations and cite some earlier results. Section 3 is devoted to the case where $|p_k - q_k|$ is estimated in terms of Δ, while in Section 4 the total variation distance is treated.

2. Preliminaries

The difference of two probability generating function belongs to the following class. Let $\mathcal{F}$ denote the set of all functions $f : [0,1] \to \mathbb{R}$ such that

$$f(z) = \sum_{k=0}^{\infty} a_k z^k, \quad \sum_{k=0}^{\infty} |a_k| \le 2, \quad f(1) = 0. \tag{1}$$

On the other hand, every $f \in \mathcal{F}$ can be obtained as the difference of two probability generating functions, namely, $f(z) = g_1(z) - g_2(z)$, where g_1 and g_2 can be constructed in the following way. Let $0 \le b_k \le 1 - |a_k|$, $k = 0, 1, \ldots$ (note that $|a_k| \le 1$ for $f \in \mathcal{F}$) such that

$$\sum_{k=0}^{\infty} b_k = 1 - \frac{1}{2}\sum_{k=0}^{\infty} |a_k| = 1 - \sum_{k=0}^{\infty} a_k^{+} = 1 - \sum_{k=0}^{\infty} a_k^{-}, \tag{2}$$

where $x^+ = \max\{x, 0\}$, $x^- = (-a_k)^+$. Then set

$$g_1(z) = \sum_{k=0}^{\infty} (a_k^+ + b_k) z^k,$$
$$g_2(z) = \sum_{k=0}^{\infty} (a_k^- + b_k) z^k. \tag{3}$$

For $f \in \mathcal{F}$ let $\Delta = \Delta(f) = \max_{0 \le z \le 1} |f(z)|$, and $L = \log(1/\Delta)$. For $d \in (0, 1]$ and $k = 0, 1, \ldots$ define

$$\varphi_k(d) = \sup\{|a_k| : f \in \mathcal{F},\ \Delta(f) = d\}. \tag{4}$$

In [2] it was shown that, for every $k = 0, 1, \ldots$,

$$C_k \Delta L^{2k} \le \varphi_k(\Delta) \le \Delta \exp\left(2(kL)^{4/5}\right) \tag{5}$$

holds with a suitable positive constant C_k, if Δ is sufficiently small. In fact, the upper estimate is valid for every $\Delta \in (0, 1]$, but it becomes trivial for $\Delta \ge \exp(-32k^4)$ by elementary calculus. Though in both the upper and lower bounds the multiplier of Δ is slowly varying as $\Delta \to 0$, they are not of the same order of magnitude.

It is easy to see that $\sum_{k=0}^{\infty} |a_k|$ cannot be estimated in a nontrivial way, because for arbitrary $\Delta \in (0, 1]$ there exists a function $f \in \mathcal{F}$, such that $\Delta(f) = \Delta$ and $\sum_{k=0}^{\infty} |a_k| = 2$, maximal. Indeed, let $n > 1/\Delta$ and

$$\begin{aligned} f(z) &= \Delta - \Delta z + (1 - \Delta) z^n - (1 - \Delta) z^{n+1} \\ &= \Delta - \Delta z + z(1 - \Delta)\left(z^{n-1} - z^n\right). \end{aligned} \tag{6}$$

Then $0 \le f(z) \le \Delta$ for $z \in [0, 1]$, because

$$z^{n-1} - z^n \le \left(\frac{n-1}{n}\right)^{n-1} \frac{1}{n} < \frac{1}{n-1} < \frac{\Delta}{1 - \Delta} \tag{7}$$

by the choice of n, provided $\Delta < 1$. For $\Delta = 1$ the estimation obviously holds. Hence $\Delta(f) = f(0) = \Delta$.

However, if $f = g_{\mathbf{p}} - g_{\mathbf{q}}$, and one of the distributions, say $\mathbf{p}$, is fixed (as in the case of Poisson approximation), then the class of feasible functions is smaller; thus the bounds for $|a_k|$ may decrease, and even the total variation distance can be estimated nontrivially. The following results can be found in [2].

Let $\mathbf{p}$ be a fixed discrete probability distribution such that $p_k > 0$ for every $k = 0, 1, \ldots$, and

$$\limsup_{k \to \infty} \frac{1}{k} \log \frac{1}{p_k} < \infty. \tag{8}$$

Let k be a positive integer and C a sufficiently small positive constant. Then for every sufficiently small positive Δ there exists a discrete probability distribution $\mathbf{q}$, such that $\Delta(g_{\mathbf{p}} - g_{\mathbf{q}}) = \Delta$, and

$$|p_k - q_k| > C \Delta L^{2k}. \tag{9}$$

If the tail of $\mathbf{p}$ is lighter than exponential, the lower estimate decreases.

Instead of (8), suppose that

$$\limsup_{k \to \infty} \frac{1}{h(k)} \log \frac{1}{p_k} \tag{10}$$

is positive and finite, where h is a positive, continuous, and increasing function, regularly varying at infinity, and $\lim_{k \to \infty} h(k)/k = \infty$. Let k be a positive integer and C a sufficiently small positive constant. Then for every sufficiently small positive Δ there exists a discrete probability distribution $\mathbf{q}$, such that $\Delta(g_{\mathbf{p}} - g_{\mathbf{q}}) = \Delta$, and

$$|p_k - q_k| > C \Delta \left(h^{-1}(L)\right)^{2k}. \tag{11}$$

Particularly, when $\mathbf{p}$ is the Poisson distribution with parameter λ, it follows that for every sufficiently small positive Δ there exists a discrete probability distribution $\mathbf{q}$, such that $\Delta(g_{\mathbf{p}} - g_{\mathbf{q}}) = \Delta$, and

$$|p_k - q_k| > C_k \Delta \left(\frac{L}{\log L}\right)^{2k}. \tag{12}$$

(The constant C_k does not depend on λ and $\mathbf{q}$. The parameter λ only appears in the bounds implicit in the phrase "sufficiently small.")

Let us turn to the case of total variation. In [2] for every fixed $\mathbf{p}$ an increasing function $\psi : [0, 1] \to \mathbb{R}$ was constructed in such a way that $\lim_{t \to 0} \psi(t) = 0$ and $\|\mathbf{p} - \mathbf{q}\| \le \psi(\Delta)$ for arbitrary $\mathbf{q}$. However, apart from the case where the tail of $\mathbf{p}$ was extremely light, the function ψ proved to be slowly varying at 0, which is just a little bit better than nothing. For example, in the case of Poissonian $\mathbf{p}$, the following inequality was obtained:

$$\|\mathbf{p} - \mathbf{q}\| = o\left(\exp\left(-L^{1/4}\right)\right), \tag{13}$$

as $\mathbf{q}$ varies in such a way that $\Delta \to 0$.

Since $\|\mathbf{p} - \mathbf{q}\| \ge |p_k - q_k|$, every lower estimate obtained for fixed k will do for the total variation. However, if p_k does not decrease faster than exponential, that is, condition (8) is fulfilled, there is a lower estimate of the form

$$\|\mathbf{p} - \mathbf{q}\| \ge \Delta^C \tag{14}$$

with an exponent $C \in (0, 1)$ depending on $\mathbf{p}$.

When the tail of $\mathbf{p}$ is lighter than exponential, namely, condition (10) holds, then for every sufficiently small positive Δ the following lower estimate is valid.

$$\|\mathbf{p} - \mathbf{q}\| \ge \Delta \exp\left(C h^{-1}(L)\right), \tag{15}$$

with a constant C depending on $\mathbf{p}$. Particularly, in the case where $\mathbf{p}$ is Poisson,

$$\|\mathbf{p} - \mathbf{q}\| > \Delta \exp\left(1.76 \frac{L}{\log L}\right) \tag{16}$$

was proved.

3. Estimation for the Difference of kth Terms

The following important result can be traced back to Markoff, 1892 [4], who dealt with the extremal properties of Chebyshev polynomials over the interval $[-1, 1]$; see Chapter 2 of [5]. The proof can be found in [6] or in [7].

Theorem 1. *Let f be a polynomial of degree less than or equal to n, and $0 \le k \le n$. Then*

$$|a_k| \le \frac{(2n)^{2k}}{(2k)!}\Delta. \tag{17}$$

Using this result an upper bound can be proved without any restriction on f, which is of the same order as the lower bound in the left-hand side of (5)

Theorem 2. *Let $0 \le k \le 2L$. Then*

$$|a_k| \le \frac{e^{2k+1}}{(2k)!}\Delta L^{2k} \le \frac{e^3}{2}\Delta L^{2k}. \tag{18}$$

Proof. Suppose first that $\Delta \le e^{-2k}$. Let $c = e^{-2}$ and $n = \lfloor L/2 \rfloor$. Then $n \ge k$, and

$$\begin{aligned} \max_{0\le y\le 1}\left|\sum_{j=0}^{n} a_j c^j y^j\right| &= \max_{0\le t\le c}\left|\sum_{j=0}^{n} a_j t^j\right| \\ &\le \max_{0\le t\le 1}\left|\sum_{j=0}^{\infty} a_j t^j\right| + \max_{0\le t\le c}\left|\sum_{j=n+1}^{\infty} a_j t^j\right| \\ &\le \Delta + \sum_{j=n+1}^{\infty} c^j = \Delta + \frac{1}{1-c}e^{-2(n+1)}. \end{aligned} \tag{19}$$

Here $e^{-2(n+1)} < e^{-L} = \Delta$; therefore the right-hand side is less than $e\Delta$. By Theorem 1 we then have

$$|a_k c^k| \le \frac{(2n)^{2k}}{(2k)!}e\Delta \le \frac{e}{(2k)!}\Delta L^{2k}, \tag{20}$$

as claimed.

If $e^{-2k} < \Delta \le e^{-k/2}$, then the upper bound, being greater than 1, is trivial. Indeed, in that interval, ΔL^{2k} is decreasing, hence the upper bound attains its minimum at $\Delta = e^{-k/2}$. Then its value is

$$\frac{e^{2k+1}}{(2k)!}\Delta L^{2k} = \frac{1}{(2k)!}\left(\frac{k}{2}\right)^{2k} e^{1+3k/2}. \tag{21}$$

Here the right-hand side is equal to e for $k = 0$ and to $(1/8)e^{5/2} > 1$ for $k = 1$. Stepping further from k to $k+1$, the right-hand side gets multiplied by

$$\begin{aligned} &\frac{1}{(2k+1)(2k+2)} \cdot \frac{(k+1)^{2(k+1)}}{k^{2k}} \cdot \frac{e^{3/2}}{4} \\ &> \left(\frac{k+1}{k}\right)^{2k} \frac{e^{3/2}}{16} > \frac{e^{3/2}}{4} > 1, \end{aligned} \tag{22}$$

where we used the fact that $((k+1)/k)^k$ is increasing; hence it is not less than 2.

Finally, it is easy to see that $e^{2k+1}/(2k)!$ is decreasing for $k \ge 1$, from which the second inequality follows.

If $\mathbf{p}$ satisfies (8), that is, p_k cannot decrease faster than exponential, then (9) implies that the estimate of Theorem 2 is sharp in the order of magnitude.

Theorem 3. *Let $a_k = p_i - q_k$, where $\mathbf{p} = (p_k)_{k\ge 0}$ and $\mathbf{q} = (q_k)_{k\ge 0}$ are discrete probability distributions. Suppose that*

$$\sum_{k=n}^{\infty} p_k \le F(n), \tag{23}$$

where F is a positive, continuous, and strictly decreasing function tending to zero. Then

$$\begin{aligned} |p_k - q_k| &\le 3\frac{(8e+4)^k}{(2k)!}\Delta\left(F^{-1}(\Delta)\right)^{2k} \\ &\le 2(2e+1)^2\Delta\left(F^{-1}(\Delta)\right)^{2k} \end{aligned} \tag{24}$$

for every $k < F^{-1}(\Delta)$.

Proof. Let $n = \lceil F^{-1}(\Delta) \rceil$, and introduce

$$\begin{aligned} f(z) &= \sum_{k=0}^{\infty}(q_k - p_k)z^k, \\ g(z) &= \sum_{k=0}^{\infty} q_k z^k - \sum_{k=0}^{n-1} q_k z^k, \\ h(z) &= \sum_{k=0}^{\infty} q_k z^k, \end{aligned} \tag{25}$$

$$0 \le z \le 1.$$

Then

$$|f(z) - g(z)| \le \sum_{k=n}^{\infty} p_k \le F(n) \le \Delta. \tag{26}$$

Hence

$$\sup_{0\le z\le 1}|g(z)| \le 2\Delta. \tag{27}$$

Let $c = 2e/(2e+1)$, and $z_i = (1-c) + c(i/n)$, $i = 0, \ldots, n$. Let us approximate $g(z)$ with its Lagrange interpolation polynomial $p_n(z)$ over the uniform grid $\{z_i : 0 \le i \le n\}$. That is,

$$p_n(z) = \sum_{i=0}^{n} g(z_i)\prod_{j\ne i}^{n}\frac{z - z_j}{z_i - z_j}. \tag{28}$$

Then $g(z) - p_n(z)$ has $n+1$ zeros in the interval $[1-c; 1]$; hence, there exists a $\xi \in [1-c; 1]$ such that

$$\begin{aligned} g^{(n)}(\xi) &= p_n^{(n)}(\xi) = \sum_{i=0}^{n} g(z_i)\binom{n}{i}\left(\frac{n}{c}\right)^n (-1)^{n-i} \\ &\le 2\Delta\left(\frac{2n}{c}\right)^n < \Delta\left(\frac{2e}{c}\right)^n n! = \Delta(1-c)^{-n} n!. \end{aligned} \tag{29}$$

Since $h^{(n)}$ is increasing in the interval $[0;1]$, and it is equal to $g^{(n)}$, we have that $0 \le h^{(n)}(z) \le \Delta(1-c)^{-n}n!$ for all $z \in [0;1-c]$. Moreover, $h(0) = h'(0) = \cdots = h^{(n-1)}(0) = 0$; thus

$$0 \le h(z) \le \Delta\left(\frac{z}{1-c}\right)^n, \quad 0 \le z \le 1-c. \tag{30}$$

From all these it follows that

$$\left|\sum_{k=0}^{n-1}(p_k - q_k)z^k\right| \le 3\Delta \tag{31}$$

for $0 \le z \le 1-c$. Applying Theorem 1 in the same way as in the proof of Theorem 2 we get that

$$\begin{aligned}|p_k - q_k| &\le \frac{3\Delta}{(1-c)^k}\frac{(2(n-1))^{2k}}{(2k)!} \\ &\le 3\frac{(8e+4)^k}{(2k)!}\Delta\left(F^{-1}(\Delta)\right)^{2k},\end{aligned} \tag{32}$$

if $k < F^{-1}(\Delta)$. Investigating the ratio of consecutive terms one can easily see that

$$3\frac{(8e+4)^k}{(2k)!} \le 3\frac{(8e+4)^2}{4!} = 2(2e+1)^2. \tag{33}$$

Remark 4. Let us write F in exponential form: $F(n) = e^{-h(n)}$, where h is nonnegative, continuous, and tends increasingly to infinity. Then $F^{-1}(\Delta) = h^{-1}(L)$. The estimate of Theorem 3 is better than that of Theorem 2 in the order of magnitude only if the distribution $\mathbf{p}$ has a tail lighter than exponential; that is, $h(n)/n \to \infty$.

Remark 5. If $1/h(k)\log(1/p_k)$ is bounded away from zero and infinity as well, then both the upper estimate of Theorem 3 and the lower estimate of (11) are applicable and they are of the same order of magnitude. Use the fact that for regularly varying functions h of order $\alpha > 0$ we have $(ch)^{-1}(z) \sim c^{-1/\alpha}h^{-1}(z)$ as $z \to \infty$.

Remark 6. Reference [6] proved similar bounds with different constants, but of the same order of magnitude. However, they imposed conditions similar to (10) on the sequence $a_k = p_k - q_k$, rather than on $\mathbf{p}$, which is less useful for applications in probability. Besides, for the estimate of Theorem 2, which is true without any restriction on the coefficients, they needed exponential decay of the sequence (a_k).

Remark 7. If h is linear, that is, $\sum_{k\ge n} p_k \le e^{-\nu n}$, and it holds with a sufficiently large ν ($\nu \ge 2$ will do), the upper bound of Theorem 3 is better for $k > 0$ than that of Theorem 2. (For $k = 0$ the bound $|p_0 - q_0| \le \Delta$ is obviously the best possible.)

Particularly, let $\mathbf{p}$ be the Poisson distribution with mean λ. Then, for $n > \lambda$, we have

$$\sum_{k=n}^{\infty} p_k \le e^{-\lambda}\left(\frac{\lambda e}{n}\right)^n \tag{34}$$

(see Theorem A.15 in [8]); hence $h(x) = x\log x + O(x)$, and

$$h^{-1}(x) = \frac{x}{\log x - \log\log x + O(1)} \sim \frac{x}{\log x} \tag{35}$$

as $x \to \infty$. In addition, $h^{-1}(x) > x/\log x$ eventually. Let us plug this back into Theorem 3 to get the following estimate.

Corollary 8. *If $\mathbf{p}$ is Poisson, then, uniformly in $\mathbf{q}$, one has*

$$|p_k - q_k| \le 4\frac{(8e+4)^k}{(2k)!}\Delta\left(\frac{L}{\log L}\right)^{2k} \tag{36}$$

for every $k < L/\log L$, if Δ is sufficiently small.

Note that the order of this upper bound is the same as that of the lower bound (12).

4. Estimation for the Total Variation Distance

Let again $a_k = p_k - q_k$, where $\mathbf{p} = (p_k)_{k\ge0}$ and $\mathbf{q} = (q_k)_{k\ge0}$ are discrete probability distributions. As we have seen, if nothing is known about $\mathbf{p}$ and $\mathbf{q}$, it is impossible to give a nontrivial upper bound for the total variation distance

$$\|\mathbf{p} - \mathbf{q}\| = \sum_{k=0}^{\infty}|p_k - q_k|. \tag{37}$$

However, when $\mathbf{p}$ is fixed, the situation is completely different.

Theorem 9. *Let $a_k = p_k - q_k$, where $\mathbf{p} = (p_k)$ satisfies condition (23). Then*

$$\|\mathbf{p} - \mathbf{q}\| < 8\Delta\exp\left(\kappa F^{-1}\Delta\right), \tag{38}$$

where $\kappa = 2 + \log(2e+1) = 3.861\cdots$.

The right-hand side tends to 0 as $\Delta \to 0$ only if the tail of $\mathbf{p}$ is not too heavy; namely, $F(x) = o(e^{-\kappa x})$.

The method of proof will be applied a couple of times in the sequel with different parameters. Therefore we formulate its essence in a separate lemma as a master inequality.

Lemma 10. *Let A, B, and C be positive real numbers; $C \le B$. Suppose*

$$|p_k - q_k| \le A\frac{(2B)^{2k}}{(2k)!}, \quad 0 \le k < C. \tag{39}$$

Then

$$\|\mathbf{p} - \mathbf{q}\| \le 2A\left(\frac{eB}{C}\right)^{2C} + 2F(C). \tag{40}$$

Proof. Let $n = \lceil C \rceil$, then $n - 1 < C \le n$. Clearly,

$$\begin{aligned}\|\mathbf{p} - \mathbf{q}\| &= \sum_{k=0}^{\infty}|p_k - q_k| \\ &= 2\sum_{k=0}^{\infty}(p_k - q_k)^+ + \sum_{k=0}^{\infty}(p_k - q_k) \\ &= 2\sum_{k=0}^{\infty}(p_k - q_k)^+ \le 2\sum_{k=0}^{n-1}|p_k - q_k| + 2\sum_{k=n}^{\infty}p_k.\end{aligned} \tag{41}$$

By supposition we have

$$(2B)^{2k} = (2C)^{2k}\left(\frac{B}{C}\right)^{2k} \le (2C)^{2k}\left(\frac{B}{C}\right)^{2C}. \tag{42}$$

Hence

$$\begin{aligned}\|\mathbf{p}-\mathbf{q}\| &\le 2A\sum_{k=0}^{n-1}\frac{(2B)^k}{(2k)!} + 2F(n) \\ &\le 2A\left(\frac{B}{C}\right)^{2C}\sum_{k=0}^{n-1}\frac{(2C)^{2k}}{(2k)!} + 2F(n) \\ &\le 2A\left(\frac{eB}{C}\right)^{2C} + 2F(C).\end{aligned} \tag{43}$$

Proof of Theorem 9. Starting from Theorem 3, let us apply Lemma 10 with $A = 3\Delta$, $B = (2e+1)^{1/2}F^{-1}(\Delta)$, and $C = F^{-1}(\Delta)$. We get that

$$\begin{aligned}\|\mathbf{p}-\mathbf{q}\| &\le 6\Delta\left(e(2e+1)^{1/2}\right)^{2F^{-1}(\Delta)} + 2\Delta \\ &\le 8\Delta\exp\left(\kappa F^{-1}(\Delta)\right).\end{aligned} \tag{44}$$

Particularly, let $\mathbf{p}$ be the Poisson distribution with mean λ. As we have seen in (35),

$$F^{-1}(\Delta) = h^{-1}(L) = (1+o(1))\frac{L}{\log L}. \tag{45}$$

Let us plug this into Theorem 9. Writing 4 in place of κ in the exponent we can get rid of the term $1+o(1)$, and even the multiplier 8 gets eventually absorbed. Thus we obtain the following estimate.

Corollary 11. *Let* $\mathbf{p}$ *be Poisson; then, uniformly in* $\mathbf{q}$, *one has*

$$\|\mathbf{p}-\mathbf{q}\| \le \Delta\exp\left(\frac{4L}{\log L}\right), \tag{46}$$

if Δ *is sufficiently small.*

This is already similar to the lower bound (16), and is much better than (13).

If the tail of $\mathbf{p}$ is subexponential, that is, $\sum_{k\ge n}p_k = \exp(-o(n))$, then the estimate of Theorem 3 is useless: it tends to infinity as $\Delta \to 0$. However, with suitably chosen parameters in the master inequality, a reasonable upper bound can be obtained: not really sharp, but at least not trivial.

Theorem 12. *Suppose* $F(n) = e^{-o(n)}$. *Then*

$$\begin{aligned}&\sup\left\{\|\mathbf{p}-\mathbf{q}\| : \max_{0\le z\le 1}\left|g_{\mathbf{p}}(z) - g_{\mathbf{q}}(z)\right| = \Delta\right\} \\ &= O\left(F\left(\frac{L}{8}\right)\right), \quad \text{as } \Delta \longrightarrow 0.\end{aligned} \tag{47}$$

Proof. This time we use Theorem 2. Let $A = e\Delta$, $B = eL/2$, and $C = L/8$. Then Lemma 10 gives that

$$\|\mathbf{p}-\mathbf{q}\| \le 2e\Delta\left(4e^2\right)^{L/4} + 2F\left(\frac{L}{8}\right). \tag{48}$$

Since $(4e^2)^{1/4} = 2.331\cdots < e$, the first term in the right-hand side can be estimated by a positive power of Δ, while the second term decreases slower than any positive power of Δ, as $\Delta \to 0$; thus it will eventually dominate.

Corollary 13. *If the* α*th moment of* $\mathbf{p}$ *is finite, then* $\sum_{k=n}^{\infty}p_k = o(n^{-\alpha})$; *thus*

$$\sup\left\{\|\mathbf{p}-\mathbf{q}\| : \max_{0\le z\le 1}\left|g_{\mathbf{p}}(z) - g_{\mathbf{q}}(z)\right| = \Delta\right\} = o\left(L^{-\alpha}\right), \quad \text{as } \Delta \longrightarrow 0. \tag{49}$$

Finally, let us deal with the case of exponentially decaying tails. Let $F(n) = e^{-\nu n}$, with $\nu \ge 0$. Then the total variation $\|\mathbf{p}-\mathbf{q}\|$ can be estimated by a positive power of Δ. This follows from Theorem 9, but only for $\nu > \kappa$. The following theorem is valid for all positive ν.

Theorem 14. *Suppose* $F(n) \le e^{-\nu n}$, *with* $\nu > 0$. *Then*

$$\|\mathbf{p}-\mathbf{q}\| \le 8\Delta^{\nu/(\nu+7)}. \tag{50}$$

As not sharp, the concrete form of the exponent holds no interest; what really matters is that it is positive and tends to 1 as $\nu \to \infty$. Note that if p_k tends to 0 at an exponential rate, (14) provides a lower bound which is also a positive power of Δ (from Theorem 2.4 of [2] it follows that $C = (\nu+1/4)/(\nu+2)$ can be used for exponent).

Proof. Firstly, suppose $\nu \le 1.9$. Based on Theorem 2, apply Lemma 10 with $A = e\Delta$, $B = eL/2$, and $C = L/(\nu+7)$:

$$\begin{aligned}\|\mathbf{p}-\mathbf{q}\| &\le 2e\Delta\left(\frac{e^2(\nu+7)}{2}\right)^{2L/(\nu+7)} \\ &\quad + 2\exp\left(-\frac{\nu L}{\nu+7}\right).\end{aligned} \tag{51}$$

Let us deal with the first term on the right-hand side

$$\begin{aligned}&\Delta\left(\frac{e^2(\nu+7)}{2}\right)^{2L/(\nu+7)} \\ &= \exp\left(-L\left[1 - \frac{2}{\nu+7}\left(2 + \log\frac{\nu+7}{2}\right)\right]\right) \\ &= \exp\left(-\frac{L}{\nu+7}\left(\nu + 3 - 2\log\frac{\nu+7}{2}\right)\right).\end{aligned} \tag{52}$$

Since $3 - 2\log((\nu+7)/2) \ge 3 - 2\log 4.45 > 0$, it follows that

$$\|\mathbf{p}-\mathbf{q}\| \le 2(e+1)\Delta^{\nu/(\nu+7)}. \tag{53}$$

Secondly, let $\nu > 1.9$. Now we apply Theorem 3. Note that $F^{-1}(\Delta) = L/\nu$, and set $A = 3\Delta$, $B = (2e+1)^{1/2}L/\nu$, and $C = L/(\nu+7)$. Then we obtain that

$$\|\mathbf{p}-\mathbf{q}\| \le 6\Delta\left(\frac{e(2e+1)^{1/2}(\nu+7)}{\nu}\right)^{2L/(\nu+7)} + 2\exp\left(-\frac{\nu L}{\nu+7}\right). \tag{54}$$

Again, the first term can be estimated by the second one, because

$$\begin{aligned}\Delta&\left(\frac{e(2e+1)^{1/2}(\nu+7)}{\nu}\right)^{2L/(\nu+7)} \\ &= \exp\left(-L\left[1-\frac{2}{\nu+7}\left(\frac{\kappa}{2}+\log\frac{\nu+7}{\nu}\right)\right]\right) \\ &= \exp\left(-\frac{L}{\nu+7}\left(\nu+7-\kappa-2\log\frac{\nu+7}{\nu}\right)\right),\end{aligned} \tag{55}$$

and $7-\kappa-2\log((\nu+7)/\nu) > 7-3.862-2\log 4.685 > 0$. Thus, it follows that

$$\|\mathbf{p}-\mathbf{q}\| \le 8\Delta^{\nu/(\nu+7)} \tag{56}$$

for $\nu > 1.9$, as needed.

Competing Interests

The author declares that there are no competing interests regarding the publication of this paper.

Acknowledgments

This research has been supported by the Hungarian Scientific Research Fund OTKA, Grant no. K 108615. The idea of the proof of Theorem 3 is due to Gábor Halász, whom the author is indebted to for his invaluable help.

References

[1] A. Rényi, "A general method for proving theorems in probability theory and some of its applications," in *Selected Papers of A. Rényi*, vol. 2, pp. 581–602, Akadémiai Kiadó, Budapest, Hungary, 1976.

[2] T. F. Móri, "Deviation of discrete distributions—positive and negative results," *Statistics & Probability Letters*, vol. 79, no. 8, pp. 1089–1096, 2009.

[3] T. F. Móri, "Bonferroni inequalities and deviations of discrete distributions," *Journal of Applied Probability*, vol. 33, no. 1, pp. 115–121, 1996.

[4] W. Markoff and J. Grossmann, "Über Polynome, die in einem gegebenen Intervalle möglichst wenig von Null abweichen," *Mathematische Annalen*, vol. 77, no. 2, pp. 213–258, 1916.

[5] P. Borwein and T. Erdélyi, *Polynomials and Polynomial Inequalities*, vol. 161 of *Graduate Texts in Mathematics*, Springer, New York, NY, USA, 1995.

[6] L. Antal and T. F. Móri, "How large can the coefficients of a power series be?" *Annales Universitatis Scientiarum Budapestinensis de Rolando Eötvös Nominatae. Sectio Computatorica*, vol. 34, pp. 25–32, 2011.

[7] K. S. Berenhaut, J. V. Baxley, and R. G. Lyday, "Deviations of discrete distributions and a question of Móri," *Statistics & Probability Letters*, vol. 81, no. 12, pp. 1940–1944, 2011.

[8] N. Alon and J. H. Spencer, *The Probabilistic Method*, John Wiley & Sons, New York, NY, USA, 1989.

16

Detecting Spatial Clusters via a Mixture of Dirichlet Processes

Meredith A. Ray,[1] **Jian Kang**,[2] **and Hongmei Zhang**[3]

[1] *Division of Epidemiology, Biostatistics, and Environmental Health, University of Memphis, 220 Robison Hall, Memphis, TN 38152, USA*
[2] *Department of Biostatistics, University of Michigan, 1415 Washington Heights, Ann Arbor, MI 48109, USA*
[3] *Division of Epidemiology, Biostatistics, and Environmental Health, University of Memphis, 224 Robison Hall, Memphis, TN 38152, USA*

Correspondence should be addressed to Meredith A. Ray; maray@memphis.edu

Guest Editor: Yichuan Zhao

We proposed an approach that has the ability to detect spatial clusters with skewed or irregular distributions. A mixture of Dirichlet processes (DP) was used to describe spatial distribution patterns. The effects of different batches of data collection efforts were also modeled with a Dirichlet process. To cluster spatial foci, a birth-death process was applied due to its advantage of easier jumping between different numbers of clusters. Inferences of parameters including clustering were drawn under a Bayesian framework. Simulations were used to demonstrate and assess the method. We applied the method to an fMRI meta-analysis dataset to identify clusters of foci corresponding to different emotions.

1. Introduction

This work was motivated by a study aiming to detect centers of activated foci from a function magnetic resonance imaging (fMRI) metadataset. In summary, fMRI metadata is a collection of fMRI studies identifying areas of the brain that are significantly activated by stimuli to examine a specific outcome. FMRIs are expensive which leads to small sample sizes and therefore can use metadata to increase sample size and power. To identify spatial clusters, finite mixture models are generally implemented [1–3]. Mixture components, representing a cluster, typically share a common parametric family with each component containing respective parameters [1, 2]. Each component also has a mixing proportion or weight that is respective to the frequency of data in that component [1]. Because of the model's ease of implementation, this allows various applications such as pattern recognition, computer vision, signal and image analysis, and machine learning to list a few [4].

One commonly used distribution in the aforementioned finite mixtures is the normal distribution [1, 5], appreciating its established properties and, in the Bayesian context, conjugacy. However, when it comes to clustering, assuming normality for each part of the mixture can potentially lead to oversensibility, e.g., when one cluster is formed by a mixture of two normal distributions but with rather close centers. This type of oversensibility in many research fields should be avoided; one example is emotional foci inferred from brain imaging data, where a certain emotion is covered by a wide region.

To infer the number of clusters, under the Bayesian framework, different methods have been proposed. Reversible jump Markov chain Monte Carlo has been commonly used to infer the number of clusters [6, 7]. At each iteration, a decision is made between splitting one cluster to two, combining two clusters to one, or no movement. One potential difficulty of this approach is the risk of being trapped at a local maximum. Recently, the Dirichlet process (DP) has been used often to estimate the number of clusters [8, 9]. This process has the ability to capture irregular patterns. DP has the ability to detect clusters without being burdened about additional clustering parameters. However, this feature of clustering also has an inherit weakness in that it tends to produce more clusters, making interpretations more difficult.

To overcome the aforementioned gaps, we implemented a mixture of Dirichlet processes (DP). These processes have the ability to describe irregular patterns [8, 9] and by using it as our common parametric family allows the model to identify more complex patterns than the normal distribution would be able to identify. Furthermore, motivated from our previous work with the spatial Cox process application in [10], we elected to incorporate the birth-death process to statistically determine the number of clusters. Compared to other clustering approaches noted earlier, the birth-death process has the advantage of quick convergence and, by controlling birth rate, embraces a potential of generating redundant clusters.

The remainder of the article is organized as follows. In Section 2, we introduce the model structure, notation, and priors and hyperpriors, simulations are presented in Section 3, the application of the model to an fMRI meta-analysis dataset is in Section 4, and conclusions and discussion are in Section 5.

2. Methods

2.1. The Model. We let s_{rj} denote the (x, y, z) coordinate of a spatial point in a three-dimensional space, in particular, point j in group (study) r, $j = 1, \ldots, J_r$ and $r = 1, \ldots, R$. We have $\sum_{r=1}^{R} J_r = n$, where n is the total number of observed points. It straightforwardly follows that $\mathbf{s} = (s_{1,1} \cdots s_{R,J_R})$ represents all points in the study. We model s_{rj} as

$$s_{rj} = p_r + \theta_{rj} + \epsilon, \tag{1}$$

where p_r denotes the effect of group r, while θ_{rj} represents the mean of s_{rj} for the jth point in group r after adjusting for group effects, p_r, and ϵ denotes some random error. By modeling the random error as a standard multivariate normal distribution, the distribution of s_{rj} satisfies

$$s_{rj} \sim MVN_3\left(p_r + \theta_{rj}, \Sigma\right) \tag{2}$$

with

$$P\left(s_{rj} \mid p_r, \theta_{rj}, \Sigma\right) = (2\pi)^{-3/2} |\Sigma|^{-1/2} \cdot \exp\left[-\frac{1}{2}\left(s_{rj} - p_r - \theta_{rj}\right)' \Sigma^{-1} \left(\left(s_{rj} - p_r - \theta_{rj}\right)\right)\right], \tag{3}$$

where $\Sigma = \sigma^2 I_3$ is the covariance matrix.

2.2. Prior and Hyperprior Distributions. We start from the prior distribution of θ_{rj}. To detect underlying clusters of s_{rj} due to similarities of θ_{rj}, we describe the prior of θ_{rj} as a mixture of distribution G_k, $k = 1, \ldots, K$. Common choices of G_k are normal distributions. To improve flexibility, we relax such normalizing assumptions in the mixture and assume G_k is generated from a DP, i.e., $G_k \sim DP(\alpha, G_{0k})$, where α is the precision parameter and G_{0k} is a base distribution and taken as $G_{0k} = MVN_3(\boldsymbol{\mu_k}, \Sigma_k)$. In particular, $\theta_{rj} \sim \sum_{k=1}^{K} \pi_k G_k$, where $0 < \pi_k < 1$ such that $\sum_{k=1}^{K} \pi_k = 1$. For the number of clusters K, we assign a truncated Poisson prior distribution to K, $P(K) = (\lambda^K/K!) \exp(-\lambda)$, $K = 1, \ldots, n$. We assign a Dirichlet distribution with parameter 1 to $\boldsymbol{\pi} = (\pi_1, \ldots, \pi_K)$, which implies that $\boldsymbol{\pi}$ is k-dimensional uniformly distributed. The prior distribution of $\boldsymbol{\mu_k}$ from the base distribution G_{0k} is chosen to be $\boldsymbol{\mu_k} \sim MVN_3(\boldsymbol{\xi}, \kappa^{-1})$, where $\boldsymbol{\xi} = (\xi_1, \xi_2, \xi_3)$, which are known and set as the midpoint of observed interval of variation of the data. Parameter κ is set as

$$\kappa = \begin{bmatrix} \frac{1}{R_1^2} & 0 & 0 \\ 0 & \frac{1}{R_2^2} & 0 \\ 0 & 0 & \frac{1}{R_3^2} \end{bmatrix}, \tag{4}$$

where R_1^2, R_2^2, and R_3^2 are the range of the data for each dimension. This prior setting is adopted from [1] and we feel it is reasonable for this setting given the fact that the number and location of clusters are unknown. We let $\Sigma_k = \sigma_k I_3$ with $\sigma_k \sim IG(20, 0.5)$ since the range of the observed data is small. For group effect, p_r, we assume it is small with $p_r \sim G_p$, where $G_p \sim DP(\alpha_p, G_{0p})$ with α_p specified later and $G_{0p} = TMVN_3(\boldsymbol{\mu_p}, \Sigma_p, lower = -l, upper = l)$, where the lower and upper limits are defined as 10% of the absolute range of the data. We let $\boldsymbol{\mu_p} \sim MVN_3(\mathbf{0}, I_3)$ and $\Sigma_p = \sigma_p^2 I_3$, with $\sigma_p^2 \sim IG(5, 0.5)$. The variance component of the random error, $\Sigma = \sigma^2 I_3$, σ^2 is assumed to follow a relatively noninformative Inverse Gamma (IG) distribution, $\sigma^2 \sim IG(0.5, 0.5)$. The precision parameters α and α_p are selected by minimizing the deviance information criterion (DIC) [11, 12].

2.3. Conditional Posterior Distributions and Posterior Computing. Sampling parameter estimates from their posterior distributions can be achieved via Gibbs sampler, in which the statistical inference on the number of clusters is modeled using the birth-death process. The birth-death process is one type of continuous-time Markov chain originally introduced in [13]. This process is often used to simulate realizations of point processes as they can be difficult to directly sample from [1]. These realizations are further used for likelihood inferences for model parameters [1]. The birth-death scheme allows events to randomly occur throughout the chain; these events are either a "birth" or "death." If a birth occurs, the number of components increases by one, while if a death occurs, the number of components decreases by one.

Recall considering a finite mixture prior for θ_{rj} such that all θ_{rj} are assumed independently distributed with each generated from one of K Dirichlet processes denoted as G_k, i.e.,

$$P\left(\theta_{rj} \mid K, \boldsymbol{\pi}, \boldsymbol{\phi}, \alpha\right) = \sum_{k=1}^{K} \pi_k G_k\left(\alpha, G_{0k}\right), \tag{5}$$

where each $G_k(\alpha, G_{0k})$ represents a DP but K is unknown, $\boldsymbol{\pi} = (\pi_1, \ldots, \pi_K)$ are the mixing proportions, and $\boldsymbol{\phi} = (\boldsymbol{\mu}_1, \Sigma_k, \ldots, \boldsymbol{\mu}_K, \Sigma_K)$ are the component specific parameters for each DP. For cluster assignment, we introduce an index variable Z_{rj} that indicates the assignment of observation rj and Z_{rj} takes the values of 1 to K. Denoted by $z_{rj} \in \mathbf{z}$, where $\mathbf{z} = (z_{1,1}, \ldots, z_{R,J_R})$ represents the realization of independently and identically distributed discrete random variables $\mathbf{Z} = (Z_{1,1}, \dots, Z_{R,J_R})$ with probability mass function,

$$P\left(Z_{rj} = k \mid \boldsymbol{\pi}, \boldsymbol{\phi}, \eta\right) = \pi_k \quad (k = 1, \dots, K). \tag{6}$$

The joint posterior distribution is proportional to

$$\begin{aligned} P(\boldsymbol{\vartheta} \mid \mathbf{s}) \propto & \prod_r^R P\left(p_r \mid G_p\right) P\left(G_p \mid \alpha, G_{0p}\right) \\ & \cdot G_{0p}\left(p_r; \mu_p, \sigma_p^2\right) P\left(\mu_p\right) P\left(\sigma_p^2\right) \\ & \cdot \prod_j^{J_r} P\left(s_{rj} \mid \phi\right) \\ & \times P\left(\theta_{rj} \mid \mathbf{z}, \boldsymbol{\pi}, G_1, \dots, G_K\right) \\ & \cdot \prod_{k=1}^{K} P\left(G_k \mid \alpha, G_{0k}\right) G_{0k}\left(\theta_{rj}; \mu_k, \sigma_k^2\right) \\ & \cdot P\left(\mu_k\right) P\left(\sigma_k^2\right) \times P(\boldsymbol{\pi}) P(K) P(\mathbf{Z}) P\left(\sigma^2\right), \end{aligned} \tag{7}$$

where $\boldsymbol{\vartheta} = (p_r, \theta_{rj}, \sigma^2, \alpha_p, \boldsymbol{\mu}_p, \sigma_p^2, K, \boldsymbol{\mu}_1, \dots, \boldsymbol{\mu}_K, \sigma_1^2, \dots, \sigma_K^2, \alpha, \boldsymbol{\pi})$ is a vector of all estimable parameters. From here, the birth-death algorithm and Markov chain can be described for introducing and assigning clusters for θ_{rj}:

(1) Starting with the initial model $y = \{(\pi_1, \phi_1), \dots, (\pi_K, \phi_K)\}$, let (π_k, ϕ_k) represent the mixing proportion and cluster specific parameters for the unique θ_{rj} clusters. Let the birth rate be $\beta(y) = \lambda_b$.

(2) Calculate the death rate for each component:

$$\delta_k(y) = \frac{L(y \setminus (\pi_k, \phi_k))}{L(y)} \frac{P(K-1 \mid \alpha, \cdot)}{KP(K \mid \alpha, \cdot)} \quad (k = 1, \dots, K). \tag{8}$$

(3) Calculate the total death rate $\delta(y) = \sum_{k=1}^{K} \delta_k(y)$. To quicken convergence, we elected not to model the time to next jump as exponential and allowed an event to occur at each iteration of the Markov chain.

(4) Simulate the type of event, birth or death, with the respective probabilities:

$$\begin{aligned} P(birth) &= \frac{\beta(y)}{\beta(y) + \delta(y)}, \\ P(death) &= \frac{\delta(y)}{\beta(y) + \delta(y)}. \end{aligned} \tag{9}$$

(5) Adjust the model y to reflect the birth or death by the following:

(i) Birth: Simulate a new component (π_{K+1}, ϕ_{K+1}) from each parameter's respective (independent) prior distributions, π_{K+1} from $K(1-\pi)^{K-1}$ and ϕ_{K+1} from base distribution, such that the model becomes $y = \{(\pi_1, \phi_1), \dots, (\pi_K, \phi_K), (\pi_{K+1}, \phi_{K+1})\}$. It can be mentioned that $K(1-\pi)^{K-1}$ is the Beta distribution with parameters $(1, K)$ and can be easily simulated from $Y_1 \sim G(1,1)$ and $Y_2 \sim G(K,1)$ such that $\pi_{K+1} = Y_1/(Y_1 + Y_2)$.

(ii) Death: Select a component to die with the probabilities $\delta_k(y)/\delta(y)$ for $k = 1, \dots, K$ such that the model becomes $y = \{(\pi_1, \phi_1), \dots, (\pi_{K-1}, \phi_{K-1})\}$.

(6) Given the current state of the model at time t, simulate values for all remaining parameters.

(7) Go to step (2).

Incorporating the birth-death process into our model, we need to further define the likelihood for removing cluster i:

$$\begin{aligned} P(\boldsymbol{\Phi} \setminus (\pi_i, G_i) \mid \mathbf{s}) \propto & \prod_r^R P\left(p_r \mid G_p\right) P\left(G_p \mid \alpha, G_{0p}\right) \\ & \cdot G_{0p}\left(p_r; \mu_p, \sigma_p^2\right) P\left(\mu_p\right) P\left(\sigma_p^2\right) \prod_j^{J_r} P\left(s_{rj} \mid \phi\right) \\ & \times P\left(\theta_{rj} \mid \mathbf{z}, \boldsymbol{\pi}, G_1, \dots, G_K\right) \prod_{k=1}^{K^{(i)}} P\left(G_k \mid \alpha, G_{0k}\right) \\ & \cdot G_{0k}\left(\theta_{rj}; \mu_k, \sigma_k^2\right) \times P\left(\mu_k\right) P\left(\sigma_k^2\right) P(\boldsymbol{\pi}) P(K) P(\mathbf{Z}) \\ & \cdot P\left(\sigma^2\right), \end{aligned} \tag{10}$$

where $K^{(i)} = i \notin (1, \dots, K)$. The birth-death process is conditional on the prespecified birth rate, λ_b. By setting this birth rate, which controls how often a "birth" of a new component occurs, equal to λ as suggested and done in [1], this computationally allows the death rates to be a likelihood ratio absent of λ. In other words, the likelihood of the data drives the death rates and ultimately the decision of a new cluster. Given that decision is a "birth," the new cluster's parameters $\boldsymbol{\mu}_{K+1}$, σ_{K+1}^2, and π_{K+1} are sampled from their prior distributions:

$$\begin{aligned} \boldsymbol{\mu}_{K+1} &\sim MVN_3\left(\boldsymbol{\xi}, \kappa^{-1}\right) \\ \sigma_{K+1}^2 &\sim IG(20, 0.5) \\ \pi_{K+1} &\sim K(1-\pi)^{K-1}. \end{aligned} \tag{11}$$

The mixing proportions are adjusted by multiplying all current proportions by $(1-\pi_{K+1})$ if a birth occurs or dividing by $(1-\pi_i)$ if a death occurs.

To simulate values for all remaining parameters and hyperparameters, we implement the Gibbs sampler. Conditional posterior distributions are listed below. Note that "·" denotes data and other parameters not listed. The conditional posterior of Z_{rj} is

$$\begin{aligned} & P\left(Z_{rj} = k \mid \boldsymbol{\pi}, \theta_{rj}, G_k, \cdot\right) \\ & \quad \propto P\left(\theta_{rj} \mid Z_{rj} = k, G_k, \cdot\right) P\left(G_k \mid \cdot\right) P\left(Z_{rj} = k\right) \end{aligned}$$

$$= \left\{ \frac{\alpha MVN_3(\boldsymbol{\mu_k}, \Sigma_k)}{\alpha + n_k - 1} + \frac{\sum_{c=1}^{C_k} \delta_c(\theta_{rj})}{\alpha + n_k - 1} \right\} \pi_k, \tag{12}$$

where $c = 1, \ldots, C_k$ is the number of subclusters for cluster $k \in (1, \ldots, K)$, n_k is the number of foci in cluster k, and $\delta_c(\theta_{rj})$ denotes the unit point mass,

$$\boldsymbol{\pi} \mid \mathbf{Z} \sim Dir(n_1 + 1, \ldots, n_K + 1), \tag{13}$$

and

$$\begin{aligned} P(\theta_{rj} \mid k, \cdot) &\propto \prod_{Z_{rj} \in k} P(s_{rj} \mid p_r, \theta_{rj}, \cdot) \\ &\cdot P(\theta_{rj} \mid Z_{rj} = k, G_k) P(G_k \mid G_{0k}, \alpha) \\ &\cdot G_{0k}(\theta_{rj}; \boldsymbol{\mu_k}, \sigma_k^2) = \prod_{Z_{rj} \in k} \{MVN_3(p_r + \theta_{rj}, \Sigma)\} \\ &\cdot \left\{ \frac{\alpha MVN_3(\boldsymbol{\mu_k}, \Sigma_k)}{\alpha + n_k - 1} + \frac{\sum_{q=1, q \neq rj}^{n_k} \delta_{\theta_q}(\theta_{rj})}{\alpha + n_k - 1} \right\}, \end{aligned} \tag{14}$$

which is the distribution of a DP with $\delta_{\theta_q}(\theta_{rj})$ being the unit point mass and n_k the number of foci in some cluster $k \in (1, \ldots, K)$,

$$\begin{aligned} \boldsymbol{\mu_k} \mid \cdot &\sim MVN_3\left(\left(\kappa + n_k \sigma_k^{-2} I_3\right)^{-1} \right. \\ &\left. \cdot \left(\kappa \boldsymbol{\xi} + n_k \sigma_k^{-2} I_3 \overline{\theta}_{rj}\right), \left(\kappa + n_k \sigma_k^{-2} I_3\right)^{-1} \right), \end{aligned} \tag{15}$$

where $\overline{\theta}_{rj} = \sum_{rj \in k} \theta_{rj} / n_k$ and denotes the average of all θ_{rj} in cluster k,

$$\begin{aligned} \sigma_k^2 \mid \cdot &\sim IG\left(\frac{3 \times n_k}{2} + 20, 0.5 \right. \\ &\left. + \frac{1}{2} \sum_{Z_{rj} \in k} (\theta_{rj} - \boldsymbol{\mu_k})' (\theta_{rj} - \boldsymbol{\mu_k}) \right), \end{aligned} \tag{16}$$

where n_k is the number of foci in cluster k, $k = 1, \ldots, K$. The conditional posterior of p_r is generated from a DP again:

$$\begin{aligned} P(p_r \mid \cdot) &\propto \prod_{r=1}^{R} P(s_{rj} \mid p_r, \theta_{rj}, \cdot) P(\theta_{rj} \mid \cdot) P(p_r \mid G_p) \\ &\cdot P(G_p \mid G_{0p}, \alpha_p) G_{0p}(p_r; \mu_p, \sigma_p^2) \\ &= \prod_{r=1}^{R} \{MVN_3(p_r + \theta_{rj}, \Sigma)\} \\ &\cdot \left\{ \frac{\alpha_p TMVN_3(\boldsymbol{\mu_p}, \Sigma_p, -l, l)}{\alpha_p + R - 1} \right. \\ &\left. + \frac{\sum_{q=1, q \neq r}^{R} \delta_{p_q}(p_r)}{\alpha_p + R - 1} \right\}, \end{aligned} \tag{17}$$

where $\delta_{p_q}(p_r)$ is the unit point mass. The conditional posterior distributions for the related hyperparameters are

$$\begin{aligned} \boldsymbol{\mu_p} \mid \cdot &\sim MVN_3\left(\left(1 + R\sigma_p^{-2}\right)^{-1} \right. \\ &\left. \cdot \left(R\sigma_p^{-2} (\overline{p_r})\right), \left(1 + R\sigma_p^{-2}\right)^{-1} I_3 \right), \end{aligned} \tag{18}$$

where $\overline{p_r} = \sum_{r \in R} p_r / R$ and represents the average group effect and

$$\sigma_p^2 \mid \cdot \sim IG\left(\frac{3 \times R}{2} + 5, \frac{\sum_{r=1}^{R} (p_r - \boldsymbol{\mu_p})^2 + 1}{2} \right). \tag{19}$$

Lastly, the sampling distribution for Σ is

$$\begin{aligned} \sigma^2 \mid \cdot &\sim IG\left(\frac{3 \times n + 1}{2}, \right. \\ &\left. \frac{1 + \sum_{Z_{rj} \in n} (s_{rj} - p_r - \theta_{rj})' (s_{rj} - p_r - \theta_{rj})}{2} \right). \end{aligned} \tag{20}$$

The sampling of unique values for θ_{rj} and p_r can be performed using Neal's algorithm 8 [14]. It works by introducing m auxiliary parameters that are independent to other parameters to represent potential values for θ_{rj} and p_r [14]. Algorithm 8 for updating clustering assignments, denoted as c, is as follows:

(i) The state of the Markov chain consists of $c = \{c_1, \ldots, c_n\}$ and $\Phi = (\phi_c; c \in c_1, \ldots, c_n)$ with ϕ_c denoting cluster parameters, e.g., θ_c in our application. Repeatedly sample as follows:

(ii) For $i = 1, \ldots, n$, let k^- be the number of distinct c_l for $l \neq i$, and $h = k^- + m$. Label these c_l with values in $\{1, \ldots, k^-\}$. If $c_i = c_l$ for some $l \neq i$, draw values independently from base distribution G_0 for those ϕ_c for which $k^- < c \leq h$. If $c_i \neq c_l$ for all $l \neq i$, let c_i have the label $k^- + 1$, and draw values independently from G_0 for those ϕ_c for which $k^- + 1 < c \leq h$. Draw a new value for c_i from $\{1, \ldots, h\}$ using the following probabilities:

$$\begin{aligned} P(c_i = c \mid c_{-i}, y_i, \phi_1, \ldots, \phi_h) \\ = \begin{cases} \dfrac{n_{-i,c}}{n - 1 + \alpha} F(y_i, \phi_c) & \text{for } 1 \leq c \leq k^- \\ \dfrac{(\alpha/m)}{n - 1 + \alpha} F(y_i, \phi_c) & \text{for } k^- < c \leq h, \end{cases} \end{aligned} \tag{21}$$

where $F(y_i, \theta_c)$ is the likelihood with θ_c and observation i, y_i, involved. In our case, it is s_{ij},

(iii) where $n_{-i,c}$ is the number of c_l for $l \neq i$ that are equal to c and b is the appropriate normalizing constant. Change the state to contain only those ϕ_c that are now associated with one or more observation.

(iv) For all $c \in \{c_1, \ldots, c_n\}$, draw new values from $\phi_c \mid y_i$ such that $c_i = c$, or perform some other update to ϕ_c that leaves this distribution invariant [14].

Thus, the value for θ_{rj} for those foci in cluster k and subcluster c may be sampled from

$$\theta_{k,c} \mid \cdot \sim MVN_3\left(\left(\sigma_k^{-2} + n_{k,c}\sigma^{-2}\right)^{-1} \cdot \left(\sigma_k^{-2}\boldsymbol{\mu}_k + n_{k,c}\sigma^{-2}\overline{\left(s_{rj} - p_r\right)}\right), \left(\sigma_k^{-2} + n_k\sigma^{-2}\right)^{-1} \cdot I_3\right), \quad (22)$$

where $n_{k,c}$ are the number of foci in some cluster k and subcluster c and $\overline{(s_{rj} - p_r)}$ denotes the mean of the observed data after adjusting for group effect. The value for p_r for those groups (studies) in group cluster c (different c notation than subclusters above) may be sampled from

$$p_c \mid \cdot \sim TMVN_3\left(\left(\sigma_p^{-2} + n_c\sigma^{-2}\right)^{-1} \cdot \left(\sigma_p^{-2}\boldsymbol{\mu}_p + n_c\sigma^{-2}\overline{\left(s_{rj} - \theta_{rj}\right)}\right), \left(\sigma_p^{-2} + n_c\sigma^{-2}\right)^{-1} I_3, -l, l\right), \quad (23)$$

where n_c is the number of groups in cluster c and similarly $\overline{(s_{rj} - \theta_{rj})}$ denotes the mean of the observed data after adjusting for individual effect.

2.4. Determining the Clusters. To estimate the number of clusters and the center of each cluster and cluster assignment, we implement the same least-squared Euclidean distance method introduced in [15] and used previously in our work in [10] and reiterated below. This method draws inferences on clusters based on a set of converged MCMC iterations and chooses one iteration as the final estimates for the clusters and related parameters. This final MCMC iteration is selected due to its smallest Euclidean distance to the expected cluster assignments estimated based on a set of independent converged MCMC iterations. This approach incorporates all clustering information in the MCMC sampling process [15]:

(1) After the prespecified number of MCMC burn-ins, let the MCMC simulations continue for an additional W iterations. An averaged clustering matrix is then created, denoted as A, and is an $n \times n$ matrix with each block or $(i, j)^{th}$ entry denoting how often foci i and j $(i, j = 1, \ldots, n)$ are in the same cluster. Specifically, each $(i, j)^{th}$ entry is the proportion of the W iterations that two foci are in the same cluster.

(2) Let the MCMC run additional F iterations, where, for each iteration,

(a) create an $n \times n$ matrix using indicators to denote which foci are clustered together; i.e., let the $(i, j)^{th}$ entry denote a 1 if foci i and j are in one cluster and 0 otherwise.

(b) use Euclidean distance to determine the similarity between this indicator matrix and the averaged clustering matrix A.

(3) Among the F iterations, select the iteration and respective clustering pattern, number of clusters, and parameters that produce the smallest Euclidean distance.

3. Inference

3.1. Simulation Settings. Simulations were utilized to illustrate and assess the proposed method. We assumed an fMRI metadata study setting including 50 studies, with each study containing 10 foci. Collectively these foci were simulated from three clusters centered as (x,y,z) talairach coordinates at $(1,1,1)^T$, $(2,2,2)^T$, and $(4,4,4)^T$ containing 150, 150, and 200 foci, respectively. It was also assumed that half the data, 250 foci or half from each individual cluster, came from one study cluster centered at $(0.1,0.1,0.1)^T$ and the remaining data from a second study cluster centered at $(0.4,0.4,0.4)^T$. These study clusters were linear shifts to cluster centers. For example, 75 foci in cluster one were centered at $(1.1,1.1,1.1)^T$ and the other half were centered at $(1.4,1.4,1.4)^T$; the study effect was a linear shift to all three dimensions from the cluster center. We made various alterations to this general setting:

(1) Normal setting: We simulated data from a multivariate normal for each cluster with respective means described above and a variance of $\Sigma = 0.002 I_3$. This creates spheres with little variation and we expect the method to have the ability to correctly identify the clusters.

(2) Chi-squared (skewed) setting: The method's ability to cluster in the presence of abnormal patterns is an important factor in spatial clustering. For this setting, we applied the same scenario as in the normal setting in respect of clusters 1 and 2 but simulated cluster 3 using a chi-squared distribution with 4 degrees of freedom.

(3) Large variance setting: The last scenario is designed to assess the robustness of the method with respect to the distance between and among clusters. To this end, we applied the normal setting but considered increasing levels (referred to large1, large2, large3, and large4 settings, resp.) of Σ: $\Sigma = 0.01 I_3, 0.05 I_3, 0.1 I_3$, and $0.2 I_3$, representing gradually closer distances among clusters.

For each setting, we implemented a grid search for a single dataset to estimate values of α_p and α based on the minimization of DIC. We let precision parameter values be 0.01, 0.05, 0.1, 0.5, 1, 2, and 5. Based on α_p and α estimates, 100 MC datasets were generated with 2500 burn-in iterations, 500 working iterations to calculate the probability matrix for determining the clusters, and 100 additional iterations

TABLE 1: Simulation assessments for three foci-level clusters.

Scenario (α_p, α)	Median Num. of Clusters (SD)∗	Cluster index	Average Sensitivity (SD)∗	Average Specificity (SD)∗	Average Accuracy (%) (SD)∗
Normal (0.1, 1.0)	∗∗IC: 3 (0.88)	1	0.96 (0.13)	1 (0)	0.96 (0.13)
		2	0.96 (0.13)	1 (0)	
		3	0.96 (0.13)	1 (0)	
	∗∗SC: 2 (0.32)	1	0.93 (0.26)	1 (0.04)	0.96 (0.13)
		2	1 (0.04)	0.93 (0.26)	
Chi-squared (0.05, 0.1)	IC: 19 (2.17)	1	0.94 (0.09)	0.91 (0.15)	0.54 (0.13)
		2	0.74 (0.43)	0.96 (0.02)	
		3	0.09 (0.03)	0.94 (0.16)	
	SC: 2 (0.56)	1	0.76 (0.37)	0.9 (0.1)	0.82 (0.18)
		2	0.88 (0.11)	0.78 (0.36)	
Large 1 (0.05, 0.1)	IC: 3 (0.82)	1	0.96 (0.13)	1 (0)	0.96 (0.13)
		2	0.96 (0.13)	1 (0)	
		3	0.96 (0.13)	1 (0)	
	SC: 2 (0.29)	1	0.92 (0.27)	1 (0)	0.96 (0.14)
		2	1 (0)	0.92 (0.27)	
Large 2 (0.01, 0.5)	IC: 3 (0.77)	1	0.99 (0.04)	1 (0)	0.99 (0.02)
		2	0.99 (0.03)	1 (0)	
		3	1 (0.01)	1 (0)	
	SC: 2 (0.14)	1	1 (0.02)	1 (0.01)	1 (0.01)
		2	1 (0.01)	1 (0)	
Large 3 (0.05, 0.05)	IC: 5 (1.25)	1	0.91 (0.13)	1 (0)	0.92 (0.07)
		2	0.88 (0.16)	1 (0)	
		3	0.96 (0.09)	1 (0)	
	SC: 2 (0.37)	1	0.97 (0.06)	0.98 (0.03)	0.97 (0.04)
		2	0.98 (0.03)	0.98 (0.03)	
Large 4 (0.01, 2.0)	IC: 9 (1.3)	1	0.66 (0.14)	0.99 (0.01)	0.62 (0.09)
		2	0.33 (0.08)	0.99 (0.01)	
		3	0.8 (0.17)	1 (0)	
	SC: 1 (0.36)	1	0.04 (0.17)	0.99 (0.05)	0.51 (0.08)
		2	0.98 (0.07)	0.05 (0.18)	

∗SD: standard deviation across 100 MC replicates; ∗∗IC: individual foci cluster; SC: study effect clusters.

to infer the number of clusters and individual foci cluster centers.

Model assessment consists of three evaluations: sensitivity, specificity, and accuracy. Sensitivity and specificity are defined by their generic definitions, the proportion of foci that are correctly assigned to their simulated cluster, and the proportion of foci that are correctly not assigned their nonsimulated cluster. Accuracy is defined as the percentage of foci that are correctly clustered. Note that the definition of accuracy takes into account both true positive and true negatives. In addition to our methodology, we applied a very common existing clustering approach for continuous data, K-means, to our simulation settings. Although this method cannot adjust for additional covariates, it allows for a comparison to existing methods. Lastly, to highlight the advantage of using a mixture of DPs over existing clustering approaches, we applied our approach, a revised version of our approach using a mixture of multivariate normal distributions rather than DPs, and Kmeans to the normal and chi-squared simulation scenarios. As the emphasis was on clustering performance, the group effect was assumed to be known for these two settings.

3.2. Simulation Results. Table 1 summarizes the findings on the three foci-level cluster identifications and the quality of the identified clusters. The proposed method gives high sensitivities and specificities across all scenarios. The accuracy of cluster assignment overall is higher than 90% when the variation in the data is relatively small and only dropping once clusters were large enough to overlap (scenario Large 4). The proposed methodology was also accurate at identifying the correct number of clusters as indicated by the median

TABLE 2: Kmeans simulation assessments for three foci-level clusters.

Scenario:	Median Num. of Clusters (SD)*	Cluster index	Average Sensitivity (SD)*	Average Specificity (SD)*	Average Accuracy (%) (SD)*
		1	0.73 (0.29)	0.98 (0.09)	
Normal	5 (0.95)	2	0.72 (0.3)	0.98 (0.08)	0.7 (0.12)
		3	0.67 (0.26)	1 (0)	
		1	0.58 (0.19)	0.99 (0.01)	
Chi-squared	10 (0.5)	2	0.86 (0.28)	0.92 (0.03)	0.5 (0.09)
		3	0.16 (0.08)	0.98 (0.11)	
		1	0.68 (0.33)	1 (0.04)	
Large 1	5 (0.98)	2	0.76 (0.26)	0.96 (0.12)	0.7 (0.13)
		3	0.67 (0.26)	1 (0)	
		1	0.61 (0.27)	0.99 (0.06)	
Large 2	7 (1.29)	2	0.61 (0.27)	0.99 (0.06)	0.58 (0.11)
		3	0.54 (0.27)	1 (0)	
		1	0.49 (0.19)	1 (0)	
Large 3	9 (0.94)	2	0.47 (0.17)	1 (0)	0.46 (0.08)
		3	0.43 (0.24)	1 (0)	
		1	0.43 (0.16)	1 (0.01)	
Large 4	10 (0.57)	2	0.42 (0.14)	1 (0.01)	0.41 (0.07)
		3	0.38 (0.2)	1 (0)	

*SD: standard deviation across 100 MC replicates.

number of individual and study clusters. In comparison, results from the Kmeans approach indicate relatively lower statistics in sensitivity and specificity (Table 2). The accuracy is around 70%, when the variations in the data are relatively small. However, compared to the proposed method, the Kmeans approach often inferred a higher number of clusters as indicated by the larger median number of clusters. The computation time of a single dataset for the mixture of DPs took, on average, 7-8 hours on a high performance computer (Dell cluster with 88 compute nodes, 3120 total central processing unit cores, 20664 Giga-bytes of RAM, and 61440 total graphic processing unit cores).

After removing study effects, the comparison between the mixture of DPs, mixture of normal, and Kmeans is as expected (Table 3). Both mixtures performed exceptionally well at identifying the three normally distributed clusters while the Kmeans performance was adequate with an overall accuracy of 80% (compared to 100% for both mixtures). Once the data deviated from normality, the mixture of normals approach was unable to differentiate the clusters resulting in low accuracy (32%). The Kmeans approach performed similarly to the mixture of DPs with both approaches having low sensitivities for the third cluster which was skewed but with the mixture of DPs resulting in superior accuracy. The mixture of DPs was able to differentiate clusters 1 and 2 as indicated by 99% sensitivity and 100% specificity measures but tended to "overcluster" cluster 3 into smaller clusters as indicated by the large median number of clusters, 14% sensitivity, and 100% specificity. In regard to overall accuracy, the mixture of DPs outperformed the mixture of normals and Kmeans when the data are skewed.

4. Real Data Application

For this application, we applied the proposed method to a meta-analysis dataset. Constructed originally in [16], this data consists of a total of 162 neuroimaging publications, of which 57 were PET and 105 were fMRI. Among these 162 publications, there were 437 contrasts or studies. Only those foci that were deemed significantly activated by their study specific criteria were included for a total of 2,478 foci. Summary statistics for this data can be seen in Tables 4 and 5.

As with the simulation studies, grid search and DIC were used to estimate values for α_p and α. Potential precision parameters values were 0.01, 0.05, 0.1, 0.5, 1, 2, 5, and 7.5. Each combination was performed over 2,600 iterations, 2,000 of those for burn-in, 500 for the probability matrix calculation, and final 100 to infer individual clusters and their centers. To assist with the magnitude of the likelihood calculations, the data was scaled down by 10.

It was found that the precision parameter combination of $\alpha_p = 0.05$ and $\alpha = 1$ produced the smallest DIC. Convergence, with the initial 2,000 discarded, was checked based on trace plots. Based on the proposed method, we identified four study clusters and 14 individual foci clusters. A single DIC setting for this data took, on average, 72 hours to run on the HPC. The break down of the 14

TABLE 3: Comparison of approaches for selected simulated settings assuming study effect is known.

Setting	Approach	Median Num. of Clusters (SD)∗	Cluster index	Average Sensitivity (SD)∗	Average Specificity (SD)∗	Average Accuracy (%) (SD)∗
Normal	DP	3 (0)	1	1 (0)	1 (0)	1 (0)
			2	1 (0)	1 (0)	
			3	1 (0)	1 (0)	
	Mixture	3 (0)	1	1 (0)	1 (0)	1 (0)
			2	1 (0)	1 (0)	
			3	1 (0)	1 (0)	
	Kmeans	3 (0.6)	1	0.81 (0.36)	0.93 (0.16)	0.8 (0.22)
			2	0.79 (0.38)	0.94 (0.15)	
			3	0.79 (0.25)	1 (0)	
Chi-squared	DP	16 (1.76)	1	0.99 (0.01)	0.99 (0.01)	0.65 (0.01)
			2	0.99 (0.01)	0.97 (0.01)	
			3	0.14 (0.03)	1 (0)	
	Mixture	1 (0.29)	1	0.06 (0.24)	0.98 (0.11)	0.32 (0.1)
			2	0.95 (0.22)	0.06 (0.24)	
			3	0.05 (0.21)	1 (0)	
	Kmeans	10 (0.22)	1	0.99 (0.06)	0.98 (0.01)	0.62 (0.1)
			2	0.87 (0.34)	0.93 (0.03)	
			3	0.16 (0.06)	0.94 (0.17)	

∗SD: standard deviation across 100 MC replicates.

TABLE 4: Descriptive statistics∗.

	Min.	1st Qu.	Median	Mean	3rd Qu.	Max.
Number of foci per pub.	1.00	5.75	10.00	15.11	17.25	110.00
Number of foci per study	1.00	2.00	4.00	5.67	7.00	47.00
Number of subjects per pub.	4.00	9.00	11.00	12.26	14.00	40.00
Number of studies per pub.	1.000	1.000	2.000	2.67	4.000	12.000

∗Min: minimum, 1st Qu: 25% percentile, 3rd Qu: 75% percentile, Max: maximum, pub: publication.

TABLE 5: Frequency of emotions.

Emotions	Frequency of studies (% of total studies)	Frequency of foci (% of total foci)
aff∗	175 (40.05%)	881 (35.55%)
anger	26 (5.95 %)	166 (6.7%)
disgust	44 (10.07%)	337 (13.6%)
fear	68 (15.56%)	367 (14.81%)
happiness	36 (8.24%)	178 (7.18%)
mixed	41 (9.38%)	195 (7.87%)
sadness	45 (10.3%)	348 (14.04%)
surprise	2 (0.46%)	6 (0.24%)
Total	437	2478

∗aff: affective.

individual foci clusters by center location, brain location, foci frequency, and study frequency can be seen in Table 6. The frequency of each foci-associated emotion within each of the 14 clusters can be seen in Table 7. The affective emotion was dominating in all clusters with fear being the second dominating emotion in clusters 1, 2, 3, 11, and 13, disgust in clusters 5 and 14, sadness in clusters 6 and 10, and a mixture of emotions in the remaining clusters. When only focusing on those foci that fell within known brain regions of interest, as seen in Table 8, the dominating emotion, other

TABLE 6: Meta-data cluster results.

Cluster Centers	Brian Regions	Cluster Index	# of foci per cluster (% of total foci)	# of studies per cluster (% of all studies)
(-9.11, -14.45, 1.4)	Temporal Mid L	1	1615 (65.17)	386 (88.33)
(-0.83, -4.64, 0.54)	Temporal Mid L	2	188 (7.59)	139 (31.81)
(16.75, -10.63, -11.33)	Temporal Mid R	3	148 (5.97)	117 (26.77)
(3.48, -5.52, -2.15)	Temporal Inf L	4	102 (4.12)	84 (19.22)
(2.08, -4.61, -0.14)	Temporal Mid R	5	77 (3.11)	67 (15.33)
(0.87, -7.7, -0.63)	Temporal Pole Sup L	6	67 (2.7)	59 (13.5)
(0.99, -6.33, 2.62)	Cerebelum 6 R	7	55 (2.22)	43 (9.84)
(2.3, -5.89, 0.32)	NA	8	54 (2.18)	46 (10.53)
(0.43, -6.59, 0.6)	Postcentral L	9	41 (1.65)	36 (8.24)
(-0.18, -4.99, 0.11)	Cerebelum 6 R	10	39 (1.57)	38 (8.7)
(1.69, -5.25, 1.43)	Temporal Sup R	11	38 (1.53)	34 (7.78)
(-0.64, -5.93, 1.77)	Postcentral L	12	22 (0.89)	20 (4.58)
(1.29, -5.28, 1.23)	Precentral R	13	18 (0.73)	17 (3.89)
(0.45, -5.31, -0.87)	Occipital Inf R	14	14 (0.56)	14 (3.2)

R: right hemisphere, L: left hemisphere.

than affective, was sadness, fear, and disgust, respectively. When compared to the number of clusters identified by the spatial Cox Point process (53) and Kmeans (20) performed in [10], fewer clusters were identified with our current application. It should be noted that this particular data does not visually indicate distinct clusters and is closer to a more uniform distribution throughout the brain which may lead to an inaccurate number of identifiable clusters. However, given the results from our previous analysis in [10] and findings in simulation studies, it is possible that the clusters formed were rather subtle and actually might not be distinct enough.

5. Conclusion and Discussion

Modeling the realization of observed foci as a linear association of study effect and individual foci cluster effect with a multivariate normal random error was motivated by the limitation of the spatial Cox process to statistically distinguish between a cluster and a mode or peak of a cluster. The overall aim remained to identify activated regions within the brain using fMRI coordinate-based metadata. By modeling the data in this fashion, it was hopeful that the distribution could statistically differentiate between clusters and modes of clusters while retaining the flexibility and robustness to mimick the behavior of the data.

Simulation studies demonstrated that the method can fit data generated from normal or abnormal distributions. Furthermore, it was able to identify clusters within covariates while retaining the integrity to identify individual clusters. Both the proposed method and Kmeans were unable to correctly identify clusters when they were large and overlapped and both the mixture of normals and mixture of DPs performed poorly at identifying a cluster severely skewed.

When applied to a fMRI metadataset, the method identified a relatively low number of clusters. Given the low sensitivity findings in the simulated study with high noise, it can be concluded that this data had a high likelihood of being too broad. When the same data was analyzed with the spatial Cox process, the difference in the results was extreme. Not only were the number of clusters substantially less, but also none of the cluster centers identified from the proposed method came close to those identified in the first method. It is worth mentioning that the meta-analysis data is not distinctly grouped and is more uniformly distributed throughout the brain and perhaps the model used did not provide the best fit.

The primary advantage to this method, besides its flexibility, is its ability to describe irregular spatial patterns and its sampling design to statistically differentiate clusters. Because of its adaptable nature, this model can also adjust for any covariate(s) of interest. However, based on simulation studies and the fMRI metadata application, the proposed method tends to be too insensitive and has a difficult time identifying clusters when data are not distinctly differentiated. A potential limitation in the approach is that each DP within the mixture was assumed to have the same precision parameter. It was noted that, during the simulation studies when the mixture of DPs was attempting to fit the Chi-squared simulations (without study effect), it was overclustering the skewed cluster. However, when the precision parameter was smaller, the identifiability of cluster 3 became more accurate but became more inaccurate for clusters 1 and 2. Thus, to further improve flexibility and accuracy for this approach when data is skewed, each DP potentially requires its own unique precision parameter. Furthermore, this method's clustering ability is limited by the identification of study effects which may be improved by implementing stronger restrictions or could be an effect of having multiple DPs. Our future work will focus on these issues, allowing study effect to be random rather than a fixed effect, and identifying if a large number of DPs within the model is indeed a limitation.

Table 7: Breakdown of emotions and their frequencies by individual foci cluster∗.

Cluster Index: Total foci in that cluster Emotion Frequency of emotion (% of total cluster foci)					
Cluster: 1	1615	**Cluster: 2**	188	**Cluster: 3**	148
aff	562 (34.8)	aff	57 (30.32)	aff	53 (35.81)
anger	110 (6.81)	anger	10 (5.32)	anger	12 (8.11)
disgust	225 (13.93)	disgust	18 (9.57)	disgust	22 (14.86)
fear	234 (14.49)	fear	35 (18.62)	fear	31 (20.95)
happiness	121 (7.49)	happiness	17 (9.04)	happiness	8 (5.41)
mixed	131 (8.11)	Mixed	18 (9.57)	mixed	8 (5.41)
sadness	228 (14.12)	sadness	33 (17.55)	sadness	13 (8.78)
surprise	4 (0.25)			surprise	1 (0.68)
Cluster: 4	102	**Cluster: 5**	77	**Cluster: 6**	67
aff	44 (43.14)	aff	35 (45.45)	aff	22 (32.84)
anger	6 (5.88)	anger	3 (3.9)	anger	4 (5.97)
disgust	15 (14.71)	disgust	13 (16.88)	disgust	5 (7.46)
fear	9 (8.82)	fear	5 (6.49)	fear	14 (20.9)
happiness	9 (8.82)	happiness	5 (6.49)	happiness	6 (8.96)
mixed	4 (3.92)	Mixed	7 (9.09)	Mixed	4 (5.97)
sadness	15 (14.71)	sadness	9 (11.69)	sadness	11 (16.42)
				surprise	1 (1.49)
Cluster: 7	55	**Cluster: 8**	54	**Cluster: 9**	41
aff	21 (38.18)	aff	19 (35.19)	aff	13 (31.71)
anger	7 (12.73)	anger	2 (3.7)	anger	5 (12.2)
disgust	7 (12.73)	disgust	9 (16.67)	disgust	8 (19.51)
fear	6 (10.91)	fear	7 (12.96)	fear	8 (19.51)
happiness	3 (5.45)	mixed	9 (16.67)	happiness	2 (4.88)
mixed	4 (7.27)	sad	8 (14.81)	mixed	1 (2.44)
sadness	7 (12.73)			sad	4 (9.76)
Cluster: 10	39	**Cluster: 11**	38	**Cluster: 12**	22
aff	15 (38.46)	aff	20 (52.63)	aff	9 (40.91)
anger	2 (5.13)	anger	1 (2.63)	anger	2 (9.09)
disgust	5 (12.82)	disgust	2 (5.26)	disgust	3 (13.64)
fear	3 (7.69)	fear	7 (18.42)	fear	3 (13.64)
happiness	3 (7.69)	happiness	2 (5.26)	mixed	2 (9.09)
mixed	4 (10.26)	sad	6 (15.79)	sad	3 (13.64)
sadness	7 (17.95)				
Cluster: 13	18	**Cluster: 14**	14		
aff	5 (27.78)	aff	6 (42.86)		
anger	2 (11.11)	disgust	3 (21.43)		
disgust	2 (11.11)	happiness	1 (7.14)		
fear	5 (27.78)	mixed	2 (14.29)		
happiness	1 (5.56)	sad	2 (14.29)		
mixed	1 (5.56)				
sadness	2 (11.11)				

Table 8: Breakdown of emotions and their frequencies by individual foci cluster for ROI∗.

Cluster Index: Total foci in that cluster Emotion Frequency of emotion (% of total cluster foci)					
Cluster: 1	489	**Cluster: 2**	63	**Cluster: 3**	54
aff	217 (44.38)	aff	29 (46.03)	aff	21 (38.89)
anger	20 (4.09)	anger	3 (4.76)	anger	3 (5.56)
disgust	68 (13.91)	disgust	8 (12.7)	disgust	4 (7.41)
fear	63 (12.88)	fear	13 (20.63)	fear	8 (14.81)
happiness	28 (5.73)	happiness	2 (3.17)	happiness	6 (11.11)
mixed	22 (4.5)	mixed	4 (6.35)	mixed	5 (9.26)
sadness	71 (14.52)	sadness	4 (6.35)	sadness	7 (12.96)
Cluster: 4	34	**Cluster: 5**	25	**Cluster: 6**	20
aff	15 (44.12)	aff	13 (52)	aff	8 (40)
anger	1 (2.94)	anger	1 (4)	disgust	1 (5)
disgust	4 (11.76)	disgust	4 (16)	fear	4 (20)
fear	3 (8.82)	fear	1 (4)	happiness	1 (5)
happiness	1 (2.94)	happiness	2 (8)	mixed	1 (5)
mixed	1 (2.94)	mixed	2 (8)	sadness	5 (25)
sadness	9 (26.47)	sadness	2 (8)		
Cluster: 7	19	**Cluster: 8**	17	**Cluster: 9**	16
aff	6 (31.58)	aff	6 (35.29)	aff	9 (56.25)
anger	1 (5.26)	anger	2 (11.76)	fear	3 (18.75)
disgust	4 (21.05)	disgust	3 (17.65)	happiness	2 (12.5)
mixed	2 (10.53)	happiness	1 (5.88)	sadness	2 (12.5)
sadness	6 (31.58)	mixed	1 (5.88)		
		sadness	4 (23.53)		
Cluster: 10	15	**Cluster: 11**	9	**Cluster: 12**	7
aff	6 (40)	aff	3 (33.33)	aff	3 (42.86)
anger	1 (6.67)	anger	1 (11.11)	anger	1 (14.29)
disgust	2 (13.33)	disgust	2 (22.22)	disgust	1 (14.29)
fear	2 (13.33)	fear	1 (11.11)	fear	2 (28.57)
happiness	1 (6.67)	sadness	2 (22.22)		
mixed	2 (13.33)				
sadness	1 (6.67)				
Cluster: 13	6	**Cluster: 14**	5		
aff	3 (50)	aff	3 (60)		
disgust	1 (16.67)	happiness	1 (20)		
sadness	2 (33.33)	sadness	1 (20)		

∗ROI: region of interest; aff: affective.

Conflicts of Interest

No conflicts of interest exist for any author.

Acknowledgments

Dr. Kang's effort was supported by an NIH Grant 1R01MH105561. Dr. Zhang and Dr. Ray's efforts were supported by their start funds provided by the University of Memphis.

References

[1] M. Stephens, "Bayesian analysis of mixture models with an unknown number of components—an alternative to reversible jump methods," *The Annals of Statistics*, vol. 28, no. 1, pp. 40–74, 2000.

[2] M. Aitkin and R. Healey, "Estimation and hypothesis testing in finite mixture models. Journal of the Royal Statistical Society," *Journal of the Royal Statistical Society. Series B (Methodological)*, pp. 67–75, 1985.

[3] G. J. McLachlan and K. E. Basford, *Mixture Models: Applications to Clustering*, Marcel Dekker, New York, NY, USA, 1988.

[4] M. A. T. Figueiredo and A. K. Jain, "Unsupervised learning of finite mixture models," *IEEE Transactions on Pattern Analysis and Machine Intelligence*, vol. 24, no. 3, pp. 381–396, 2002.

[5] J.-M. Marin, K. Mengersen, and C. . Robert, "Bayesian modelling and inference on mixtures of distributions," in *Bayesian thinking: modeling and computation*, vol. 25 of *Handbook of Statist.*, pp. 459–507, Elsevier/North-Holland, Amsterdam, 2005.

[6] P. J. Green, "Reversible jump Markov chain Monte Carlo computation and Bayesian model determination," *Biometrika*, vol. 82, no. 4, pp. 711–732, 1995.

[7] P. J. Green and D. I. Hastie, "Reversible jump mcmc," *Genetics*, vol. 1550, no. 3, pp. 1391–1403, 2009.

[8] M. D. Escobar and M. West, "Bayesian density estimation and inference using mixtures," *Journal of the American Statistical Association*, vol. 90, no. 430, pp. 577–588, 1995.

[9] C. E. Antoniak, "Mixtures of Dirichlet processes with applications to Bayesian nonparametric problems," *The Annals of Statistics*, vol. 2, pp. 1152–1174, 1974.

[10] M. Ray, J. Kang, and H. Zhang, "Identifying Activation Centers with Spatial Cox Point Processes Using fMRI Data," *IEEE Transactions on Computational Biology and Bioinformatics*, vol. 13, no. 6, pp. 1130–1141, 2016.

[11] P. Congdon, *Bayesian statistical modelling*, vol. 704, John Wiley & Sons, 2007.

[12] D. J. Spiegelhalter, N. G. Best, B. P. Carlin, and A. van der Linde, "Bayesian measures of model complexity and fit," *Journal of the Royal Statistical Society: Series B (Statistical Methodology)*, vol. 64, no. 4, pp. 583–639, 2002.

[13] C. Preston, "Spatial birth-and-death processes," *Advances in applied probability*, vol. 70, no. 03, pp. 405–408, 1975.

[14] R. M. Neal, "Markov chain sampling methods for Dirichlet process mixture models," *Journal of Computational and Graphical Statistics*, vol. 9, no. 2, pp. 249–265, 2000.

[15] D. B. Dahl, "Model-based clustering for expression data via a dirichlet process mixture model," *Bayesian Inference for Gene Expression and Proteomics*, pp. 201–218, 2006.

[16] H. Kober, L. F. Barrett, J. Joseph, E. Bliss-Moreau, K. Lindquist, and T. D. Wager, "Functional grouping and cortical-subcortical interactions in emotion: A meta-analysis of neuroimaging studies," *NeuroImage*, vol. 42, no. 2, pp. 998–1031, 2008.

17

Polyhedral Star-Shaped Distributions

Wolf-Dieter Richter and Kay Schicker

Institute of Mathematics, University of Rostock, Ulmenstr. 69, Haus 3, 18057 Rostock, Germany

Correspondence should be addressed to Wolf-Dieter Richter; wolf-dieter.richter@uni-rostock.de

Academic Editor: Ramón M. Rodríguez-Dagnino

A new method of probabilistic modelling of polyhedrally contoured sample clouds is presented and applied to statistical reasoning for a real dataset. Various representations of the new class of polyhedral star-shaped distributions are derived and basic properties of the moments as well as characteristic and moment generating functions of these distributions are studied. Along with location-scale transformations, estimating and hypothesis testing are dealt with.

1. Introduction

One of nowadays challenges of statistical modelling is the construction of flexible multivariate probability distributions given a dataset. In [1], disparities in premature mortality between high- and low-income US counties are dealt with on the basis of descriptive statistics. For a subsequent step of statistical reasoning, a method of constructing a suitable probabilistic model is needed. Analyzing the correlation between mortality and income cannot be done in a common linear regression model because of absence of homoscedasticity. The present paper provides therefore a new method of constructing flexible multivariate distribution laws being well adapted to practical problems characterized by polyhedral contours of their sample clouds. Moreover, we will particularly suggest a specific model and further statistical reasoning for the premature mortality-income data mentioned above.

Several basic methods and results from the research area of constructing multivariate distributions are surveyed, for example, in [2] and the numerous papers mentioned there. Multivariate densities with given contours were already introduced in [3] and further studied in [4, 5]. Constructing star-shaped distributions on using Minkowski functionals and generalized uniformly distributed random vectors on generalized spheres, proving geometric measure representations of such distributions and stochastic representations of correspondingly distributed random vectors are to be found in [6]. Constructions and representations in the special cases of norm and antinorm contoured distributions are recently dealt with in [7]. Numerous applications of geometric measure representations are surveyed in the last mentioned two papers and in [8]. In [9] more recent applications to extreme values are presented. More general order statistics are dealt with elsewhere.

The aim of the present paper is to deal with another important special case of star-shaped distributions where the contours are the topological boundaries of polyhedra.

Polyhedral star-shaped distributions can be considered being a subclass of the class of star-shaped distributions. Before studying specific properties of polyhedral star-shaped distributions, we start therefore with a short introduction to the general theory of star-shaped distributions. To this end, we follow [6].

Let $K \subset \mathbb{R}^n$ denote a star body, that is, a nonempty and compact star-shaped set being equal to the closure of its interior and having the origin 0_n in its interior. The functional $h_K : \mathbb{R}^n \to [0, \infty)$ defined by

$$h_K(x) = \inf\{r > 0 : x \in rK\}, \quad x \in \mathbb{R}^n \tag{1}$$

is known as the Minkowski functional of K. Here, we assume that h_K is positive homogeneous; that is, $h_K(\lambda x) = \lambda h_K(x)$, $\lambda > 0$, and consider $K(r) = rK = \{x \in \mathbb{R}^n : h_K(x) \le r\}$ and its boundary $S(r) = rS = \{x \in \mathbb{R}^n : h_K(x) = r\}$ being the star ball and star sphere of Minkowski radius $r > 0$, respectively. Since h_K unambiguously defines the considered star ball, it is possible to study subclasses of star bodies by specifying h_K. One can choose, for example, h_K to be a norm or an antinorm. For the latter notion we refer to [10]. Here, specific representations of h_K are considered later in this paper, if h_K denotes a star-shaped polyhedron. For simplicity, consideration will be restricted throughout this paper to star bodies having the following property.

A countable collection $\mathfrak{F} = \{C_1, C_2, \ldots\}$ of pairwise disjoint cones C_j with vertex being the origin 0_n and $\mathbb{R}^n = \bigcup_j C_j$ is called a fan. Let $S_j = S \cap C_j$, $S_j \cap \mathfrak{B}_n = \mathfrak{B}_{S,j}$, where $\mathfrak{B}_n$ denotes the Borel-σ-field in $\mathbb{R}^n$, and $\mathfrak{B}_S = \sigma\{\mathfrak{B}_{S,1}, \mathfrak{B}_{S,2}, \ldots\}$. In what follows, the star body K and a set $A \in \mathfrak{B}_S$ are chosen such that for every j the set

$$G(A \cap S_j) = \left\{\vartheta \in \mathbb{R}^{n-1} : \exists \eta \text{ with } \theta = (\vartheta^T, \eta)^T \in A \cap S_j\right\} \tag{2}$$

is well defined and for every $\vartheta = (\vartheta_1, \ldots, \vartheta_{n-1})^T \in G(A \cap S_j)$ there is uniquely determined $\eta > 0$ satisfying $h_K((\vartheta_1, \ldots, \vartheta_{n-1}, \eta)^T) = 1$.

A function $g : \mathbb{R}^+ \to \mathbb{R}^+$ satisfying the assumptions $0 < \mathcal{I}(g) < \infty$ where $\mathcal{I}(g) = \int_0^\infty r^{n-1} g(r) dr$ is called a density generating function (dgf),

$$\varphi_{g,K}(x) = C(g, K)\, g(h_K(x)), \quad x \in \mathbb{R}^n \tag{3}$$

a star-shaped density and K its contour defining star body. Such densities are studied in [3, 6, 11]. A probability measure having the density $\varphi_{g,K}$ will be denoted by $\Phi_{g,K}$. Note that the normalizing constant $C(g, K)$ allows the representation

$$C(g, K) = \frac{1}{n\mu(K)\mathcal{I}(g)}, \tag{4}$$

where μ denotes the Lebesgue measure in $\mathbb{R}^n$. For examples of density generating functions, see [12] or Table 1. Moreover, the definition of specific star-shaped densities appears already in earlier work where h_K is prespecified. If h_K is the Euclidean-norm, one considers the class of spherically symmetric distributions; see [12–14] and many other contributions. The class of convex contoured $l_{n,p}$-symmetric distributions is considered if h_K is the $l_{n,p}$-norm, $p \geq 1$. This class of distributions was introduced in [15] and studied, for example, in [16–19]. In [20] stochastic and geometric representations are derived also for the case which can be called according to [10] the antinorm contoured case. General norm contoured distributions in $\mathbb{R}^2$ are studied in [8] by choosing h_K in terms of a norm in $\mathbb{R}^2$. Minkowski functionals of ellipsoids and even p-generalized ellipsoids are used in the (re-)construction of common and p-generalized elliptically contoured distributions in [21] and in [6], respectively.

The introduction of h_K as generalized radius functional is closely connected with the notion of generalized surface content measure which turns out to be a suitably chosen non-Euclidean surface content measure on a star sphere. These notions and their properties are very useful in the consideration of nonspherical distributions; see [6, 7, 20, 21].

Here, we recall the local definition of a generalized surface content measure on star spheres from [22], noting that there exists an equivalent integral approach introduced for various special classes of star bodies in [6, 7, 20, 21]. For $A \in \mathfrak{B}_S$, we introduce the central projection cone $\mathrm{CPC}(A) = \{x \in \mathbb{R}^n : x/h_K(x) \in A\}$ and the star sector of star radius $\rho > 0$, $\mathrm{sector}(A, \rho) = \mathrm{CPC}(A) \cap K(\rho)$. The star-generalized surface measure is then defined by

$$\mathfrak{O}_S(A) = f'(\rho), \quad \text{where } f(\rho) = \mu(\mathrm{sector}(A, \rho)), \; A \in \mathfrak{B}_S. \tag{5}$$

The star-generalized uniform probability distribution on the Borel-σ-field $\mathfrak{B}_S$ is defined as $\omega_S(A) = \mathfrak{O}_S(A)/\mathfrak{O}_S(S)$. With these notations, we can recall the geometric measure representation formula of $\Phi_{g,K}(B)$ for every $B \in \mathfrak{B}_n$,

$$\begin{aligned}\Phi_{g,K}(B) &= C(g, K) \int_0^\infty r^{n-1} g(r)\, \mathfrak{O}_S\left(\left[\frac{1}{r}B\right] \cap S\right) dr \\ &= C(g, K)\, \mathfrak{O}_S(S) \int_0^\infty r^{n-1} g(r)\, \mathfrak{F}_S(B, r)\, dr,\end{aligned} \tag{6}$$

where $r \to \mathfrak{F}_S(B, r) = \omega_S([(1/r)B] \cap S)$ denotes the star intersection-proportion function (ipf) of the set B. Furthermore, a star-shaped distributed random vector $Y \sim \Phi_{g,K}$ allows the stochastic representation

$$Y \stackrel{d}{=} R_g U_S, \tag{7}$$

where R_g and U_S are stochastically independent, $U_S \sim \omega_S$, and R_g follows the density $f(r) = r^{n-1}g(r)/\mathcal{I}(g)$, $r > 0$. Furthermore, we recall from [7] integral representations of $\mathfrak{O}_S(A)$, $A \in \mathfrak{B}_S^+$, where $\mathfrak{B}_S^+$ denotes the Borel-σ-field on the upper half-sphere of S, if h_k is a norm and an antinorm, respectively. It holds

$$\begin{aligned}\mathfrak{O}_S(A) &= \int_{G(A)} h_{K^*}(N(\vartheta))\, d\vartheta, \\ \mathfrak{O}_S(A) &= \int_{G(A)} h_{K^\circ}(N^\circ(\vartheta))\, d\vartheta,\end{aligned} \tag{8}$$

where $G(A) = \{\vartheta \in \mathbb{R}^{n-1} : \exists \eta = \eta(\vartheta) \text{ with } (\vartheta^T, \eta)^T \in A\}$, h_{K^*} denotes the dual norm of h_K, if h_K is a norm, and h_{K° denotes the Minkowski functional of the antipolar set K° of K if h_k is an antinorm with K being an element of the class of antinorm balls AN1 defined in [7]. Here, $N(\vartheta)$ is the outer normal vector to S at $(\vartheta^T, \eta)^T$ and $N^\circ(\vartheta)$ is the inner normal vector to S at $(\vartheta^T, \eta)^T$, respectively. For particular representations of h_{K^*} and h_{K° if K is a norm or antinorm, respectively, generated by a star-shaped polyhedron, we refer already here to Theorem 6 (d) and (e) below. In this context, we will say that K belongs to the class of antinorm balls AN2 if h_K is an antinorm and $K = \mathbb{P}$ is a star-shaped polyhedron. We notice that convex geometry and certain types of polytopes play also a basic role in [23] where max-stable distributions and all norms that give rise to strictly max-stable distributions are studied.

TABLE 1: Density generating functions, normalizing constants, and moments of R_g.

Type	$g(r)$	$\mathcal{I}(g)$	$\mathbb{E}(R_g)$	$\mathbb{E}(R_g^2)$	$\mathbb{E}(R_g^k)$
Kotz	$r^{2M-2}e^{-\beta r^{2\gamma}},\quad M,\beta,\gamma\in(0,\infty)$	$\dfrac{\Gamma((2M+n-2)/2\gamma)}{2\gamma\beta^{(2M+n-2)/2\gamma}}$	$\dfrac{\Gamma((2M+n-1)/2\gamma)}{\beta^{1/2\gamma}\Gamma((2M+n-2)/2\gamma)}$	$\dfrac{\Gamma((2M+n)/2\gamma)}{\beta^{1/\gamma}\Gamma((2M+n-2)/2\gamma)}$	$\dfrac{\Gamma((2M+n+k-2)/2\gamma)}{\beta^{k/2\gamma}\Gamma((2M+n-2)/2\gamma)}$
Multinormal	$e^{-r^2/2}$	$\dfrac{\Gamma(n/2)}{2^{1-n/2}}$	$\dfrac{\Gamma((n+1)/2)\sqrt{2}}{\Gamma(n/2)}$	n	$\dfrac{\Gamma((n+k)/2)\,2^{k/2}}{\Gamma(n/2)}$
Pearson VII	$\left(1+\dfrac{r^2}{m}\right)^{-M},\quad M>\dfrac{n}{2}, m>0$	$\dfrac{\Gamma(n/2)\Gamma(M-n/2)}{2\Gamma(M)m^{-n/2}}$	$\dfrac{\Gamma((n+1)/2)\Gamma(M-(n+1)/2)\sqrt{m}}{\Gamma(n/2)\Gamma(M-n/2)}$	$\dfrac{mn}{2M-n-2}$	$\dfrac{\Gamma((n+k)/2)\Gamma(M-(n+k)/2)}{m^{-k/2}\Gamma(n/2)\Gamma(M-n/2)}$
Multivariate t	$\left(1+\dfrac{r^2}{m}\right)^{-m/2-n/2},\quad m\in\mathbb{N}^{>0}$	$\dfrac{\Gamma(n/2)\Gamma(m/2)}{2\Gamma((m+n)/2)m^{-n/2}}$	$\dfrac{\Gamma((n+1)/2)\Gamma((m-1)/2)\sqrt{m}}{\Gamma(n/2)\Gamma(m/2)}$	$\dfrac{mn}{m-2}$	$\dfrac{\Gamma((n+k)/2)\Gamma((m-k)/2)}{m^{-k/2}\Gamma(n/2)\Gamma(m/2)}$
Multivariate Cauchy	$(1+r^2)^{-(n+1)/2}$	$\dfrac{\sqrt{\pi}\Gamma(n/2)}{2\Gamma((n+1)/2)}$	—	—	—
Pearson II	$I_{[0,1]}(r)(1-r^2)^m,\quad m>-1$	$\dfrac{\Gamma(n/2)\Gamma(m+1)}{2\Gamma(n/2+m+1)}$	$\dfrac{\Gamma((n+1)/2)\Gamma(n/2+m+1)}{\Gamma(n/2+m+3/2)\Gamma(n/2)}$	$\dfrac{n}{2m+n+2}$	$\dfrac{\Gamma((n+k)/2)\Gamma((n/2)+m+1)}{\Gamma((n+k)/2+m+1)\Gamma(n/2)}$
Uniform	$I_{[0,1]}(r)$	$\dfrac{1}{n}$	$\dfrac{n}{n+1}$	$\dfrac{n}{n+2}$	$\dfrac{n}{n+k}$

The paper is now organized as follows. We present the Minkowski functionals of star-shaped polyhedra in various kinds in Section 2 and derive corresponding representations of the polyhedral star-shaped surface and uniform measures in Section 3. This allows an extension of the ball number function in Section 3.2. The new classes of general polyhedral star-shaped distributions and specific $\mathbb{W}$-star contoured distributions are considered then in Sections 4.1 and 4.2, respectively. Moments, characteristic, and moment generating functions are studied in Section 5, and simulation with a rejection method is discussed in Section 6. Location-scale transformations considered in Section 7 give rise to considering estimating of parameters and testing of hypothesis in Section 8. In Section 8, moreover, we present the probabilistic modelling and further statistical reasoning for the premature mortality-income data mentioned at the beginning of this introduction. We discuss some open problems in Section 9. Appendix mainly shows how various theoretical results apply in one and the same situation.

2. Minkowski Functionals of Star-Shaped Polyhedra

In this section, we specify the representation of h_K if the star ball K is a star-shaped polyhedron. Let $\mathbb{P} \subset \mathbb{R}^n$ be a star-shaped polyhedron. We call $\mathfrak{F} = \{C_1, \ldots, C_m\}$ a convex polyhedral fan for $\mathbb{P}$ if $\mathbb{P} \cap C_j$ is for every j, $j \in \{1, \ldots, m\}$ a convex polyhedron. Due to the considerations of convex polyhedra in the broad literature of convex geometry, it is possible to represent a convex polyhedron in two different but equivalent ways; see [24]. A convex polyhedron $\mathbb{P} \subset \mathbb{R}^n$ can be given by the set of its vertices $\{p_1, \ldots, p_l\}$,

$$\mathbb{P} = \operatorname{conv}(\{p_1, \ldots, p_l\}), \qquad (\mathscr{V}\mathscr{P})$$

where $\operatorname{conv}(M)$ denotes the convex hull of the point set M. Alternatively, it is possible to consider a convex polyhedron as the intersection of suitably chosen closed half-spaces. In this case, there exist a matrix $A \in \mathbb{R}^{k\times n}$ and a vector $b \in \mathbb{R}^k$ such that

$$\mathbb{P} = \{x \in \mathbb{R}^n : Ax \le b\}, \qquad (\mathscr{H}\mathscr{P})$$

where "$\le$" is declared componentwise. The following proposition concludes that star-shaped polyhedra allow analogous representations.

Proposition 1. *If $\mathbb{P} \subset \mathbb{R}^n$ is a star-shaped polyhedron and $\mathfrak{F} = \{C_1, \ldots, C_m\}$ a convex polyhedral fan for $\mathbb{P}$ then $\mathbb{P}$ always allows each of the following two equivalent representations.*

(a) *There exist integer numbers $l_1, \ldots, l_m$ and points $p_{j,1}, \ldots, p_{j,l_j}$ from $\mathbb{R}^n$, $j \in \{1, \ldots, m\}$ such that $\mathbb{P} \cap C_j = \operatorname{conv}(\{0_n, p_{j,1}, \ldots, p_{j,l_j}\})$ and*

$$\mathbb{P} = \bigcup_{j=1}^{m} \operatorname{conv}\left(\{0_n, p_{j,1}, \ldots, p_{j,l_j}\}\right). \qquad (\mathscr{S}\mathscr{V}\mathscr{P})$$

(b) *There exist matrices $A_j \in \mathbb{R}^{k_j\times n}$ and vectors $b_j \in \mathbb{R}_+^{k_j}$ with positive real components such that*

$$\mathbb{P} = \bigcup_{j=1}^{m} \{x \in \mathbb{R}^n : A_j x \le b_j\}. \qquad (\mathscr{S}\mathscr{H}\mathscr{P})$$

The following lemma is basic for proving the main results of this paper. It makes use of both representations in Proposition 1. For any $C \subset \mathbb{R}^n$ and $x \in \mathbb{R}^n$, let $I_C : \mathbb{R}^n \to \{0, 1\}$ be the indicator function defined by

$$I_C(x) = \begin{cases} 1, & \text{if } x \in C, \\ 0, & \text{otherwise}, \end{cases} \qquad (9)$$

$\operatorname{pos} x = \{\alpha x \mid \alpha \ge 0\}$ and $\operatorname{pos} C = \operatorname{conv}(\bigcup_{x\in C} \operatorname{pos} x) = \operatorname{conv}(\bigcup_{\alpha\ge 0} \alpha C)$.

Lemma 2. *Let $\mathbb{P}$ and $\mathfrak{F}$ be as in Proposition 1.*

(a) *With notation as in $(\mathscr{S}\mathscr{V}\mathscr{P})$, $h_{\mathbb{P}}$ can be equivalently represented as follows:*

(a1) $h_{\mathbb{P}}(x) = \sum_{j=1}^m I_{C_j}(x) \inf\{r > 0 : x \in \operatorname{conv}(\{0_n, rp_{j,1}, \ldots, rp_{j,l_j}\})\}$.

(a2) $h_{\mathbb{P}}(x) = \sum_{j=1}^m I_{C_j}(x) \inf\{r > 0 : x \in \{\sum_{i=1}^{l_j} \lambda_i p_{j,i}, \lambda_i \ge 0, \sum_{i=1}^{l_j} \lambda_i = r\}\}$.

(a3) $h_{\mathbb{P}}(x) = \sum_{j=1}^m I_{C_j}(x) h_{\mathbb{P}}^{(j)}(x)$, *where $h_{\mathbb{P}}^{(j)}(x)$ denotes the solution of the following linear optimization problem (LOP): minimize $\lambda_1 + \cdots + \lambda_{l_j}$ subject to the conditions $A^{(j)}(\lambda_1, \ldots, \lambda_{l_j})^T = x$ and $\lambda_i \ge 0$, $i \in \{1, \ldots, l_j\}$, where $A^{(j)} = (p_{j,1}, \ldots, p_{j,l_j}) \in \mathbb{R}^{n\times l_j}$, $j \in \{1, \ldots, m\}$.*

(a4) *If additionally for every j, $j \in \{1, \ldots, m\}$, $\mathbb{P} \cap C_j$ is an n-dimensional simplex then*

$$h_{\mathbb{P}}(x) = \sum_{j=1}^{m} \left(I_{C_j}(x)\, \mathbb{1}_n^T \left(\left(A^{(j)}\right)^{-1} x\right)\right), \qquad (10)$$

where $\mathbb{1}_n = (1, \ldots, 1)^T$ denotes the n-dimensional vector of ones, $C_j = \{x \in \mathbb{R}^n : (A^{(j)})^{-1} x \ge 0\}$, and $\ge$ reads componentwise.

(b1) *With notation as in $(\mathscr{S}\mathscr{H}\mathscr{P})$, where $\mathbb{P}$ is assumed to have m facets H_j and $C_j = \operatorname{pos} H_j$ for every j, $j \in \{1, \ldots, m\}$, there exist reals $a_{j1}, \ldots, a_{jn}$, $j = 1, \ldots, m$, such that $h_{\mathbb{P}}$ allows the representation*

$$h_{\mathbb{P}}(x) = \sum_{j=1}^{m} I_{C_j}(x) \left(a_{j1}x_1 + a_{j2}x_2 + \cdots + a_{jn}x_n\right), \quad x = (x_1, \ldots, x_n)^T \in \mathbb{R}^n. \qquad (11)$$

(b2) *If $\mathbb{P}$ is additionally convex then*

$$h_{\mathbb{P}}(x) = \max\{A_1' x, \ldots, A_m' x\}, \qquad (12)$$

with $A' = (A'_1, \dots, A'_m)^T = (A_1/b_1, \dots, A_m/b_m)^T$, *where the quantities* $A = (A_1, \dots, A_m)^T$ *and* $b = (b_1, \dots, b_m)^T$ *are equal to those in* $(\mathscr{HP})$ *and* A_j/b_j, $j \in \{1, \dots, m\}$ *is declared componentwise.*

For details on the application of the latter representation, we refer to [25]. This representation is used there and elsewhere to study ball numbers, circle numbers, and generalized uniform distributions on the boundaries of platonic bodies and regular convex polygons, respectively.

Proof. Representation (a1) follows immediately by applying the definition of $h_{\mathbb{P}}$ in (1) and $(\mathscr{SVP})$. Representation (a2) follows then by applying (a1) and the well-known representation

$$\operatorname{conv}(\{x_1, \dots, x_l\}) = \left\{\sum_{i=1}^{l} \lambda_i x_i : \lambda_i \geq, \sum_{i=1}^{l} \lambda_i = 1\right\} \quad (13)$$

for a point set $\{x_1, \dots, x_l\}$ in $\mathbb{R}^n$. From (a2), it follows that $\sum_{i=1}^{l_j} \lambda_i$ has to be minimal where $\lambda_i \geq 0$ and $x \in \mathbb{R}^n$ has to be represented by the linear combination $\sum_{i=1}^{l_j} \lambda_i p_{j,i}$. From this, (a3) follows. We derive now representation (a4) from (a3). If all $\mathbb{P} \cap C_j$ are n-dimensional simplizes then $A^{(j)} \in \mathbb{R}^{n \times n}$ and $A^{(j)}(\lambda_1, \dots, \lambda_n)^T = x$ is uniquely determined since $A^{(j)}$ is built up by affinely independent points thus satisfying $\det(A^{(j)}) \neq 0$. In turn, the minimum of $\sum_{i=1}^{n} \lambda_i$ can be calculated directly by $\mathbb{1}_n^T((A^{(j)})^{-1}x)$. Furthermore, if a given vector x is an element of the simplex $\mathbb{P} \cap C_j$, all $\lambda_1, \dots, \lambda_n$ have to be nonnegative in this case. Thus the vector $\lambda = (A^{(j)})^{-1}x$ satisfies $x \in (\mathbb{P} \cap C_j)$ iff all components of λ are nonnegative, yielding the definition of I_{C_j}.

Representation (b1) can be proved as follows: if H_j, $j = 1, \dots, m$ are the facets of $\mathbb{P}$, each H_j is a subset of the boundary of an n-dimensional, closed half-space. Thus, there exist reals $a_{j1}, \dots, a_{jn}$, $j = 1, \dots, m$, such that

$$H_j = \{x \in \mathbb{P} : a_{j1}x_1 + a_{j2}x_2 + \cdots + a_{jn}x_n = 1\},$$
$$C_j \cap \mathbb{P} = \{x \in C_j : a_{j1}x_1 + a_{j2}x_2 + \cdots + a_{jn}x_n \leq 1\} \quad (14)$$

since $C_j = \operatorname{pos} H_j$.

From this, it follows

$$\begin{aligned} h_{\mathbb{P}}(x) &= \inf\left\{r > 0 : \frac{x}{r} \in \mathbb{P}\right\} = \inf\left\{r > 0 : \frac{x}{r} \in \bigcup_{j=1}^{m} (C_j \cap \mathbb{P})\right\} = \inf\left\{r > 0 : \frac{x}{r} \right. \\ &\left. \in (\{x \in \mathbb{R}^n : x \in C_1, a_{11}x_1 + \cdots + a_{1n}x_n \leq 1\} \cup \cdots \cup \{x \in \mathbb{R}^n : x \in C_m, a_{m1}x_1 + \cdots + a_{mn}x_n \leq 1\})\right\} \\ &= \inf\left\{r > 0 : \sum_{j=1}^{m} I_{C_j}(x)\left(a_{j1}x_1 + a_{j2}x_2 + \cdots + a_{jn}x_n\right) \leq r\right\} = \sum_{j=1}^{m} I_{C_j}(x)\left(a_{j1}x_1 + a_{j2}x_2 + \cdots + a_{jn}x_n\right), \quad \forall x = (x_1, \dots, x_n)^T \in \mathbb{R}^n. \end{aligned} \quad (15)$$

Representation (b2) follows by applying $(\mathscr{HP})$ and using that

$$\begin{aligned} \mathbb{P} &= \{x \in \mathbb{R}^n : A'x \leq \mathbb{1}_n\} \\ &= \{x \in \mathbb{R}^n : A'_1 x \leq 1, A'_2 x \leq 1, \dots, A'_m x \leq 1\} \quad (16) \\ &= \{x \in \mathbb{R}^n : \max\{A'_1 x, A'_2 x, \dots, A'_m x\} \leq 1\}. \end{aligned}$$

3. Polyhedral Generalized Surface Content Measure

3.1. Specific Representations. In this section, specific representations of the star-generalized surface content measure $\mathfrak{O}_{\mathbb{S}}$ are considered in case that $K = \mathbb{P}$ is a star-shaped polyhedron and $S = \mathbb{S}$ its boundary. Prior to this, we consider a generalized polar coordinate transformation allowing one of the specific representations derived later on. For another powerful application of these generalized polar coordinates, we refer already here to Section 5.2.

Definition 3. Let $\mathbb{P} \subset \mathbb{R}^n$ be a star-shaped polyhedron. The polyhedral star-shaped polar coordinate transformation $SPOL_{\mathbb{P}} : M_n \to \mathbb{R}^n$, $M_n = [0, \infty) \times M_n^*$, $M_n^* = [0, \pi)^{\times(n-2)} \times [0, 2\pi)$ is defined by

$$\begin{aligned} x_1 &= \frac{r}{N(\varphi_1, \dots, \varphi_{n-1})} \cos(\varphi_1) \\ x_2 &= \frac{r}{N(\varphi_1, \dots, \varphi_{n-1})} \sin(\varphi_1)\cos(\varphi_2) \\ &\vdots \\ x_{n-1} &= \frac{r}{N(\varphi_1, \dots, \varphi_{n-1})} \sin(\varphi_1), \dots, \sin(\varphi_{n-2}) \cdot \cos(\varphi_{n-1}) \\ x_n &= \frac{r}{N(\varphi_1, \dots, \varphi_{n-1})} \sin(\varphi_1), \dots, \sin(\varphi_{n-2}) \cdot \sin(\varphi_{n-1}), \end{aligned} \quad (17)$$

where $N(\varphi_1, \dots, \varphi_{n-1}) = h_{\mathbb{P}}(\cos(\varphi_1), \sin(\varphi_1)\cos(\varphi_2), \dots, \sin(\varphi_1), \dots, \sin(\varphi_{n-1}))$.

We will later also use the notation $x = x(\varphi, r) = (x_1(r, \varphi), \dots, x_n(r, \varphi))$ if $x = SPOL_{\mathbb{P}}(r, \varphi)$.

Lemma 4. *The map $SPOL_{\mathbb{P}}$ is almost one-to-one, for $x \neq 0$, its inverse $SPOL_{\mathbb{P}}^{-1}$ is given by*

$$r = h_{\mathbb{P}}(x_1, \ldots, x_n),$$

$$\varphi_i = \arccos\left(\frac{x_i}{\left(\sum_{j=i}^{n} x_j^2\right)^{1/2}}\right), \quad i = 1, \ldots, n-2,$$

$$\varphi_{n-1} = \begin{cases} \operatorname{arccot}\left(\dfrac{x_{n-1}}{x_n}\right), & x_n > 0, \\ \operatorname{arccot}\left(\dfrac{x_{n-1}}{x_n}\right) + \pi, & x_n < 0 \end{cases} \tag{18}$$

and its Jacobian is

$$J(SPOL_{\mathbb{P}})(r, \varphi) = r^{n-1} J^*(SPOL_{\mathbb{P}})(\varphi), \quad (r, \varphi) \in M_n \text{ where } J^*(SPOL_{\mathbb{P}})(\varphi_1, \ldots, \varphi_{n-1}) = \frac{1}{N(\varphi)^n} \prod_{i=1}^{n-1} \sin(\varphi_i)^{n-i-1}. \tag{19}$$

Proof. It holds

$$h_{\mathbb{P}}(x_1, \ldots, x_n) = \frac{r}{N(\varphi_1, \ldots, \varphi_{n-1})} \cdot \underbrace{h_{\mathbb{P}}(\cos(\varphi_1), \ldots, \sin(\varphi_1), \ldots, \sin(\varphi_{n-1}))}_{=N(\varphi_1, \ldots, \varphi_{n-1})} = r \tag{20}$$

and the relations of φ_i, $i = 1, \ldots, n-1$, are the same as those in the case of usual polar coordinates. For calculating the Jacobian, we refer to the proof of Theorem 3.1 in [25] that can be extended to the n-dimensional case. □

Let us denote the cdf of a random vector Z by F_Z.

Theorem 5. *Let $(R, \Phi_1, \ldots, \Phi_{n-1})^T = SPOL_{\mathbb{P}}^{-1}(X)$ be the random polyhedral star-shaped coordinates of X, where $X \sim \Phi_{g,\mathbb{P}}$; then R and $(\Phi_1, \ldots, \Phi_{n-1})^T$ are stochastically independent, $F_R(t) = \int_0^t (r^{n-1}/\mathcal{I}(g)) g(r) dr$, and with $\varphi = (\varphi_1, \ldots, \varphi_{n-1})$*

$$F_{\Phi_i}(\alpha) = \frac{1}{n\mu(\mathbb{P})} \cdot \underbrace{\int_0^{\pi} \cdots \int_0^{\pi}}_{i-1 \text{ times}} \int_0^{\alpha} \underbrace{\int_0^{\pi} \cdots \int_0^{\pi}}_{n-i-2 \text{ times}} \int_0^{2\pi} J^*(SPOL_{\mathbb{P}})(\varphi) \, d(\varphi), \tag{21}$$

$\forall i = 1, \ldots, n-2$, $\alpha \in [0, \pi)$ *and*

$$F_{\Phi_{n-1}}(\alpha) = \frac{1}{n\mu(\mathbb{P})} \underbrace{\int_0^{\pi} \cdots \int_0^{\pi}}_{n-2 \text{ times}} \int_0^{\alpha} J^*(SPOL_{\mathbb{P}})(\varphi) \, d(\varphi), \quad \alpha \in [0, 2\pi). \tag{22}$$

Proof. Applying the described method in Remark 11 in [6] for elliptically contoured distributions analogously here yields the representation of F_R. Furthermore, $P(\Phi_i < \alpha) = \Phi_{g,\mathbb{P}}(A(\alpha))$ for every $i = 1, \ldots, n-1$, where $A(\alpha) = \{(r, \varphi_1, \ldots, \varphi_{n-1}) \in [0, \infty) \times M_n^* : \varphi_i \in [0, \alpha)\}$. Thus with $\varphi = (\varphi_1, \ldots, \varphi_{n-1})$

$$P(\Phi_i < \alpha) = \underbrace{C(g, \mathbb{P})}_{1/n\mathcal{I}(g)\mu(\mathbb{P})} \underbrace{\int_0^{\infty} r^{n-1} g(r) \, dr}_{=\mathcal{I}(g)} \underbrace{\int_0^{\pi} \cdots \int_0^{\pi}}_{i-1 \text{ times}} \int_0^{\alpha} \underbrace{\int_0^{\pi} \cdots \int_0^{\pi}}_{n-i-2 \text{ times}} \int_0^{2\pi} J^*(SPOL_{\mathbb{P}})(\varphi) \, d(\varphi). \tag{23}$$

Let us denote by $SPOL_{\mathbb{P}}^*(\varphi) = SPOL_{\mathbb{P}}(1, \varphi)$ the restriction of $SPOL_{\mathbb{P}}$ to the case $r = 1$ and, analogously, by $SPOL_{\mathbb{P}}^{*-1}$ the inverse of the map $SPOL_{\mathbb{P}}^* : M_n^* \to \mathbb{S}$. Now, the following representations of $\mathfrak{O}_{\mathbb{S}}$ can be proved.

Theorem 6. *Let $\mathbb{P} \subset \mathbb{R}^n$ be a star-shaped polyhedron and $B \in (\mathbb{S} \cap \mathfrak{B}_n)$.*

(a) *$\mathfrak{O}_{\mathbb{S}}$ satisfies the polyhedral star-shaped polar coordinate representation*

$$\mathfrak{O}_{\mathbb{S}}(B) = \int_{SPOL_{\mathbb{P}}^{*-1}(B)} J^*(SPOL_{\mathbb{P}})(\varphi) \, d\varphi. \tag{24}$$

(b1) *With notation as in ($\mathcal{SHP}$) and Lemma 2 (b1), $\mathfrak{O}_{\mathbb{S}}$ satisfies the star-spherical coordinate representation*

$$\mathfrak{O}_{\mathbb{S}}(B) = \sum_{j=1}^{m} \int_{G(B \cap (C_j \cap \mathbb{P}))} \frac{1}{|a_{jn}|} d\vartheta. \tag{25}$$

(b2) *Let additionally vol_{n-1} denote the volume in $\mathbb{R}^{n-1}$, $B = \bigcup_{j=1}^{m} \mathcal{B}_j$ where $\mathcal{B}_j \subseteq (C_j \cap \mathbb{P})$, $\operatorname{int}(\mathcal{B}_j) \cap \operatorname{int}(\mathcal{B}_i) \neq \emptyset$ if $i \neq j$, $i, j \in \{1, \ldots, m\}$; then $\mathfrak{O}_{\mathbb{S}}$ satisfies the facet-content representation*

$$\mathfrak{O}_{\mathbb{S}}(B) = \sum_{j=1}^{m} \frac{vol_{n-1}(\mathcal{B}_j)}{\sqrt{a_j^T a_j}}, \tag{26}$$

where $a_j = (a_{j1}, \ldots, a_{jn})^T$. *Note that* $\mathfrak{D}_{\mathbb{S}}$ *is proportional to the Lebesgue measure in* $\mathbb{R}^{n-1}$ *in every sector* $C_j \cap \mathbb{P}$, $j = 1, \ldots, m$.

(c) *Let additionally* $\mathbb{P} \cap C_j$, $j \in \{1, \ldots, m\}$ *be n-dimensional simplizes and* $\mathbb{P} \cap C_j$ *and* $B \cap C_j$, $j \in \{1, \ldots, m\}$ *represented according to* $(\mathscr{V}\mathscr{P})$. *Then there exist integers* k_j *such that* $B \cap C_j = \bigcup_{i=1}^{k_j} \mathrm{conv}(\{b_{(j),i1}, \ldots, b_{(j),in}\})$ *and* $\mathrm{int}(\mathrm{conv}(\{b_{(j),i_1 1}, \ldots, b_{(j),i_1 n}\})) \cap \mathrm{int}(\mathrm{conv}(\{b_{(j),i_2 1}, \ldots, b_{(j),i_2 n}\})) = \emptyset$, $i_1 \neq i_2$, *where* $\mathrm{conv}(\{b_{(j),i1}, \ldots, b_{(j),in}\})$ *are* $(n-1)$*-dimensional simplizes. Then* $\mathfrak{D}_{\mathbb{S}}$ *satisfies the simplicial representation*

$$\mathfrak{D}_{\mathbb{S}}(B) = \frac{1}{(n-1)!} \sum_{j=1}^{m} V\left(B \cap C_j\right), \tag{27}$$

where

$$V\left(B \cap C_j\right) = \begin{cases} \sum_{i=1}^{k_j} \left|\det\left(b_{(j),i1}, \ldots, b_{(j),in}\right)\right|, & \textit{if } B \cap C_j \neq \emptyset, \\ 0, & \textit{otherwise}. \end{cases} \tag{28}$$

(d) *If additionally* $h_{\mathbb{P}}$ *denotes a norm and* $N(\vartheta) = (\nabla\eta(\vartheta), -1)^T$ *is the outer normal vector to the norm sphere* $\mathbb{S}$ *at the point* $(\vartheta^T, \eta(\vartheta))^T$, *then, with notations as in* $(\mathscr{S}\mathscr{V}\mathscr{P})$, $\mathfrak{D}_{\mathbb{S}}$ *satisfies the dual norm representation*

$$\mathfrak{D}_{\mathbb{S}}(A) = \int_{G(A)} \max\left\{p_1^T N(\vartheta), \ldots, p_l^T N(\vartheta)\right\} d\vartheta, \quad A \in \mathfrak{B}_{\mathbb{S}}^{+}, \tag{29}$$

$$G(A) = \left\{\vartheta \in \mathbb{R}^{n-1} : \exists \eta = \eta(\vartheta) \text{ with } \left(\vartheta^T, \eta\right)^T \in A\right\}.$$

(e) *If additionally* $h_{\mathbb{P}}$ *denotes an antinorm and* $\mathbb{P} \in AN2$ *and* $\overline{N}(\vartheta) = (\nabla\eta(\vartheta), 1)^T$ *is the inner normal vector to the antinorm sphere* $\mathbb{S}$ *at the point* $(\vartheta^T, \eta(\vartheta))^T$, *then with notations as in* $(\mathscr{S}\mathscr{H}\mathscr{P})$ *the Minkowski functional of the antipolar set* $\mathbb{P}^{\circ}$ *of* $\mathbb{P}$ *defined by* $h_{\mathbb{P}^{\circ}}(u) = \sum_{i=1}^{m} I_{C_i}(u) \inf\{u^T y : y \in \mathbb{S} \cap C_i\}$, $u \in \mathbb{R}^n$ *and* $\mathfrak{D}_{\mathbb{S}}$ *satisfies the (antipolar set) representation*

$$\mathfrak{D}_{\mathbb{S}}(B) = \int_{G(B)} h_{\mathbb{P}^{\circ}}\left(\overline{N}(\vartheta)\right) d\vartheta. \tag{30}$$

Proof. Since

$$\mu\left(\mathrm{sector}_{\mathbb{P}}(B, \rho)\right) = \int_{\mathrm{sector}_{\mathbb{P}}(B,\rho)} dx = \int_0^{\rho} \int_{SPOL_{\mathbb{P}}^{*-1}(B)} r^{n-1} J^*\left(SPOL_{\mathbb{P}}\right)(\varphi)\, d(\varphi, r), \tag{31}$$

(5) yields (a). Analogously,

$$\mu\left(\mathrm{sector}_{\mathbb{P}}(B, \rho)\right) = \int_{\mathrm{sector}_{\mathbb{P}}(B,\rho)} dx = \int_0^{\rho} r^{n-1}\, dr \sum_{i=1}^{m} \int_{G(B \cap (C_i \cap \mathbb{P}))} J_j^*(\vartheta)\, d\vartheta, \tag{32}$$

where, for every $\vartheta \in G(\mathbb{S}_j)$, $J_j^*(\vartheta)$, $j = 1, \ldots, m$ is the value of the Jacobian of the star-spherical coordinate transformation; see [6]. Since rH_j, $j = 1, \ldots, m$ can be represented by star-spherical coordinates as

$$rH_j = r \cdot \left\{\left(\vartheta, \eta_j(\vartheta)\right)^T \in \mathbb{P} : \eta_j(\vartheta) = \frac{1}{a_{jn}}\left(1 - \sum_{i=1}^{n-1} a_{ji}\vartheta_i\right)\right\}, \quad j = 1, \ldots, m, \tag{33}$$

it follows by applying Lemma 1 in [6] that $J_j^*(\vartheta) = 1/|a_{jn}|$. Thus, (5) yields (b1). Considering part (b2), it holds $\mu(\mathrm{sector}_{\mathbb{P}}(B, 1)) = \sum_{j=1}^{m} \mu(\mathrm{sector}_{\mathbb{P}}(\mathscr{B}_j, 1))$, where every $\mathrm{sector}_{\mathbb{P}}(\mathscr{B}_j, 1)$, $j = 1, \ldots, m$ is a convex cone with base $\mathscr{B}_j$ and cusp in the origin. Thus $\mu(\mathrm{sector}_{\mathbb{P}}(\mathscr{B}_j, 1)) = vol_{n-1}(\mathscr{B}_j)/(n\sqrt{a_j^T a_j})$, where $1/\sqrt{a_j^T a_j}$ denotes the Euclidean length of the height of $\mathrm{sector}_{\mathbb{P}}(\mathscr{B}_j, 1)$. According to Remark 2 in [6], it holds $\mathfrak{D}_{\mathbb{S}}(B) = \mathfrak{D}_{\mathbb{S}}(\mathbb{S})\mu(\mathrm{sector}_{\mathbb{P}}(B, 1))/\mu(\mathbb{P})$. Thus

$$\mathfrak{D}_{\mathbb{S}}(B) = \underbrace{\frac{\mathfrak{D}_{\mathbb{S}}(\mathbb{S})}{n\mu(\mathbb{P})}}_{=1} \sum_{j=1}^{m} \frac{vol_{n-1}\left(\mathscr{B}_j\right)}{\sqrt{a_j^T a_j}}, \tag{34}$$

and (b2) is proved. Considering part (c), it holds

$$\mu\left(\mathrm{sector}_{\mathbb{P}}\left(B \cap C_j, 1\right)\right) = \sum_{i=1}^{k_j} \mu\left(\mathrm{conv}\left(\left\{0_n, b_{(j),i1}, \ldots, b_{(j),in}\right\}\right)\right), \quad j = 1, \ldots, m, \tag{35}$$

and since $\mu(\mathrm{conv}(\{0_n, b_{(j),i1}, \ldots, b_{(j),in}\})) = (1/n!)|\det(b_{(j),i1}, \ldots, b_{(j),in})|$ it follows

$$\mathfrak{D}_{\mathbb{S}}(B) = \underbrace{\frac{\mathfrak{D}_{\mathbb{S}}(\mathbb{S})}{\mu(\mathbb{P})}}_{=n} \mu\left(\mathrm{sector}_{\mathbb{P}}(B, 1)\right) = \frac{1}{(n-1)!} \sum_{j=1}^{m} V\left(B \cap C_j\right). \tag{36}$$

Considering part (d) and applying (8), it remains to prove that $h_{\mathbb{P}^*}(N(\vartheta)) = \max\{p_1^T N(\vartheta), \ldots, p_l^T N(\vartheta)\}$. According to [26] it is well known that $h_{\mathbb{P}^*}$ equals the support function of $\mathbb{P}$; thus $h_{\mathbb{P}^*}(u) = \sup\{u^T y : y \in \mathbb{P}\}$, $u \in \mathbb{R}^n$, and it follows $h_{\mathbb{P}^*}(u) = \max\{p_1^T u, \ldots, p_l^T u\}$.

Considering part (e) we can use the results from Lemma 1 in [6], according to which it holds

$$\mathfrak{O}_{\mathbb{S}}(B) = \int_{G(B)} \left|\left(\vartheta^T, \eta(\vartheta)\right) \overline{N}(\vartheta)\right| d\vartheta. \quad (37)$$

With notations as in ($\mathscr{SHP}$), it follows for every $i = 1, \ldots, m$ that $\mathbb{S} \cap C_i = \{x \in \mathbb{R}^n : a_{i1}x_1 + \cdots + a_{in}x_n = 1\}$, $\eta(\vartheta) = (1/a_{in})(1 - \sum_{j=1}^{n-1} a_{ij}\vartheta_j)$ and $\overline{N}(\vartheta) = (a_{i1}/a_{in}, \ldots, a_{i(n-1)}/a_{in}, 1)^T$. Thus, for every $(\vartheta^T, \eta(\vartheta))^T \in \mathbb{S} \cap C_i$,

$$\begin{aligned}\left|\left(\vartheta^T, \eta(\vartheta)\right) \overline{N}(\vartheta)\right| &= \left|\frac{1}{a_{in}}\right| \\ &= \inf\left\{\overline{N}(\vartheta)\, y : y \in \mathbb{S} \cap C_i\right\}.\end{aligned} \quad (38)$$

Since the antisupport function $\mathfrak{h}_{\mathbb{P}}^{\mathfrak{F}}$ of $\mathbb{P}$ with respect to $\mathfrak{F}$ is defined by

$\mathfrak{h}_{\mathbb{P}}^{\mathfrak{F}}(u) = \sum_{i=1}^{m} I_{C_i}(u) \inf\{u^T y : y \in \mathbb{S} \cap C_i\}$, $u \in \mathbb{R}^n$ and it is shown in Lemma 10 in [7] that $\mathfrak{h}_{\mathbb{P}}^{\mathfrak{F}} = h_{\mathbb{P}^\circ}$, part (e) is proved.

According to [7] it is possible to represent the antipolar set $\mathbb{P}^\circ$ of a $\mathbb{P} \in$ AN2 by

$$\mathbb{P}^\circ = \left\{\lambda(u)\, u : 0 \le \lambda(u)\, h_{\mathbb{P}^\circ}(u) \le 1,\ u \in S_E^{(n-1)}\right\}, \quad (39)$$

where $S_E^{(n-1)}$ denotes the Euclidean unit sphere in $\mathbb{R}^n$. For an illustration, we refer to Section 4.2, where antipolar sets of $\mathbb{W}$-stars are considered. For an application of certain representations of Theorem 6, we refer to Example A.1 in the Appendix.

3.2. Extension of the Ball Number Function. The circumference and area content properties of Euclidean circles which motivate the generalization of the circle number π have been discussed to a certain extent first in [27] for $l_{2,p}$-circles. These considerations were later extended to ellipses and general star-circles. Related multivariate studies started by introducing the generalized surface and volume properties of $l_{n,p}$-balls and were followed up in [6, 21]. General ball numbers being values attained by the ball number function are defined in the literature, and it is also stated there as a challenging problem to extend the ball number function to as many as possible further classes of generalized balls. In this respect, an extension of the ball number function to the class of regular convex polygons can be found in [25], and an extension to platonic bodies is considered elsewhere by the authors. Furthermore, in Section 7 of [25], several other possible extensions of the circle number function one could think about are discussed.

Note that the equation

$$\mu(\mathbb{P}(r)) = \int_0^r \mathfrak{O}_{\mathbb{S}}(\mathbb{S}(\rho))\, d\rho \quad (40)$$

reflects a certain generalization of the method of indivisibles of Cavalieri and Torricelli and that the ratios

$$\frac{\mu(\mathbb{P}(r))}{r^n} = \frac{\mathfrak{O}_{\mathbb{S}}(\mathbb{S}(r))}{nr^{n-1}} \quad (41)$$

not only do coincide but are even independent of r. This motivates the following definition.

Definition 7. The polyhedral star ball number $\pi_{\mathbb{P}}$ is defined by $\pi_{\mathbb{P}} = \mu(\mathbb{P})$.

Note that $\pi_{\mathbb{P}} = \mathfrak{O}_{\mathbb{S}}(\mathbb{S})/n$. For concrete values of $\pi_{\mathbb{P}}$ in the special case of regular convex polygons, see [25]. Ball numbers of platonic bodies are dealt with elsewhere. Apart from this extension of the domain of the ball number function, $\pi_{\mathbb{P}}$ will be used in this paper as kind of normalizing constant for polyhedral star-shaped contoured distributions.

3.3. Polyhedral Star-Shaped Generalized Uniform Distributions. The notion of a polyhedral generalized surface content measure makes it possible to define the polyhedral star-shaped generalized uniform distribution on $\mathbb{S}$ as

$$\omega_{\mathbb{S}}(A) = \frac{\mathfrak{O}_{\mathbb{S}}(A)}{\mathfrak{O}_{\mathbb{S}}(\mathbb{S})}, \quad A \in \mathfrak{B}_{\mathbb{S}}. \quad (42)$$

Note that plugging in any of the representations of Theorem 6 into (42) yields the polyhedral star-shaped polar coordinate, star-spherical coordinate, and facet-content, simplicial and dual norm representations of the generalized uniform distribution, respectively. A numerical example of the application of $\omega_{\mathbb{S}}$ can be found in Example A.1 in the Appendix.

Let $(\Omega, \mathfrak{A}, P)$ be a probability space and G a random vector being defined on Ω and taking values in $\mathbb{R}^n$. Assume further that G follows the uniform probability distribution on $\mathbb{P}$; that is, $P(G \in A) = \mu(A)/\mu(\mathbb{P})$, for $A \in (\mathbb{P} \cap \mathfrak{B}_n)$, and put $Y = G/h_{\mathbb{P}}(G)$, where division is defined componentwise. The proof of the following result can be done analogously to that in case of platonically generalized uniform distributions on platonic spheres.

Theorem 8. *The random vector Y follows the polyhedral star-shaped generalized uniform distribution on* $\mathbb{S}$.

4. Polyhedral Star-Shaped Distributions

4.1. Representations. The results in the previous section can be used for representing general polyhedral star-shaped distributions. Doing this, we follow the considerations in [6–8, 12, 20, 21, 28], where stochastic representations like (7) were repeatedly exploited.

A random vector $X : \Omega \to \mathbb{R}^n$ is said to follow a polyhedral star-shaped distribution if there exists a random variable $R : \Omega \to [0, \infty)$ such that X satisfies the stochastic representation

$$X \stackrel{d}{=} R \cdot U_{\mathbb{S}}, \quad (43)$$

where $U_{\mathbb{S}} \sim \omega_{\mathbb{S}}$ is polyhedral star-shaped generalized uniformly distributed on $\mathbb{S}$ and $U_{\mathbb{S}}$ and R are stochastically independent. The set of all polyhedral star-shaped distributions on $\mathfrak{B}_n$ will be denoted by $\mathrm{PStSh}^{(n)}$ and its subset of continuous distributions by

$$\mathrm{CPStSh}^{(n)} = \{\Phi_{g,\mathbb{P}} : \mathbb{P} \text{ is a star-shaped polyhedron}, \; g \text{ is a dgf}\}. \quad (44)$$

Applying (6) and the representations of Theorem 6, we can state the following specific geometric measure representation formulae of continuous polyhedral star-shaped distributions. For the application of geometric measure representation formulae of other classes of distributions, we refer to [28–30].

Proposition 9. *Let the assumptions from Theorem 6 be fulfilled and* $B \in \mathfrak{B}_n$. *Then* $\Phi_{g,\mathbb{P}}(B)$ *allows correspondingly*

(a) *polyhedral star-shaped polar coordinate representation*

$$\Phi_{g,\mathbb{P}}(B) = \frac{1}{n\pi_{\mathbb{P}}\mathcal{I}(g)} \int_0^\infty \int_{SPOL_{\mathbb{P}}^{*-1}([(1/r)B]\cap\mathbb{S})} r^{n-1} g(r) \cdot J^*(SPOL_{\mathbb{P}})(\varphi)\, d(\varphi)\, dr, \quad (45)$$

(b1) *star-spherical coordinate representation*

$$\Phi_{g,\mathbb{P}}(B) = \frac{1}{n\pi_{\mathbb{P}}\mathcal{I}(g)} \int_0^\infty r^{n-1} g(r) \cdot \sum_{j=1}^m \int_{G(([(1/r)B]\cap\mathbb{S})\cap(C_j\cap\mathbb{P}))} \frac{1}{|a_{jn}|} d\vartheta\, dr, \quad (46)$$

(b2) *facet-content representation*

$$\Phi_{g,\mathbb{P}}(B) = \frac{1}{n\pi_{\mathbb{P}}\mathcal{I}(g)} \int_0^\infty r^{n-1} g(r) \sum_{j=1}^m \frac{vol_{n-1}(B_j(r))}{\sqrt{a_j^T a_j}} dr, \quad (47)$$

if additionally $[(1/r)B] \cap \mathbb{S} = \bigcup_{j=1}^m \mathcal{B}_j(r)$ *where* $\mathcal{B}_j(r) \subseteq (C_j \cap \mathbb{P})$ *and* $\mathrm{int}(\mathcal{B}_i(r)) \cap \mathrm{int}(\mathcal{B}_j(r)) = \emptyset$, *if* $i \neq j$, $i, j \in \{1, \ldots, m\}$,

(c) *simplicial representation*

$$\Phi_{g,\mathbb{P}}(B) = \frac{1}{n\pi_{\mathbb{P}}\mathcal{I}(g)(n-1)!} \cdot \int_0^\infty r^{n-1} g(r) \sum_{j=1}^m V\left(\left(\left[\frac{1}{r}B\right] \cap \mathbb{S}\right) \cap C_j\right) dr, \quad (48)$$

(d) *dual norm representation*

$$\Phi_{g,\mathbb{P}}(B) = \frac{1}{n\pi_{\mathbb{P}}\mathcal{I}(g)} \int_0^\infty \int_{G([(1/r)B]\cap\mathbb{S})} r^{n-1} g(r) \cdot \max\{p_1^T N(\vartheta), \ldots, p_l^T N(\vartheta)\}\, d(\vartheta)\, dr, \quad (49)$$

(e) *(antipolar set) representation*

$$\Phi_{g,\mathbb{P}}(B) = \frac{1}{n\pi_{\mathbb{P}}\mathcal{I}(g)} \cdot \int_0^\infty \int_{G([(1/r)B]\cap\mathbb{S})} r^{n-1} g(r)\, h_{\mathbb{P}^\circ}(\overline{N}(\vartheta))\, d\vartheta\, dr, \quad (50)$$

(f) *sector measure representation*

$$\Phi_{g,\mathbb{P}}(B) = \frac{1}{\pi_{\mathbb{P}}\mathcal{I}(g)} \cdot \int_0^\infty r^{n-1} g(r)\, \mu\left(\mathrm{sector}_{\mathbb{P}}\left(\left[\frac{1}{r}B\right] \cap \mathbb{S}, 1\right)\right) dr. \quad (51)$$

Note that for the sector measure also the notion of cone measure is used; see [17, 31]

Finally note that if $X \sim \Phi_{g,\mathbb{P}}$ then we write (43) as

$$X \overset{d}{=} R_g \cdot U_{\mathbb{S}}, \quad (52)$$

where the nonnegative random variable R_g is independent of $U_{\mathbb{S}}$.

4.2. The Class of $\mathbb{w}$*-Star Contoured Distributions in* $\mathbb{R}^2$. The polyhedral star-shaped distributions considered in this section may be alternatively norm or antinorm contoured. For simplicity, consideration is restricted here to the two-dimensional case. Let $\mathbb{P}_{\mathbb{w}} \subset \mathbb{R}^2$ be a $\mathbb{w}$ (vertices) driven star-shaped polygon defined for arbitrary $\mathbb{w} \in (0,\infty)$ by the vertices $q_1 = (1,0)$, $q_2 = (\mathbb{w},\mathbb{w})$, $q_3 = (0,1)$, $q_4 = (-\mathbb{w},\mathbb{w})$, $q_5 = (-1,0)$, $q_6 = (-\mathbb{w},-\mathbb{w})$, $q_7 = (0,-1)$, and $q_8 = (\mathbb{w},-\mathbb{w})$. We call $\mathbb{P}_{\mathbb{w}}$ a $\mathbb{w}$-star. For an illustration of different $\mathbb{w}$-stars, see Figure 1. Note that a $\mathbb{w}$-star appears to be convex if $\mathbb{w} \in [1/2, 1]$ and radially concave if $\mathbb{w} \in (0, 1/2) \cup (1, \infty)$. For representing the Minkowski functional $h_{\mathbb{P}_{\mathbb{w}}}$ of a $\mathbb{w}$-star, it is convenient to make use of the fans $\mathfrak{C}_1 = \bigcup_{i=1}^4 C_{1,i}$ and $\mathfrak{C}_2 = \bigcup_{i=1}^4 C_{2,i}$ if $\mathbb{w} \in (0,1)$ or $\mathbb{w} \geq 1$, respectively; see again Figure 1. If $\mathbb{w} \in (0,1)$ then

$$h_{\mathbb{P}_{\mathbb{w}}}(x,y) = I_{C_{1,1}}(x,y)\left(x + \frac{1-\mathbb{w}}{\mathbb{w}}|y|\right) + I_{C_{1,2}}(x,y) \cdot \left(\frac{1-\mathbb{w}}{\mathbb{w}}|x| + y\right) + I_{C_{1,3}}(x,y) \cdot \left(-x + \frac{1-\mathbb{w}}{\mathbb{w}}|y|\right) + I_{C_{1,4}}(x,y) \cdot \left(\frac{1-\mathbb{w}}{\mathbb{w}}|x| - y\right) \quad (53)$$

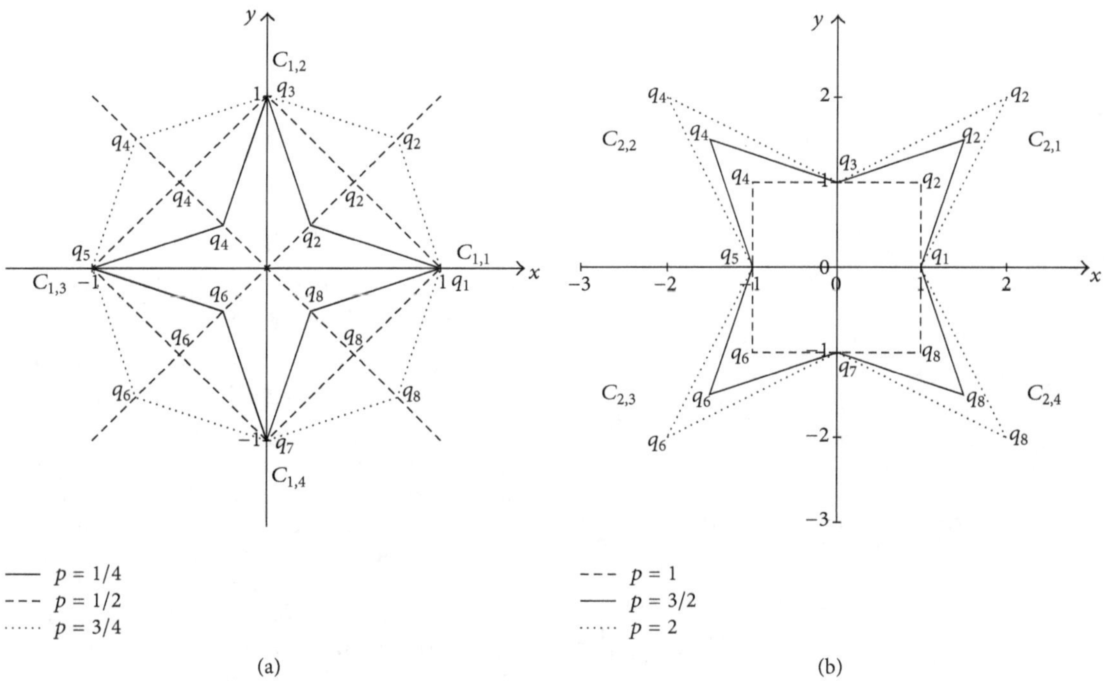

FIGURE 1: w-stars in case $0 < \mathbb{w} < 1$ (a) and in case $\mathbb{w} \geq 1$ (b).

and if $\mathbb{w} \geq 1$ then

$$h_{\mathbb{P}_{\mathbb{w}}}(x,y) = I_{C_{2,1}}(x,y) \cdot \left(\max\{x,y\} + \frac{1-\mathbb{w}}{\mathbb{w}}\min\{x,y\}\right) + I_{C_{2,2}}(x,y) \cdot \left(\max\{-x,y\} + \frac{1-\mathbb{w}}{\mathbb{w}}\min\{-x,y\}\right) + I_{C_{2,3}}(x,y) \cdot \left(\max\{-x,-y\} + \frac{1-\mathbb{w}}{\mathbb{w}}\min\{-x,-y\}\right) + I_{C_{2,4}}(x,y)\left(\max\{x,-y\} + \frac{1-\mathbb{w}}{\mathbb{w}}\min\{x,-y\}\right). \quad (54)$$

Surprisingly enough, it follows that $h_{\mathbb{P}_{\mathbb{w}}}(x,y) = \max\{|x|,|y|\} + ((1-\mathbb{w})/\mathbb{w})\min\{|x|,|y|\}$ for every $\mathbb{w} \in (0,\infty)$. This result enables us to introduce the w-star contoured densities

$$\varphi_{g,\mathbb{P}_{\mathbb{w}}}(x,y) = \frac{1}{8\mathbb{w}\mathcal{I}(g)} \cdot g\left(\max\{|x|,|y|\} + \frac{1-\mathbb{w}}{\mathbb{w}}\min\{|x|,|y|\}\right), \quad (x,y)^T \in \mathbb{R}^2. \quad (55)$$

Note that since $\mu(\mathbb{P}_{\mathbb{w}}) = 4\mathbb{w}$ the normalizing constant applies by (4). If $\mathbb{w} \in [1/2, 1]$ then $h_{\mathbb{P}_{\mathbb{w}}}$ denotes a norm. Thus, applying Theorem 6 (d), we can represent $\mathfrak{O}_{\mathbb{S}_{\mathbb{w}}}$ of a set $A \in \mathfrak{B}^+_{\mathbb{S}_{\mathbb{w}}}$ by the dual norm representation as

$$\mathfrak{O}_{\mathbb{S}_{\mathbb{w}}}(A) = \int_{G(A)} \max\{|N_1(\vartheta)|, |N_2(\vartheta)|, \mathbb{w}(|N_1(\vartheta)| + |N_2(\vartheta)|)\}\, d\vartheta, \quad (56)$$

where $N(\vartheta) = (N_1(\vartheta), N_2(\vartheta))^T$. Note that further representations of the star-generalized surface content measure can be found in [7] for a special class of radially concave star bodies and in [25] for regular polygons, especially in the case that these polygons have an odd number of vertices and thus are not symmetric.

If $\mathbb{w} \in (0, 1/2) \cup (1, \infty)$ then $h_{\mathbb{P}_{\mathbb{w}}}$ denotes an antinorm. Thus, applying Theorem 6 (e), we can represent $\mathfrak{O}_{\mathbb{S}_{\mathbb{w}}}$ of a set $B \in \mathfrak{B}_{\mathbb{S}_{\mathbb{w}}}$ by the (antipolar set) representation

$$\mathfrak{O}_{\mathbb{S}_{\mathbb{w}}}(B) = \int_{G(B)} \sum_{i=1}^{8} I_{C_i}\left(\left(\frac{a_{i1}}{a_{i2}}, 1\right)^T\right)\left(\mathbb{w}\left(\frac{a_{i1}}{a_{i2}} + 1\right)\right) d\vartheta = \int_{G(B)} \mathbb{w}\left(|\overline{N}_1(\vartheta)| + |\overline{N}_2(\vartheta)|\right) d\vartheta, \quad (57)$$

where the sectors C_i, $i = 1, \ldots, 8$, are equivalent to those in Example A.2, a_{i1} and a_{i2}, $i = 1, \ldots, 8$, are equivalent to the entries of the matrix A in Example A.2, $\overline{N}(\vartheta) = (\overline{N}_1(\vartheta), \overline{N}_2(\vartheta))^T$, and the antipolar set $\mathbb{P}^\circ_{\mathbb{w}}$ of $\mathbb{P}_{\mathbb{w}}$ is the l_1-ball of Minkowski radius $1/\mathbb{w}$, $\mathbb{P}^\circ_{\mathbb{w}} = \{(x_1, x_2)^T \in \mathbb{R}^2 : |x_1| + |x_2| \leq 1/\mathbb{w}\}$.

5. Expectations

5.1. Moments of Polyhedral Star-Shaped Distributions. Let $G = (G_1, \dots, G_n)^T$ be uniformly distributed on a star-shaped polyhedron $\mathbb{P}$. Such vector has dgf $g(r) = I_{[0,1]}(r)$. Since $\mathcal{I}(g) = 1/n$ the density of G is

$$\varphi_{g,\mathbb{P}}(x) = \frac{I_{[0,1]}(h_{\mathbb{P}}(x))}{\mu(\mathbb{P})} = \frac{I_{\mathbb{P}}(x)}{\mu(\mathbb{P})}, \quad x \in \mathbb{R}^n. \tag{58}$$

With $\mathbb{E}(G_i^k) = (1/\mu(\mathbb{P})) \int_{\mathbb{P}} x_i^k d\mu$, $i = 1, \dots, n$, the vector of moments of order k is defined as $\mathbb{E}(G^k) = (\mathbb{E}(G_1^k), \dots, \mathbb{E}(G_n^k))^T$. Mixed moments of order $k = k_1 + \cdots + k_n$, $k_j \geq 0$, $j \in \{1, \dots, n\}$ are

$$\mathbb{E}\left(G_1^{k_1} \cdot \dots \cdot G_n^{k_n}\right) = \frac{1}{\mu(\mathbb{P})} \int_{\mathbb{P}} x_1^{k_1} \cdot \dots \cdot x_k^{k_n} \, d\mu. \tag{59}$$

The numerical computation of these moments of G requires the integration of polynomials over star-shaped polyhedra. To solve this computational problem, especially in higher dimensions, one can use the software package "LattE Integrale" which computes integrals of polynomials over convex polyhedra. The polyhedra can be given by $(\mathcal{VP})$ or $(\mathcal{HP})$, and the results are exact if the entries (vertices in case of $(\mathcal{VP})$ or A and b in case of $(\mathcal{HP})$) are rational numbers; see [32–34]. Note that star-shaped polyhedra can always be divided into disjoint convex polyhedra $\mathbb{P} \cap C_j$, $j \in \{1, \dots, m\}$, say. Moreover, $\mu(\mathbb{P})$ can also be computed by integrating the function $f(x) = 1$ over $\mathbb{P}$.

Theorem 10. *Let $\mathbb{P} \subset \mathbb{R}^n$ be a star-shaped polyhedron, $G = (G_1, \dots, G_n)^T$ uniformly distributed on $\mathbb{P}$ and $X = (X_1, \dots, X_n)^T \sim \Phi_{g,\mathbb{P}}$.*

(a) *If the moment $\mathbb{E}(R_g^k)$ exists then $\mathbb{E}(X^k) = ((n + k)/n)\mathbb{E}(R_g^k)\mathbb{E}(G^k)$.*

(b) *If $\mathbb{E}(R_g^k)$ exists then, for all $k_j \geq 0$, $j \in \{1, \dots, n\}$ with $k = k_1 + \cdots + k_n$,*

$$\mathbb{E}\left(X_1^{k_1} \cdot \dots \cdot X_n^{k_n}\right) = \frac{n+k}{n} \mathbb{E}\left(R_g^k\right) \mathbb{E}\left(G_1^{k_1} \cdot \dots \cdot G_n^{k_n}\right). \tag{60}$$

(c) *In case of existence, the covariance matrix $\operatorname{cov}(X)$ of X allows the representation*

$$\operatorname{cov}(X) = \frac{n+2}{n} \mathbb{E}\left(R_g^2\right) \mathbb{E}\left(GG^T\right) - \left(\frac{n+1}{n}\right)^2 \left(\mathbb{E}\left(R_g\right)\right)^2 \mathbb{E}(G)\, \mathbb{E}(G)^T. \tag{61}$$

Proof. By the stochastic representation in (52), we have $\mathbb{E}(X^k) = \mathbb{E}(R_g^k \cdot U_{\mathbb{S}}^k) = \mathbb{E}(R_g^k)\mathbb{E}(U_{\mathbb{S}}^k)$. According to Theorem 8, $G \overset{d}{=} h_{\mathbb{P}}(G) \cdot U_{\mathbb{S}}$ where $h_{\mathbb{P}}(G)$ and $U_{\mathbb{S}}$ are stochastically independent, $h_{\mathbb{P}}(G)$ follows the density function $f_{(h_{\mathbb{P}}(G))}(t) = I_{[0,1]}(t) n t^{n-1}$, and

$$\mathbb{E}\left(\left(h_{\mathbb{P}}(G)\right)^k\right) = \int_{-\infty}^{\infty} t^k f_{(h_{\mathbb{P}}(G))}(t)\, dt = n \int_0^1 t^{n+k-1}\, dt = \frac{n}{n+k}. \tag{62}$$

Since $\mathbb{E}(G^k) = \mathbb{E}((h_{\mathbb{P}}(G))^k)\mathbb{E}(U_{\mathbb{S}}^k)$ it follows (a). Parts (b) and (c) can be proved analogously.

For basic types of dgf g and representations of the moments of R_g, we refer to [12] and to Table 1.

5.2. Characteristic Functions. Integral representations of characteristic functions of arbitrary star-shaped distributed random vectors are derived in [6]. Specific representations are proved for spherically distributed random vectors in [35], for $l_{n,1}$-symmetrically distributed random vectors in [36] and for $l_{n,p}$-symmetrically distributed random vectors in [37]. We present a specific integral representation for continuous polyhedral star-shaped distributed random vectors by applying the polyhedral star-shaped polar coordinates to the arbitrary representations from [6]. This change of the coordinate system is very fruitful for the visualization of characteristic functions of continuous polyhedral star-shaped distributions by numerical methods; see Figures 2 and 3.

Theorem 11. *If $X \sim \Phi_{g,\mathbb{P}}$ then its characteristic function allows the representation*

$$\begin{aligned}\phi_X(t) = C_{g,\mathbb{P}}(n) &\cdot \left(\int_0^\infty \left[\int_{M_n^*} \cos\left(\sum_{i=1}^n t_i x_i(r,\varphi)\right) J^*(SPOL_{\mathbb{P}})(\varphi)\, d\varphi\right] \cdot r^{n-1} g(r)\, dr\right.\\ &\left. + i \int_0^\infty \left[\int_{M_n^*} \sin\left(\sum_{i=1}^n t_i x_i(r,\varphi)\right) J^*(SPOL_{\mathbb{P}})(\varphi)\, d\varphi\right] \cdot r^{n-1} g(r)\, dr\right), \quad t \in \mathbb{R}^n.\end{aligned} \tag{63}$$

Proof. Applying the stochastic representation (52) and Theorem 1.1.6 in [38], it is proved in Theorem 9 in [6] that

$$\phi_X(t) = \frac{1}{\mathcal{I}(g)} \int_0^\infty \phi_{U_{\mathbb{S}}}(rt)\, r^{n-1} g(r)\, dr, \quad t \in \mathbb{R}^n. \tag{64}$$

Applying the representation of $\phi_{U_{\mathbb{S}}}(t)$ in Remark 10 (a) of [6] and the polyhedral star-shaped polar coordinate transformation to (64) yields (63).

Note that if $U_{\mathbb{S}}$ is symmetrically distributed with respect to the origin, $U_{\mathbb{S}} \overset{d}{=} -U_{\mathbb{S}}$, the imaginary part in equation (63) vanishes. To be specific, let $(X,Y)^T$ follow the $\mathbb{w}$-star contoured distribution $\Phi_{g,\mathbb{P}_{\mathbb{w}}}$. The characteristic function of $(X,Y)^T$ is

$$\phi_{g,\mathrm{w}}(t_1,t_2)=\frac{1}{8\mathrm{w}\mathcal{I}(g)}\cdot\int_0^\infty\int_0^{2\pi}\cos\left(\frac{t_1r\cos(\varphi)+t_2r\sin(\varphi)}{\max\{|\cos(\varphi)|,|\sin(\varphi)|\}+((1-\mathrm{w})/\mathrm{w})\min\{|\cos(\varphi)|,|\sin(\varphi)|\}}\right)\cdot\frac{rg(r)}{(\max\{|\cos(\varphi)|,|\sin(\varphi)|\}+((1-\mathrm{w})/\mathrm{w})\min\{|\cos(\varphi)|,|\sin(\varphi)|\})^2}d\varphi\,dr. \tag{65}$$

Note that the use of polyhedral star-shaped polar coordinates yields an integral representation of the characteristic function where the range of integration is much easier to describe than if one makes use of other types of coordinates. We exploit this fact for drawing Figures 2 and 3 where this characteristic function is visualized for the particular dgf $g_0(r) = (1+r^2)^{-2}$.

5.3. Moment Generating Functions. Considering a random vector $X \sim \Phi_{g,\mathbb{P}}$ the moment generating function $M_X : \mathbb{R}^n \to \mathbb{R}$ can be formally represented by

$$M_X(t_1,\ldots,t_n)=C(g,\mathbb{P})\cdot\int_{\mathbb{R}^n}e^{(t_1x_1+\cdots+t_nx_n)}g(h_{\mathbb{P}}(x_1,\ldots,x_n))\,d(x_1,\ldots,x_n). \tag{66}$$

It is possible to specify representation (66) given a certain type of dgf and a certain representation of $\mathbb{P}$. The results are shown in the following theorem.

Theorem 12. *Let $X \sim \Phi_{g,\mathbb{P}}$ and the special Kotz-type dgf $g(r)=e^{-\beta r}$, $\beta>0$ be given. If $h_{\mathbb{P}}$ is represented by Lemma 2 (b1) and it exists an $\epsilon>0$ such that*

$$h_{\mathbb{P}(t)}(x)=\sum_{i=1}^{m}I_{C_i}(x)\left(\left(a_{i1}-\frac{t_1}{\beta}\right)x_1+\cdots+\left(a_{in}-\frac{t_n}{\beta}\right)x_n\right) \tag{67}$$

for all $t=(t_1,\ldots,t_n)^T\in[-\epsilon,\epsilon]^{\times n}$ is the Minkowski functional of the star-shaped polyhedron $\mathbb{P}(t)=\{x\in\mathbb{R}^n : h_{P(t)}(x)\le 1\}$ then

$$M_X(t)=\frac{\pi_{\mathbb{P}(t)}}{\pi_{\mathbb{P}}},\quad t=(t_1,\ldots,t_n)^T\in[-\epsilon,\epsilon]^{\times n}. \tag{68}$$

Proof. Since $C(g,\mathbb{P})=\beta^n/n\pi_{\mathbb{P}}\Gamma(n)$ it holds

$$M_X(t)=\frac{\beta^n}{n\pi_P\Gamma(n)}\int_{\mathbb{R}^n}e^{\langle t,x\rangle}e^{-\beta h_P(x)}dx=\frac{\beta^n}{n\pi_P\Gamma(n)}\cdot\sum_{i=1}^{m}\int_{C_i}e^{-\beta((a_{i1}-t_1/\beta)x_1+\cdots+(a_{in}-t_n/\beta)x_n)}d(x_1,\ldots,x_n)=\frac{\beta^n}{n\pi_P\Gamma(n)}\underbrace{\int_{\mathbb{R}^n}e^{-\beta h_{P(t)}(x)}dx}_{=n\pi_{P(t)}(\Gamma(n)/\beta^n)}=\frac{\pi_{P(t)}}{\pi_P}. \tag{69}$$

The specific representation (68) of the moment generating function can be applied to compute vectors of moments of polyhedral star-shaped distributions. For an illustration, we refer to Example A.2 in the Appendix.

6. Simulation of Polyhedral Star-Shaped Distributions

The stochastic representations of Theorem 8 and (52) offer the possibility to generate polyhedral star-shaped distributed random points for a given contour defining star-shaped polyhedron $\mathbb{P}\subset\mathbb{R}^n$ and a dgf g. By generating a random number R_g from the density $f_{R_g}(r)=\mathcal{I}(g)^{-1}r^{n-1}g(r)$ and independently of this a polyhedral star-shaped generalized uniformly distributed random point $U_{\mathbb{S}}$ on $\mathbb{S}$, the product $R_gU_{\mathbb{S}}$ gives the desired polyhedral star-shaped distributed random point. $U_{\mathbb{S}}$ can be simulated by generating a uniformly on $\mathbb{P}$ distributed random point G and applying Theorem 8; that is, $U_{\mathbb{S}}=G/h_{\mathbb{P}}(G)$. It remains the simulation of G. To this end, we use a cuboid $\mathrm{Q}\subset\mathbb{R}^n$ satisfying $\mathbb{P}\subseteq\mathrm{Q}$ and apply the rejection method. For basic details on rejection sampling, we refer to [39]. Applications of the rejection method to the simulation of the p-generalized Gaussian distribution can be found in [29] and to the simulation of the uniform distribution on platonic bodies elsewhere. All single steps are summarized in Algorithm 1.

The sampling method in step (1) has to be specified in accordance with the chosen dgf g. For the dgfs in Table 1, sampling methods can be found in [12]. Proving that G is uniformly distributed on $\mathbb{P}$ and the stopping time of step (2) is finite can be done analogously to Appendix 1 of [29]. A suitable cuboid Q can be found as follows. If $\mathbb{P}$ is represented by $(\mathcal{SVP})$, one can put

$$q_{\min,i}=\min\{p_{j,k}[i],\ j=1,\ldots,m,\ k=1,\ldots,l_j\}\quad i=1,\ldots,n,$$
$$q_{\max,i}=\max\{p_{j,k}[i],\ j=1,\ldots,m,\ k=1,\ldots,l_j\}\quad i=1,\ldots,n, \tag{70}$$

where $p_{j,k}[i]$ denotes the ith component of the vertex $p_{j,k}$. If $\mathbb{P}$ is represented by $(\mathcal{SHP})$, one can solve for every $i=1,\ldots,n$, the optimization problems: minimize $q_{\min,i}=x_i$ (maximize $q_{\max,i}=x_i$) subject to $h_{\mathbb{P}}(x_1,\ldots,x_n)\le 1$. Since it appears to be easier to find a suitable Q in case of $(\mathcal{SVP})$, note that it is possible to transform $\mathbb{P}$ from $(\mathcal{SHP})$ to $(\mathcal{SVP})$ according to [40, 41].

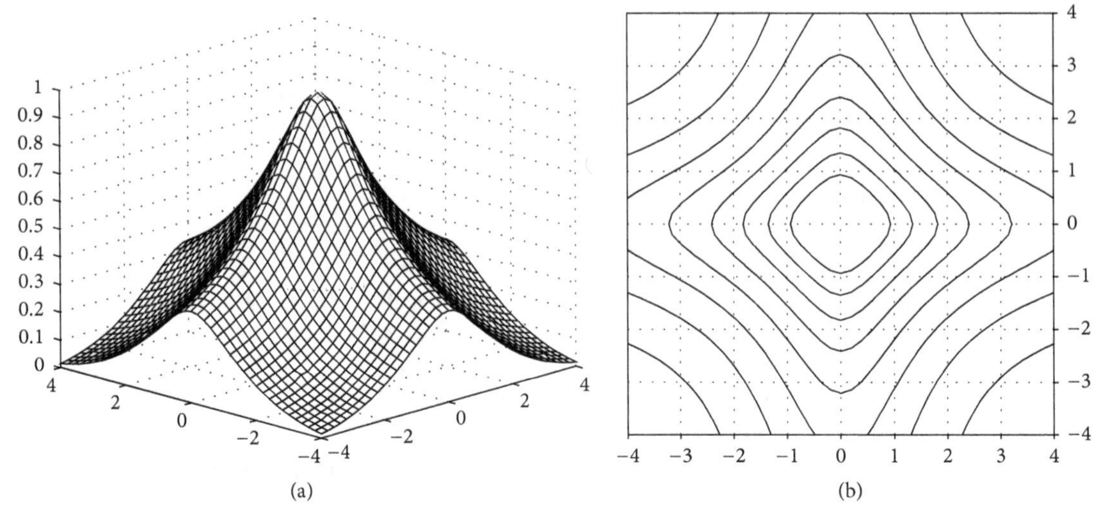

Figure 2: $\phi_{g_0,\mathrm{w}}$ with $\mathrm{w} = 0.25$ (a) and its level sets (b).

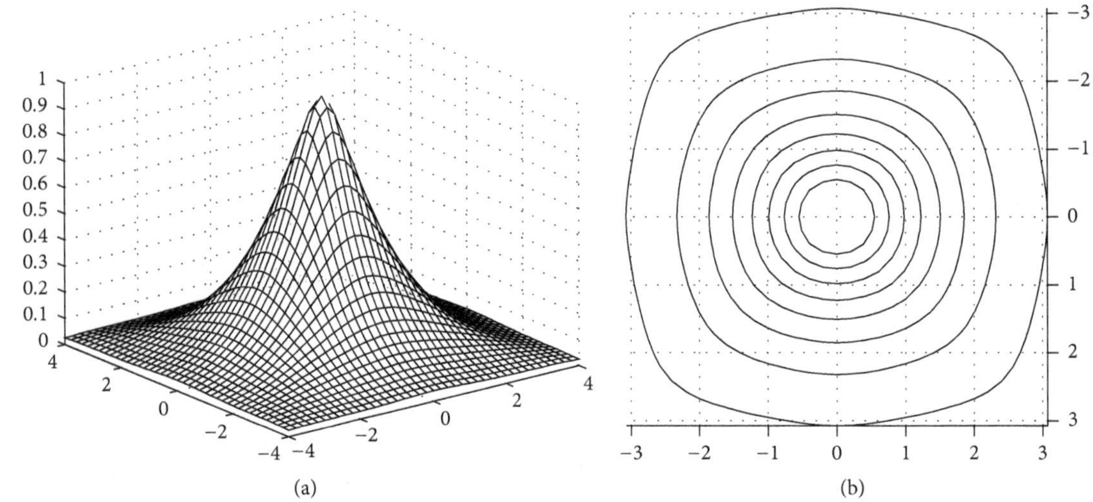

Figure 3: $\phi_{g_0,\mathrm{w}}$ with $\mathrm{w} = 0.8$ (a) and its level sets (b).

Input: $g, h_{\mathbb{P}}, Q = [q_{\min,1}, q_{\max,1}] \times \cdots \times [q_{\min,n}, q_{\max,n}]$
Output: X from $\Phi_{g,\mathbb{P}}$
Algorithm
(1) Sample R_g from $f_{R_g}(r) = \mathcal{I}(g)^{-1} r^{n-1} g(r)$.
(2) Sample G uniformly from Q until $h_{\mathbb{P}}(G) \le 1$.
(3) Build $U_{\mathbb{S}} = G/h_{\mathbb{P}}(G)$ componentwise.
(4) Build $X = R_g \cdot U_{\mathbb{S}}$ componentwise.
Return X

Algorithm 1: Sampling algorithm for polyhedral star-shaped distributions $\Phi_{g,\mathbb{P}}$.

Example 13. Let X be w-star contoured distributed, $X \sim \Phi_{g,\mathbb{P}_{\mathrm{w}}}$ with the Kotz-type dgf $g(r) = g_{K(M,\beta,\gamma)}(r) = r^{2M-2} e^{-\beta r^{2\gamma}}$ where $M, \beta, \gamma \in (0, \infty)$. Figures 4, 5, and 6 show 2000 points simulated independently according to $\Phi_{g,\mathbb{P}_{\mathrm{w}}}$.

7. Location-Scale Transformations of Continuous Polyhedral Star-Shaped Distributions

Throughout this section, we assume that R_g has a finite second-order moment and use the notation $X \sim \Phi_{g,\mathbb{P},\vartheta,\Psi}$, where ϑ and Ψ denote expectation vector and covariance matrix of X, respectively. We consider location and scale transformations of X and study transformed density level sets.

Theorem 14. *Let $X \sim \Phi_{g,\mathbb{P},\vartheta,\Psi}$ with symmetric and positive definite matrix Ψ. If $\Sigma \in \mathbb{R}^{n\times n}$ is another symmetric and*

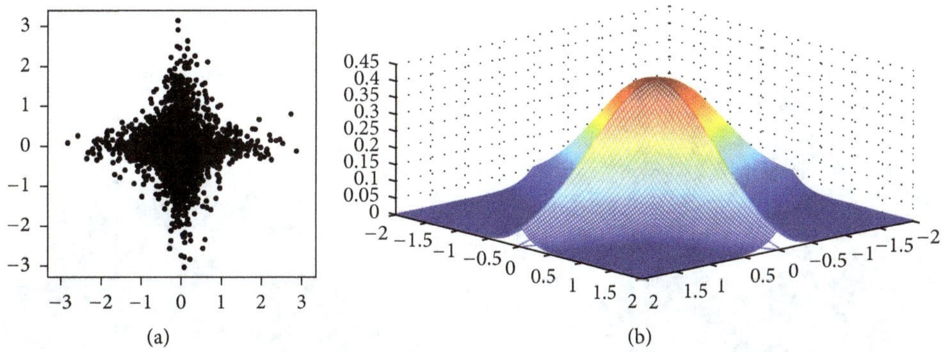

FIGURE 4: 2000 points (a) simulated according to $\varphi_{g_{K(1,0.5,1)},\mathbb{P}_{0.3}}$ (b).

positive definite matrix and $\mu \in \mathbb{R}^n$ then there exist regular lower triangular matrices D and V from $\mathbb{R}^{n\times n}$ such that $DD^T = \Psi$, $VV^T = \Sigma$ and the random vector $Z = VD^{-1}(X - \vartheta) + \mu$ follows the distribution $\Phi_{g,\mathbb{P}^{LT},\mu,\Sigma}$, where $\mathbb{P}^{LT} = VD^{-1}(\mathbb{P} - \vartheta) + \mu$, and the density is represented as

$$\varphi_{g,\mathbb{P}^{LT}}(z) = \frac{\det(D)\,C(g,\mathbb{P})}{\det(V)} g\left(h_{\mathbb{P}}\left(DV^{-1}(z-\mu)+\vartheta\right)\right), \quad z \in \mathbb{R}^n. \tag{71}$$

Proof. For the existence of D and V, see [42]. Furthermore, $\mathbb{E}(Z) = VD^{-1}(\mathbb{E}(X) - \vartheta) + \mu = \mu$, and $\operatorname{cov}(Z) = VD^{-1}\Psi(D^{-1})^T V^T = VV^T = \Sigma$. The representation of $\varphi_{g,\mathbb{P}}(z)$ follows by the density transformation formula.

For an example of the application of the linear transformation in Theorem 14, see Example A.3 in the Appendix. Note that the linear transformation does not only transform location and scale of the given random vector but also transforms the contour defining star-shaped polyhedron $\mathbb{P}$ into a polyhedron being generally not congruent with $\mathbb{P}$. Referring to this, see again Figures 7 and 8. Since noncongruent change of shape is not always a desired property, because $\mathbb{P}$ may be strictly chosen, we introduce a second, orthogonal transformation by using Givens rotations and Givens matrices. For further information about Givens rotations and matrices, see [42].

Definition 15. Let $l \in \{1,\ldots,n-1\}$ and $k \in \{l+1,\ldots,n\}$. The n-dimensional (l,k)-Givens matrix $S_n(l,k;\alpha) = (S_n(l,k;\alpha)_{(i,j)})_{i,j\in\{1,\ldots,n\}} \in \mathbb{R}^{n\times n}$ is defined by

$$S_n(l,k;\alpha)_{(i,j)} = \begin{cases} \cos(\alpha), & i=j=l \vee i=j=k, \\ \sin(\alpha), & i=l \wedge j=k, \\ -\sin(\alpha), & i=k \wedge j=l, \\ 1, & i=j \wedge i\neq l \wedge i \neq k, \\ 0, & \text{otherwise.} \end{cases} \tag{72}$$

Note that, for every dimension n, there exist $n(n-1)/2$ Givens matrices. A Givens rotation of a vector $x \in \mathbb{R}^n$ defined by $S_n(l,k;\alpha)x$ is an orthogonal transformation and rotates x within the l-k-plane of $\mathbb{R}^n$ by angle α. It follows that the application of all possible matrices $S_n(l,k;\alpha)$ to x allows an orthogonal transformation in all $n(n-1)/2$ two-dimensional planes of $\mathbb{R}^n$. For this reason, the product of all $n(n-1)/2$ n-dimensional Givens-matrices is defined by

$$S_n\left(\alpha_1,\ldots,\alpha_{(1/2)n(n-1)}\right) := \prod_{i=1}^{n-1}\left(\prod_{j=i+1}^{n} S_n\left(i,j;\alpha_{(1/2)(i-1)(2n-i)+j-i}\right)\right). \tag{73}$$

By now, it is possible to transform a random vector $X \sim \Phi_{g,\mathbb{P},\vartheta,\Psi}$ orthogonally with given angles $\alpha_1,\ldots,\alpha_{n(n-1)/2}$. The transformed contour defining polyhedron then represents a rotated but congruent representative of given $\mathbb{P}$. The proof of the following theorem can be done analogously to that of Theorem 14.

Theorem 16. *Let $X \sim \Phi_{g,\mathbb{P},\vartheta,\Psi}$. Given the angles $\alpha_1,\ldots,\alpha_{n(n-1)/2}$, the random vector $Z = S_n(\alpha_1,\ldots,\alpha_{n(n-1)/2})X$ is $\Phi_{g,\mathbb{P}^{OT},\mu,\Sigma}$ distributed, where $\mathbb{P}^{OT} = S_n(\alpha_1,\ldots,\alpha_{n(n-1)/2})\mathbb{P}$, $\Sigma = S_n(\alpha_1,\ldots,\alpha_{n(n-1)/2})\Psi S_n(\alpha_1,\ldots,\alpha_{n(n-1)/2})^T$, $\mu = S_n(\alpha_1,\ldots,\alpha_{n(n-1)/2})\vartheta$, and $\varphi_{g,\mathbb{P}^{OT}}(z) = C(g,\mathbb{P})g(h_{\mathbb{P}}(S_n(\alpha_1,\ldots,\alpha_{n(n-1)/2})^T z))$ is the density of Z.*

For an illustration of how Theorem 16 applies, see Example A.4 in the Appendix. In case $n = 2$, the Givens rotation reduces to a case considered in [43] for a geometric parametrization of the two-dimensional Gaussian law.

8. Statistics in Location-Scale Families

8.1. Estimating and Testing. Estimating the parameters μ and Σ from a $\Phi_{g,\mathbb{P},\mu,\Sigma}$-distributed sample can be done using

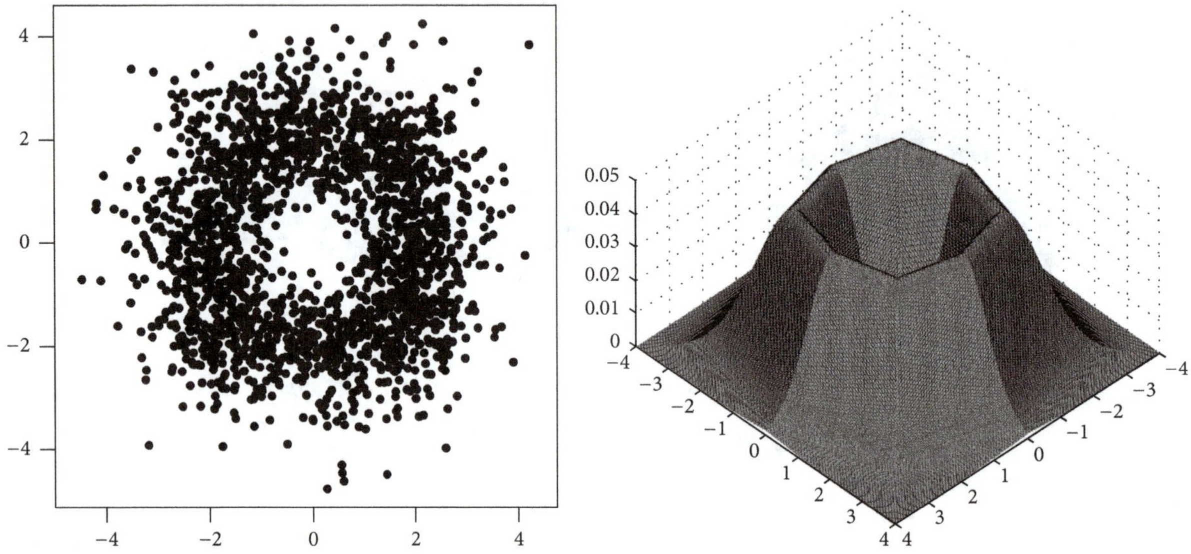

(a) (b)

FIGURE 5: 2000 points (a) simulated according to $\varphi_{g_{K(3,0.5,1)},\mathbb{P}_{0.8}}$ (b).

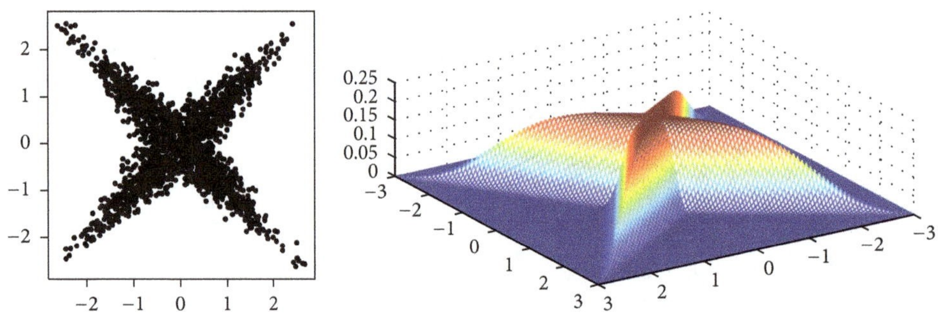

(a) (b)

FIGURE 6: 2000 points (a) simulated according to $\varphi_{g_{K(1,5,2)},\mathbb{P}_{3}}$ (b).

different methods. In a reasonable way, classical empirical moments

$$\hat{\mu} = \frac{1}{k}\sum_{r=1}^{k} X_r,$$
$$\hat{\Sigma} = \frac{1}{k}\sum_{r=1}^{k} \left(X_r - \hat{\mu}\right)\left(X_r - \hat{\mu}\right)^T \tag{74}$$

turn out to be method of moments estimators. To find maximum-likelihood estimators of the lower triangular matrix $V \in \mathbb{R}^{n \times n}$ with $v_{ii} \neq 0$, $\forall i$, and the vector $\rho \in \mathbb{R}^n$ in a $\Phi_{g,\mathbb{P}^{LT},V\vartheta+\rho,V\Psi V^T}$-model, one has to numerically solve a nonlinear minimizing problem. Note that there are various algorithms introduced in the literature to solve such problems. To this end, we refer to [44–46] and for a global overview to [47], respectively. A concrete application of this method will be given below.

Similarly, deriving maximum-likelihood estimates of $S_n(\alpha_1, \ldots, \alpha_{n/2(n-1)})$ and ρ in a $\Phi_{g,\mathbb{P}^{OT},S_n(\alpha_1,\ldots,\alpha_{n/2(n-1)})\mu,S_n(\alpha_1,\ldots,\alpha_{n/2(n-1)})\Psi S_n(\alpha_1,\ldots,\alpha_{n/2(n-1)})}$ model needs to solve a slightly modified nonlinear problem. Moreover, robust parameter estimation following [48, 49] can be seen to successfully work in the present distribution class. An analogous remark holds true for the application in the following section.

Now we discuss how to use one-dimensional Kolmogorov-Smirnov tests to verify if a dataset is distributed according to a given polyhedral star-shaped distribution. To this end, we use the polyhedral star-shaped polar coordinate transformation and Theorem 5 from Section 3.1 to convert the random vector $X \sim \Phi_{g,\mathbb{P},\vartheta,\Psi}$ into a tuple of random variables and applying the well-known Kolmogorov-Smirnov test to each random variable. Assuming a given realization $(x_1, \ldots, x_k)$ of independent and identically distributed random vectors $X_1, \ldots, X_k$, $X_i \in \mathbb{R}^n$, $i = 1, \ldots, k$, we can apply the transformation $SPOL_{\mathbb{P}}^{-1}$ for every $i = 1, \ldots, k$ to the realization x_i to generate polyhedral star-shaped polar coordinates r_i, $\varphi_{i,1}, \ldots, \varphi_{i,n-1}$, $i = 1, \ldots, k$. For a given polyhedral star-shaped distribution $\Phi_{g,\mathbb{P},\vartheta,\Psi}$ and $X \sim \Phi_{g,\mathbb{P},\vartheta,\Psi}$ we can apply Theorem 5 and test the empirical distribution functions of the realizations against the distributions of the

TABLE 2: Model estimates.

	Estimates
Method of moments	$\widehat{\mu} = \begin{pmatrix} 39.27992 \\ 44.28225 \end{pmatrix}$, $\widehat{\Sigma} = \begin{pmatrix} 118.70887 & -61.49893 \\ -61.49893 & 66.09405 \end{pmatrix}$
Maximum-likelihood	(a) $\widehat{\rho} = \begin{pmatrix} 37.65567 \\ 42.24069 \end{pmatrix}$, $\widehat{V} = \begin{pmatrix} 0.6957135 & 0 \\ -0.1285103 & 0.6720302 \end{pmatrix}$
	(b) $\widehat{\rho} = \begin{pmatrix} 36.6896 \\ 42.0099 \end{pmatrix}$, $S_2(\widehat{\alpha}) = \begin{pmatrix} \cos(0.1343) & \sin(0.1343) \\ -\sin(0.1343) & \cos(0.1343) \end{pmatrix}$
M-estimation	$\widehat{\rho} = \begin{pmatrix} 35.7269 \\ 42.7719 \end{pmatrix}$, $S_2(\widehat{\alpha}) = \begin{pmatrix} \cos(0.1437) & \sin(0.1437) \\ -\sin(0.1437) & \cos(0.1437) \end{pmatrix}$

random variables $R, \Phi_1, \ldots, \Phi_{n-1}$. This can be done by the Kolmogorov-Smirnov test statistics

$$T_k(r) = \sup_t \left|F_k(t, r) - F_R(t)\right|,$$
$$T_k(\varphi_j) = \sup_\alpha \left|F_k(\alpha, \varphi_j) - F_{\Phi_j}(\alpha)\right|, \qquad j = 1, \ldots, n-1, \tag{75}$$

where $F_k(t, r)$ and $F_k(\alpha, \varphi_j)$ are the usual empirical distribution functions. Since we consider the marginal distributions we can reject our realization to be $\Phi_{g,\mathbb{P},\vartheta,\Psi}$ distributed if the tests are rejected for one polar coordinate. Note that the integral representations of F_R and F_{Φ_j}, $j = 1, \ldots, n-1$ are difficult to compute, especially in high dimensions. To this end, we can use Algorithm 1 to simulate $\widetilde{k}$ independently, $\Phi_{g,\mathbb{P},\vartheta,\Psi}$ distributed realizations $(\widetilde{x}_1, \ldots, \widetilde{x}_{\widetilde{k}})$, apply Theorem 5 to $(\widetilde{x}_1, \ldots, \widetilde{x}_{\widetilde{k}})$, replace the exact cdf F_R and F_{Φ_j}, $j = 1, \ldots, n-1$ by the empirical cdf $F_{\widetilde{k}}(t, \widetilde{r})$ and $F_{\widetilde{k}}(\alpha, \widetilde{\varphi}_j)$, $j = 1, \ldots, n-1$, and perform two sample Kolmogorov-Smirnov tests with test statistics

$$T_{k,\widetilde{k}}(r, \widetilde{r}) = \sup_t \left|F_k(t, r) - F_{\widetilde{k}}(t, \widetilde{r})\right|,$$
$$T_{k,\widetilde{k}}(\varphi_j, \widetilde{\varphi}_j) = \sup_\alpha \left|F_k(\alpha, \varphi_j) - F_{\widetilde{k}}(\alpha, \widetilde{\varphi}_j)\right|, \qquad j = 1, \ldots, n-1. \tag{76}$$

8.2. Application. To conclude our considerations we model a polyhedral star-shaped distribution based on real data using the methods we described in this paper. To be more specific, we consider premature mortality and the median household income in all counties of the state Georgia (USA) and model their joint distribution to describe how premature mortality is influenced by household income in that particular state. Doing this is inspired by the descriptive statistical analysis in [1].

Premature mortality is defined as the all-cause, age-adjusted mortality rate for all individuals based on all deaths occurring before the age of 75 that could have been prevented. The data for the sample of all 159 counties in Georgia are from the CDC Compressed Mortality database (https://wonder.cdc.gov) and from the County Health Ranking of the USA (http://www.countyhealthrankings.org) and averaged form 2008 through 2011. For an illustration of the sample, see the scatter plot in Figure 9. Testing this sample for normality via Shapiro-Wilk test is significantly high rejected with a resulting p value $p = 1.516 \cdot 10^{-7}$. Since the density of the bivariate normal distribution is characterized by elliptical level sets (see [43]), the intention to model the data is here to keep the multinormal dgf $g(r) = e^{-r^2/2}$ but to change the contours of the level sets using a specific star-shaped polygon. Inspired by the optical appearance of the data point cloud, we choose $\mathbb{P}$ to be the triangle defined by its vertices $p_1 = (-7, -6)^T$, $p_2 = (-11, 19)^T$, and $p_3 = (21, -6)^T$; see Figure 9. This results in the star-shaped polyhedral distribution $\Phi_{g,\mathbb{P},\vartheta,\Psi}$, where

$$\vartheta = \begin{pmatrix} 1.87997 \\ 4.3866 \end{pmatrix},$$
$$\Psi = \begin{pmatrix} 203.13237 & -98.91335 \\ -98.91335 & 141.42441 \end{pmatrix},$$
$$\varphi_{g,\mathbb{P},\vartheta,\Psi}(x, y) = \frac{1}{700} e^{-(1/2)(\max\{-25x/199-4y/199, 25x/333+32y/333, -y/6\})^2}, \qquad x, y \in \mathbb{R}. \tag{77}$$

For the modelling we divide the data randomly into two groups, one for the estimation and one for testing the estimated models. Applying the estimation methods from Section 8.1 for the first group of our data yields estimates that can be found in Table 2.

Applying Theorem 14 with the method of moments estimates and the maximum-likelihood estimates (a) yields $\Phi_{g,\mathbb{P}^{LT},\widehat{\mu},\widehat{\Sigma}}$ and $\Phi_{g,\mathbb{P}^{LT},\widehat{V}\vartheta+\widehat{\rho},\widehat{V}\Psi\widehat{V}^T}$, respectively. The application of Theorem 16 with the maximum-likelihood estimates (b) and the m-estimates, respectively, yields $\Phi_{g,\mathbb{P}^{OT}+\widehat{\rho},S_2(\widehat{\alpha})\vartheta+\widehat{\rho},S_2(\widehat{\alpha})\Psi S_2(\widehat{\alpha})^T}$. We tested the second group of the data against these estimated models by using the two sample Kolmogorov-Smirnov tests from Section 8.1 with 10^5 simulated random points of the estimated models. The results are shown in Table 3. It turns out that the estimated models we generated using orthogonal transformation cannot be accepted as appropriate models of the data in this case. The choice of the contour defining polygon $\mathbb{P}$ was not good enough to model the data properly by using orthogonal transformations of $\Phi_{g,\mathbb{P},\vartheta,\Psi}$. Since the greatest p values of our tests are received with the maximum-likelihood estimates (a), we prefer $\Phi_{g,\mathbb{P}^{LT},\widehat{V}\vartheta+\widehat{\rho},\widehat{V}\Psi\widehat{V}^T}$ as distribution model of the data, where

$$\widehat{V}\vartheta + \widehat{\rho} = \begin{pmatrix} 38.96359 \\ 44.94702 \end{pmatrix},$$

$$\widehat{V}\Psi\widehat{V}^T = \begin{pmatrix} 98.31958 & -64.40732 \\ -64.40732 & 84.31032 \end{pmatrix},$$

$$\varphi_{g,\mathbb{P}^{LT}}(x,y) = 0.003 \cdot \exp\left(\frac{1}{2}\left(\max\left(-0.1818x - 0.0312y + 8.161, -0.0449x - 0.2585y + 12.611, 0.1313x + 0.149y - 11.2397\right)\right)^2\right), \quad x, y \in \mathbb{R}. \tag{78}$$

For an illustration of the empirical cdf of the marginal distributions used for the two sample Kolmogorov-Smirnov tests of our preferred model, see Figure 10. The density and its level sets of the model are visualized in Figure 11. Since the value of the correlation of our estimated model is −0.7074 we can conclude a strong negative correlation between premature mortality and household income which is a first hint to that lower income results can be accompanied by higher premature mortality rates in Georgia. Furthermore, in a simulation of 10^6 independent $\Phi_{g,\mathbb{P}^{LT},\widehat{V}\vartheta+\widehat{\rho},\widehat{V}\Psi\widehat{V}^T}$ distributed random points, we observed that 36.7% of the points were related to an income higher than 38.96359 and a mortality rate lower than 44.94702, and 36.2% of the points were related to an income lower than 38.96359 and a mortality rate higher than 44.94702. Even when choosing a tolerance level of a half standard deviation $\sigma/2$, the rates are approximately equal, namely, 20.3% (income higher than $38.96359 + \sigma/2$ and a mortality rate lower than $44.94702 - \sigma/2$) and 19.6% (income lower than $38.96359 - \sigma/2$ and a mortality rate higher than $44.94702 + \sigma/2$). This is a second hint to that a higher household income is accompanied by a lower premature mortality rate in Georgia.

9. Discussion and Outlook

In Sections 8.1 and 8.2 we considered m-estimates assuming fully specified dgf g. Referring to Table 1 where density generating functions are given with unknown parameters, one could ask for a simultaneous estimation of these parameters and m-estimates. The simultaneous m-estimation of $\alpha_1, \dots, \alpha_{n(n-1)/2}$, ρ, and parameters of g failed so far by numerical problems. Solving this is postponed to future work.

In Section 8.1 we presented univariate Kolmogorov-Smirnov tests to test data against a given polyhedral star-shaped distribution $\Phi_{g,\mathbb{P},\vartheta,\Psi}$. Since with this method we test with univariate marginal distributions of $\Phi_{g,\mathbb{P},\vartheta,\Psi}$, it is an open problem to develop a multivariate test for the class of polyhedral star-shaped distributions to validate also the dependencies in the model. Doing this, however, is postponed to future work.

Appendix

Example A.1. Let $\mathbb{P}$ be a pentagon given by the vertices $p_1 = (3,0)$, $p_2 = (1,1)$, $p_3 = (-1,3)$, $p_4 = (-3,0)$, and $p_5 = (2,-3)$ and $\mathfrak{C} = \{C_1, C_2, C_3, C_4, C_5\}$ a convex polyhedral fan for $\mathbb{P}$, where $C_1 = \{(x,y)^T \in \mathbb{R}^2 : y > 0,\ y \le x\}$, $C_2 = \{(x,y)^T \in \mathbb{R}^2 : y > x, -3x \le y\}$, $C_3 = \{(x,y)^T \in \mathbb{R}^2 : y < -3x,\ y \ge 0\}$, $C_4 = \{(x,y)^T \in \mathbb{R}^2 : y < 0,\ y \le -3/2x\}$, and $C_5 = \{(x,y)^T \in \mathbb{R}^2 : y > -3/2x,\ y \le 0\}$; see Figure 12. Applying Lemma 2 (b1), where $H_1 = \overline{p_1p_2}$, $H_2 = \overline{p_2p_3}$, $H_3 = \overline{p_3p_4}$, $H_4 = \overline{p_4p_5}$, and $H_5 = \overline{p_5p_1}$ the Minkowski functional of $\mathbb{P}$ can be represented as

$$\begin{aligned} h_{\mathbb{P}}(x,y) &= I_{C_1}(x,y)\left(\frac{1}{3}x + \frac{2}{3}y\right) \\ &+ I_{C_2}(x,y)\left(\frac{1}{2}x + \frac{1}{2}y\right) \\ &+ I_{C_3}(x,y)\left(-\frac{1}{3}x + \frac{2}{9}y\right) \\ &+ I_{C_4}(x,y)\left(-\frac{1}{3}x - \frac{5}{9}y\right) \\ &+ I_{C_5}(x,y)\left(\frac{1}{3}x - \frac{1}{9}y\right), \quad (x,y)^T \in \mathbb{R}^2. \end{aligned} \tag{A.1}$$

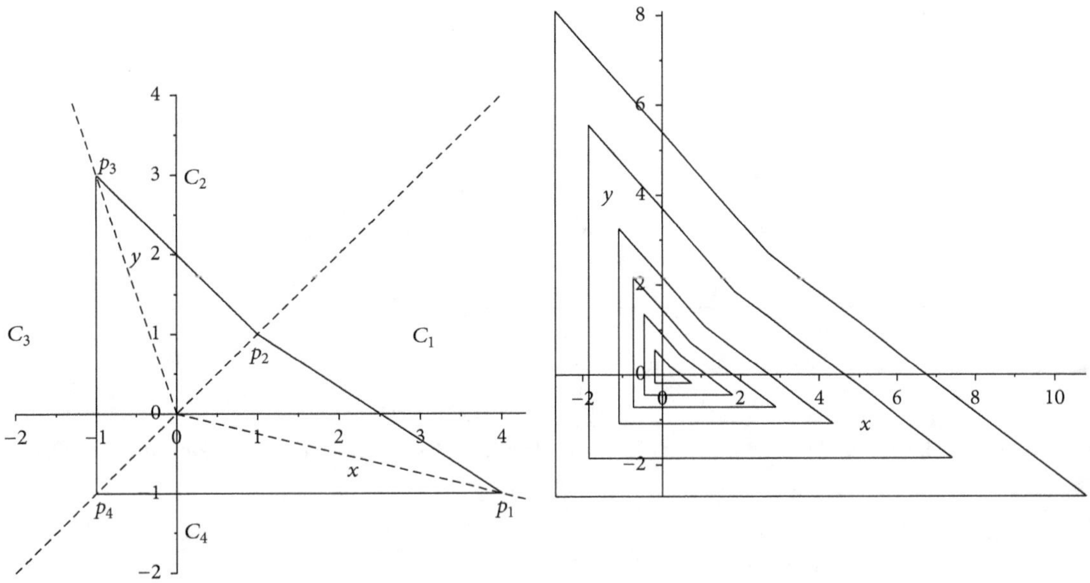

(a) (b)

FIGURE 7: Tetragon $\mathbb{P}$ (a) and level sets of the density of X (b) in Example A.3.

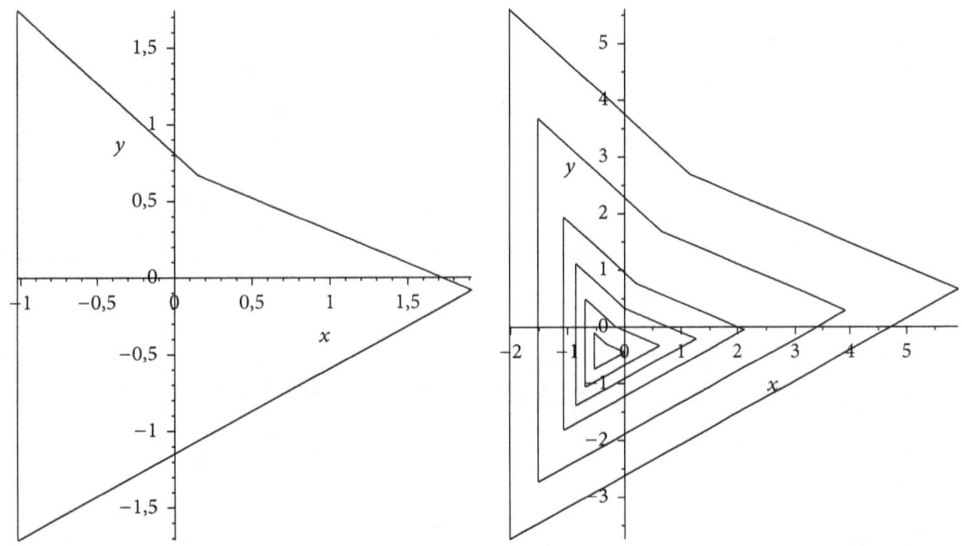

(a) (b)

FIGURE 8: Linear transformed contour defining tetragon $\mathbb{P}^{LT}$ (a) and density level sets of the density of Y (b) in Example A.3.

Let $q = (-2, 1.5)$ and $B = \{x \in (\mathrm{conv}(p_1, p_2) \cup \mathrm{conv}(p_2, p_3) \cup \mathrm{conv}(p_3, q))\}$.

Note that $\mathfrak{O}_{\mathbb{S}}(B)$ can be calculated applying

(i) the polyhedral star-shaped polar coordinate representation as

$$\begin{aligned}
\mathfrak{O}_{\mathbb{S}}(B) &= \int_{SPOL_P^{*(-1)}(B)} J^*(SPOL_P(\varphi))\,d\varphi \\
&= \int_0^{\pi+\arctan(-3/4)} \frac{1}{h_P(\cos(\varphi), \sin(\varphi))^2}\,d\varphi \\
&= \int_0^{\pi/4} \frac{1}{((1/3)\cos(\varphi) + (2/3)\sin(\varphi))^2}\,d\varphi \\
&\quad + \int_{\pi/4}^{\pi+\arctan(-3)} \frac{1}{((1/2)\cos(\varphi) + (1/2)\sin(\varphi))^2}\,d\varphi \\
&\quad + \int_{\pi+\arctan(-3)}^{\pi+\arctan(-3/4)} \frac{1}{(-(1/3)\cos(\varphi) + (2/9)\sin(\varphi))^2}\,d\varphi \\
&= 3 + 4 + 4.5 = 11.5,
\end{aligned} \tag{A.2}$$

(ii) the star-spherical coordinate representation as

$$\begin{aligned}
\mathfrak{O}_{\mathbb{S}}(B) &= \int_1^3 \frac{3}{2}\,d\vartheta + \int_{-1}^{1} 2\,d\vartheta + \int_{-2}^{-1} \frac{9}{2}\,d\vartheta \\
&= 3 + 4 + 4.5 = 11.5,
\end{aligned} \tag{A.3}$$

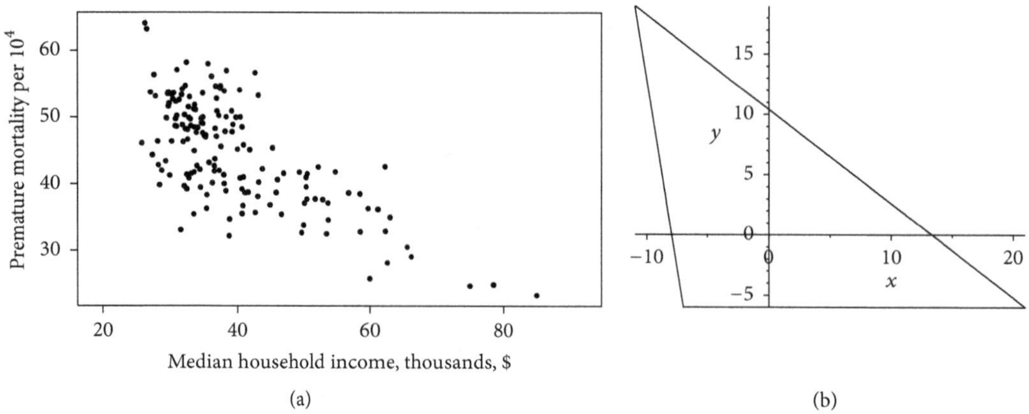

FIGURE 9: Median household income and premature mortality in the counties of Georgia, averaged from 2008 through 2011 (a) and contour defining triangle $\mathbb{P}$ (b).

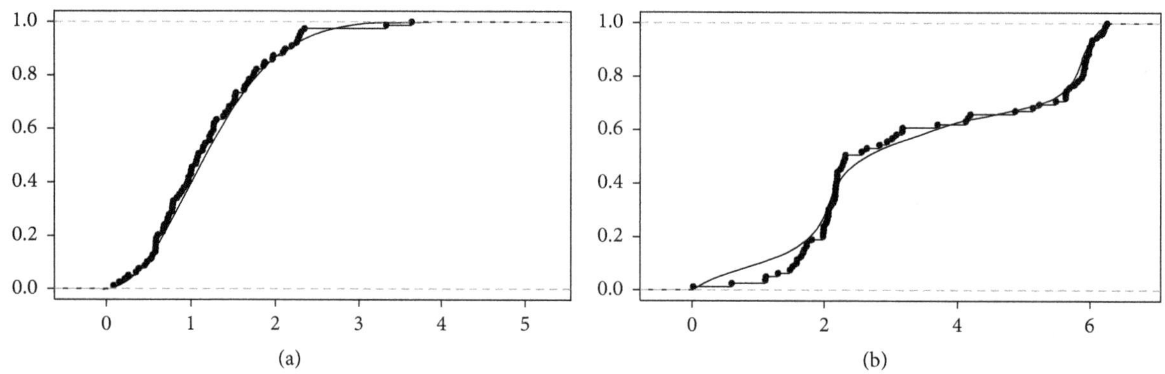

FIGURE 10: Simulated empirical cdf of the marginal distributions of R (a) and Φ (b) of the second group of data and $\Phi_{g,\mathbb{P}^{LT},\bar{\nabla}\vartheta+\bar{\rho},\bar{\nabla}\Psi\bar{\nabla}^T}$.

(iii) the facet-content representation by calculating $\sqrt{a_1 a_1^T} = 3/\sqrt{5}$, $\sqrt{a_2 a_2^T} = 2/\sqrt{2}$, $\sqrt{a_3 a_3^T} = 9/\sqrt{13}$, $vol_1(\mathscr{B}_1) = \sqrt{2^2+1^2} = \sqrt{5}$, $vol_1(\mathscr{B}_2) = \sqrt{2^2+2^2} = 2\sqrt{2}$, $vol_1(\mathscr{B}_3) = \sqrt{1^2+1.5^2} = \sqrt{13}/2$, and

$$\mathfrak{O}_{\mathbb{S}}(B) = \frac{3}{\sqrt{5}}\sqrt{5} + \frac{2}{\sqrt{2}}2\sqrt{2} + \frac{9}{\sqrt{13}}\frac{1}{2}\sqrt{13} = 11.5 \text{ or} \quad (A.4)$$

(iv) the simplicial representation as

$$\mathfrak{O}_{\mathbb{S}}(B) = \frac{1}{(2-1)!}\left(\left|\begin{pmatrix} 3 & 1 \\ 0 & 1 \end{pmatrix}\right| + \left|\begin{pmatrix} 1 & -1 \\ 1 & 3 \end{pmatrix}\right| + \left|\begin{pmatrix} -1 & -2 \\ 3 & \frac{3}{2} \end{pmatrix}\right|\right) = 11.5. \quad (A.5)$$

Applying the definition of the polyhedral star-shaped generalized uniform distribution (42) we can calculate $\omega_{\mathbb{S}}(B)$ by $\omega_{\mathbb{S}}(B) = \mathfrak{O}_{\mathbb{S}}(B)/\mathfrak{O}_{\mathbb{S}}(\mathbb{S}) = (23/2)/34 = 23/68$. Note that it is also possible to calculate this ratio using the sector measure and cone measure, respectively, since $\mu(\text{sector}_{\mathbb{P}}(B,1))/\mu(\mathbb{P}) = (23/4)/17 = 23/68$.

TABLE 3: Kolmogorov-Smirnov tests of the data against the estimated distribution models.

	p value of the K. S. test for R	p value of the K. S. test for Φ
Method of moments	0.7782463	0.4186046
Maximum-likelihood	(a) 0.9205786 (b) $6.992589 \cdot 10^{-8}$	(a) 0.6398323 (b) 0.1791245
M-estimation	$3.928482 \cdot 10^{-8}$	0.2240325

Example A.2. In this example we consider again the class of $\mathbb{W}$-stars from Section 4.2. Apart from the representation of $h_{\mathbb{P}_{\mathbb{W}}}$ in the mentioned section it is also possible to apply Lemma 2 (b2) and represent $h_{\mathbb{P}_{\mathbb{W}}}$ by

$$h_{\mathbb{P}_{\mathbb{W}}}(x) = \sum_{i=1}^{8} I_{C_i}(x,y)(a_{i1}x + a_{i2}y), \quad (A.6)$$

where a_{i1} and a_{i2} are for all $i = 1, \ldots, 8$ the elements of the matrix

$$A = \begin{pmatrix} 1 & \frac{1-w}{w} & -\frac{1-w}{w} & -1 & -1 & -\frac{1-w}{w} & \frac{1-w}{w} & 1 \\ \frac{1-w}{w} & 1 & 1 & \frac{1-w}{w} & -\frac{1-w}{w} & -1 & -1 & -\frac{1-w}{w} \end{pmatrix}^T \quad (A.7)$$

and $C_i = C_{1,(\lfloor i/2 \rfloor + 1 \bmod 4)} \cap C_{2,\lceil i/2 \rceil}$, $i = 1, \ldots, 8$, where $C_{1,j}$ and $C_{2,j}$, $j = 1, \ldots, 4$ are equivalent to those in Section 4.2. Since $\mathbb{P}(t_1, t_2) = \{(x, y) \in \mathbb{R}^2 : \sum_{i=1}^{8} I_{C_i}(x, y)((a_{i1} - t_1)x + (a_{i2} - t_2)y)\}$, $(t_1, t_2)^T \in [-\epsilon, \epsilon]^{\times 2}$, $\epsilon > 0$ it follows

$$\begin{aligned}
M_X(t_1, t_2) &= \frac{1}{8}\sum_{i=0}^{3} \frac{1/\left(1 + (-1)^i t_1\right) + 1/\left(1 + (-1)^{i!} t_2\right)}{1 + (-1)^i w t_1 + (-1)^{i!} w t_2}, \\
&\qquad (t_1, t_2)^T \in [-\epsilon, \epsilon]^{\times 2}, \\
\frac{\partial}{\partial t_j} M_X(t_1, t_2) &= \frac{1}{8} \\
&\cdot \sum_{i=0}^{3} \left[\frac{(-1)^{i+1} w \left(1/\left(1 + (-1)^i t_1\right) + 1/\left(1 + (-1)^{i!} t_2\right)\right)}{\left(1 + (-1)^i w t_1 + (-1)^{i!} w t_2\right)^2} \right. \\
&\left. + \frac{(-1)^{i+1}}{\left(1 + (-1)^i w t_1 + (-1)^{i!} w t_2\right)\left(1 + (-1)^i t_j\right)} \right], \\
&\qquad j = 1, 2, \\
\frac{\partial^2}{\partial t_1} M_X(t_1, t_2) &= \frac{1}{4} \\
&\cdot \sum_{i=0}^{3} \left[\frac{w^2 \left(1/\left(1 + (-1)^i t_1\right) + 1/\left(1 + (-1)^{i!} t_2\right)\right)}{\left(1 + (-1)^i w t_1 + (-1)^{i!} w t_2\right)^3} \right. \\
&+ \frac{w}{\left(1 + (-1)^i w t_1 + (-1)^{i!} w t_2\right)^2 \left(1 + (-1)^i t_1\right)^2} \\
&\left. + \frac{1}{\left(1 + (-1)^i w t_1 + (-1)^{i!} w t_2\right)\left(1 + (-1)^i t_1\right)^3} \right], \\
\frac{\partial^2}{\partial t_2} M_X(t_1, t_2) &= \frac{1}{4} \\
&\cdot \sum_{i=0}^{3} \left[\frac{w^2 \left(1/\left(1 + (-1)^i t_1\right) + 1/\left(1 + (-1)^{i!} t_2\right)\right)}{\left(1 + (-1)^i w t_1 + (-1)^{i!} w t_2\right)^3} \right. \\
&+ \frac{w}{\left(1 + (-1)^i w t_1 + (-1)^{i!} w t_2\right)^2 \left(1 + (-1)^{i!} t_2\right)^2} \\
&\left. + \frac{1}{\left(1 + (-1)^i w t_1 + (-1)^{i!} w t_2\right)\left(1 + (-1)^{i!} t_2\right)^3} \right].
\end{aligned} \quad (A.8)$$

Thus we can calculate the first and second vector of moments by

$$\begin{aligned}
\mathbb{E}(X) &= \begin{pmatrix} \frac{\partial}{\partial t_1} M_X(t_1, t_2)\Big|_{(t_1,t_2)=(0,0)} \\ \frac{\partial}{\partial t_2} M_X(t_1, t_2)\Big|_{(t_1,t_2)=(0,0)} \end{pmatrix} = \begin{pmatrix} 0 \\ 0 \end{pmatrix}, \\
\mathbb{E}\left(X^2\right) &= \begin{pmatrix} \frac{\partial^2}{\partial t_1} M_X(t_1, t_2)\Big|_{(t_1,t_2)=(0,0)} \\ \frac{\partial^2}{\partial t_2} M_X(t_1, t_2)\Big|_{(t_1,t_2)=(0,0)} \end{pmatrix} \\
&= \begin{pmatrix} 2w^2 + w + 1 \\ 2w^2 + w + 1 \end{pmatrix}.
\end{aligned} \quad (A.9)$$

Example A.3. Let $\mathbb{P} \subset \mathbb{R}^2$ be the tetragon defined by the vertices $p_1 = (4, -1)$, $p_2 = (1, 1)$, $p_3 = (-1, 3)$, and $p_4 = (-1, -1)$. Considering the convex polyhedral fan $\mathfrak{C} = \{C_1, C_2, C_3, C_4\}$ for $\mathbb{P}$, where

$$\begin{aligned}
C_1 &= \left\{(x, y)^T \in \mathbb{R}^2 : x > y,\ -y \le \frac{1}{4}x\right\}, \\
C_2 &= \left\{(x, y)^T \in \mathbb{R}^2 : x \le y,\ -3x < y\right\}, \\
C_3 &= \left\{(x, y)^T \in \mathbb{R}^2 : -3x \ge y,\ y > x\right\}, \\
C_4 &= \left\{(x, y)^T \in \mathbb{R}^2 : y \le x,\ -y > \frac{1}{4}x\right\},
\end{aligned} \quad (A.10)$$

as shown in Figure 7(a), we can represent $h_{\mathbb{P}}$ applying Lemma 2 (b1) as $h_{\mathbb{P}}(x, y) = I_{C_1}(x, y)((2/5)x + (3/5)y) + I_{C_2}(x, y)(x/2 + y/2) - I_{C_3}(x, y)x - I_{C_4}(x, y)y$, $(x, y)^T \in \mathbb{R}^2$.

Assuming the dgf $g(r) = (1 + r^2/2)^{-4}$, it follows $\mathcal{I}(g) = 1/3$ and $C(g, \mathbb{P}) = 1/6$, since $\pi_{\mathbb{P}} = 9$. The random vector $X \sim \Phi_{g,\mathbb{P},\vartheta,\Psi}$ follows the density $f_{g,\mathbb{P}}(x, y) = 1/6 \cdot (1 + h_{\mathbb{P}}(x, y)^2/2)^{-4}$ and first- and second-order moments are

$$\begin{aligned}
\vartheta &= \frac{\pi\sqrt{2}}{96}\begin{pmatrix} 16 \\ 7 \end{pmatrix}, \\
\Psi &= \begin{pmatrix} 2.896 & -1.0917 \\ -1.0917 & 1.7469 \end{pmatrix}.
\end{aligned} \quad (A.11)$$

If we wish to transform X by a linear transformation into a random vector $Y \sim \Phi_{g,\mathbb{P}^{LT},0_2,E_2}$ where E_2 denotes the two-dimensional identity matrix, we can apply Theorem 14. It

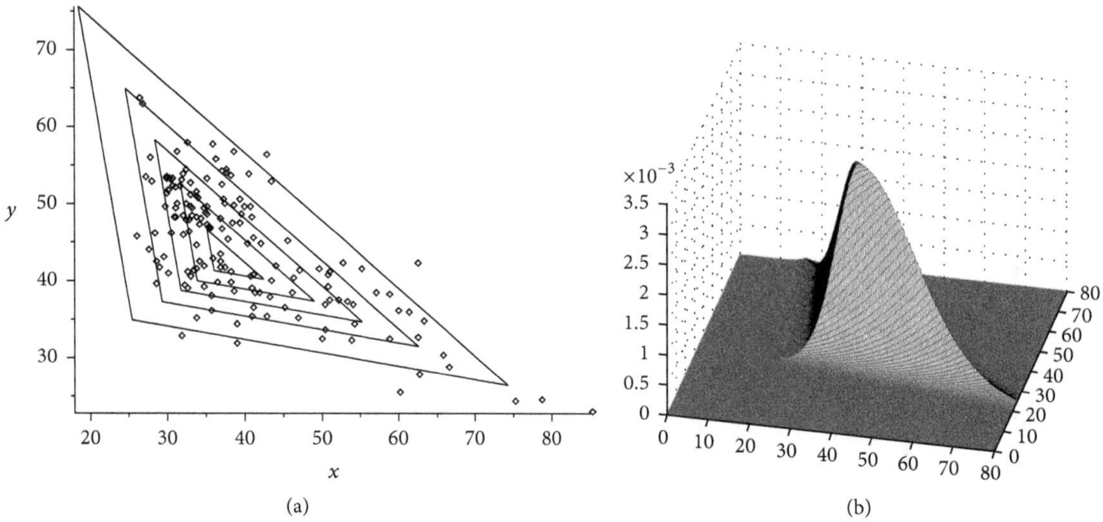

FIGURE 11: Density levels sets (a) and density (b) of $\Phi_{g,\mathbb{P}^{LT},\tilde{V}\vartheta+\tilde{\rho},\tilde{V}\Psi\tilde{V}^T}$.

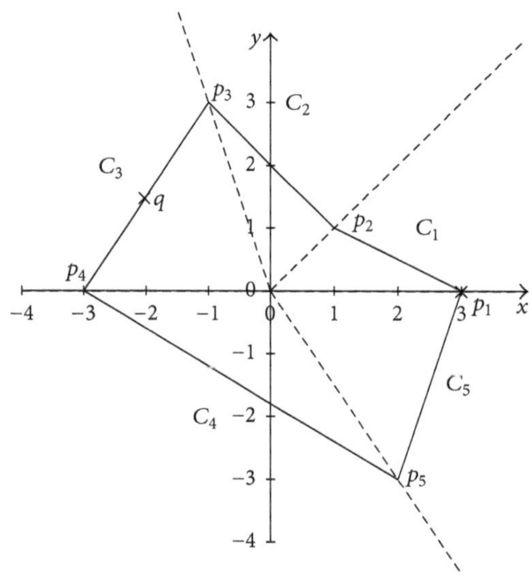

FIGURE 12: Pentagon $\mathbb{P}$ and the convex polyhedral fan $\mathfrak{C}$ for $\mathbb{P}$ in Example A.1.

follows then $Y = D^{-1}(X - \vartheta)$, where D can be calculated by Cholesky decomposition of Ψ; thus

$$D = \begin{pmatrix} 1.7018 & 0 \\ -0.6415 & 1.1556 \end{pmatrix}, \qquad D^{-1} = \begin{pmatrix} 0.5876 & 0 \\ 0.3262 & 0.86537 \end{pmatrix}. \tag{A.12}$$

Y then follows the density

$$f_{g,\mathbb{P}^{LT}}(x, y) = 1.96656 \cdot C(g, \mathbb{P}) \cdot \left(1 + \frac{1}{2} h_{\mathbb{P}}\left(D(x, y)^T + \vartheta\right)^2\right)^{-4}. \tag{A.13}$$

For an illustration of the transformed contour defining tetragon $\mathbb{P}^{LT}$ and the density level sets of the transformed density, see Figure 8.

Example A.4. Assume that the polyhedron $\mathbb{P} \subset \mathbb{R}^2$, the dgf g, and the random vector $X \sim \Phi_{g,\mathbb{P},\vartheta,\Psi}$ are the same as in Example A.3. Applying Theorem 16 with

$$S_2(5.74) = \begin{pmatrix} 0.856 & -0.5168 \\ 0.5168 & 0.856 \end{pmatrix}, \tag{A.14}$$

it follows that $Y = S_2(5.74)X$ is $\Phi_{g,\mathbb{P}^{OT},S_2(5.74)\vartheta,\mathrm{diag}(3.555,1.0878)}$ distributed, and the resulting density function allows the representation

$$f_{g,\mathbb{P}^{OT}}(x, y) = C(g, \mathbb{P}) \left(1 + \frac{1}{2} h_{\mathbb{P}}\left(S_2(\alpha)^T (x, y)^T\right)\right)^{-4}. \tag{A.15}$$

For an illustration of the transformed contour defining tetragon $\mathbb{P}^{OT}$ and the density level sets of the transformed density, see Figure 13.

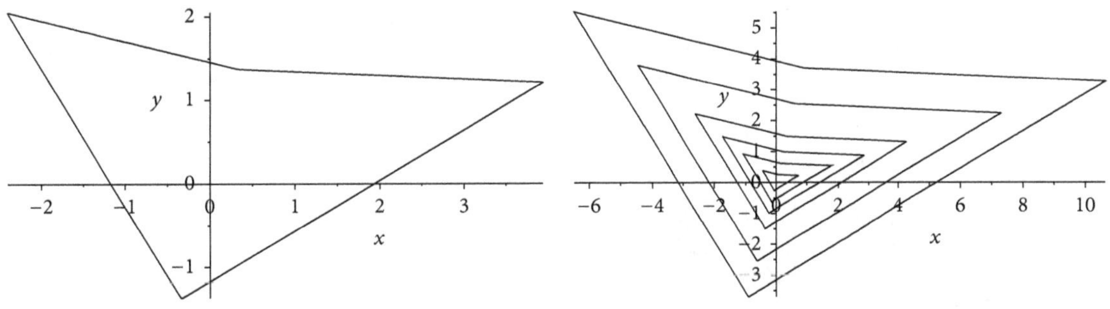

FIGURE 13: Orthogonal transformed contour defining tetragon $\mathbb{P}^{OT}$ (a) and density level sets of the density of Y (b) in Example A.4.

Competing Interests

The authors declare further that there is no conflict of interests regarding the publication of this paper.

References

[1] E. R. Cheng and D. A. Kindig, "Disparities in premature mortality between high- and low-income us counties," *Preventing Chronic Disease*, vol. 9, no. 3, Article ID 110120, 2012.

[2] J. M. Sarabia and E. Gómez-Déniz, "Construction of multivariate distributions: a review of some results," *SORT*, vol. 32, no. 1, pp. 3–46, 2008.

[3] C. Fernandez, J. Osiewalski, and M. F. J. Steel, "Modeling and inference with vspherical distributions," *Journal of the American Statistical Association*, vol. 90, pp. 1331–1340, 1995.

[4] H. Kamiya, A. Takemura, and S. Kuriki, "Star-shaped distributions and their generalizations," *Journal of Statistical Planning and Inference*, vol. 138, no. 11, pp. 3429–3447, 2008.

[5] B. C. Arnold, E. Castillo, and J. M. Sarabia, "Multivariate distributions defined in terms of contours," *Journal of Statistical Planning and Inference*, vol. 138, no. 12, pp. 4158–4171, 2008.

[6] W.-D. Richter, "Geometric disintegration and star-shaped distributions," *Journal of Statistical Distributions and Applications*, vol. 1, article 20, 2014.

[7] W.-D. Richter, "Convex and radially concave contoured distributions," *Journal of Probability and Statistics*, vol. 2015, Article ID 165468, 12 pages, 2015.

[8] W.-D. Richter, "Norm contoured distributions in R2," in *Lecture Notes of Seminario Interdisciplinare di Matematica*, vol. 12, pp. 179–199, Seminario Interdisciplinare di Matematica (S.I.M.), University of Basilicata,, Potenza, Italy, 2015.

[9] K. Müller and W.-D. Richter, "Extreme value distributions for dependent jointly ln,p-symmetrically distributed random variables," *Dependence Modeling*, vol. 4, no. 1, pp. 30–62, 2016.

[10] M. Moszyńska and W.-D. Richter, "Reverse triangle inequality. Antinorms and semi-antinorms," *Studia Scientiarum Mathematicarum Hungarica*, vol. 49, no. 1, pp. 120–138, 2012.

[11] G. Balkema and N. Nolde, "Asymptotic independence for unimodal densities," *Advances in Applied Probability*, vol. 42, no. 2, pp. 411–432, 2010.

[12] K.-T. Fang, S. Kotz, and K.-W. Ng, *Monographson Statistics and Applied Probability: Symmetric Multivariate and Related Distributions*, vol. 36, Chapman & Hall, New York, NY, USA, 1990.

[13] E. Hashorva, S. Kotz, and A. Kume, "L_p-norm generalised symmetrised dirichlet distributions," *Albanian Journal of Mathematics*, vol. 1, no. 1, pp. 31-56, 2007.

[14] N. Balakrishnan and E. Hashorva, "On pearson-Kotz dirichlet distributions," *Journal of Multivariate Analysis*, vol. 102, no. 5, pp. 948–957, 2011.

[15] J. Osiewalski and M. F. J. Steel, "Robust bayesian inference in lq-spherical models," *Biometrika*, vol. 80, no. 2, pp. 456–460, 1993.

[16] A. K. Gupta and D. Song, "L_p-norm spherical distribution," *Journal of Statistical Planning and Inference*, vol. 60, no. 2, pp. 241–260, 1997.

[17] G. Schechtman and J. Zinn, "On the volume of the intersection of two ln p balls," *Proceedings of the American Mathematical Society*, vol. 110, no. 1, pp. 217–224, 1990.

[18] S. T. Rachev and L. Rüschendorf, "Approximate independence of distributions on spheres and their stability properties," *The Annals of Probability*, vol. 19, no. 3, pp. 1311–1337, 1991.

[19] P. J. Szabłowski, "Uniform distributions on spheres in finite dimensional L_α and their generalizations," *Journal of Multivariate Analysis*, vol. 64, no. 2, pp. 103–117, 1998.

[20] W.-D. Richter, "Continuous $l_{n,p}$-symmetric distributions," *Lithuanian Mathematical Journal*, vol. 49, no. 1, pp. 93–108, 2009.

[21] W.-D. Richter, "Geometric and stochastic representations for elliptically contoured distributions," *Communications in Statistics. Theory and Methods*, vol. 42, no. 4, pp. 579–602, 2013.

[22] W.-D. Richter, "Generalized spherical and simplicial coordinates," *Journal of Mathematical Analysis and Applications*, vol. 336, no. 2, pp. 1187–1202, 2007.

[23] I. Molchanov, "Convex geometry of max-stable distributions," *Extremes*, vol. 11, no. 3, pp. 235–259, 2008.

[24] G. M. Ziegler, *Lectures on Polytopes*, Springer, Berlin, Germany, 1995.

[25] W.-D. Richter and K. Schicker, "Circle numbers of regular convex polygons," *Results in Mathematics*, vol. 69, no. 3, pp. 521–538, 2016.

[26] R. Webster, *Convexity*, Oxford University Press, New York, NY, USA, 1994.

[27] W.-D. Richter, "On $l_{2,p}$-circle numbers," *Lithuanian Mathematical Journal*, vol. 48, no. 2, pp. 228–234, 2008.

[28] V. Henschel and W.-D. Richter, "Geometric generalization of the exponential law," *Journal of Multivariate Analysis*, vol. 81, no. 2, pp. 189–204, 2002.

[29] S. Kalke and W.-D. Richter, "Simulation of the p-generalized Gaussian distribution," *Journal of Statistical Computation and Simulation*, vol. 83, no. 4, pp. 641–667, 2013.

[30] S. Kalke, W.-D. Richter, and F. Thauer, "Linear combinations, products and ratios of simplicial or spherical variates," *Communications in Statistics—Theory and Methods*, vol. 42, no. 3, pp. 505–527, 2013.

[31] A. Naor, "The surface measure and cone measure on the sphere of ℓ_p^n," *Transactions of the American Mathematical Society*, vol. 359, no. 3, pp. 1045–1080, 2007.

[32] V. Baldoni, N. Berline, J. A. De Loera, M. Köppe, and M. Vergne, "How to integrate a polynomial over a simplex," *Mathematics of Computation*, vol. 80, no. 273, pp. 297–325, 2011.

[33] J. A. De Loera, B. Dutra, M. Köppe, S. Moreinis, G. Pinto, and J. Wu, "Software for exact integration of polynomials over polyhedra," *Computational Geometry*, vol. 46, no. 3, pp. 232–252, 2013.

[34] V. Baldoni, N. Berline, J. A. De Loera et al., "A User's Guide for LattE integrale v1.7.1," URL software package LattE, 2013, https://www.math.ucdavis.edu/~latte/.

[35] I. J. Schoenberg, "Metric spaces and completely monotone functions," *Annals of Mathematics*, vol. 39, no. 4, pp. 811–841, 1938.

[36] K. W. Ng and G.-L. Tian, "Characteristic functions of L_1-spherical and L_1-norm symmetric distributions and their applications," *Journal of Multivariate Analysis*, vol. 76, no. 2, pp. 192–213, 2001.

[37] S. Kalke, *Geometrische Strukturen l_n,p-symmetrischer charakteristischer Funktionen [Ph.D. thesis]*, Universität Rostock, 2013.

[38] Z. Sasvári, *Multivariate characteristic and correlation functions*, De Gruyter Studies in Mathematics, Walter de Gruyter & Co., Berlin, Germany, 2013.

[39] J. von Neumann, "Various techniques used in connection with random digits. monte carlo methods," *National Bureau of Standards*, vol. 12, pp. 36–38, 1951.

[40] K. Fukuda and A. Prodon, "Double description method revisited," in *Combinatorics and computer science*, vol. 1120 of *Lecture Notes in Computer Science*, pp. 91–111, Springer, Berlin, 1996.

[41] D. Avis, "lrs: a revised implementation of the reverse search vertex enumeration algorithm," in *Polytopes—Combinatorics and Computation*, pp. 177–198, Birkhäuser, Basel, Switzerland, 2000.

[42] G. H. Golub and C. F. van Loan, *Matrix Computations*, The John Hopkins University Press, Baltimore, Md, USA, 4th edition, 2013.

[43] T. Dietrich, S. Kalke, and W.-D. Richter, "Stochastic representations and a geometric parametrization of the two-dimensional Gaussian law," *Chilean Journal of Statistics*, vol. 4, no. 2, pp. 27–59, 2013.

[44] R. Fletcher and C. M. Reeves, "Function minimization by conjugate gradients," *The Computer Journal*, vol. 7, no. 2, pp. 149–154, 1964.

[45] J. A. Nelder and R. Mead, "A simplex method for function minimization," *The Computer Journal*, vol. 7, no. 4, pp. 308–313, 1965.

[46] C. J. P. Bélisle, "Convergence theorems for a class of simulated annealing algorithms on Rd," *Journal of Applied Probability*, vol. 29, no. 4, pp. 885–895, 1992.

[47] J. Nocedal and S. J. Wright, *Numerical optimization*, Springer Series in Operations Research and Financial Engineering, Springer, New York, USA, Second edition, 2006.

[48] P. J. Huber, "The behavior of maximum likelihood estimates under nonstandard conditions," in *Proceedings of the 5th Berkeley Symposium on Mathematical Statistics and Probability*, vol. 1, pp. 221–233, 1967.

[49] F. R. Hampel, E. M. Ronchetti, P. J. Rousseeuw, and A. Stahel, *Robust Statistics—The Approach Based on Influence Functions*, John Wiley & Sons, New York, NY, USA, 1986.

18

An Approach of Randomness of a Sample Based on Its Weak Ergodic Limit

Jaime A. Londoño

Departamento de Matemáticas y Estadística, Universidad Nacional de Colombia, Manizales, Colombia

Correspondence should be addressed to Jaime A. Londoño; jaime.a.londono@gmail.com

Academic Editor: Ramón M. Rodríguez-Dagnino

For a Polish Sample Space with a Borel σ-field with a surjective measurable transformation, we define an equivalence relation on sample points according to their ergodic limiting averages. We show that this equivalence relation partitions the subset of sample points on measurable invariant subsets, where each limiting distribution is the unique ergodic probability measure defined on each set. The results obtained suggest some natural objects for the model of a probabilistic time-invariant phenomenon are uniquely ergodic probability spaces. As a consequence of the results gained in this paper, we propose a notion of randomness that is weaker than recent approaches to Schnorr randomness.

1. Introduction

To understand the meaning of randomness several approaches have been undertaken. Some of them include Kolmogorov [1], Martin-Löf [2], Zvonkin and Levin [3], and Schnorr [4] among many others.

On the initial attempts, researchers were trying to identify when a sequence of numbers was random. The sequence of digits of π or sequences of computer-generated numbers produced by an algorithm that seems to exhibit a "random behavior" motivated the original problem. In any case of interest, the sequences are elements of $\{0, 1, \dots, 9\}^{\mathbb{N}}$ or $\mathbb{R}^{\mathbb{N}}$ and researchers would like to give a criterion that decides for a sequence to be "random." In general, sample spaces that allow a Polish Space structure and a surjective transformation (the shifting transformation) embrace the interesting cases.

In this paper, we show that a natural approach for a sample space is to consider all samples that describe the same physical time-invariant phenomenon. We would demonstrate that this method allows for a richer probability structure. The latter introduces a different view for randomness, in a way ... two sample objects are equivalent if they describe the ...bability phenomenon (in a way that we determine ...r), and, as a consequence, we give an alternative ...blem of trying to define randomness.

The Ergodic Theorem and the Ergodic Decomposition Theorem motivate our definition. Indeed, our approach goes hand in hand with Gray [5] and extends some results of Shields [6]. Some recent works which study the relation between Ergodic Theory and randomness are V'yugin [7], Gács et al. [8], and Galatolo et al. [9].

The Birkhoff-Kinchin Ergodic Theorem (See Billingsley [10]) states that if θ is a measure preserving transformation defined on a probability space $(\Omega, \mathcal{F}, P)$ and $X : (\Omega, \mathcal{F}) \to (\mathbb{R}^d, \mathcal{B}(\mathbb{R}^d))$ is an integrable random vector, then

$$\lim_n \frac{1}{n} \sum_{k=1}^{n} X \circ \theta^{k-1} = E\left[X \mid \mathcal{F}^{\theta}\right] \quad \text{a.e.,} \tag{1}$$

where $\mathcal{F}^{\theta} = \{F \in \mathcal{F} \mid \theta^{-1}(F) = F\}$ is the σ-field of invariant sets (under θ). For another interpretation of this fact, we define for each n

$$P_n^{\theta}(\omega, F) = \frac{1}{n} \sum_{k=1}^{n} 1_F \circ \theta^{k-1}(\omega) \quad \text{for } F \in \mathcal{F}. \tag{2}$$

It follows that $P_n^{\theta} : \Omega \times \mathcal{F} \to [0,1]$ defines a (finite) kernel, where $P_n^{\theta}(\omega, \cdot)$ is an ergodic measure. Moreover for any measurable random vector $X : (\Omega, \mathcal{F}) \rightarrow (\mathbb{R}^d, \mathcal{B}(\mathbb{R}^d))$

$$\int_\Omega X \, dP_n^\theta (\omega, \cdot) = \frac{1}{n} \sum_{k=1}^{n} X \circ \theta^{k-1} (\omega) . \tag{3}$$

The Birkhoff Ergodic Theorem implies that whenever θ is an ergodic measure preserving transformation and X is an integrable random vector there exists a set $N_X \in \mathcal{F}$ with $P(N_X) = 0$ such that for $\omega \notin N_X$

$$\int_\Omega X \, dP_n^\theta (\omega, \cdot) \longrightarrow \int_\Omega X \, dP. \tag{4}$$

Using the terminology of Gács et al. [8], if ω satisfies (4), ω is said to be θ typical with respect to the observable X. Moreover, if the last equation holds for any bounded continuous function X, then ω is said to be θ typical. The latter is equivalent to the weak convergence of $P_n^\theta(\omega, \cdot) \rightarrow P$. For a review on weak convergence see Billingsley [11]. Moreover, if ω is θ typical for any effective ergodic surjective transformation, then it is said that ω is typical. For a definition of effective ergodic transformation see Gács et al. [8].

For any computable measure P, defined on $\Omega = \{0, 1\}^{\mathbb{N}}$, with Borel σ-field $\mathcal{F}$, V'yugin [7] showed that any Martin-Löf random sequence $\omega \in \Omega$ is typical. For a review on Martin-Löf randomness see Martin-Löf [2]. Galatolo et al. [9] generalize the last result to computable probability spaces and P-computable observables. For a discussion of the definition and consequences of computable probability, spaces see Gács et al. [8].

Moreover Gács et al. [8] extend the concept of Schnorr randomness to computable probability spaces, and they show that the idea of Schnorr randomness is weaker than the notion of Martin-Löf randomness in this context. For a definition of Schnorr randomness see Schnorr [4]. Gács et al. [8] also showed that on an arbitrary computable probability space $(\Omega, \mathcal{F}, P)$ space with no atoms $\omega \in \Omega$ is Schnorr random if and only if ω is θ typical for every mixing endomorphism θ.

In this paper, we obtain another approach to randomness. Assume that (Ω, d) is a Polish Space, with metric d, and assume the Borel σ-field $\mathcal{F} = \mathcal{B}_d(\Omega)$ on Ω. Moreover assume a measure preserving transformation θ defined on Ω. For most problems related to finding a reasonable definition for randomness it is sufficient to assume that $\Omega = \mathbb{R}^{\mathbb{N}}$ is the sample space of real valued sequences with the metric $d(\omega, \omega') = \sum_n 2^{-n} \min(|\omega(n) - \omega'(n)|, 1)$, the shifting transformation $\theta(\omega)(n) = \omega(n + 1)$, and the sequence of processes $X_n(\omega) = \omega(n)$ and with a probability that makes X_n independent with the same distribution P_{X_1} defined as $P_{X_1}(F) = P(X_1^{-1}(F))$ for all $F \in \mathcal{B}(\mathbb{R})$ Borel set.

With the help of the Ergodic Decomposition Theorem (see Proposition 1) it is possible to show that there exists a set $N \in \mathcal{F}$ with $P(N) = 0$ where, for any $\omega \notin N$, $P_n^\theta(\omega, \cdot)$ are probability measures that converge *weakly*, and the limiting distribution is an ergodic invariant probability measure $Q(\omega, \cdot)$ on $\mathcal{F}$.

The latter suggests, in a manner to be precise by Theorem 3, that only samples that describe a physical phenomenon (those whose ergodic limit is a "reasonable" probability) should be considered as samples for any physical "reasonable probability space." It is also apparent, as shown by Theorem 3, that a stronger theory can be established if other samples are not considered.

We propose to partition the sample space according to their weak ergodic limit: assume a measurable space $(\Omega, \mathcal{F})$, as above, and assume a surjective measurable function $\theta : (\Omega, \mathcal{F}) \rightarrow (\Omega, \mathcal{F})$. We define an equivalent relation on Ω, by $\omega \cong \omega'$ iff for any bounded continuous function $X : \Omega \rightarrow \mathbb{R}$

$$\lim_n \left| \int_\Omega X \, dP_n^\theta (\omega, \cdot) - \int_\Omega X \, dP_n^\theta (\omega', \cdot) \right| = 0. \tag{5}$$

If $\omega \cong \omega'$ we say that ω has the *same stochastic type* of ω'. Clearly $\cong$ is an equivalence relation on elements on Ω. Moreover if $P_n^\theta(\omega, \cdot) \rightarrow P$ weakly and $\omega \cong \omega'$ then $P_n^\theta(\omega', \cdot) \rightarrow P$ weakly, and if P is a θ null invariant probability (see Definition 2) then it is a θ invariant probability (see Theorem 3).

We notice that the requirement of weak convergence of $P_n^\theta(\omega, \cdot)$ as $n \rightarrow \infty$ to a probability measure is quite natural. The latter is because for any probability space defined on a Polish Space with a measure preserving transformation for almost all ω the mentioned weak convergence holds (see Proposition 1). In the case where $P_n^\theta(\omega, \cdot) \rightarrow P$ we say that *stochastic type of ω is P.*

The main contribution of this paper is Theorem 3. Roughly speaking, Theorem 3 states that any reasonable space is divided into nonoverlapping subspaces, where on each subspace the statistics behavior of all the points is identical, and where these spaces have a simple statistical structure (they are uniquely ergodic). In fact Shields [6] [Section I.4.c] suggested in the case of finite alphabets ($\Omega = \{1, \ldots, n\}^{\mathbb{Z}}$ for some n) that, as a consequence of the Ergodic Decomposition Theorem, such decomposition is natural.

As a result of Theorem 3, for any physical "reasonable probability space" only the samples that describe a physical phenomenon (those whose ergodic limit is a "reasonable" probability) are sufficient as realizations of the sample space to study the physical phenomenon (the other realizations are unnecessary).

In the current stay of affairs, for any sample realization of a probability space and any statistical application, only a finite set of values $X_n(\omega)$ are observed, due to our technical limitations (the amount of time required to gather data is finite). Therefore, the assumption that *any* ω belongs to the sample space (for instance in the case that $\Omega = \mathbb{R}^{\mathbb{N}}$) is a technical assumption that it is motivated as long as it provides a rich and consistent theory. However we are showing in Theorem 3 that a richer theory is obtained if we restrict the sample space to those ω whose weak ergodic limit is a θ null invariant probability. The gain achieved in the modeling structure arises from the fact that the limiting probabilities describing the physical phenomenon are uniquely ergodic.

Therefore, our conclusions suggest further studies could point out to consider the uniquely ergodic prob...

as natural spaces for the modeling of any physical time-invariant phenomenon.

Also, as a consequence of Theorem 3, and results obtained by Gács et al. [8], a natural definition for a ω to be random is that $P_n^\theta(\omega,\cdot)$ converges weakly to a θ null invariant ergodic measure (see Definition 2). In fact, it follows that our definition of randomness extends to noncomputable probability spaces, and by Gács et al. [8] on computable probability spaces our notion of randomness is weaker than Schnorr randomness (and therefore weaker than Martin-Löf randomness).

In our approach, we do not consider any possible mixing endomorphism. Contrary to existing methods we fix a surjective transformation θ. Our approach works on any Polish Space, and it works for any probability measure (instead of computable probability spaces).

Issues that remain open include studying the consequences of modeling probability spaces as uniquely ergodic spaces. On the side of randomness, it remains open to examining definitions of randomness for discrete sequences that model non-ergodic Dependent Processes and a definition of randomness for functions defined on an interval (in the ergodic and non-ergodic cases). Finally, it remains open to provide specific sequences that are random (or whose limiting distribution is a particular distribution) and to provide examples of sequences that are random in the sense we give in this paper and are not random in the Schnorr sense (and therefore not random in the Martin-Löf sense).

2. Randomness

The first proposition says that on "reasonable spaces," for any reasonable physical realization, *always* the ergodic averages converge weakly to ergodic invariant measures.

Proposition 1. *Assume a measure preserving transformation θ defined on a probability space $(\Omega,\mathcal{F},P)$, where (Ω,d) is a Polish Space, and $\mathcal{F}=\mathcal{B}_d(\Omega)$ is the Borel σ-field. Let $\mathcal{F}^\theta$ be the σ-field of invariant sets (with respect to θ). Define $P_n^\theta(\omega,\cdot)$ by (2).*

There exist a set $N\in\mathcal{F}$ with $P(N)=0$ where for any $\omega\notin N$, $P_n^\theta(\omega,\cdot)$ are probability measures that converge weakly to the ergodic invariant probability measure $P_{\mathcal{F}|\mathcal{F}^\theta}(\omega,\cdot)$ where $P_{\mathcal{F}|\mathcal{F}^\theta}(\cdot,\cdot)$ is the regular conditional probability kernel with respect to $\mathcal{F}^\theta$ of the Ergodic Decomposition Theorem.

Proof. Assume that D is a countable dense subset of Ω. We notice that the class of sets $\mathcal{A}$ that are finite intersections of balls with a rational radius centered at points in D is a countable π-system with $\sigma(\mathcal{A})=\mathcal{F}$. Let $\mathcal{F}_0$ be the countable field generated by $\mathcal{A}$. By the Ergodic Decomposition Theorem there exists a set N, with $P(N)=0$ such that, for $F\in\mathcal{F}_0$,

$$\lim_n \frac{1}{n}\sum_{k=1}^{n} 1_F\circ\theta^{k-1}(\omega)=P_{\mathcal{F}|\mathcal{F}^\theta}(\omega,F)\quad \text{for } \omega\notin N, \tag{6}$$

where $P_{\mathcal{F}|\mathcal{F}^\theta}(\omega,\cdot)$ is the ergodic invariant probability measure corresponding to the regular conditional probability kernel with respect to $\mathcal{F}^\theta$ of the Ergodic Decomposition Theorem.

Assume an open set U, then there exist $A_i\in\mathcal{A}$, with $U=\bigcup_{i=1}^\infty A_i=\bigcup_{i=1}^\infty B_i$, where $B_i\in\mathcal{F}_0$ can be assumed disjoint sets with $\bigcup_{i=1}^m B_i=\bigcup_{i=1}^m A_i$ for each $m\geq 1$. It follows that for $\omega\notin N$

$$\begin{aligned}\liminf_n P_n(\omega,U)&=\liminf_n \frac{1}{n}\sum_{k=1}^{n}1_U\circ\theta^{k-1}(\omega)\\ &=\liminf_n\sum_{k=1}^{n}\sum_{i=1}^{\infty}\frac{1}{n}1_{B_i}\circ\theta^{k-1}(\omega)\\ &=\liminf_n\sum_{i=1}^{\infty}\sum_{k=1}^{n}\frac{1}{n}1_{B_i}\circ\theta^{k-1}(\omega)\\ &\geq\sum_{i=1}^{\infty}\lim_n\sum_{k=1}^{n}\frac{1}{n}1_{B_i}\circ\theta^{k-1}(\omega)\\ &=\sum_{i=1}^{\infty}P_{\mathcal{F}|\mathcal{F}^\theta}(\omega,B_i)=P_{\mathcal{F}|\mathcal{F}^\theta}(\omega,U),\end{aligned}\tag{7}$$

where the second equality follows by absolute convergence of $\sum_{k=1}^n\sum_{i=1}^\infty(1/n)1_{B_i}\circ\theta^{k-1}(\omega)$, and the third equality follows Fatou's Lemma. Finally the result follows by the Portmanteau Theorem (see Billingsley [11]).

The latter proposition suggests that, for any reasonable physical realization ($\omega\notin N$), the ergodic average $P_n(\omega,\cdot)$ converges weakly to an ergodic θ invariant probability measure. It turns out that a weaker concept for invariability is sufficient to describe the limiting ergodic behavior of any reasonable "physical realization."

Definition 2. Given a measurable space $(\Omega,\mathcal{F})$ and a surjective transformation $\theta:(\Omega,\mathcal{F})\to(\Omega,\mathcal{F})$, we say that a probability measure P defined on $\mathcal{F}$ is a θ *null invariant probability* if $P(F)=0$ implies that $P(\theta^{-1}(F))=0$.

Under conditions of Proposition 1 we obtain a necessary requirement for the limit probabilities of P_n^θ to be null invariant. In Theorem 3 we prove that the latter condition is sufficient demand for the limiting probabilities to converge weakly to an invariant probability measure.

In the next theorem, we prove that for any physical "reasonable probability space" only the realizations (or samples) that describe a physical phenomenon (those whose ergodic limit is a "reasonable." probability) are sufficient as realizations of the sample space to study the physical phenomenon (the other samples are unnecessary) and that those realizations (or samples) partition the whole space according to the physical phenomenon that they describe. Moreover, we show that when we restrict ourselves to those samples that describe the same physical phenomenon (those whose limiting ergodic behavior is the same), the mathematical structure of the space defined is richer (in the sense that the probability spaces are uniquely invariant). The above conclusions make a strong point to consider the unique invariant probability spaces as natural spaces for the modeling of a physical time-invariant phenomenon.

Theorem 3. *Assume a measurable surjective function $\theta : (\Omega, \mathscr{F}) \to (\Omega, \mathscr{F})$ defined on a measurable space $(\Omega, \mathscr{F})$, where (Ω, d) is a Polish Space and $\mathscr{F} = \mathscr{B}_d(\Omega)$ is the Borel σ-field, where d is a metric for which Ω is a complete separable space. Let $\omega \in \Omega$ and assume that $P_n^\theta(\omega, \cdot) \to P$ weakly, where P is a θ null invariant probability. Then P is a θ invariant probability measure defined on $\mathscr{F}$. Moreover if*

$$\Omega_P = \left\{\omega' \mid P_n^\theta\left(\omega', \cdot\right) \longrightarrow P\right\} \tag{8}$$

then Ω_P is a measurable invariant set with $\theta(\Omega_P) = \Omega_P$ where either $P(\Omega_P) = 0$ or $P(\Omega_P) = 1$ and P is ergodic iff $P(\Omega_P) = 1$. Moreover, there exist a countable field $\mathscr{F}_0$ where any open can be written as a countable union of elements of $\mathscr{F}_0$, such that, for any $A \in \mathscr{F}_0$, and $\omega' \in \Omega_P$, $P_n^\theta(\omega', A) \to P(A)$ (and in particular $\sigma(\mathscr{F}_0) = \mathscr{F}$).

Moreover, if P is an ergodic probability measure there is only one ergodic θ invariant measure on $(\Omega_P, \mathscr{F} \cap \Omega_P = \{F \cap \Omega_P,\ F \in \mathscr{F}\})$ (namely, P).

Proof. Let D be a countable dense subset of Ω. For each $d \in D$, define $\mathscr{A}(d)$ to be a countable set of balls with center at d, with radius $r(d,n) \downarrow 0$ when $n \to \infty$ such that $P(\partial B_{r(d,n)}(d)) = 0$, for $n = 1, 2, \ldots$, where $x \in D$, $B_r(x) = \{y \mid d(y,x) < r\}$, and ∂F denotes the boundary of given set F. The set $\mathscr{A}(d)$ with the desired properties exists by a standard argument. Let $\mathscr{F}_0$ be the field generated by $\bigcup_{d\in D} \mathscr{A}(d)$. Clearly any open set can be written as a countable union of elements in $\mathscr{F}_0$. By the Portmanteau Theorem $P_n^\theta(\omega, A) \to P(A)$ for any $A \in \mathscr{F}_0$.

For any set $A \in \mathscr{F}_0$, it is clear that $\lim_n P_n^\theta(\omega, A) = \lim_n P_n^\theta(\omega, \theta^{-1}(A))$, and since the class of sets where $P(F) = P(\theta^{-1}(F))$ is a λ-system that contains $\mathscr{A}$ since P is a null invariant probability, it follows by the Dynkin's π-λ theorem that $P(F) = P(\theta^{-1}(F))$ for any $F \in \mathscr{F}$, proving that P is θ-invariant.

Clearly $\lim_n P_n^\theta(\omega, A) = \lim_n P_n^\theta(\theta(\omega), A)$, for any $A \in \mathscr{F}_0$, and therefore using the Dinkyn π-λ theorem the limiting measure associated with ω and $\theta(\omega)$ is the same, proving that $\theta(\Omega_P) \subset \Omega_P$. We also notice that, for $\omega \in \Omega_P$, if $\omega' \in \Omega$ is an element such that $\theta(\omega') = \omega$, then $\lim_n P_n^\theta(\omega', A) = \lim_n P_n^\theta(\omega, A)$, for any $A \in \mathscr{F}_0$, and then by the Dynkin's π-λ theorem $\omega' \cong \omega$ proving that $\theta(\Omega_P) = \Omega_P$. Next we prove that Ω_P is a measurable set. For the latter, let $\psi : \mathbb{N} \to \mathscr{F}_0$ be an enumeration of $\mathscr{F}_0$. Define $\underline{\varphi}, \overline{\varphi} : (\Omega, \mathscr{F}) \to (\mathbb{R}^{\mathbb{N}}, \mathscr{B}(\mathbb{R}^{\mathbb{N}}))$ by

$$\begin{aligned} \overline{\varphi}(\omega)(m) &= \limsup_n P_n^\theta(\omega, \psi(m)), \\ \underline{\varphi}(\omega)(m) &= \liminf_n P_n^\theta(\omega, \psi(m)). \end{aligned} \tag{9}$$

Then by the Portmanteau Theorem $\Omega_P = \{\omega \mid \overline{\varphi}(\omega) = \underline{\varphi}(\omega) = \omega_P\}$, where $\omega_P \in \mathbb{R}^{\mathbb{N}}$ is defined by $\omega_P(m) = \lim_n P_n^\theta(\omega, \psi(m))$, and therefore Ω_P is measurable.

By the Ergodic Decomposition Theorem, there exist a kernel $P_{\mathscr{F}|\mathscr{F}^\theta} : \Omega \times \mathscr{F} \to [0,1]$ and set N, with $P(N) = 0$ with the property that for all $\omega \notin N$ $\lim_n P_n^\theta(\omega, A) \to P_{\mathscr{F}|\mathscr{F}^\theta}(\omega, A)$ for all $A \in \mathscr{F}_0$, $P_{\mathscr{F}|\mathscr{F}^\theta}(\omega, \cdot)$ is an ergodic measure, and

$$P(G \cap F) = \int_G P_{\mathscr{F}|\mathscr{F}^\theta}(\omega, F)\, dP_{\mathscr{F}^\theta} \quad F \in \mathscr{F},\ G \in \mathscr{F}^\theta, \tag{10}$$

where $P_{\mathscr{F}^\theta}$ is the restriction to $\mathscr{F}^\theta$ of P and $\mathscr{F}^\theta$ is the σ-field of invariant sets (under θ). It follows that for $A \in \mathscr{F}_0$

$$\begin{aligned} P(\Omega_P \cap A) &= \int_{\Omega_P} P_{\mathscr{F}|\mathscr{F}^\theta}(\omega, A)\, dP_{\mathscr{F}^\theta} \\ &= \int_{\Omega_P} P(A)\, dP_{\mathscr{F}^\theta} = P(A)\, P(\Omega_P). \end{aligned} \tag{11}$$

Since the class of sets $F \in \mathscr{F}$ where $P(\Omega_P \cap F) = P(F)P(\Omega_P)$ is a λ system, by the Dynkin's π-λ theorem it follows that for all $F \in \mathscr{F}$, $P(\Omega_P \cap F) = P(F)P(\Omega_P)$. In particular $P(\Omega_P) = P^2(\Omega_P)$, proving that $P(\Omega_P) = 0$ or $P(\Omega_P) = 1$.

Assume that $P(\Omega_P) = 1$, then by the Ergodic Decomposition Theorem, there exist a set N where $P(N) = 0$ and $\omega \notin N$, $P_n^\theta(\omega, F) \to P_{\mathscr{F}|\mathscr{F}^\theta}(\omega, F) = P(F)$, for $F \in \mathscr{F}_0$ where $P_{\mathscr{F}|\mathscr{F}^\theta}(\omega, \cdot)$ is ergodic. It follows that P is ergodic.

Finally assume that $P(\Omega_P) = 1$, and assume an ergodic probability measure Q on Ω_P. It is clear that Q can be extended to $(\Omega, \mathscr{F})$, in a natural way as $Q(F) = Q(F \cap \Omega_P)$. We also denote the extension as Q. Moreover the extension to $\mathscr{F}$ is also ergodic. By the Ergodic Decomposition Theorem, it follows that

$$\begin{aligned} Q(A) = Q(\Omega_P \cap A) &= \int_{\Omega_P} P(A)\, dQ_{\mathscr{F}^\theta} \\ &= P(A)\, Q(\Omega_P) \quad \text{for } A \in \mathscr{A}. \end{aligned} \tag{12}$$

It follows that, on $\mathscr{F}_0$, $Q = P$ and by the Dinkyn's π-λ Theorem $Q = P$ on $\mathscr{F}$, and therefore $Q = P$ on $\mathscr{F} \cap \Omega_P$. □

Probably the most interesting consequence of the previous theorem is the uniform convergence of ergodic averages for all sample points. If Ω_P is a compact set then, for any continuous bounded function X defined on Ω_P (for the induced topology), *for all* $\omega \in \Omega_P$

$$\frac{1}{n}\sum_{k=1}^{n} X \circ \theta^{k-1}(\omega) \longrightarrow \int_{\Omega_P} X\, dP \tag{13}$$

uniformly on Ω_P. The latter is the content of the Oxtoby Theorem [12] [Theorem 6.19].

The consequence of Theorem 3 that, for any $\omega \in \Omega_P$, P is ergodic iff $P(\Omega_P) = 1$ has been observed in the case of finite alphabets by Shields [6] [Section I.9.b]. In fact Shields [6] [Sections I.4.c and I.9.b] suggests a similar approach in the case of finite alphabets, although uniquely ergodicity is not obtained.

Finally we observe some elementary facts of our definition.

Proposition 4. *Assume a measurable space $(\Omega, \mathscr{F})$ and a surjective transformation $\theta : (\Omega, \mathscr{F}) \to (\Omega, \mathscr{F})$; then:*

(i) *If θ is periodic on ω in the sense that there exist n with $\theta^n(\omega) = \omega$, with period k (namely, $k = \min\{n \mid \theta^n(\omega) = \omega\}$, then $P_n^{\theta}(\omega, \cdot) \to P$), where P is a probability measure supported on a finite set.*

(ii) *If there exist n, m such that $\theta^n(\omega) = \theta^m(\omega')$, then $\omega \cong \omega'$. In particular it is impossible to tell from a finite set of values of ω any information about the distribution of ω.*

(iii) *For any n, $\omega \cong \theta^n(\omega)$.*

Conflicts of Interest

The author declares that there are no conflicts of interest regarding the publication of this paper.

References

[1] A. N. Kolmogorov, "Three approaches to the quantitative definition of information," *International Journal of Computer Mathematics*, vol. 2, no. 1-4, pp. 157–168, 1968.

[2] P. Martin-Löf, "The definition of random sequences," *Information and Control*, vol. 9, pp. 602–619, 1966.

[3] A. K. Zvonkin and L. A. Levin, "The complexity of finite objects and the development of the concepts of information and randomness by means of the theory of algorithms," *Russian Mathematical Surveys*, vol. 25, no. 6, pp. 83–124, 1970.

[4] C.-P. Schnorr, *Zufälligkeit und Wahrscheinlichkeit. Eine algorithmische Begründung der Wahrscheinlichkeitstheorie*, vol. 218 of *Lecture Notes in Mathematics*, Springer-Verlag, Cham, Switzerland, 1971.

[5] R. M. Gray, *Probability, Random Processes, and Ergodic Properties*, Springer, Dordrecht, Netherlands, 2nd edition, 2009.

[6] P. C. Shields, *The Ergodic Theory of Discrete Sample Paths*, vol. 13 of *Graduate Studies in Mathematics*, American Mathematical Society, Providence, RI, USA, 1996.

[7] V. V. V'yugin, "Effective convergence in probability and an ergodic theorem for individual random sequences," *Theory of Probability & its Applications*, vol. 420, no. 1, Article ID S0040585, pp. 39–50, 1998.

[8] P. Gács, M. Hoyrup, and C. Rojas, "Randomness on computable probability spaces–a dynamical point of view," *Theory of Computing Systems*, vol. 48, no. 3, pp. 465–485, 2011.

[9] S. Galatolo, M. Hoyrup, and C. Rojas, "Effective symbolic dynamics, random points, statistical behavior, complexity and entropy," *Information and Computation*, vol. 208, no. 1, pp. 23–41, 2010.

[10] P. Billingsley, *Probability and Measure*, John Wiley & Sons, Inc., Hoboken, NJ, USA, 3rd edition, 1995.

[11] P. Billingsley, *Convergence of Probability Measures*, John Wiley & Sons, New York, NY, USA, 2nd edition, 1999.

[12] P. Walters, *An Introduction to Ergodic Theory*, vol. 79, Springer, New York, NY, USA, 1st edition, 2000.

19

Scan Statistics for Detecting High-Variance Clusters

Lionel Cucala

Institut Montpelliérain Alexander Grothendieck, 34000 Montpellier, France

Correspondence should be addressed to Lionel Cucala; lionel.cucala@univ-montp2.fr

Academic Editor: Dejian Lai

Scan statistics are mostly used to detect spatial areas or time intervals in which the mean level of a given variable is more important. Sometimes, when this variable is continuous, there is an interest in looking for clusters in which its variability is more important. In this paper, two scan statistics are introduced for identifying clusters of values exhibiting higher variance. Like many classical scan statistics, the first one relies on a generalized likelihood ratio test whereas the second one is based on ratios of empirical variances. These methods are useful to identify spatial areas or time intervals in which the variability of a given variable is more important. In an application of the new methods, I look for geographical clusters of high-variability income in France and then for residuals exhibiting higher variance in a linear regression context.

1. Introduction

Cluster detection has become a very fruitful research subject since the earlier work of Naus [1] looking for an unusual clustering of random points on the line. Thereafter, it was extended to identify time intervals or spatial areas where the observations of a given random variable are different than elsewhere. These methods are nowadays very popular in disease surveillance for the detection of disease clusters, and they are also used in many other fields, such as forestry, astronomy, and criminology. A thorough review is given by Glaz et al. [2].

Most of the cluster detection methods are designed for count data, that is, point processes made of the random coordinates of n events observed in S, a bounded subset of $\mathbb{R}^d$: the goal is to identify, if they exist, the areas in which the concentration of events is abnormally high. According to the article by Cressie [3], the scan statistic denotes the maximal concentration observed on a collection of potential clusters. Originally, the size of all the potential clusters had to be the same, so that the scan statistic was just the maximum number of events in a window of size d, d being fixed a priori. This major drawback vanished when Kulldorff [4] introduced the scan statistic based on generalized likelihood ratio in either Poisson model or Bernoulli model. This scan statistic is defined to analyse point processes with binary marks, such as case/control data. Later on, Kulldorff et al. [5] introduced the Gaussian model scan statistic which allows analysing point processes with continuous marks.

This scan method is useful to detect high-mean clusters, that is, areas where the marks are significantly higher than elsewhere. However, when analysing income inequalities, for example, it might be useful to look for high-variance clusters: in these areas, inequalities are more obvious and this could generate more violence or more crimes. Thus, existing scan methods need to be modified in order to detect such clusters.

In this paper I introduce two scan statistics for detecting high-variance clusters. Section 2 describes the scan statistics and their computational aspects in the framework of marked point processes. Their performances are compared through a simulation study in Section 3. In Section 4, they are applied to two real data sets: on the first one, which describes the spatial distribution of incomes in France, the high-variance scan statistics are directly applied; on the second one, which illustrates the link between expenditure on public schools and per capita income, a linear regression model is fitted and its residuals are analysed through high-variance scan statistics. The paper is concluded with a discussion.

2. Two High-Variance Scan Statistics

Let $\{(x_i, s_i),\ i = 1, \ldots, n\}$ denote the realization of a marked point process, where $s_i \in S$ is the location of the event and

$x_i \in \mathbb{R}$ its associated mark. The area $S \subset \mathbb{R}^d$ is the observation domain and the locations can be either one-dimensional ($d = 1$) or spatial ($d = 2$ or $d = 3$). Our goal is to detect the area $Z \subset S$ in which the marks exhibit a significantly higher variance than elsewhere.

From now on, let us consider that the locations are spatial. The setting of one-dimensional locations will be handled in Section 4.2. Most of the spatial cluster detection methods rely on likelihood ratio between two hypotheses depending on a potential cluster: the scan statistic is nothing but the maximum of all these likelihood ratios. Thus the two questions to answer are how to choose the potential clusters and which likelihood ratio should be used?

Concerning the potential clusters, I will focus on circular clusters, such as Kulldorff [4]. The set of potential clusters, denoted by $\mathcal{D}$, is the set of discs (if $d = 2$) or balls (if $d = 3$) centered on a location and passing through another one:

$$\mathcal{D} = \left\{D_{i,j},\ 1 \le i \le n,\ 1 \le j \le n\right\}, \tag{1}$$

where $D_{i,j}$ is the disc (or the ball) centered on s_i and passing through s_j: the number of potential clusters is n^2.

2.1. A Generalized Likelihood Ratio Scan Statistic. As said in the Introduction, Kulldorff et al. [5] introduced a Gaussian-based scan statistic to detect clusters exhibiting higher means than elsewhere. Their concentration index is based on the likelihood ratio between two hypotheses. In order to detect high-variance clusters instead of high-mean clusters, I decided to modify their alternative hypothesis.

Let $X_1, \ldots, X_n$ denote the random variables associated with the marks, which are assumed to be independent. Under the null hypothesis H_0, corresponding to the absence of any cluster, all the marks come from the same Gaussian distribution:

$$X_1, \ldots, X_n \sim \mathcal{N}\left(\mu^*, \sigma^{*2}\right), \tag{2}$$

where (μ^*, σ^{*2}) is the maximum-likelihood (ML) estimate associated with this distribution; that is,

$$\left(\mu^*, \sigma^{*2}\right) = \underset{(\mu,\sigma^2)}{\arg\max} \prod_{i=1}^{n} f_{(\mu,\sigma^2)}\left(x_i\right), \tag{3}$$

$f_{(\mu,\sigma^2)}(\cdot)$ being the density function associated with the Gaussian distribution with mean μ and variance σ^2.

Let the spatial area $Z \subset S$ be a potential cluster and $\overline{Z} = S \setminus Z$ its complement. Under the alternative hypothesis $H_{1,Z}$, corresponding to a high-variance cluster in Z, the marks come from a mixture Gaussian distribution with equal means but different variances:

$$\begin{aligned}\left(X_i : s_i \in Z\right) &\sim \mathcal{N}\left(\mu^*_{Z,\overline{Z}}, \sigma^{*2}_{Z}\right),\\ \left(X_i : s_i \in \overline{Z}\right) &\sim \mathcal{N}\left(\mu^*_{Z,\overline{Z}}, \sigma^{*2}_{\overline{Z}}\right),\end{aligned} \tag{4}$$

where $(\mu^*_{Z,\overline{Z}}, \sigma^{*2}_{Z}, \sigma^{*2}_{\overline{Z}})$ is the ML estimate associated with this mixture distribution; that is,

$$\begin{aligned}&\left(\mu^*_{Z,\overline{Z}}, \sigma^{*2}_{Z}, \sigma^{*2}_{\overline{Z}}\right)\\ &\quad = \underset{(\mu_{Z,\overline{Z}},\sigma^2_Z,\sigma^2_{\overline{Z}})}{\arg\max} \prod_{i:s_i\in Z} f_{(\mu_{Z,\overline{Z}},\sigma^2_Z)}\left(x_i\right) \prod_{i:s_i\in \overline{Z}} f_{(\mu_{Z,\overline{Z}},\sigma^2_{\overline{Z}})}\left(x_i\right).\end{aligned} \tag{5}$$

Both of these ML estimates have closed forms. Let

$$\begin{aligned}n_Z &= \sum_{i=1}^{n} \mathbb{1}\quad\left(s_i \in Z\right),\\ \overline{X}_Z &= \frac{1}{n_Z}\sum_{i=1}^{n} X_i \mathbb{1}\quad\left(s_i \in Z\right),\\ \overline{X^2}_Z &= \frac{1}{n_Z}\sum_{i=1}^{n} X_i^2 \mathbb{1}\quad\left(s_i \in Z\right)\end{aligned} \tag{6}$$

denote, respectively, the number and the mean of marks and the mean of squares of marks in Z. The ML estimate under H_0 is

$$\left(\mu^*, \sigma^{*2}\right) = \left(\overline{X}_S, \overline{X^2}_S - \left(\overline{X}_S\right)^2\right), \tag{7}$$

the vector containing the mean and the (biased) empirical variance of the marks in the whole domain. As shown in the Appendix, the ML estimate under $H_{1,Z}$ satisfies

$$\begin{aligned}\sigma^{*2}_{Z} &= \overline{X^2}_Z - 2\mu^*_{Z,\overline{Z}}\overline{X}_Z + \left(\mu^*_{Z,\overline{Z}}\right)^2,\\ \sigma^{*2}_{\overline{Z}} &= \overline{X^2}_{\overline{Z}} - 2\mu^*_{Z,\overline{Z}}\overline{X}_{\overline{Z}} + \left(\mu^*_{Z,\overline{Z}}\right)^2\end{aligned} \tag{8}$$

and $\mu^*_{Z,\overline{Z}}$ is the root of a cubic function which can be obtained using, for example, Cardano's method [6]. The log-likelihood under the null hypothesis is

$$\begin{aligned}\mathrm{LL}_0 &= -\frac{n}{2}\log(2\pi) - \frac{n}{2}\log\left(\sigma^{*2}\right)\\ &\quad - \frac{1}{2\sigma^{*2}}\sum_{i=1}^{n}\left(X_i - \mu^*\right)^2\\ &= -\frac{n}{2}\log(2\pi) - \frac{n}{2}\log\left(\sigma^{*2}\right) - \frac{n}{2}\end{aligned} \tag{9}$$

and the one under the alternative hypothesis is

$$\begin{aligned}\mathrm{LL}_{1,Z} &= -\frac{n_Z}{2}\log(2\pi) - \frac{n_Z}{2}\log\left(\sigma^{*2}_Z\right)\\ &\quad - \frac{1}{2\sigma^{*2}_Z}\sum_{s_i\in Z}\left(X_i - \mu^*_{Z,\overline{Z}}\right)^2 - \frac{n_{\overline{Z}}}{2}\log(2\pi)\\ &\quad - \frac{n_{\overline{Z}}}{2}\log\left(\sigma^{*2}_{\overline{Z}}\right) - \frac{1}{2\sigma^{*2}_{\overline{Z}}}\sum_{s_i\in \overline{Z}}\left(X_i - \mu^*_{Z,\overline{Z}}\right)^2\\ &= -\frac{n}{2}\log(2\pi) - \frac{n_Z}{2}\log\left(\sigma^{*2}_Z\right) - \frac{n_{\overline{Z}}}{2}\log\left(\sigma^{*2}_{\overline{Z}}\right)\\ &\quad - \frac{n_Z}{2} - \frac{n_{\overline{Z}}}{2},\end{aligned} \tag{10}$$

so that the log-likelihood ratio between both hypotheses reduces to

$$\begin{aligned}\mathrm{LLR}_Z &= \mathrm{LL}_{1,Z} - \mathrm{LL}_0 \\ &= \frac{1}{2}\left[n\log\left(\sigma^{*2}\right) - n_Z\log\left(\sigma_Z^{*2}\right) - n_{\overline{Z}}\log\left(\sigma_{\overline{Z}}^{*2}\right)\right].\end{aligned} \tag{11}$$

The next step is only the maximization of this log-likelihood ratio on the set of potential clusters previously defined. The generalized likelihood ratio (GLR) high-variance scan statistic is

$$\lambda = \max_{Z\in\mathscr{D}} \mathrm{LLR}_Z \tag{12}$$

and the potential cluster for which this maximum is obtained,

$$\widehat{C} = \arg\max_{Z\in\mathscr{D}} \mathrm{LLR}_Z, \tag{13}$$

is called the most likely cluster.

Once the scan statistic is computed, I need to evaluate its significance. Unfortunately, the null distribution of λ is untractable due to the dependence between the log-likelihood ratios: indeed, the log-likelihood ratios LLR_Z and $\mathrm{LLR}_{Z'}$ associated with two potential clusters Z and Z' are independent only if $n_{Z\cap Z'} = 0$. Another solution, chosen by Kulldorff [4], would be to simulate random data sets under the null hypothesis. However, by doing the randomization this way, the correct alpha level will not be maintained if the marks do not truly come from a normal distribution. Thus I decided to run a technique used by Kulldorff et al. [5] called random labelling: a simulated data set is obtained by randomly associating the marks with the spatial locations. Let T denote the number of simulated data sets and let $\lambda^{(1)},\ldots,\lambda^{(T)}$ be the observations of the scan statistic associated with these data sets: these simulated scan statistics must be compared to λ, the scan statistic observed on the real data set. According to Dwass [7], the Monte Carlo based p value of the scan statistic λ is $R/(T+1)$, where R is the rank of λ in the $(T+1)$-sample $(\lambda^{(1)},\ldots,\lambda^{(T)},\lambda)$. Note that this p value is unbiased in the sense that under the null hypothesis, the probability of observing a p value less than or equal to p is exactly p. According to the classical test theory, the most likely cluster $\widehat{C}$ is said to be significant if the associated p value is less than type I error α.

2.2. A Generalized Variance-Ratio Scan Statistic. Even if the likelihood-based scan statistics are broadly used, Cucala [8] proposed, in a different context, an alternative scan statistic showing better results than the one based on likelihood ratio: the power was slightly higher, such as the ability to detect small clusters. Here I propose a method to build a high-variance scan statistic not relying on likelihood ratio.

A classical test for equality of variances between the marks in Z and in $\overline{Z}$, usually known as F-test, relies on the ratio of (unbiased) empirical variances

$$R_Z = \frac{S_Z^{*2}}{S_{\overline{Z}}^{*2}}, \tag{14}$$

where

$$S_Z^{*2} = \frac{1}{n_Z - 1}\sum_{i=1}^{n}\left(X_i - \overline{X}_Z\right)^2 \mathbb{1}\left(s_i \in Z\right). \tag{15}$$

Note that, since the X_i's are assumed to be independent, the empirical variances S_Z^{*2} and $S_{\overline{Z}}^{*2}$ are also independent. Moreover, under Gaussian assumption, the equality of variances of marks in Z and $\overline{Z}$ ensures that R_Z follows, by definition, a Fisher distribution $F(n_Z-1, n_{\overline{Z}}-1)$, as mentioned by Saporta [9]. Let me introduce the high-variance index for potential cluster Z

$$I_Z = F_{n_Z-1,n_{\overline{Z}}-1}\left(R_Z\right), \tag{16}$$

where $F_{n_Z-1,n_{\overline{Z}}-1}(\cdot)$ denotes the cumulative distribution function associated with the Fisher distribution $F(n_Z-1, n_{\overline{Z}}-1)$. If the variances are equal, the distribution of I_Z is uniform $(0,1)$ and does not depend anymore on n_Z: this method is called the probability integral transform [10]. Moreover, a value of I_Z close to 1 shows that the variance in Z is significantly larger than in $\overline{Z}$. Therefore, the generalized variance-ratio (GVR) scan statistic is

$$\widetilde{\lambda} = \max_{Z\in\mathscr{D}} I_Z. \tag{17}$$

The way I evaluate the significance of $\widetilde{\lambda}$ is completely similar to the method for λ.

3. A Simulation Study

I decided to run a simulation study in order to compare the results of the tests based on the scan statistics previously introduced, λ and $\widetilde{\lambda}$. I generate artificial data sets according to three different models. Whatever the model, the geographical locations are the locations of the 94 French departments. Note that the location associated with each department is the location of its capital city. The true cluster, called C, is a set of eight departments around Paris called Ile-de-France: in this area, the variance of the marks is larger than elsewhere. In the first model, denoted by homogeneous Gaussian model, the associated marks are independent and follow a Gaussian distribution with equal means and different variances:

$$X_i \sim \begin{cases}\mathscr{N}(0,r) & \text{if } s_i \in C,\\ \mathscr{N}(0,1) & \text{otherwise.}\end{cases} \tag{18}$$

In the second model, denoted by Exponential model, the associated marks are independent and are derived from exponentially distributed random variables with different rate parameters, $X_i = Y_i - \mathbb{E}(Y_i)$, where

$$Y_i \sim \begin{cases}\mathscr{E}\left(r^{-1/2}\right) & \text{if } s_i \in C,\\ \mathscr{E}(1) & \text{otherwise,}\end{cases} \tag{19}$$

so that the means are all equal to 0. In the third model, denoted by heterogeneous Gaussian model, the associated

Table 1: Homogeneous Gaussian model.

Cluster intensity r		Results of the following λ	$\tilde{\lambda}$
1.5	Power	0.080	**0.118**
	TP	0.567	**0.741**
	FP	3.702	**3.590**
2.0	Power	0.265	**0.322**
	TP	1.723	**2.009**
	FP	3.998	**3.400**
2.5	Power	0.528	**0.553**
	TP	3.492	**3.500**
	FP	2.844	**2.118**

Table 2: Exponential model.

Cluster intensity r		Results of the following λ	$\tilde{\lambda}$
1.5	Power	0.056	**0.057**
	TP	0.191	**0.352**
	FP	**1.441**	2.384
2.0	Power	0.070	**0.071**
	TP	0.349	**0.472**
	FP	**1.806**	2.195
2.5	Power	0.074	**0.093**
	TP	0.332	**0.659**
	FP	**1.538**	2.987

Table 3: Heterogeneous Gaussian model.

Cluster intensity r	Mean lag l		Results of the following λ	$\tilde{\lambda}$
1.0	1.0	Power	**0.082**	0.075
		TP	**0.612**	0.504
		FP	5.499	**3.480**
1.5	1.0	Power	**0.168**	0.135
		TP	**1.196**	0.960
		FP	4.548	**4.236**
2.0	1.0	Power	**0.390**	0.345
		TP	**2.597**	2.317
		FP	4.822	**4.712**
2.5	1.0	Power	**0.603**	0.571
		TP	**3.970**	3.800
		FP	**2.618**	3.354

marks are independent and follow a Gaussian distribution with different means and different variances:

$$X_i \sim \begin{cases} \mathcal{N}(l, r) & \text{if } s_i \in C, \\ \mathcal{N}(0, 1) & \text{otherwise.} \end{cases} \quad (20)$$

Note that, for every model, the parameter r is the ratio between the variance inside the cluster and the variance outside the cluster: I should call it the cluster intensity. For the third model, the parameter l, which denotes the difference of means inside the cluster and outside of the cluster, is called the mean lag.

For each model and each value of the cluster intensity r, I generated 1000 simulated data sets. The high-variance scan statistics λ and $\tilde{\lambda}$ have been computed, using the set of circular potential clusters described in Section 2, and their p values are estimated based on $T = 999$ permutations. In both methods, the most likely cluster $\widehat{C}$ is said to be significant if the associated p value is less than $\alpha = 5\%$.

When applied to such data sets exhibiting a true cluster, scan methods are useful when they manage simultaneously to identify that there is a significant cluster and to recover as precisely as possible the true cluster. Therefore, in order to compare the different scan methods, I computed three different criterions. The first one is the power of the method, that is, the proportion of data sets exhibiting a significant cluster. The second one is the mean number of true positive (TP) departments, that is, the departments included both in the most likely cluster $\widehat{C}$ and in the true cluster C. The third one is the mean number of false positive (FP) departments, that is, the departments included in the most likely cluster $\widehat{C}$ but not in the true cluster C. Note that the sum of TP and FP is the mean number of departments in the most likely cluster $\widehat{C}$. Table 1 gives the results obtained with the homogeneous Gaussian model, Table 2 the results obtained with the Exponential model, and Table 3 the results obtained with the heterogeneous Gaussian model.

The bold values are the best results obtained in each procedure. As expected, the power of the methods increases with the cluster intensity. Under the homogeneous Gaussian model, the GVR method is always more powerful than the GLR one, whatever the cluster intensity is. These results are very similar to the ones obtained by Cucala [8] stating that, for detecting clusters in count data, GLR methods are not always the most powerful ones. However, although these power results are quite satisfactory for the homogeneous Gaussian model, they are quite poor when the marks are exponentially distributed: even when the variance is 2.5 times larger in the cluster, both methods hardly detect it. The results in Table 3 also reveal that, even if those scan statistics are designed for detecting high-variance clusters, both are sensitive to a difference of means. Indeed, the cluster intensity being equal, the powers of both methods increases when there is a mean lag. Finally, looking at the FP results, I also conclude that the GLR method tends to exhibit larger significant clusters than the GVR one, which leads to a larger number of departments exhibited by mistake.

4. Applications

4.1. An Application to Economic Data. For many years now, the spatial analysis of income inequality is an interesting scheme for many economists; for example, Shelnutt and Yao [11] investigated the income inequalities in the different counties of Arkansas and the interaction with economic growth, whereas Atems [12] analysed the Gini index of the incomes in the 3109 American counties. Among others, Deutsch et al.

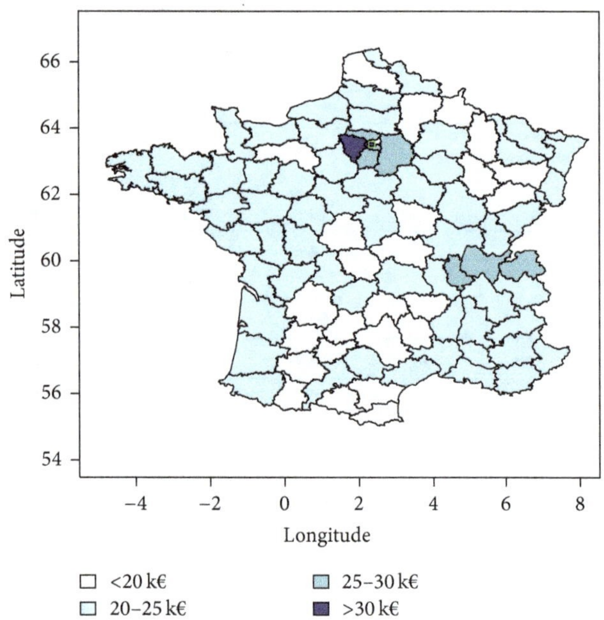

FIGURE 1: Median income.

[13] compared the spatial distribution of crime and economic inequalities in the US and found a correlation. These studies confirm that knowing precisely the geographic areas in which inequalities are more significant is of great interest. However, this search is usually done in a nonmathematical way. In this subsection, I show how the high-variance scan methods introduced later on can be useful for such a purpose. I apply the scan methods to an economic data set provided by a French institute, l'Institut National de la Statistique et des Etudes Economiques [14]. For each of the 94 departments in France, the median income for year 2009 has been computed and Figure 1 illustrates the results.

This figure clearly exhibits a cluster of wealthy departments in the North of France, next to Paris (whose location is given by the light square), whose significance has been assessed by Cucala [15] using the classical Gaussian-based scan statistic. However, we may wonder whether there is any high-variance cluster, enlightening a region where inequalities between departments are significant. To answer this question, I applied both high-variance scan statistics to this data set, once again using the set of circular potential clusters described in Section 2. Their associated p values are also estimated based on $T = 9999$ permuted samples. The most likely clusters are given by Figure 2. Note that, for convenience, the clusters I highlighted on this figure are not the circular areas exhibited by the scan methods, but the set of departments whose capital cities are included in those circular areas.

The most likely cluster exhibited by GLR scan statistic λ is exactly the area around Paris called Ile-de-France. This is not surprising since it includes 7 wealthy departments but also the department called Seine-Saint-Denis which is one of the French departments in which the incomes are much lower. This cluster is quite significant since its p value equals 0.0001: none of the 9999 likelihood-based scan statistics computed from the permuted samples was larger than the likelihood-based scan statistic computed from the observed sample. The cluster exhibited by GVR scan statistic $\widetilde{\lambda}$ is a bit less significant: its p value equals 0.0075. It contains 7 departments in Ile-de-France but also 6 neighboring departments in which the incomes are much lower.

4.2. An Application to Residuals Analysis. Even if the high-variance scan statistics have been introduced in a spatial setting, they can also be adapted to data with one-dimensional locations, like high-mean scan statistics. This includes data indexed by time but also any kind of data indexed along an axis. In order to illustrate this, I apply the high-variance scan methods to a set of residuals in a regression framework.

The data set I use results from an investigation of public school spending in the United States: the sample consists of 51 observations of per capita expenditure on public schools and per capita income for each state and the District of Columbia in 1979 [16]. As recommended by Greene [17], a linear regression model is fitted to explain the expenditure: the explanatory variables are the income and the squared income. A classical way to check the homoscedasticity assumption, that is, that the errors variance is constant across all the observations, is the analysis of the plot of the studentized residuals versus one of the explanatory variables, or alternatively the fitted values. In Figure 3, the studentized residuals are plotted versus the income. Note that, since the other explanatory variable of the model is the squared income, the plot of the residuals versus this explanatory variable or versus the fitted values would have been quite similar since the ordering of the residuals would have been exactly the same.

The variance of the residuals seems to be constant, except for larger values of the income. A few methods have been introduced in order to check that the residuals have equal variance: the Breusch-Pagan [18] test and the White test [19] are powerful to do so but they do not mention which of the residuals exhibit higher variance. The use of high-variance scan methods is of great interest to obtain such information.

Let $x_1, \ldots, x_n$ denote the studentized residuals, ordered such that x_i corresponds to the ith observation of the income in ascending order. For any one-dimensional cluster detection method, the potential clusters are all the sets of observations $\{x_i, \ldots, x_j\}$, $1 \le i \le j \le n$. Here, since the variance of the marks needs to be estimated inside and outside of the potential cluster, intervals containing one observation $(j = i + 1)$ or $n - 1$ observations $(j = i + n - 2)$ are excluded. I computed both high-variance scan statistics on these data. The most likely cluster given by λ corresponds to the four last residuals, from solid line in Figure 3. The most likely cluster given by $\widetilde{\lambda}$ only contains the two last residuals, from dotted line in Figure 3. This matches the conclusion of the simulation study: on this specific data set, the most likely cluster exhibited by the likelihood-based scan statistic is smaller than the one exhibited by the variance-ratio scan statistic. Based on $T = 9999$ permuted samples, their p values are, respectively, 0.0535 and 0.0426: this clearly indicates that the regression model is not valid when the income overcomes a certain level. Note that the Breusch-Pagan test and the White test both gave the same conclusion.

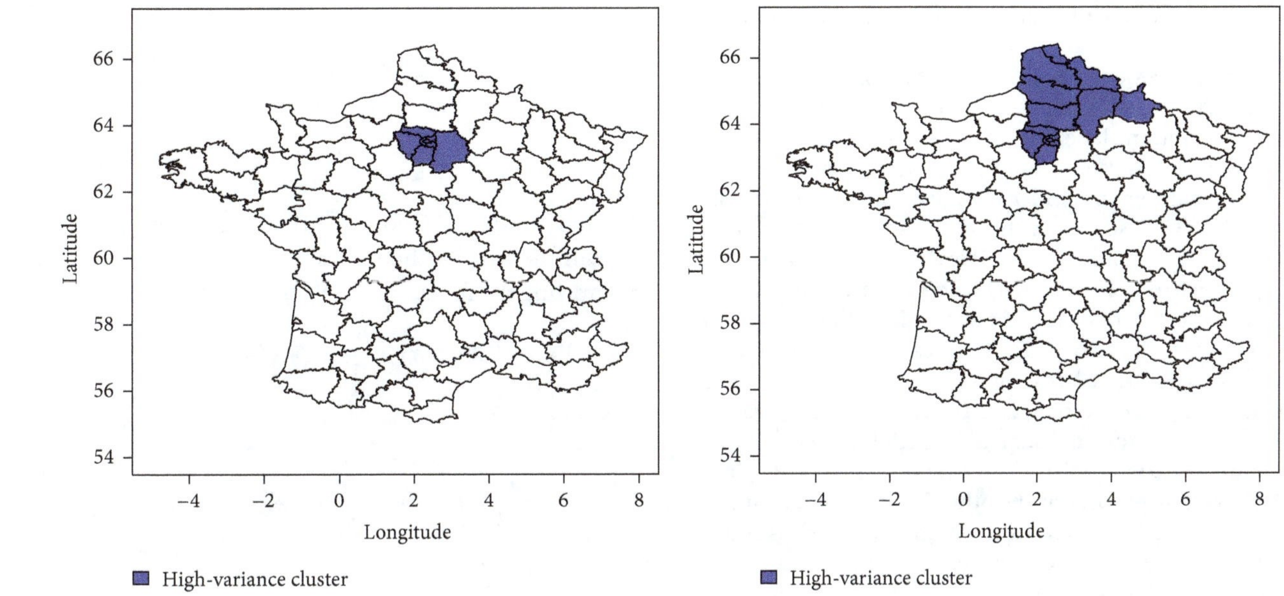

(a) (b)

FIGURE 2: (a) Most likely cluster given by λ. (b) Most likely cluster given by $\tilde{\lambda}$.

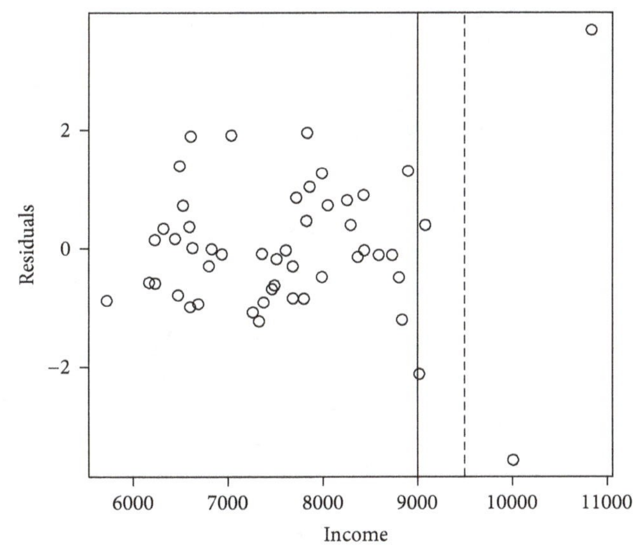

FIGURE 3: Residuals versus income.

5. Discussion

The scan statistics introduced in this paper allow one to detect high-variance clusters in marked point processes without any parameter to set up. According to the simulation study, the GVR scan statistic is slightly more powerful than the GLR one against any homogeneous clustering alternative: this result confirms that GLR scan statistics are not always the most efficient, even when the distribution of the marks is known. It also seems that, such as in the regression residuals analysis, the GLR scan statistic hardly detects clusters containing very few observations. Note that the GLR scan methods compare likelihood ratios which are, under the null hypothesis, not exactly equally distributed but only asymptotically when n tends to infinity, as stated by Wilks [20]. On the other hand, the GVR scan method compares the I_Z's which are equally distributed under the null hypothesis. Therefore, I should recommend using the GVR scan statistic instead of the GLR one.

In Section 4, I gave two examples of real data analysis through high-variance scan statistics: the search for inequalities hotspots on economic data and a homoscedasticity checking procedure on regression residuals. In the first example, the scan statistics clearly indicate a significant high-variance cluster, which is not very surprising, but also indicate precisely where social programs should be focused on in order to reduce the inequalities. Note also that a finer analysis could be done using the median income in each city instead of each department. In the second example, a new possibility is given by the high-variance scan methods to analyse regression residuals: they lead to the rejection of homoscedasticity hypothesis and focus on the individuals for which the linear regression model seems to be nonvalid. Let me mention that, among others, another application possibility would be the analysis of meteorological time series such as annual rainfall records in a specific station. Indeed, a high-variance scan test could determine whether the amplitude of rainfall increased in the last decades, because of global warming, in certain regions, leading to a higher frequency of very dry and very wet years.

The GVR scan statistic I introduce is based on the most famous test for equality of variances, the F-test. However, many other tests are available in the same purpose, as mentioned by Gartside [21]. For example, a scan statistic derived from the squared rank test introduced by Conover [22] could be set up, similar to the Mann-Whitney scan statistic for high-mean clusters defined by Cucala [23]. It might be more suitable to marks whose distribution is not Gaussian.

The randomization procedure used to estimate the significance of both high-variance scan statistics is the most basic one, that is, random labelling. I focused on this procedure because it is the only one for which type I error remains equal to α whatever the underlying distribution of the data is. Since the GVR scan statistic itself is also distribution-free, this choice sounds natural. However, if the labels are spatially autocorrelated, this may lead to overestimating the significance of the detected clusters. Note that the same problem arises with other GLR based scan statistics when the significance is estimated through Monte Carlo simulation. As mentioned by Haining [24], restricted randomization procedures taking into account this spatial autocorrelation are usually applied to global clustering tests such as Ripley's K and derived methods. On the other hand, for local cluster detection tests such as scan statistics, this approach is much less frequent, except in a few articles including the ones by Loh and Zhu [25] and Zhang et al. [26]. Since the methods they introduced are designed for high-mean cluster detection, it could be interesting to adapt them for high-variance cluster detection: this might be the subject of a future work.

In the last few years, there have been more and more papers dealing with spatiotemporal cluster detection, such as Viel et al. [27]. It should be noted that the high-variance scan statistics can also be applied to spatiotemporal data using the spatiotemporal distance introduced by Demattei and Cucala [28].

In this work I only focused on the most likely cluster but the detection of secondary clusters is straightforward using the method proposed by Zhang et al. [29]: once a significant cluster is found, remove the data included in that cluster and restart the analysis.

Finally, I should emphasize that the high-variance scan tests could be adjusted for any continuous covariate adjustment as proposed by Klassen et al. [30], such as the age of an underlying population. This could be done by modeling a regression function of the marks depending on the adjusted covariates and then analysing the corresponding residuals.

Appendix

Computing the ML Estimate under the Alternative Hypothesis

The ML estimate under $H_{1,Z}$ is the value for $(\mu_{Z,\overline{Z}}, \sigma_Z^2, \sigma_{\overline{Z}}^2)$ maximizing the log-likelihood

$$\begin{aligned} & -\frac{n_Z}{2}\log\left(2\pi\sigma_Z^2\right) - \frac{1}{2\sigma_Z^2}\sum_{i:s_i\in Z}\left(x_i - \mu_{Z,\overline{Z}}\right)^2 \\ & -\frac{n_{\overline{Z}}}{2}\log\left(2\pi\sigma_{\overline{Z}}^2\right) - \frac{1}{2\sigma_{\overline{Z}}^2}\sum_{i:s_i\in \overline{Z}}\left(x_i - \mu_{Z,\overline{Z}}\right)^2. \end{aligned} \tag{A.1}$$

When $\mu_{Z,\overline{Z}}$ is fixed, the value for σ_Z^2 maximizing

$$-\frac{n_Z}{2}\log\left(2\pi\sigma_Z^2\right) - \frac{1}{2\sigma_Z^2}\sum_{i:s_i\in Z}\left(x_i - \mu_{Z,\overline{Z}}\right)^2 \tag{A.2}$$

is

$$\begin{aligned} \sigma_Z^{*2} &= \frac{1}{n_Z}\sum_{i:s_i\in Z}\left(x_i - \mu_{Z,\overline{Z}}\right)^2 \\ &= \overline{X^2}_Z - 2\mu_{Z,\overline{Z}}\overline{X}_Z + \left(\mu_{Z,\overline{Z}}\right)^2. \end{aligned} \tag{A.3}$$

The same result holds for $\sigma_{\overline{Z}}^{*2}$. Thus, inserting these expressions in the log-likelihood function, $\mu_{Z,\overline{Z}}^*$ is the value for μ maximizing the function

$$\begin{aligned} g(\mu) &= -\frac{n_Z}{2}\log(2\pi) - \frac{n_Z}{2}\log\left(\overline{X^2}_Z - 2\mu\overline{X}_Z + \mu^2\right) \\ &\quad - \frac{n_Z}{2} \\ &= -\frac{n_{\overline{Z}}}{2}\log(2\pi) - \frac{n_{\overline{Z}}}{2}\log\left(\overline{X^2}_{\overline{Z}} - 2\mu\overline{X}_{\overline{Z}} + \mu^2\right) \\ &\quad - \frac{n_{\overline{Z}}}{2}. \end{aligned} \tag{A.4}$$

The derivative of this function is

$$\begin{aligned} & g'(\mu) \\ &= -\frac{n_Z}{2}\frac{-2\overline{X}_Z + 2\mu^2}{\overline{X^2}_Z - 2\mu\overline{X}_Z + \mu^2} \\ &\quad - \frac{n_{\overline{Z}}}{2}\frac{-2\overline{X}_{\overline{Z}} + 2\mu^2}{\overline{X^2}_{\overline{Z}} - 2\mu\overline{X}_{\overline{Z}} + \mu^2} \\ &= \frac{n}{2}\frac{h(\mu)}{\left(\overline{X^2}_Z - 2\mu\overline{X}_Z + \mu^2\right)\left(\overline{X^2}_{\overline{Z}} - 2\mu\overline{X}_{\overline{Z}} + \mu^2\right)}, \end{aligned} \tag{A.5}$$

where

$$\begin{aligned} h(\mu) &= \mu^3 + \frac{-n_Z\overline{X}_Z - 2n_Z\overline{X}_{\overline{Z}} - n_{\overline{Z}}\overline{X}_{\overline{Z}} - 2n_{\overline{Z}}\overline{X}_Z}{n}\mu^2 \\ &\quad + \frac{n_Z\overline{X^2}_{\overline{Z}} + 2n\overline{X}_Z\overline{X}_{\overline{Z}} + n_{\overline{Z}}\overline{X^2}_Z}{n}\mu \\ &\quad + \frac{-n_Z\overline{X}_Z\overline{X^2}_{\overline{Z}} - n_{\overline{Z}}\overline{X}_{\overline{Z}}\overline{X^2}_Z}{n}. \end{aligned} \tag{A.6}$$

Following Cardano's method [6], the three roots of this cubic function $h(\cdot)$ can be computed. Therefore, $\mu_{Z,\overline{Z}}^*$ is either the only real root (if the two others are complex) or the one of the three roots in which the function $g(\cdot)$ is maximized (if the three roots are real).

Conflict of Interests

The author declares that there is no conflict of interests regarding the publication of this paper.

References

[1] J. Naus, *Clustering of random points in the line and plane [Ph.D. thesis]*, Rutgers University, New Brunswick, NJ, USA, 1963.

[2] J. Glaz, J. Naus, and S. Wallenstein, *Scan Statistics*, Springer, New York, NY, USA, 2001.

[3] N. Cressie, "On some properties of the scan statistic on the circle and the line," *Journal of Applied Probability*, vol. 14, no. 2, pp. 272–283, 1977.

[4] M. Kulldorff, "A spatial scan statistic," *Communications in Statistics. Theory and Methods*, vol. 26, no. 6, pp. 1481–1496, 1997.

[5] M. Kulldorff, L. Huang, and K. Konty, "A scan statistic for continuous data based on the normal probability model," *International Journal of Health Geographics*, vol. 8, article 58, 2009.

[6] N. Jacobson, *Basic Algebra*, Dover, 2009.

[7] M. Dwass, "Modified randomization tests for nonparametric hypotheses," *Annals of Mathematical Statistics*, vol. 28, pp. 181–187, 1957.

[8] L. Cucala, "A hypothesis-free multiple scan statistic with variable window," *Biometrical Journal*, vol. 50, no. 2, pp. 299–310, 2008.

[9] G. Saporta, *Probabilités, Analyse des Données et Statistique*, Technip, Paris, France, 2011.

[10] Y. Dodge, *The Oxford Dictionary of Statistical Terms*, Oxford University Press, 2003.

[11] J. Shelnutt and V. Yao, "A spatial analysis of income inequality in Arkansas at the county level: evidence from tax and commuting data," *Regional Economic Development*, vol. 1, pp. 52–65, 2005.

[12] B. Atems, "A note on the differential regional effects of income inequality: empirical evidence using U.S. county-level data," *Journal of Regional Science*, vol. 53, no. 4, pp. 656–671, 2013.

[13] J. Deutsch, U. Spiegel, and J. Templeman, "Crime and income inequality: an economic approach," *Atlantic Economic Journal*, vol. 20, no. 4, pp. 46–54, 1992.

[14] Institut National de la Statistique et des Études Économiques, http://www.insee.fr/.

[15] L. Cucala, "A distribution-free spatial scan statistic for marked point processes," *Spatial Statistics*, vol. 10, pp. 117–125, 2014.

[16] US Department of Commerce, *Statistical Abstract of the United States*, US Government Printing Office, Washington, DC, USA, 1979.

[17] W. Greene, *Econometric Analysis*, Macmillan Publishing Company, New York, NY, USA, 2nd edition, 1993.

[18] T. S. Breusch and A. R. Pagan, "A simple test for heteroscedasticity and random coefficient variation," *Econometrica*, vol. 47, no. 5, pp. 1287–1294, 1979.

[19] H. White, "A heteroskedasticity-consistent covariance matrix estimator and a direct test for heteroskedasticity," *Econometrica*, vol. 48, no. 4, pp. 817–838, 1980.

[20] S. S. Wilks, "The large-sample distribution of the likelihood ratio for testing composite hypotheses," *The Annals of Mathematical Statistics*, vol. 9, no. 1, pp. 60–62, 1938.

[21] P. S. Gartside, "A study of methods for comparing several variances," *Journal of the American Statistical Association*, vol. 67, no. 338, pp. 342–346, 1972.

[22] W. Conover, *Practical Nonparametric Statistics*, Wiley, New York, NY, USA, 1980.

[23] L. Cucala, "A Mann-Whitney scan statistic for continuous data," *Communications in Statistics. Theory and Methods*, In press.

[24] R. Haining, *Spatial Data Analysis: Theory and Practice*, Cambridge University Press, 2003.

[25] J. M. Loh and Z. Zhu, "Accounting for spatial correlation in the scan statistic," *The Annals of Applied Statistics*, vol. 1, no. 2, pp. 560–584, 2007.

[26] T. Zhang, Z. Zhang, and G. Lin, "Spatial scan statistics with overdispersion," *Statistics in Medicine*, vol. 31, no. 8, pp. 762–774, 2012.

[27] J.-F. Viel, N. Floret, and F. Mauny, "Spatial and space-time scan statistics to detect low rate clusters of sex ratio," *Environmental and Ecological Statistics*, vol. 12, no. 3, pp. 289–299, 2005.

[28] C. Demattei and L. Cucala, "Multiple spatio-temporal cluster detection for case event data: an ordering-based approach," *Communications in Statistics. Theory and Methods*, vol. 40, no. 2, pp. 358–372, 2011.

[29] Z. Zhang, M. Kulldorff, and R. Assunção, "Spatial scan statistics adjusted for multiple clusters," *Journal of Probability and Statistics*, vol. 2010, Article ID 642379, 11 pages, 2010.

[30] A. C. Klassen, M. Kulldorff, and F. Curriero, "Geographical clustering of prostate cancer grade and stage at diagnosis, before and after adjustment for risk factors," *International Journal of Health Geographics*, vol. 4, article 1, 2005.

20

Gram-Charlier Processes and Applications to Option Pricing

Jean-Pierre Chateau[1] and Daniel Dufresne[2]

[1] *Faculty of Business Administration, University of Macau, Macau*
[2] *Montreal, QC, Canada*

Correspondence should be addressed to Daniel Dufresne; dufresne@unimelb.edu.au

Academic Editor: Aera Thavaneswaran

A Gram-Charlier distribution has a density that is a polynomial times a normal density. For option pricing this retains the tractability of the normal distribution while allowing nonzero skewness and excess kurtosis. Properties of the Gram-Charlier distributions are derived, leading to the definition of a process with independent Gram-Charlier increments, as well as formulas for option prices and their sensitivities. A procedure for simulating Gram-Charlier distributions and processes is given. Numerical illustrations show the effect of skewness and kurtosis on option prices.

1. Introduction

Gram-Charlier series are expansions of the form

$$f(x) = \phi(a, b; x) \cdot \left[1 + c_1 He_1\left(\frac{x-a}{b}\right) + c_2 He_2\left(\frac{x-a}{b}\right) + \cdots\right], \tag{1}$$

where

$$\phi(a, b; x) = \frac{1}{b\sqrt{2\pi}} e^{-(1/2b^2)(x-a)^2}, \quad x \in \mathbf{R}, \tag{2}$$

is the usual normal density and He_k is the Hermite polynomial of order k. The expression within square brackets in (1) is an orthogonal polynomial expansion for the ratio $f(x)/\phi(x)$; given an arbitrary function the expansion may or may not converge to the true value of $f(x)$. In this paper we focus on the properties of the Gram-Charlier distributions, obtained by truncating the series after a finite number of terms. What is obtained is a family of distributions parametrized by $a, b, c_0, \ldots, c_N$, as is explained in detail below.

This paper has three main goals: (1) define and study the properties of the family of Gram-Charlier distributions; (2) define a Gram-Charlier process and derive its basic properties; (3) apply those to European options. The formulas we give for European option prices and Greeks apply to Gram-Charlier distributions of any order, and we use four- and six-parameter Gram-Charlier distributions in our examples. Two numerical illustrations show how option prices are affected by the skewness and kurtosis of returns. This paper can a reference for those using Gram-Charlier distributions in option pricing but also in statistics.

Most previous applications to option pricing have assumed that $c_1 = c_2 = 0$. We believe this restriction is not necessary in a general theory of Gram-Charlier distributions, as there may well be situations where the extra degrees of freedom given by c_1 and c_2 will be useful. We discuss it in detail at the end of Section 2.

It has been observed that option prices have nonconstant implied volatilities, meaning that log returns do not have a normal distribution under the risk-neutral measure. There is a wide literature on modelling log returns to fit observed option prices, the main alternatives to Brownian motion being stochastic volatility models (where the parameter σ in Black-Scholes is replaced with a continuous-time stochastic process), GARCH time series, and Lévy processes. Gram-Charlier distributions are mathematically simpler than the models just mentioned, while allowing a better fit to data than the normal distribution. Several authors have used Gram-Charlier distributions in option pricing, as a model for log returns, among others [1–10]. The majority of previous authors assumed a density of the form

$$\phi(x)\left(1+\frac{\bar{s}}{6}He_3(x)+\frac{\bar{k}}{24}He_4(x)\right),$$
$$\phi(x)=\phi(0,1,x)=\frac{1}{\sqrt{2\pi}}e^{-x^2/2}, \tag{3}$$

for the normalized log return. (In our notation this is a $\mathbf{GC}(0,1;0,0,\bar{s}/6,\bar{k}/24)$ distribution; see Section 2.) The notation emphasizes that in this case the coefficient of $He_3(x)$ turns out to be Pearson's skewness coefficient $\bar{s}$ divided by 6, and the coefficient of $He_4(x)$ the excess kurtosis coefficient $\bar{k}$ divided by 24. The distribution of log returns is then a four-parameter Gram-Charlier distribution (since there are two other parameters, the mean and variance, besides $\bar{s}$ and $\bar{k}$). This distribution allows nonzero skewness and excess kurtosis, unlike the normal distribution found in Black-Scholes. In (3) the parameters $(\bar{s},\bar{k})$ are restricted to a specific region (see Figure 3), because outside that region the function in (3) becomes negative for some values of x (see Section 2.1).

Jurczenko et al. [8] specify the martingale restriction

$$S_0=e^{-rT}\mathbf{E}^{\mathbf{Q}}(S_T) \tag{4}$$

that the four-parameter Gram-Charlier density must satisfy in pricing options (previous authors had not taken it into account). The martingale condition for general Gram-Charlier distributions is described in Section 5. Our Gram-Charlier distribution with parameters $a,b,c_1,\ldots,c_N$, denoted by $\mathbf{GC}(a,b;c_1,\ldots,c_N)$, has density

$$\phi(a,b,x)\cdot\left[1+c_1He_1\left(\frac{x-a}{b}\right)+\cdots+c_NHe_N\left(\frac{x-a}{b}\right)\right]. \tag{5}$$

As stated above we do not set $c_1=c_2=0$, and N can be any even positive integer. The question whether (5) is nonnegative for all x is an important one. Several authors have disregarded this issue and, in fact, some have come up with parameters that do not yield a true probability density function. One might argue that this is the price to pay for truncating an infinite Gram-Charlier series. However, if the same log return distribution is used to price many options, then a true probability density function is the only safe choice, because otherwise there might be inconsistencies among option prices. For instance, if the function used as density is negative over the interval (α,β), then a digital option that pays off only when the log return is in that interval will have a negative price. In this paper we consider Gram-Charlier *distributions*, not *expansions*, and insist that the densities integrate to one and be nonnegative. Our goal is to define a family of proper probability distributions; nevertheless, our formulas do apply to truncated Gram-Charlier expansions as well.

The paper by León et al. [11] presents an alternative to the general Gram-Charlier distributions we study in this paper. Those authors consider the subclass of Gram-Charlier distributions consisting of densities

$$f_X(x)=\phi(a,b,x)\,p(x), \tag{6}$$

where the polynomial $p(x)$ is the square of another polynomial, $p(x)=q(x)^2$. This has the obvious advantage that the nonnegativity restriction on $f_X(\cdot)$ is automatically satisfied. We discuss that subclass of "squared" Gram-Charlier distributions in the Conclusion.

Almost all previous authors have used Gram-Charlier distributed log returns over a single time period. This has an obvious downside, in that it becomes tricky, if not impossible, to preserve consistency between the prices of options with different maturities. Section 3 shows that a Lévy process with Gram-Charlier increments does not exist; however, it also shows that the sum of independent Gram-Charlier distributed variables also has a Gram-Charlier distribution. This opens the way for multiperiod Gram-Charlier option pricing, using a discrete-time random walk model for which the log return over any period has a Gram-Charlier distribution. The Gram-Charlier distribution of the multiperiod return has a larger number of parameters, though the model is still simpler than almost any (if not all) alternative stochastic volatility models. There is no problem computationally, since we give explicit formulas for options under Gram-Charlier distributions with an arbitrary number of parameters.

The layout of the paper is as follows. In Section 2, we extend the study of Gram-Charlier distributions to all possible polynomials $p(\cdot)$ and derive their properties (moments, cumulants, moment determinacy, properties of the set of valid parameters, tail, and so on). Some of the formulas and properties in Theorem 2 appear to be new. In Section 3 we show that there is no Lévy process with Gram-Charlier distributed increments, apart from Brownian motion, and we define a discrete-time process with independent Gram-Charlier increments that is suitable for option pricing. In Section 4 we show that the log Gram-Charlier distribution is not determined by its moments, just like the lognormal. Next, Section 5 gives formulas for European call and put prices when the log-price returns of the underlying have a general Gram-Charlier distribution; the previous literature mostly considered the $\mathbf{GC}(a,b;0,0,c_3,c_4)$ family and the squared Gram-Charlier distributions (an exception is [10], where it is assumed that $c_1=c_2=0$). In particular, we derive a change of measure formula that extends the Cameron-Martin formula for the normal distribution; the latter is used in pricing European options in the Black-Scholes model. We also derive formulas for the sensitivities (Greeks) of those option prices with respect to all parameters. A technique for simulating Gram-Charlier distributions is described. Parts (c), (d), (i), (j) (k), and (l) of Theorem 2 and part (b) of Theorem 3 and Theorems 6, 7, 9, and 11 appear to be new. Theorem 3(a) is for the first time formulated for general Gram-Charlier distributions. Theorems 3(a), 4, and 8 are given, more or less explicitly, for the subclass of squared

Gram-Charlier distributions in [11], and some of the Greeks in Theorem 9 had been calculated for the $\mathbf{GC}(a, b; 0, 0, c_3, c_4)$ distribution by previous authors.

In Section 6 we give two applications that show how option prices depend on skewness and kurtosis of the log returns (this of course cannot be done in the classical Black-Scholes setting, while it appears quite complicated to do so in stochastic volatility or Lévy driven models). The first example is equity indexed annuities (EIAs in the sequel) premium options. The pricing of EIAs has been studied by many authors, including Hardy [12], Gaillardetz and Lin [13], and Boyle and Tian [14]. The second example is lookback options, which illustrates the use of the simulation method in Section 2.5. Lookback options have been studied by numerous authors, but there is no closed form formula for the price of discretely monitored lookback options within the Black-Scholes model. We refer the reader to Kou [15]. In those examples all parameters are estimated by maximum likelihood, with the range of the parameters restricted to where they correspond to a true probability distribution. Moment estimation is of course possible when $c_1 = c_2 = 0$, but it is not trustworthy, because of the range restriction on the parameters $c_3, c_4, \ldots$; for instance, the empirical skewness and excess kurtosis have a positive probability of falling outside the feasible region (see Section 6). Theorem 2(f) shows that the first k moments of the Gram-Charlier distribution depend on more than k parameters, which probably rules out moment estimation, unless one a priori fixes two of the parameters (which again is not trustworthy). We apply maximum likelihood to the six-parameter $\mathbf{GC}(a, b; c_1, c_2, c_3, c_4)$ in Section 6, but we do realize that maximum likelihood estimation for higher order Gram-Charlier distributions poses computational problems, which are an interesting avenue for further research. (A referee pointed out that fixing a and b would make estimation easier; our guess is that if one wishes to simplify maximum likelihood estimation then fixing a and b might be a better idea than fixing c_1 and c_2, though we have not looked at this in any depth. A two-step process might be imagined, whereby the data first give information about a and b, and then maximum likelihood is applied using that preliminary information. This would help by constraining the optimization to a smaller region, while agreeing with the intuitive idea that a and b are location and scale parameters, resp.)

Notation 1. The normal density function is denoted as

$$\phi(a, b, x) = \frac{1}{b\sqrt{2\pi}} e^{-(1/2b^2)(x-a)^2},$$
$$\phi(0, 1, x) = \phi(x) = \frac{e^{-x^2/2}}{\sqrt{2\pi}}, \tag{7}$$

and its distribution function is

$$\Phi(x) = \int_{-\infty}^{x} \phi(y)\, dy. \tag{8}$$

Two equivalent versions of the Hermite polynomials may be found in the literature: for $n = 0, 1, 2 \ldots$,

$$H_n(x) = (-1)^n e^{x^2} \frac{d^n}{dx^n} e^{-x^2}$$
$$He_n(x) = (-1)^n e^{x^2/2} \frac{d^n}{dx^n} e^{-x^2/2}. \tag{9}$$

The first one is common in mathematics and physics, but in probability and statistics there is an obvious advantage in using the second one. (The conversion formula is $He_n(x) = 2^{-n/2} H_n(x/\sqrt{2})$.) The first few Hermite polynomials are

$$\begin{aligned} He_0(x) &= 1, \\ He_1(x) &= x, \\ He_2(x) &= x^2 - 1, \\ He_3(x) &= x^3 - 3x, \\ He_4(x) &= x^4 - 6x^2 + 3, \\ He_5(x) &= x^5 - 10x^3 + 15x, \\ He_6(x) &= x^6 - 15x^4 + 45x^2 - 15. \end{aligned} \tag{10}$$

2. Gram-Charlier Distributions

For a fixed N, consider the class of distributions that have a pdf of the form

$$f(x) = \phi(x) \sum_{k=0}^{N} c_k He_k(x), \quad x \in \mathbf{R}, \tag{11}$$

with $c_N \neq 0$. Noting that the leading term of $He_k(x)$ is x^k, we conclude that N must necessarily be even, because if N were odd then the polynomial that multiplies $\phi(x)$ would take negative values for some x. For the same reason c_N cannot be negative.

Definition 1. Let $a \in \mathbf{R}$, $b > 0$, $c_k \in \mathbf{R}$, $c_0 = 1$, and $N \in \{0, 2, 4, \ldots\}$. We write $Y \sim \mathbf{GC}(a, b; c_1, \ldots, c_N)$ (or $Y \sim \mathbf{GC}(a, b; \overline{c})$) if the variable $(Y - a)/b$ has probability density function

$$\phi(x) \sum_{k=0}^{N} c_k He_k(x). \tag{12}$$

This will be called a *Gram-Charlier distribution* with parameters $a, b, \overline{c}$, with $\overline{c} = (c_1, \ldots, c_N)$. The largest N such that $c_N > 0$ is called the *order* of the Gram-Charlier distribution. The normal distribution with mean a and variance b^2 is a $\mathbf{GC}(a, b; (0, \ldots 0))$ (or $\mathbf{GC}(a, b; -)$) with order 0.

The class of Gram-Charlier distributions just defined includes all distributions with density

$$\frac{1}{b}\phi\left(\frac{y-a}{b}\right) p(y), \tag{13}$$

where $p(y)$ is a polynomial of degree N, since $p(y)$ can be rewritten as a combination of

$$He_k\left(\frac{y-a}{b}\right), \quad k = 0, 1, \ldots, N. \tag{14}$$

The condition $c_0 = 1$ ensures that function (12) integrates to one, since

$$\int_{-\infty}^{\infty} \phi(x) He_k(x)\, dx = (-1)^k \int_{-\infty}^{\infty} \frac{d^k}{dx^k}\phi(x)\, dx = 0, \quad k = 1, 2, \ldots. \tag{15}$$

There are no simple conditions that ensure that a polynomial remains nonnegative everywhere, though in some cases precise conditions on $\bar{c}$ are known; see below. If a vector $\bar{c}$ leads to a true Gram-Charlier pdf, then we will say that $\bar{c}$ is *valid*.

Generating functions are convenient when dealing with orthogonal polynomials. One is

$$w_1(t, x) = \sum_{n=0}^{\infty} \frac{t^n}{n!} He_n(x) = e^{tx-(t^2/2)}. \tag{16}$$

Another one is

$$\begin{aligned} w_2(t, u, x, y) &= \sum_{k=0}^{\infty}\sum_{n=0}^{\infty} \frac{t^k}{k!}\frac{u^n}{n!} \int_{-\infty}^{\infty} e^{-x^2/2} He_k(x) He_n(x+y)\, dx \\ &= \sqrt{2\pi} e^{u(t+y)}. \end{aligned} \tag{17}$$

Letting $y = 0$ leads to

$$\begin{aligned} \frac{1}{\sqrt{2\pi}} \int_{-\infty}^{\infty} e^{-x^2/2} He_k(x) He_n(x)\, dx &= \left.\frac{\partial^k}{t^k}\frac{\partial^n}{u^n} e^{tu}\right|_{t=u=0} \\ &= \begin{cases} 0 & \text{if } k \neq n \\ n! & \text{if } k = n. \end{cases} \end{aligned} \tag{18}$$

This proves the orthogonality of the Hermite polynomials and gives us the value of

$$\frac{1}{\sqrt{2\pi}} \int_{-\infty}^{\infty} e^{-x^2/2} He_n^2(x)\, dx, \tag{19}$$

which is essential in deriving Gram-Charlier series. Since $He_0 = 1$ this also implies

$$\int_{-\infty}^{\infty} \phi(x) He_k(x)\, dx = 0, \quad k = 1, 2, \ldots. \tag{20}$$

Another formula is the Laplace transform of $\phi(x)He_n(x)$, which may be found by integrating by parts n times:

$$\int_{-\infty}^{\infty} e^{tx}\phi(x) He_k(x)\, dx = t^k e^{t^2/2}, \quad k = 0, 1, \ldots. \tag{21}$$

Let us calculate the moments of a distribution with density (11). First consider

$$\begin{aligned} &\int_{-\infty}^{\infty} x^n \phi(x) He_k(x)\, dx \\ &\quad = (-1)^k \int_{-\infty}^{\infty} x^n \left(\frac{d^k}{dx^k}\phi(x)\right) dx. \end{aligned} \tag{22}$$

Integrating by parts repeatedly yields

$$\begin{aligned} &\int_{-\infty}^{\infty} x^n \phi(x) He_k(x)\, dx \\ &= \begin{cases} \dfrac{n!}{2^{(n-k)/2}((n-k)/2)!} & \text{if } n-k \text{ is even and nonnegative} \\ 0 & \text{if } n-k \text{ is odd and nonnegative.} \end{cases} \end{aligned} \tag{23}$$

Hence, the nth moment of the distribution in (11) is

$$\sum_{k=0}^{N\wedge n} c_k \frac{n!}{2^{(n-k)/2}((n-k)/2)!} \mathbf{1}_{\{n-k \text{ even}\}}. \tag{24}$$

This says in particular that the parameter c_k only affects the moments of order k and higher of the $\mathbf{GC}(a, b; \bar{c})$ distribution (see part (i) of Theorem 2). This is confirmed by the moment generating function, which may be found from (21):

$$\int_{-\infty}^{\infty} e^{tx} f(x)\, dx = e^{t^2/2} \sum_{k=0}^{N} c_k t^k. \tag{25}$$

It can be checked that differentiating this expression n times and setting $t = 0$ give the same expression found for the nth moment of (11).

Theorem 2. *Suppose $Y \sim \mathbf{GC}(a, b; \bar{c})$, $\bar{c} \in \mathbf{R}^N$ with $b > 0$, $c_0 = 1$, and $c_N > 0$. The order N of the distribution is necessarily even.*

(a)

$$\mathbf{E}(Y-a)^n = b^n \sum_{k=0}^{N\wedge n} c_k \frac{n!}{2^{(n-k)/2}((n-k)/2)!} \mathbf{1}_{\{n-k \text{ even}\}}. \tag{26}$$

(b)

$$\mathbf{E}e^{tY} = e^{at+(b^2t^2/2)} \sum_{k=0}^{N} b^k c_k t^k, \quad t \in \mathbf{R}. \tag{27}$$

(c) The representation of the $\mathbf{GC}(a, b; \bar{c})$ distribution in terms of the parameters a, b, and $\bar{c}$ is unique.

(d) All Gram-Charlier distributions are determined by their moments.

(e) The set of valid $\bar{c}$ in $\mathbf{R}^N$ includes the origin, is not reduced to a single point, and is convex.

(f) The first six moments of the $\mathbf{GC}(a,b;\tilde{c})$ *distribution are*

$$\begin{aligned}
m_1 &= a + bc_1,\\
m_2 &= a^2 + b^2 + 2\left(abc_1 + b^2c_2\right),\\
m_3 &= a^3 + 3ab^2 + 3bc_1\left(a^2 + b^2\right) + 6\left(ab^2c_2 + b^3c_3\right),\\
m_4 &= a^4 + 4a^3bc_1 + 12a^2b^2c_2 + 6a^2b^2 + 12ab^3c_1\\
&\quad + 24ab^3c_3 + 12b^4c_2 + 24b^4c_4 + 3b^4,\\
m_5 &= a^5 + 5a^4bc_1 + 20a^3b^2c_2 + 10a^3b^2 + 30a^2b^3c_1\\
&\quad + 60a^2b^3c_3 + 60ab^4c_2 + 120ab^4c_4 + 15ab^4\\
&\quad + 15b^5c_1 + 60b^5c_3 + 120b^5c_5,\\
m_6 &= a^6 + 6a^5bc_1 + 30a^4b^2c_2 + 15a^4b^2 + 60a^3b^3c_1\\
&\quad + 120a^3b^3c_3 + 180a^2b^4c_2 + 360a^2b^4c_4\\
&\quad + 45a^2b^4 + 90ab^5c_1 + 360ab^5c_3 + 720ab^5c_5\\
&\quad + 90b^6c_2 + 360b^6c_4 + 720b^6c_6 + 15b^6.
\end{aligned} \tag{28}$$

(g) The first six cumulants of the $\mathbf{GC}(a,b;\tilde{c})$ *distribution are*

$$\begin{aligned}
\kappa_1 &= a + bc_1,\\
\kappa_2 &= b^2\left(1 - c_1^2 + 2c_2\right),\\
\kappa_3 &= 2b^3\left(c_1^3 - 3c_1c_2 + 3c_3\right),\\
\kappa_4 &= -6b^4\left(c_1^4 - 4c_1^2c_2 + 2c_2^2 + 4c_1c_3 - 4c_4\right),\\
\kappa_5 &= 24b^5\left(c_1^5 - 5c_1^3c_2 + 5c_1c_2^2 + 5c_1^2c_3 - 5c_2c_3 - 5c_1c_4\right.\\
&\quad \left. + 5c_5\right),\\
\kappa_6 &= -120b^6\left(c_1^6 - 6c_1^4c_2 + 9c_1^2c_2^2 - 2c_2^3 + 6c_1^3c_3\right.\\
&\quad \left. - 12c_1c_2c_3 + 3c_3^2 - 6c_1^2c_4 + 6c_2c_4 + 6c_1c_5 - 6c_6\right).
\end{aligned} \tag{29}$$

(h) The following hold for any $\mathbf{GC}(a,b;\tilde{c})$ *distribution:*

$$\begin{aligned}
\textit{mean:} &\quad a + bc_1\\
\textit{variance:} &\quad b^2\left(1 - c_1^2 + 2c_2\right)\\
\textit{skewness coefficient:} &\quad \frac{2\left(c_1^3 - 3c_1c_2 + 3c_3\right)}{\left(1 - c_1^2 + 2c_2\right)^{3/2}}\\
\textit{excess kurtosis coefficient:} &\quad -\frac{6\left(c_1^4 - 4c_1^2c_2 + 2c_2^2 + 4c_1c_3 - 4c_4\right)}{\left(1 - c_1^2 + 2c_2\right)^2}.
\end{aligned} \tag{30}$$

(i) Suppose $X \sim \mathbf{GC}(a,b;\tilde{c}^X)$, $Y \sim \mathbf{GC}(a,b;\tilde{c}^Y)$. *Then the first K moments of X and Y are the same; that is,*

$$\mathbf{E}X^j = \mathbf{E}Y^j, \quad j = 1,\ldots,K, \tag{31}$$

if and only if $c_j^X = c_j^Y$, $j = 1,\ldots,K$. *This implies that if* c_1 *up to* c_4 *are equal to 0 then the distribution has zero skewness and excess kurtosis; hence, this shows how to construct an infinite number of distributions that share this property with the normal.*

(j) Suppose $X \sim \mathbf{GC}(a,b;\tilde{c}^X)$. *Then*

$$\begin{aligned}
a = \mathbf{E}X &\Longleftrightarrow c_1^X = 0,\\
b^2 = \mathbf{E}(X-a)^2 &\Longleftrightarrow c_2^X = 0,
\end{aligned} \tag{32}$$

$$\left\{a = \mathbf{E}X,\ b^2 = \mathbf{Var}\,X\right\} \Longleftrightarrow \left\{c_1^X = c_2^X = 0\right\}.$$

When $c_1^X = c_2^X = 0$ *the skewness and excess kurtosis coefficients of X are* $6c_3^X$ *and* $24c_4^X$, *respectively, for any* $N = 0,2,4,\ldots$.

(k) If $X \sim \mathbf{GC}(a,b;c_1,\ldots,c_N)$ *and q is a constant then* $Y = qX \sim \mathbf{GC}(aq, b|q|; c_1',\ldots,c_N')$, *where* $c_k' = (\mathrm{sign}(q))^k c_k$, $k \geq 1$. *In particular,* $-X \sim \mathbf{GC}(-a,b;-c_1,\ldots,(-1)^{N-1}c_{N-1},c_N)$.

(l) The law of the square of $X \sim \mathbf{GC}(0,1;c_1,\ldots,c_N)$ *is a combination of chi-square distributions with* $1,3,\ldots,N+1$ *degrees of freedom that has density*

$$g(y) = \frac{e^{-y/2}}{\sqrt{2\pi y}}\sum_{j=0}^{n/2}\alpha_j y^j, \qquad \text{where } \sum_{j=0}^{n/2}\alpha_j y^j = \sum_{j=0}^{n/2} c_{2j} He_{2j}\left(\sqrt{y}\right). \tag{33}$$

Proof. Parts (a) and (b) were proved above, and (c) follows directly from (b). To prove (d) it is sufficient to note the existence of $\mathbf{E}e^{tY}$ for t in an open neighbourhood of $s = 0$. For (e), if $\tilde{c} = \tilde{0} \in \mathbf{R}^N$ then the distribution is the standard normal, and this is a Gram-Charlier distribution. Since $c_N > 0$ and N is even, the polynomial

$$\sum_{k=1}^{N} He_k(x) \tag{34}$$

tends to ∞ when $|x|$ tends to ∞. Hence, there is $\epsilon_0 > 0$ such that

$$1+\epsilon\sum_{k=0}^{N} He_k(x) > 0, \quad x \in \mathbf{R},\ 0 \le \epsilon \le \epsilon_0 \tag{35}$$

(recall that $c_0 He_0(x) = 1$ for all Gram-Charlier distributions). The set of valid vectors $\overline{c}$ thus includes $(\epsilon,\dots,\epsilon) \in \mathbf{R}^N$ for each $0 \le \epsilon \le \epsilon_0$. If $\overline{c}^{(j)}$, $j = 1, 2$, are valid then $p\overline{c}^{(1)} + (1-p)\overline{c}^{(2)}$ is also valid, for any $p \in (0,1)$.

Part (f) is found by expanding the mgf in (b) as a series in s around the origin. Part (g) is found by expanding

$$t \longmapsto \log\left(e^{at+(b^2t^2/2)}\sum_{k=0}^{N} b^k c_k t^k\right). \tag{36}$$

The formulas in (h) follow (f) and (g).

For part (i), it is sufficient to consider the case $a = 0, b = 1$ only. Write

$$\begin{aligned}\psi(t) &= e^{t^2/2},\\ q_X(t) &= \sum_{k=0}^{N} c_k^X t^k,\\ q_Y(t) &= \sum_{k=0}^{N} c_k^Y t^k\end{aligned} \tag{37}$$

and calculate the moments of X and Y by successively differentiating the mgf's of X and Y (note that $c_0^X = q_X(0) = c_0^Y = q_Y(0) = 1$). The first K moments of X and Y are the same if and only if

$$\begin{gathered}\psi'(0)q_X(0)+\psi(0)q_X'(0) = \psi'(0)q_Y(0)+\psi(0)q_Y'(0),\\ \vdots\\ \sum_{j=0}^{K}\binom{K}{j}\psi^{(j)}(0)q_X^{(K-j)}(0) = \sum_{j=0}^{K}\binom{K}{j}\psi^{(j)}(0)q_Y^{(K-j)}(0).\end{gathered} \tag{38}$$

Suppose that $c_j^X = c_j^Y$, $j = 1,\dots,K$. Then $q_X^{(j)}(0) = q_Y^{(j)}(0)$, $j = 1,\dots,K$, and thus the first K moments of X and Y are the same. Conversely, suppose $\mathbf{E}X^j = \mathbf{E}Y^j$, $j = 1,\dots,K$. Then the first identity above implies that $c_1^X = q_X'(0) = q_Y'(0) = c_1^Y$, since $\psi(0)$ is not zero. The second identity implies that $c_2^X = c_2^Y$, and so on, up to $c_K^X = c_K^Y$.

Turning to (j), the first equivalence

$$a = \mathbf{E}X \Longleftrightarrow c_1^X = 0 \tag{39}$$

is an immediate consequence of property (i) with $K = 1$. To prove the second equivalence, suppose $X \sim \mathbf{GC}(0,1;\overline{c})$, and let $Y \sim \mathbf{GC}(0,1;-) = \mathbf{N}(0,1)$. Then

$$\begin{aligned}\mathbf{E}X^2 &= \psi''(0)q_X(0)+2\psi'(0)q_X'(0)+\psi(0)q_X''(0),\\ \mathbf{E}Y^2 &= \psi''(0)q_Y(0)+2\psi'(0)q_Y'(0)+\psi(0)q_Y''(0).\end{aligned} \tag{40}$$

Here $q_X(0) = q_Y(0) = 1$ and $\psi'(0) = 0$; hence, $\mathbf{E}X^2 = \mathbf{E}Y^2$ if and only if $q_X''(0) = q_Y''(0)$. The last equality is $c_2^X = c_2^Y$. For $X \sim \mathbf{GC}(a,b;\overline{c}^X)$, with a and b being arbitrary, this means that

$$\mathbf{E}\left(\frac{x-a}{b}\right)^2 = 1 \Longleftrightarrow c_2^X = 0. \tag{41}$$

For (k), observe that if $q \ne 0$ then the mgf of Y is

$$\begin{aligned}&e^{aqs+(1/2)b^2q^2s^2}\sum_{k=0}^{N} c_k b^k q^k s^k\\ &= e^{aqs+(1/2)b^2q^2s^2}\sum_{k=0}^{N} c_k\left(\frac{q}{|q|}\right)^k (b|q|)^k s^k.\end{aligned} \tag{42}$$

Finally, turn to (l). Routine calculations show that the density of the square is

$$\begin{aligned}&\frac{1}{2\sqrt{y}}\left(f_X(\sqrt{y})+f_X(-\sqrt{y})\right)\\ &= \frac{\phi(\sqrt{y})}{2\sqrt{y}}\sum_{k=0}^{N} c_k\left[He_k(\sqrt{y})+He(-\sqrt{y})\right].\end{aligned} \tag{43}$$

The Hermite polynomials of odd order are odd functions and so disappear from that expression, while the even order Hermite polynomials are even functions.

When $N = 0$ the $\mathbf{GC}(a,b;\overline{c})$ distribution is the normal distribution with mean a and variance b^2. However, part (b) of the theorem says that when $N > 0$ the parameters a and b^2 are not necessarily the mean and variance of the distribution. Simple calculations lead to the following result.

Theorem 3. *(a) If $X \sim \mathbf{GC}(0,1;c_1,\dots,c_N)$ then*

$$\begin{aligned}\mathbf{P}(X \le x) &= \Phi(x)-\phi(x)\sum_{k=1}^{N} c_k He_{k-1}(x),\\ \mathbf{P}(X > x) &= \Phi(-x)+\phi(x)\sum_{k=1}^{N} c_k He_{k-1}(x).\end{aligned} \tag{44}$$

(b) The tails of the $\mathbf{GC}(a,b;c_1,\dots,c_N)$ distribution are

$$\begin{aligned}&\mathbf{P}(X > x) \sim c_N\left(\frac{x-a}{b}\right)^{N-1}\phi\left(\frac{x-a}{b}\right) \quad \text{as } x \longrightarrow \infty,\\ &\mathbf{P}(X < x) \sim c_N\left(\frac{|x-a|}{b}\right)^{N-1}\phi\left(\frac{x-a}{b}\right) \quad \text{as } x \longrightarrow -\infty.\end{aligned} \tag{45}$$

The tails of the Gram-Charlier distributions are thicker than those of the normal distribution but are still "thin" because they are in the limit smaller than any exponential function $\exp(-\alpha x)$.

2.1. The $\mathbf{GC}(a, b; 0, 0, c_3, c_4)$ *Family.* Here the exact region for the (c_3, c_4) that lead to a true probability distribution has been found. This goes back to Barton and Dennis [16], but a more detailed explanation is given in Jondeau and Rockinger [6]. The region is shown in Figure 3 (use the correspondence $\overline{s} = 6c_3$, $\overline{k} = 24c_4$ from Theorem 2(h)). An important fact about this region is that it is not rectangular; the possible excess kurtosis values depend on skewness, and conversely.

2.2. $\mathbf{GC}(a, b; \overline{c})$ *Distributions of Order 4 and Higher.* The $\mathbf{GC}(a, b; c_1, c_2, c_3, c_4)$ family has six parameters, rather than four, and thus has more degrees of freedom in fitting data; to the authors' knowledge the general six-parameter Gram-Charlier distribution has been used in financial applications by León et al. [11] only (those authors use the subfamily consisting of polynomials $p(\cdot)$ that are squares of some second-degree polynomial). Schlögl [10] fits the six- and eight-parameter families $\mathbf{GC}(a, b; 0, 0, c_3, c_4, c_5, c_6)$ and $\mathbf{GC}(a, b; 0, 0, c_3, c_4, c_5, c_6, c_7, c_8)$ to data.

The set of (c_1, c_2, c_3, c_4) that yield true probability distributions has not been identified, but it is possible to fit the six parameters a, b, c_1, c_2, c_3, c_4 by maximum likelihood.

2.3. Why We Do Not Assume That $c_1 = c_2 = 0$. Almost all previous authors have assumed that $c_1 = c_2 = 0$, because they used normalized data:

$$Y = \frac{X - \mathbf{E}X}{\sqrt{\mathbf{Var}\, X}} \quad (46)$$

(see part (j) of Theorem 2) as explained below. In this section we explain why it is important not to restrict Gram-Charlier distributions or series in that way. The first reason for not doing so is that enlarging the parameter space can only be a good thing. The second one is that in fitting those distributions to data there may be very real advantages in letting c_1 and c_2 be different from zero. The only downside of letting c_1 and c_2 take nonzero values, and it is of no real importance, is that c_3 and c_4 lose their simple relationship with skewness and excess kurtosis (see part (h) of Theorem 2). A third reason is that after an exponential change of measure a Gram-Charlier distribution will rarely have $c_1 = c_2 = 0$; see Section 2.4.

We now show, using an example that can be worked out explicitly, that it is not always best to use normalized data when fitting Gram-Charlier distributions, because choosing another affine transformation of the data may well yield a much better fit.

The "standard" Gram-Charlier expansion for a function $f(\cdot)$ is

$$\phi(x) \sum_{k=0}^{\infty} c_k He_k, \quad (47)$$

where

$$c_k = \frac{1}{k!} \int_{-\infty}^{\infty} He_k(x) f(x)\, dx = \frac{1}{k!} \mathbf{E} He_k(X). \quad (48)$$

A classical result about Hermite series, proved by Cramér [17], is that sufficient conditions for the Gram-Charlier expansion (47) to converge to

$$\frac{1}{2}(f(x-) + f(x+)) \quad (49)$$

for all $x \in \mathbf{R}$ are that (i) $f(\cdot)$ has finite variation in every bounded interval, and (ii) satisfies

$$\int_{-\infty}^{\infty} e^{x^2/4} f(x)\, dx < \infty. \quad (50)$$

If X has density $f(\cdot)$, then the last condition is $\mathbf{E}e^{X^2/4} < \infty$. This condition cannot be improved upon, in the sense that there are cases where the Gram-Charlier series defined above diverges, although $\mathbf{E}e^{\rho X^2} < \infty$ for all $\rho < 1/4$ (it will be shown below that one such case is the normal distribution with mean 0 and variance 2).

Let us first calculate

$$h_{q,k} = \mathbf{E} He_k(X), \quad X \sim \mathbf{N}(0, q^2). \quad (51)$$

Using the generating function (16) we find

$$h_{q,k} = \begin{cases} 0 & \text{if } k \text{ is odd} \\ \dfrac{k!}{2^{k/2}(k/2)!}\left(q^2 - 1\right)^{k/2} & \text{if } k \text{ is even.} \end{cases} \quad (52)$$

The Gram-Charlier expansion corresponding to the $\mathbf{N}(0, q^2)$ density is thus

$$\phi(x) \sum_{k=0}^{\infty} \frac{h_{q,k}}{k!} He_k(x) = \phi(x) \sum_{n=0}^{\infty} \frac{1}{2^n n!} \left(q^2 - 1\right)^n He_{2n}(x). \quad (53)$$

It is possible to determine whether this converges or not when $x = 0$; the Gram-Charlier series for $f(0)$ is

$$\frac{1}{\sqrt{2\pi}} \sum_{n=0}^{\infty} (-1)^n \frac{(2n)!}{2^{2n}(n!)^2} \left(q^2 - 1\right)^n. \quad (54)$$

From Stirling's formula,

$$\frac{(2n)!}{2^{2n}(n!)^2} \sim \frac{1}{\sqrt{\pi n}} \quad (55)$$

as n tends to infinity, which implies that there are three possibilities: expansion (54) (i) converges absolutely if $0 < q^2 < 2$, (ii) converges simply if $q^2 = 2$, and diverges if $q^2 > 2$. (The preceding calculations are from Cramér [17].)

Let us now consider a random variable X_p, $0 \le p \le 1$, with density

$$g_p(x) = p\phi(x) + (1 - p)\frac{1}{p}\phi\left(\frac{x}{p}\right). \quad (56)$$

In words, X_p has a $\mathbf{N}(0,1)$ distribution with probability p and a $\mathbf{N}(0,p^2)$ distribution with probability $1-p$. Define the normalized variable

$$Y_p = \frac{X_p}{\sqrt{\mathbf{Var}\, X_p}} = \frac{X_p}{\sigma_p}, \qquad \sigma_p^2 = \mathbf{Var}\, X_p = p + p^2 - p^3. \tag{57}$$

The coefficients $c_{p,k}$ of the Gram-Charlier expansion for Y_p are

$$c_{p,k} = \frac{1}{k!}\mathbf{E}He_k\left(Y_p\right) = \frac{1}{k!}\left(ph_{1/\sigma_p,k} + (1-p)\, h_{p/\sigma_p,k}\right). \tag{58}$$

We have shown that (53) diverges if $q^2 > 2$. Here we see that $c_{p,k}$ involves $h_{q,k}$ with $q = 1/\sigma_p$ and $q = p/\sigma_p$; in the second case there is no problem, as $(p/\sigma_p)^2 < 2$ for all $p \in (0,1]$. In the first case,

$$\frac{1}{\sigma_p^2} > 2 \Longleftrightarrow p + p^2 - p^3 < \frac{1}{2}. \tag{59}$$

This condition is satisfied for p smaller than $p^* \approx 0.403$, and so the Gram-Charlier expansion for Y_p diverges for all $p < p^*$. This is a sad state of affair: a series designed to work for distributions that are "close to the normal" that fails for combinations of two normal densities!

There is an easy solution: use a different scaling for X_p. Rather than multiplying X_p by $1/\sigma_p$, use a factor r such that the series converges. In this example we know that the series converges if and only if $r < \sqrt{2}$. Say we choose $p = .3$; this implies

$$\frac{1}{\mathbf{Var}\, X_p} = \frac{1}{0.3 + 0.3^2 - 0.3^3} = 2.75482 > 2. \tag{60}$$

Figure 1 shows how the true density for $X_{0.3}$ is approximated by order 10 and 20 Gram-Charlier truncated series if $r = 1/\sigma_{0.3} \doteq 1.66$ is chosen; the graph is not surprising, the infinite Gram-Charlier expansion diverges. Figure 2 shows the same except that r is set to 1.4. The latter is smaller than $\sqrt{2}$, so the infinite expansion converges. In this example the density to be approximated is symmetrical about 0, so c_1 is always 0, while

$$q = \frac{1}{\sigma_{0.3}} \Longrightarrow c_2 = 0; \qquad q = 1.4 \Longrightarrow c_2 = -0.14426. \tag{61}$$

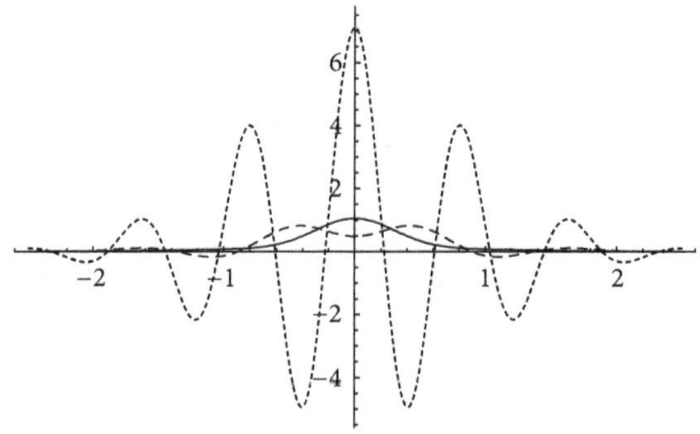

Figure 1: Gram-Charlier approximations to the density $g_{0.3}(x)$ (solid line) when $q = 1/\sigma_{0.3} \doteq 1.66$, resulting in $c_1 = c_2 = 0$. The curve with long dashes is the order 10 approximation, and the one with short dashes is the order 20 approximation.

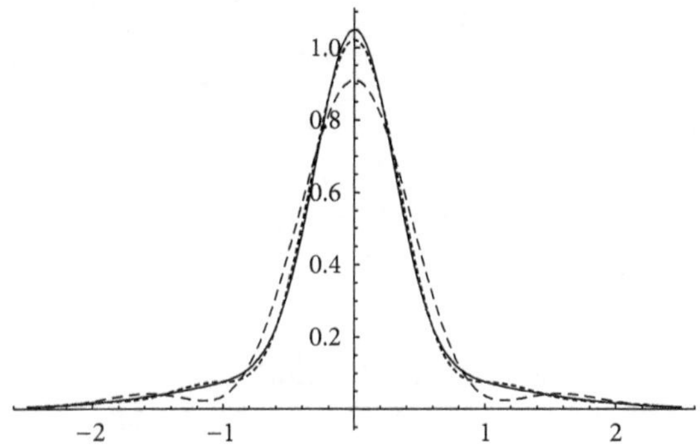

Figure 2: Gram-Charlier approximations to the density $g_{0.3}(x)$ (solid line) when $q = 1.4$. The curve with long dashes is the order 10 approximation, and the one with short dashes is the order 20 approximation.

2.4. Exponential Change of Measure. The one-dimensional Cameron-Martin formula may be stated as follows: if $X \sim \mathbf{N}(\mu,\sigma^2)$ then for $q \in \mathbf{R}$ and $f \ge 0$,

$$\mathbf{E}e^{qX} f(X) = e^{\mu q + (1/2)\sigma^2 q^2}\mathbf{E}f\left(X + \sigma^2 q\right). \tag{62}$$

The same property may be expressed in terms of a change of measure. If $X \overset{\mathbf{P}}{\sim} \mathbf{N}(\mu,\sigma^2)$ and a change of measure is defined by

$$P' = \frac{e^{qX}}{\mathbf{E}^{\mathbf{P}} e^{qX}} \cdot P, \tag{63}$$

then $X \overset{\mathbf{P}'}{\sim} \mathbf{N}(\mu + \sigma^2 q, \sigma^2)$. The next result is an extension of the Cameron-Martin formula for the normal distribution to the Gram-Charlier distributions. There are different though equivalent formulas in Schlögl [10] (for the special case $c_1 = c_2 = 0$).

Theorem 4. *Suppose that* $X \overset{\mathbf{P}}{\sim} \mathbf{GC}(a,b;c_1,\ldots,c_N)$ *and that* $\mathbf{P}'$ *is defined by (63) for* $q \in \mathbf{R}$. *Then* $X \overset{\mathbf{P}'}{\sim} \mathbf{GC}(a + b^2 q, b; c_1',\ldots,c_N')$, *where* $c_1',\ldots,c_N'$ *are found from*

$$\sum_{k=0}^{N} b^k c_k' s^k = \frac{\sum_{k=0}^{N} b^k c_k (q+s)^k}{\sum_{j=0}^{N} b^j c_j q^j}, \quad s \in \mathbf{R}, \tag{64}$$

or, more precisely,

$$c'_k = \frac{1}{\sum_{j=0}^{N} b^j c_j q^j} \sum_{\ell=k}^{N} \binom{\ell}{k} b^{\ell-k} c_\ell q^{\ell-k}. \quad (65)$$

Proof. It is sufficient to calculate the mgf of X under $\mathbf{P}'$:

$$\mathbf{E}^{P'} e^{sX} = \frac{1}{\mathbf{E}^{\mathbf{P}} e^{qX}} \mathbf{E}^{\mathbf{P}} e^{(q+s)X} = (\cdots)$$

$$= e^{(a+b^2q)s+(1/2)b^2s^2} \sum_{k=0}^{N} b^k s^k \frac{1}{\sum_{j=0}^{N} b^j c_j q^j} \cdot \sum_{\ell=k}^{N} \binom{\ell}{k} b^{\ell-k} c_\ell q^{\ell-k}. \quad (66)$$

Example 5. If $X \overset{\mathbf{P}}{\sim} \mathbf{GC}(a, b; 0, 0, c_3, c_4)$ and $q = 1$ the change of measure (63) leads to

$$c'_1 = \frac{3b^2c_3 + 4b^3c_4}{1 + b^3c_3 + b^4c_4},$$
$$c'_2 = \frac{3bc_3 + 6b^2c_4}{1 + b^3c_3 + b^4c_4},$$
$$c'_3 = \frac{c_3 + 4bc_4}{1 + b^3c_3 + b^4c_4},$$
$$c'_4 = \frac{c_4}{1 + b^3c_3 + b^4c_4}. \quad (67)$$

Hence, assuming $c_1 = c_2 = 0$ does not imply that c'_1 and c'_2 are also zero. In other words, the family $\mathbf{GC}(a, b; 0, 0, c_3, c_4)$ is not closed under a change of measure that occurs naturally in option pricing (see Section 5).

2.5. Simulating Gram-Charlier Distributions. Simulation is required for many kinds of options, and it turns out that the Gram-Charlier distributions are very easy to generate, as we now show; there is no need to invert their distribution functions. When estimating some quantity $m = \mathbf{E}g(X_1, \ldots, X_s)$ by simulation, one generates n independent vectors

$$\left(X_1^{(j)}, \ldots, X_s^{(j)}\right), \quad j = 1, \ldots, n, \quad (68)$$

with the same distribution. Suppose all the X's are independent and have the same Gram-Charlier distribution with density

$$f_X(x) = \phi(a, b, x)\, p(x), \quad (69)$$

where the polynomial $p(x)$ is given in (5). Then

$$\mathbf{E}g(X_1, \ldots, X_s) = \int_{\mathbf{R}^s} g(x_1, \ldots, x_s) \cdot \prod_{k=1}^{s} [\phi(a, b, x_k)\, p(x_k)]\, dx_1 \cdots dx_s = \mathbf{E}^{\mathbf{Q}} \left[g(X_1, \ldots, X_s) \prod_{k=1}^{s} p(X_k) \right], \quad (70)$$

where under the measure $\mathbf{Q}$ the X's have a normal distribution $\mathbf{N}(a, b^2)$. Hence, estimating $\mathbf{E}g(X_1, \ldots, X_s)$ by simulation can be performed by generating ordinary normal random variables $(X_1^{(j)}, \ldots, X_s^{(j)})$, $j = 1, \ldots, n$, and then using the estimator

$$\widehat{m}_n = \frac{1}{n} \sum_{j=1}^{n} \left[g\left(X_1^{(j)}, \ldots, X_s^{(j)}\right) \prod_{k=1}^{s} p\left(X_k^{(j)}\right) \right]. \quad (71)$$

This is an application of the likelihood ratio method.

3. Convolution of Gram-Charlier Distributions; Gram-Charlier Processes

The simplest way to find the distribution of the sum of two independent Gram-Charlier distributed variables is to multiply their moment generating functions. Suppose $X_j \sim \mathbf{GC}(a_j, b_j; c_1^{(j)}, \ldots, c_N^{(j)})$, $j = 1, 2$ are independent. Then

$$\mathbf{E}e^{z(X+Y)} = e^{(a_1+a_2)t+(b_1^2+b_2^2)(t^2/2)} \left(\sum_{k=0}^{N^{(1)}} c_k^{(1)} t^k \right) \left(\sum_{k=0}^{N^{(2)}} c_k^{(2)} t^k \right). \quad (72)$$

Expanding the product, this says that

$$X + Y \sim \mathbf{GC}\left(a_1 + a_2, \sqrt{b_1^2 + b_2^2}; c_1, \ldots, c_{N^{(1)}+N^{(2)}}\right), \quad c_k = \sum_{j=0}^{k} c_j^{(1)} c_{k-j}^{(2)}. \quad (73)$$

It is then possible to know the explicit distribution of

$$Z_n = X_1 + \cdots + X_n, \quad (74)$$

where the X's are independent and have a Gram-Charlier distribution, constituting a discrete-time Gram-Charlier process. If the X's have the same $\mathbf{GC}(a, b; c_1, \ldots, c_N)$ distribution then $\{Z_n, n \geq 0\}$ is a random walk. The derivation of the distribution of Z_n can be done recursively, using (73), or it can be done by finding the Taylor expansion of

$$\left(\sum_{k=0}^{N} c_k t^k \right)^n = \sum_{k=0}^{nN} c_k^{(n)} t^k = 1 + c_1 nt + \left(c_2 n + \frac{1}{2} c_1^2 n(n-1) \right) t^2 + \cdots + c_N^n t^{nN}, \quad (75)$$

and thus

$$Z_n \sim \mathbf{GC}\left(an, b\sqrt{n}; c_1 n, c_2 n + \frac{1}{2}c_1^2 n(n-1), \ldots, c_N^n\right). \quad (76)$$

These computations are simple using symbolic mathematics software. An example is given at the end of Section 6.1.

The above raises the question of whether there is a continuous-time process that has Gram-Charlier distributed increments. There is such a process with normal increments (Brownian motion), and it is moreover a Lévy process.

Theorem 6. *The only Lévy process with Gram-Charlier distributed increments is Brownian motion.*

Proof. It is sufficient to show that, besides the normal distribution, any Gram-Charlier distribution cannot be infinitely divisible. If X has a Gram-Charlier distribution then its moment generating function is of the form

$$M(z) = \mathbf{E}e^{zX} = e^{az+b^2(z^2/2)}\pi(z), \quad (77)$$

where $\pi(\cdot)$ is a polynomial. Suppose $X_{n,1}, \ldots, X_{n,n}$ are independent, have the same distribution, and add up to X (in law). Fubini's theorem implies that $\mathbf{E}e^{tX_{n,1}}$ is finite for all real t, and thus

$$M_n(z) = \mathbf{E}e^{zX_{n,1}} \quad (78)$$

is an analytic function of z in the whole complex plane. We then have the identity

$$M_n(z)^n = M(z), \quad z \in \mathbf{C}, \quad (79)$$

for every $n = 1, 2, 3, \ldots$. This means that the function $\pi(z)^{1/n}$, which is well defined for all z that is not a zero of π, has an analytic continuation over the whole of $\mathbf{C}$ for each $n \geq 1$, which is impossible, for instance, take n larger than the degree of π.

More precisely, this says that if we exclude Brownian motion, no increment of any Lévy process can have a Gram-Charlier distribution. Any Gram-Charlier process with independent increments must be discrete-time.

4. The Log Gram-Charlier Distribution

The distribution of the exponential of a Gram-Charlier distributed variable will naturally be called "log Gram-Charlier", as we do for the lognormal: if $Y \sim \mathbf{GC}(a, b; \overline{c})$, then $L = \exp(Y) \sim \mathbf{LogGC}(a, b; \overline{c})$. The density of L is

$$f_L(z) = \frac{1}{z} f_Y(\log z), \quad z > 0. \quad (80)$$

This distribution has all moments finite, and they are given by Theorem 2(b). The log Gram-Charlier distribution shares one property with the lognormal, it is "moment indeterminate."

Theorem 7. *The log Gram-Charlier distribution is not determined by its moments. More precisely, there is a noncountable number of other distributions that have the same moments as any particular log Gram-Charlier distribution.*

Proof. There is a well-known way to construct a family of distributions that have the same moments as the lognormal (Feller [18], p. 227); the trick works for arbitrary parameters but, for simplicity, let $L \sim \mathbf{LogN}(0, 1)$; that is,

$$f_L(z) = \frac{1}{z\sqrt{2\pi}} e^{-(1/2)\log^2 z}, \quad z > 0. \quad (81)$$

All the functions

$$f_\alpha(z) = f_L(z)\left(1 + \alpha \sin(2\pi \log z)\right), \quad -1 \leq \alpha \leq 1, \quad (82)$$

are nonnegative, integrate to 1, and have the same moments as L (as a result of the symmetry of the standard normal distribution, after an obvious change of variable). Now, suppose $L \sim \mathbf{LogGC}(0, 1; \overline{c})$ (once again the same arguments work for other values of a and b), and consider

$$g_\alpha(z) = f_L(z) + \alpha\phi(z)\sin(2\pi \log z). \quad (83)$$

By the same change of variables used for the lognormal one immediately finds that g_α integrates to one and has the same moments as L. The only difference with the case of the lognormal is that g_a is not necessarily nonnegative for $-1 \leq \alpha \leq 1$. Two cases may arise. The first one is that the polynomial

$$p(y) = \sum_{k=0}^{N} c_k He_k(y) \quad (84)$$

has no real zero. In that case its infimum is strictly greater than zero, and one can find a nonempty interval $I = (-\epsilon, \epsilon)$ such that g_α is nonnegative for all $\alpha \in I$. In the second case $p(\cdot)$ has one or more zeros and the previous argument breaks down but can be modified to yield the same conclusion, if the function $\sin(2\pi y)$ is replaced with

$$s(y) = \sin(2\pi y)\,\mathbf{1}_{\{y \notin J\}}, \quad (85)$$

where J is a collection of intervals that include the zeros of $p(\cdot)$, so defined that $s(\cdot)$ is not identically zero and satisfies

$$s(y + 2k\pi) = s(y),$$
$$s(-y) = -s(y). \quad (86)$$

(These two conditions are sufficient for

$$\int_{-\infty}^{\infty} e^{ky}\phi(y)s(y)\,dy = 0 \quad (87)$$

to hold for $k = 1, 2, \ldots$.) The details are omitted. Another way to prove that the log Gram-Charlier distribution is moment indeterminate when $p(y)$ has no zero (and thus the density $f_L(z)$ does not take the value zero for any $z > 0$) is to use a Krein condition (Stoyanov [19], p. 941), which says that a continuous distribution on $\mathbf{R}_+$ with positive density $f(\cdot)$ is not determined by its moments if

$$-\int_0^{\infty} \frac{\log f\left(z^2\right)}{1+z^2} dz < \infty. \tag{88}$$

5. Option Pricing Formulas

The formulas below hold for any vector $(c_1, \ldots, c_N)$ and thus extend those that have been derived by previous authors for the case where the log return has a $\mathbf{GC}(a, b; 0, 0, c_3, c_4)$ distribution or a "squared" Gram-Charlier distribution (León et al. [11]). Schlögl [10] derives a formula equivalent to (90) but for densities expressed as an infinite Gram-Charlier series with $c_1 = c_2 = 0$.

As previous authors have done, we consider a market with a risky security S and a risk-free security with annual return r and suppose that the log return for period $[0, T]$ (denoted as X_T) has a Gram-Charlier distribution under the risk-neutral (or "pricing") measure, which we denote as $\mathbf{Q}$. The physical measure (usually denoted as $\mathbf{P}$) is not specified (nothing says that the log return has or does not have a Gram-Charlier distribution under the physical measure).

The market model may have one or more periods, but since we consider the pricing of ordinary European puts and calls only the distribution of the log return for the whole period $[0, T]$ is required. For other types of options, in particular the applications presented in the next section, it may be necessary to use the one-period returns separately, as is done in Theorem 11.

The risky security has price S_0 at time 0, and $S_T = S_0 e^{X_T}$.

Theorem 8. *Suppose that a risky security pays dividends at a constant rate δ over $[0, T]$ and that the risk-free rate of interest is r. Suppose also that under the risk-neutral measure $\mathbf{Q}$ the log return of the risky security over $[0, T]$ is $X_T \overset{\mathbf{Q}}{\sim} \mathbf{GC}(a, b; c_1, \ldots, c_N)$, which satisfies the martingale condition*

$$e^{a+(b^2/2)} \sum_{k=0}^{N} b^k c_k = e^{(r-\delta)T}. \tag{89}$$

Then the time-0 price of a European call option with maturity T and strike price K is

$$\begin{aligned} C_0 &= S_0 e^{-\delta T} \Phi(d_1) - K e^{-rT} \Phi(d_2) + K e^{-rT} \phi(d_2) \\ &\quad \cdot \sum_{k=1}^{N} \left[c_k^* He_{k-1}(-d_1) - c_k He_{k-1}(-d_2)\right], \end{aligned} \tag{90}$$

where d_2, d_1, and $\{c_k^*\}$ are given by (97), (101), and (105). The price of the corresponding European put is

$$\begin{aligned} P_0 &= K e^{-rT} \Phi(-d_2) - S_0 e^{-\delta T} \Phi(-d_1) - K e^{-rT} \phi(d_2) \\ &\quad \cdot \sum_{k=1}^{N} \left[c_k^* He_{k-1}(-d_1) - c_k He_{k-1}(-d_2)\right]. \end{aligned} \tag{91}$$

If $N = 4$, then the above simplify to

$$\begin{aligned} C_0 &= S_0 e^{-\delta T} \Phi(d_1) - K e^{-rT} \Phi(d_2) + bK e^{-rT} \phi(d_2) \\ &\quad \cdot \left[c_2 + (b - d_2) c_3 + \left(b^2 - b d_2 + d_2^2 - 1\right) c_4\right], \\ P_0 &= K e^{-rT} \Phi(-d_2) - S_0 e^{-\delta T} \Phi(-d_1) - bK e^{-rT} \phi(d_2) \\ &\quad \cdot \left[c_2 + (b - d_2) c_3 + \left(b^2 - b d_2 + d_2^2 - 1\right) c_4\right]. \end{aligned} \tag{92}$$

Proof. Absence of arbitrage implies that $\mathbf{E}^{\mathbf{Q}} S_T = e^{(r-\delta)T} S_0$, or

$$S_0 e^{a+(b^2/2)} \sum_{k=0}^{N} b^k c_k = e^{(r-\delta)T} S_0. \tag{93}$$

This justifies (89). The price of the call is

$$\begin{aligned} C_0 &= e^{-rT} \mathbf{E}^{\mathbf{Q}} (S_T - K)_+ = C_0^+ - C_0^- \\ C_0^+ &= e^{-rT} \mathbf{E}^{\mathbf{Q}} S_T \left(\mathbf{1}_{\{S_T > K\}}\right), \\ C_0^- &= K e^{-rT} \mathbf{E}^{\mathbf{Q}} \left(\mathbf{1}_{\{S_T > K\}}\right). \end{aligned} \tag{94}$$

The second part is easier to deal with

$$C_0^- = K e^{-rT} \mathbf{Q}\left(X_T > \log\left(\frac{K}{S_0}\right)\right). \tag{95}$$

The probability of the event $X_T > \log(K/S_0)$ can be calculated explicitly by recalling that $(X_T - a)/b \overset{\mathbf{Q}}{\sim} \mathbf{GC}(0, 1; c_1, \ldots, c_N)$ and using Theorem 3:

$$\begin{aligned} &\mathbf{Q}\left(X_T > \log\left(\frac{K}{S_0}\right)\right) \\ &\quad = \mathbf{Q}\left(\frac{X_T - a}{b} > \frac{1}{b}\left(\log\left(\frac{K}{S_0}\right) - a\right)\right) \\ &\quad = \Phi(d_2) + \phi(d_2) \sum_{k=1}^{N} c_k He_{k-1}(-d_2), \end{aligned} \tag{96}$$

where

$$d_2 = \frac{1}{b}\left(\log\left(\frac{S_0}{K}\right) + a\right). \tag{97}$$

To calculate C_0^+ use the exponential change of measure formula, defining

$$\mathbf{Q}' = \frac{S_T}{\mathbf{E}^{\mathbf{Q}} S_T} \cdot \mathbf{Q} = \frac{e^{X_T}}{\mathbf{E}^{\mathbf{Q}} e^{X_T}} \cdot \mathbf{Q}. \tag{98}$$

Then

$$\mathbf{E}^{\mathbf{Q}}\left(S_T\mathbf{1}_{\{S_T>K\}}\right) = \left(\mathbf{E}^{\mathbf{Q}}S_T\right)\mathbf{E}^{\mathbf{Q}'}\left(\mathbf{1}_{\{S_T>K\}}\right). \quad (99)$$

Since $X_T \overset{\mathbf{Q}'}{\sim} \mathbf{GC}(a+b^2, b; c'_1, \dots, c'_N)$,

$$\left(\mathbf{E}^{\mathbf{Q}}S_T\right)\mathbf{E}^{\mathbf{Q}'}\left(\mathbf{1}_{\{S_T>K\}}\right) = S_0 e^{(r-\delta)T}\mathbf{Q}'\left(\frac{X_T - a - b^2}{b} > \frac{1}{b}\left(\log\left(\frac{K}{S_0}\right) - a - b^2\right)\right) = S_0 e^{(r-\delta)T}\Phi(d_1) + S_0 e^{(r-\delta)T}\phi(d_1)\sum_{k=1}^{N} c'_k He_{k-1}(-d_1), \quad (100)$$

where

$$c'_k = \frac{1}{\sum_{j=0}^{N} b^j c_j}\sum_{\ell=k}^{N}\binom{\ell}{k} b^{\ell-k} c_\ell, \quad k = 1, \dots, N,$$
$$d_1 = \frac{1}{b}\left[\log\left(\frac{S_0}{K}\right) + a + b^2\right] = d_2 + b. \quad (101)$$

Hence,

$$C_0 = S_0 e^{-\delta T}\Phi(d_1) - Ke^{-rT}\Phi(d_2) + S_0 e^{-\delta T}\phi(d_1)\sum_{k=1}^{N} c'_k He_{k-1}(-d_1) - Ke^{-rT}\phi(d_2)\sum_{k=1}^{N} c_k He_{k-1}(-d_2). \quad (102)$$

Using

$$\frac{S_0 e^{-\delta T}\phi(d_1)}{\sum_{k=0}^{N} b^k c_k} = S_0 e^{a+(b^2/2)-rT}\phi(d_1) = Ke^{-rT}\phi(d_2) \quad (103)$$

(a consequence of (89) and (101)), it is possible to write

$$S_0 e^{-\delta T}\phi(d_1) c'_k = Ke^{-rT}\phi(d_2) c^*_k, \quad (104)$$

where

$$c^*_k = c'_k \sum_{k=0}^{N} b^k c_k = \sum_{\ell=k}^{N}\binom{\ell}{k} b^{\ell-k} c_\ell, \quad k = 1, \dots, N. \quad (105)$$

This proves (90).

The price of a European put can be found from the put-call parity identity

$$P_0 = C_0 - S_0 + Ke^{-rT}. \quad (106)$$

Finally, if $N = 4$ then the summation in (90) becomes

$$\sum_{k=1}^{4} c^*_k He_{k-1}(-d_1) - \sum_{k=1}^{4} c_k He_{k-1}(-d_2) = bc_2 + \left(b^2 - bd_2\right)c_3 + \left(b^3 - b^2 d_2 + bd_2^2 - b\right)c_4. \quad (107)$$

This ends the proof.

The option price formulas are of the form "Black-Scholes plus correction term." Observe, however, that the values of d_1 and d_2 are different from what they would be in the Black-Sholes formula. More precisely, in the Black-Scholes model (where $c_k = 0$ for all $k \geq 1$) we have $a = \mathbf{E}^{\mathbf{Q}}X_T$ and $b^2 = \mathbf{Var}^{\mathbf{Q}}X_T$, but this does not happen with Gram-Charlier distributed log returns, first because of the martingale condition (89) and second because of the result in part (j) of Theorem 2.

5.1. Sensitivities ("Greeks"). The previous literature includes formulas for sensitivities of Gram-Charlier option prices, but only in the case of the four-parameter $\mathbf{GC}(a, b; 0, 0, c_3, c_4)$ Gram-Charlier distributions; see Jurczenko et al. [7], Rouah and Vainberg [20], and Chateau [3]. Below we give sensitivities of the option prices calculated above with respect to all the parameters for a general Gram-Charlier distribution, taking the martingale condition (89) into account. This means that a is a function of $\delta, b, \{c_k\}, r$ and T (we might have written $a = a(\delta, b, \{c_k\}, r, T)$). However, a is not a function of S_0.

Theorem 9. *Let C_0 be the price of the European call option described in Theorem 8. Then*

$$\Delta = \frac{\partial C_0}{\partial S_0} = e^{-\delta T}\left(\Phi(d_1) + \phi(d_1)\sum_{k=1}^{N} c'_k He_{k-1}(-d_1)\right),$$
$$\gamma = \frac{\partial^2 C_0}{\partial S_0^2} = \frac{e^{a+(b^2/2)-rT}}{bS_0}\phi(d_1)\sum_{k=0}^{N} c_k He_k(-d_2),$$
$$\rho = \frac{\partial C_0}{\partial r} = KTe^{-rT}\left(\Phi(d_2) + \phi(d_2)\sum_{k=1}^{N} c_k He_{k-1}(-d_2)\right),$$
$$\kappa = \frac{\partial C_0}{\partial b} = S_0 e^{-\delta T}\phi(d_1) \cdot \sum_{k=0}^{N}\left[c'_k - c'_1 c'_{k+1} + (k+2)c'_{k+2}\right] He_k(-d_1),$$
$$\frac{\partial C_0}{\partial c_j} = Ke^{-rT}\phi(d_2) \cdot \left(\sum_{k=1}^{j-1} b^k He_{j-1-k}(-d_2) - b^j \sum_{k=1}^{N} c'_k He_{k-1}(-d_1)\right), \quad j \geq 1. \quad (108)$$

In the formula for κ the symbols c'_{N+1} and c'_{N+2} are equal to zero.

Proof. The following lemma is obtained by elementary calculations.

Lemma 10. *For any integrable random variable U and any constant K,*

$$\frac{\partial}{\partial s}\mathbf{E}(sU-K)_+ = \mathbf{E}\left(U\mathbf{1}_{\{sU>K\}}\right). \quad (109)$$

If U has a continuous density f_U, then

$$\frac{\partial^2}{\partial s^2}\mathbf{E}(sU-K)_+ = \frac{K^2}{s^3}f_U\left(\frac{K}{s}\right). \quad (110)$$

From $S_T = S_0e^{X_T}$ and (100),

$$\Delta = \frac{\partial}{\partial S_0}\left[e^{-rT}\mathbf{E}^{\mathbf{Q}}\left(S_0e^{X_T}-K\right)_+\right] = \frac{1}{S_0}C_0^+ \\ = e^{-\delta T}\left(\Phi(d_1)+\phi(d_1)\sum_{k=1}^{N}c_k'He_{k-1}(-d_1)\right). \quad (111)$$

From the lemma,

$$\gamma = \frac{\partial^2}{\partial^2 S_0}C_0 = e^{-rT}\frac{\partial^2}{\partial^2 S_0}\mathbf{E}^{\mathbf{Q}}(S_0U-K)_+ \\ = \frac{K^2e^{-rT}}{S_0^3}f_U\left(\frac{K}{S_0}\right), \quad U=e^X, \quad (112)$$

where $X_T \overset{\mathbf{Q}}{\sim} \mathbf{GC}(a,b;c_1,\ldots,c_N)$. Now the density of U may be expressed in terms of the density of $G=(X_T-a)/b \overset{\mathbf{Q}}{\sim} \mathbf{GC}(0,1;c_1,\ldots,c_N)$:

$$f_U(u) = \frac{1}{bu}f_G\left(\frac{1}{b}(\log u - a)\right) \\ = \frac{1}{bu}\phi(y)\sum_{k=0}^{N}c_kHe_k(y)\Bigg|_{y=(1/b)(\log u-a)}. \quad (113)$$

Setting $u=K/S_0$ and then using (103), we obtain

$$\gamma = \frac{Ke^{-rT}}{bS_0^2}\phi(d_2)\sum_{k=0}^{N}c_kHe_k(-d_2) \\ = \frac{e^{a+(b^2/2)-rT}}{bS_0}\phi(d_1)\sum_{k=0}^{N}c_kHe_k(-d_2). \quad (114)$$

Next,

$$\rho = \frac{\partial}{\partial r}\mathbf{E}^{\mathbf{Q}}\left(S_0e^{a+bG-rT}-Ke^{-rT}\right)_+ = \frac{\partial}{\partial r}\mathbf{E}^{\mathbf{Q}}h(G,r), \quad (115)$$

where $G \overset{\mathbf{Q}}{\sim} \mathbf{GC}(0,1,\overline{c})$. For fixed G, the function $g(r) = h(G,r)$ is absolutely continuous and thus

$$\rho = \mathbf{E}^{\mathbf{Q}}\left[\frac{\partial}{\partial r}\left(S_0e^{a+bG-rT}-Ke^{-rT}\right)\right]\mathbf{1}_{\{S_0e^{a+bG-rT}>Ke^{-rT}\}}. \quad (116)$$

Now, from (89),

$$e^{a+bG-rT} = \frac{e^{bG-(b^2/2)-\delta T}}{\sum_{k=0}^{N}b^kc_k} \quad (117)$$

does not depend on r, so

$$\rho = \mathbf{E}^{\mathbf{Q}}\left(KTe^{-rT}\mathbf{1}_{\{S_0e^{a+bG-rT}>Ke^{-rT}\}}\right) \\ = KTe^{-rT}\left[\Phi(d_2)+\phi(d_2)\sum_{k=1}^{N}c_kHe_{k-1}(-d_2)\right]. \quad (118)$$

The same line of reasoning shows that the sensitivity to b is

$$\kappa = S_0e^{-rT}\mathbf{E}^{\mathbf{Q}}\left[\left(\frac{\partial}{\partial b}e^{a+bG}\right)\mathbf{1}_{\{S_0e^{a+bG}>K\}}\right]. \quad (119)$$

Define

$$m(b) = \mathbf{E}e^{bG-(b^2/2)} = \sum_{k=0}b^kc_k, \\ m'(b) = \sum_{k=1}kb^{k-1}c_k = c_1^*. \quad (120)$$

In order to shorten the formulas we will write (from (89))

$$e^a = \frac{e^{(r-\delta)T-(b^2/2)}}{m(b)} \quad (121)$$

and thus find an expression for

$$S_0e^{-\delta T}\mathbf{E}^{\mathbf{Q}}\left[\left(\frac{\partial}{\partial b}\frac{e^{bG-(b^2/2)}}{m(b)}\right)\mathbf{1}_{\{G>-d_2\}}\right] \\ = \frac{S_0e^{-\delta T-(b^2/2)}}{m(b)^2} \\ \cdot\mathbf{E}^{\mathbf{Q}}\left[\left(Ge^{bG}m(b)-e^{bG}\left(m'(b)+bm(b)\right)\right)\cdot\mathbf{1}_{\{G>-d_2\}}\right]. \quad (122)$$

For arbitrary $y \in \mathbf{R}$,

$$\mathbf{E}^{\mathbf{Q}}\left(e^{bG}\mathbf{1}_{\{G>-y\}}\right) = e^{b^2/2}\left[m(b)\Phi(y+b) + \phi(y+b)\sum_{k=1}^{N}c_k^*He_{k-1}(-y-b)\right]. \quad (123)$$

To calculate

$$\mathbf{E}^{\mathbf{Q}}\left(Ge^{bG}\mathbf{1}_{\{G>-y\}}\right) = \frac{\partial}{\partial b}\mathbf{E}^{\mathbf{Q}}\left(e^{bG}\mathbf{1}_{\{G>-y\}}\right) \quad (124)$$

we need (see (105))

$$\frac{\partial c_k^*}{\partial b} = \frac{\partial}{\partial b}\sum_{\ell=k}^{N}\binom{\ell}{k}b^{\ell-k}c_\ell = (k+1)c_{k+1}^*, \\ k=1,\ldots,N-1, \\ \frac{\partial c_N^*}{\partial b} = 0. \quad (125)$$

Then, noting that $m'(b) = c_1^* = c_1' m(b)$ and $(\phi He_{k-1})' = -\phi He_k$,

$$\begin{aligned}\mathbf{E}^{\mathrm{Q}}\left(Ge^{bG}\mathbf{1}_{\{G>-y\}}\right) &= b\mathbf{E}^{\mathrm{Q}}\left(e^{bG}\mathbf{1}_{\{G>-y\}}\right) + e^{b^2/2}m(b) \\ &\cdot \Bigg[c_1'\Phi(y+b) + \phi(y+b) \\ &\cdot \sum_{k=0}^{N-2}(k+2)c_{k+2}'He_k(-y-b) + \phi(y+b) \\ &\cdot \sum_{k=0}^{N} c_k' He_k(-y-b)\Bigg].\end{aligned} \tag{126}$$

Setting $y = d_2$ and putting the pieces together yield the result. When $\tilde{c} = 0$ this reduces to $S_0 e^{-\delta T}\phi(d_1)$, which is the sensitivity of the Black-Scholes call price to $b = \sigma\sqrt{T}$.

Finally, turn to the sensitivities with respect to c_j, $j \geq 1$. First, let us derive $\partial a/\partial c_j$; from (89),

$$\begin{aligned}&\frac{\partial a}{\partial c_j} e^{a+(b^2/2)} \sum_{k=0}^{N} b^k c_k + b^j e^{a+(b^2/2)} = 0 \\ &\text{or } \frac{\partial a}{\partial c_j} = -\frac{b^j}{\sum_{k=0}^{N} b^k c_k} = -b^j e^{a+(b^2/2)+(\delta-r)T}.\end{aligned} \tag{127}$$

Next,

$$\begin{aligned}&e^{-rT}\frac{\partial}{\partial c_j}\mathbf{E}^{\mathrm{Q}}\left(S_0 e^{a+bG} - K\right)_+ \\ &= e^{-rT}\frac{\partial}{\partial c_j}\int_{-d_2}^{\infty}\left(S_0 e^{a+by} - K\right)\phi(y)\sum_{k=0}^{N} c_k He_k(y)\,dy \\ &= e^{-rT}\int_{-d_2}^{\infty} S_0 \frac{\partial a}{\partial c_j} e^{a+by}\phi(y)\sum_{k=0}^{N} c_k He_k(y)\,dy \\ &+ e^{-rT}\int_{-d_2}^{\infty}\left(S_0 e^{a+by} - K\right)\phi(y) He_j(y)\,dy.\end{aligned} \tag{128}$$

The first integral reduces to $S_0\Delta\partial a/\partial c_j$ (see the derivation of Δ above). The second integral is evaluated by repeated partial integration: if $g(x)$ is continuously differentiable k times and does not grow too fast,

$$\begin{aligned}&\int_y^{\infty} g(x)\phi(x) He_j(x)\,dx \\ &= \phi(y)\sum_{k=0}^{j-1} g^{(k)}(y) He_{j-1-k}(y) \\ &+ \int_y^{\infty} g^{(j)}(x)\phi(x)\,dx.\end{aligned} \tag{129}$$

Hence the second integral in (128) becomes

$$\begin{aligned}&S_0 e^{a-bd_2-rT}\phi(d_2)\sum_{k=1}^{j-1} b^k He_{j-1-k}(-d_2) \\ &+ S_0 e^{a-rT}\int_{-d_2}^{\infty} b^j e^{by}\phi(y)\,dy \\ &= S_0 e^{a-bd_2-rT}\phi(d_2)\sum_{k=1}^{j-1} b^k He_{j-1-k}(-d_2) \\ &+ b^j S_0 e^{a-rT+(b^2/2)}\Phi(d_1).\end{aligned} \tag{130}$$

Putting all this together (using (103)) we get the formula for $\partial C_0/\partial c_j$.

6. Two Applications

6.1. Equity Indexed Annuities. Hardy [12] describes the main types of equity indexed annuities (EIAs), point-to-point, annual ratchet and high water mark, in which an embedded European call option has to be priced. We consider compound ratchet EIAs without life-of-contract guarantee to explore the dependence of ratchet premium options on skewness and kurtosis of returns.

A single premium P is paid by the policy-holder, and the benefit under the ratchet premium contract is ([12], p. 248)

$$B_n = P\prod_{t=1}^{n}\left[1 + \max\left(\alpha\left(\frac{S_t}{S_{t-1}} - 1\right), 0\right)\right]. \tag{131}$$

Here, n is the term of the contract in years, α is the participation rate (a number between 0 and 1), and S_t is the value of the equity index, usually the S&P500 index.

In [12], p. 249, the formula

$$P\left[e^{-r} + \alpha\left(e^{-\delta}\Phi(d_1) - e^{-r}\Phi(d_2)\right)\right]^n \tag{132}$$

for B_n is proved for the value of the premium option under a compound annual ratchet contract in the Black-Scholes model. The same reasoning will now be applied when index log returns have a Gram-Charlier distribution. To find a formula for the price of this option, we assume that under the risk-neutral measure the one-year log returns are independent Gram-Charlier distributed random variables

$$R_t \overset{\mathrm{Q}}{\sim} \mathbf{GC}(a, b; \tilde{c}), \quad t = 1, \ldots, n. \tag{133}$$

Independence of the variables across time implies that the value of the ratchet premium option is (if the annual risk-free rate of interest is r)

$$\begin{aligned}V_0 &= \mathbf{E}^{\mathrm{Q}}\left(e^{-rn}B_n\right) \\ &= P\prod_{t=1}^{n}\left\{e^{-r}\mathbf{E}^{\mathrm{Q}}\left[1 + \max\left(\alpha\left(\frac{S_t}{S_{t-1}} - 1\right), 0\right)\right]\right\}.\end{aligned} \tag{134}$$

As Hardy notes ([12], p. 249), each factor in the product is the price of a one-year European call option on S_t, if initial index price and strike are both equal to one.

Theorem 11. *The no-arbitrage price of the EIA ratchet premium option described above is*

$$V_0 = P\left(e^{-r} + \alpha C_0\right)^n, \tag{135}$$

where C_0 is the price of a one-year call (see (90)) with $S_0 = 1$ and $K = 1$. The first-order sensitivities are

$$\frac{\partial V_0}{\partial \xi} = \alpha P\left(e^{-r} + \alpha C_0\right)^{n-1} \frac{\partial C_0}{\partial \xi}, \tag{136}$$

with ξ replaced with one of S_0, r, b, c_j, and $\partial C_0/\partial \xi$ given in Theorem 9. The second-order sensitivity with respect to S_0 is

$$\begin{aligned}\frac{\partial^2 C_0}{\partial S_0^2} &= \alpha^2 P\left(e^{-r} + \alpha C_0\right)^{n-2} \left(\frac{\partial C_0}{\partial S_0}\right)^2 \\ &\quad + \alpha P\left(e^{-r} + \alpha C_0\right)^{n-1} \frac{\partial^2 C_0}{\partial S_0^2}.\end{aligned} \tag{137}$$

We now show the effect of skewness and kurtosis on the ratchet premium option values; this is done using the four-parameter distribution $\mathbf{GC}(a, b; 0, 0, c_3, c_4)$, for which this is especially simple. We also compute prices using the six-parameter $\mathbf{GC}(a, b; c_1, c_2, c_3, c_4)$ distribution.

Our application involves an annual ratchet, and so the option C_0 in Theorem 11 has a maturity of one year; we thus need the distribution of one-year returns. For pricing that is consistent with the (derivatives) market what is needed is the one-year distribution of returns under the risk-neutral measure, which should be obtained from observed derivative prices; see León et al. [11] or Rompolis and Tzavalis [21] for more details. For illustrative purposes, we estimated the four parameters of that distribution by maximum likelihood ([6, 22]) to obtain the parameters of the $\mathbf{GC}(a, b; 0, 0, c_3, c_4)$:

$$\begin{aligned} a &= 0.0687, \\ b &= 0.1685, \\ c_3 &= -0.1150, \\ c_4 &= 0.03598. \end{aligned} \tag{138}$$

This says that the skewness and excess kurtosis of the fitted distribution are

$$\begin{aligned} \bar{s} &= 6c_3 = -0.6898, \\ \bar{k} &= 24c_4 = 0.8634. \end{aligned} \tag{139}$$

This point is on the boundary of the feasible $(\bar{s}, \bar{k})$ region; see Figure 3.

We use the parameters a and b while varying the values of c_3, c_4 to show how EIA premiums vary with skewness and excess kurtosis. Figure 4 shows the ratchet premium option values (Theorem 11) of an EIA with participation rate $\alpha = 0.6$ and single premium $P = 100$, as a function of skewness $\bar{s}$ and excess kurtosis $\bar{k}$. The risk-free rate is 3%, the dividend rate is 2%, and the term is 7 years. The martingale condition (89) must hold and so a is replaced with a' so that (89) is satisfied. (*N.B.* The Black-Scholes model corresponds to $c_3 = c_4 = 0$, in which case $a' = (r - \delta)T - b^2/2$; this is -0.004193 with the values of r, δ, and b we are using and $T = 1$.)

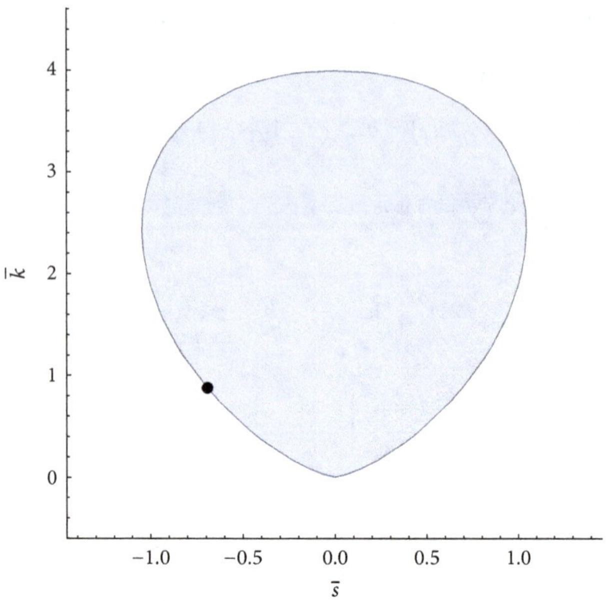

Figure 3: Feasible region for the skewness and excess kurtosis of the $\mathbf{GC}(a, b; 0, 0, c_3, c_4)$ distributions and maximum likelihood estimate.

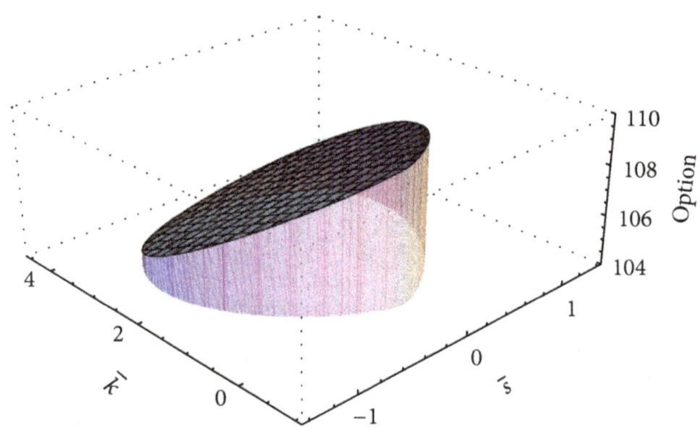

Figure 4: Ratchet premium option prices as a function of skewness and excess kurtosis, with $\mathbf{GC}(a, b; 0, 0, c_3, c_4)$ distributions.

The effect of varying (c_3, c_4) is quite significant. The highest and lowest ratchet premium options in the graph are 109.30 to 104.24, which is the range [95%, 100%] as a proportion of the Black-Scholes premium option ($109.26), which has skewness and excess kurtosis equal to zero. (The maximum value of the premium option is reached at $(c_3, c_4) = (0.0230, 0.00332567)$ and the minimum at $(-0.0836, 0.163)$; these points correspond to $(\bar{s}, \bar{k})$ equal to $(.138, .0798)$ and

$(0.023, 0.003326)$, resp.) The dependence on $(\bar{s}, \bar{k})$ is nearly linear. This is easily explained. The price of a call is

$$C_0 = e^{-rT}\mathbf{E}^{\mathbf{Q}}(S_T - K) = e^{-rT}\int_{(1/b)(\log(K/S_0)-a')}^{\infty}\left(S_0e^{a'+bx} - K\right)\phi(x) \cdot \left(\sum_{k=0}^{N} c_k He_k(x)\right)dx. \quad (140)$$

The sensitivity of C_0 to c_j $(j = 3, 4)$ is thus

$$e^{-rT}\frac{\partial a'}{\partial c_j}\mathbf{E}^{\mathbf{Q}}\left(S_T \mathbf{1}_{S_T>K}\right) + e^{-rT}\int_{(1/b)(\log(K/S_0)-a')}^{\infty}\left(S_0e^{a'+bx} - K\right)\phi(x) \cdot He_j(x)\,dx. \quad (141)$$

The first term above is relatively small, and the second one is almost constant, since a' does not change much with c_j. This carries over to the sensitivity of the ratchet premium option V_0 to changes in c_j (see the first formula in Theorem 11).

The parameters fitted by maximum likelihood for the $\mathbf{GC}(a, b; c_1, c_2, c_3, c_4)$ distribution are

$$\begin{aligned} a &= 0.1174, \\ b &= 0.1595, \\ c_1 &= -0.3053675695201066, \\ c_2 &= 0.09542079373489153, \\ c_3 &= -0.12383971126335243, \\ c_4 &= 0.06120331530131559. \end{aligned} \quad (142)$$

The martingale condition implies $a' = 0.0451$, and the EIA premium option has value 107.90. The skewness and excess kurtosis of the fitted distribution are $\bar{s} = -0.5437$ and $\bar{k} = 0.5092$. (*N.B.* The feasible region for $(\bar{s}, \bar{k})$ in the $\mathbf{GC}(a, b; c_1, c_2, c_3, c_4)$ case is naturally larger than the one for the $\mathbf{GC}(a, b; 0, 0, c_3, c_4)$; in this case $(\bar{s}, \bar{k})$ are just outside the region in Figure 3.) It was checked numerically that the parameters c_1 to c_4 lead to a nonnegative function.

Table 1 shows ratchet premium option values based on Gram-Charlier with four parameters that are lower than the Black-Scholes based ones. Rows A and F are Black-Scholes cases, row B is our estimation of the four-parameter Gram-Charlier (as explained above), row C is the four-parameter Gram-Charlier with the largest negative skewness possible (see Figure 1), row D is similar but with the largest positive skewness, and row E has maximum kurtosis but no skewness. Rows A to E have the same parameter b, the one we estimated from S&P500 for a four-parameter Gram-Charlier, while row F has the same b as the six-parameter Gram-Charlier we estimated. Skewness and kurtosis are shown, followed by the break-even (or "fair") α ("B-E α" in the table) and then the ratchet premium option values for two specific choices for α. Percentage differences with the values in row A (resp., F) are also displayed ("$\Delta A\%$" in the table). It is notable that in all cases the effect of moving away from Black-Scholes is more pronounced for the break-even participation rates than for the option values themselves. The scenario with the greatest impact (compared to Black-Scholes) is the one zero skewness and maximal excess kurtosis (row E). Insurers are unlikely to accept a cost of contract (V_0) greater than the received initial premium (\$100); they may thus wish to adjust the participation rate α down to the break-even rate (to wit, see Lin and Tan [23]). The Black-Scholes break-even participation rate in row A is 41.9%, but it increases from 44.3% up to 49.3% in rows B to E. These are significant increases, ranging from 5.5% to 17.5%. Rows F and G are based on two quite different risk-neutral probability distributions, but the ratchet premium options are surprisingly similar. The parameters we estimated may not be those that a practitioner would use, but what this example says is that skewness and kurtosis have a very real importance in the pricing of equity indexed annuities and that the fair participation rate is, at least here, more sensitive to nonzero skewness and excess kurtosis than are option values. (See also [24] regarding the separate effects of skewness and excess kurtosis on option values.)

Let us compare this with the approach in [11]. That paper is about a subclass of the Gram-Charlier distributions described in this paper; the authors call their distributions "seminonparametric" and have density (6), where $p(\cdot)$ is the square of the polynomial $q(\cdot)$ If $q(x)$ is of order m, then it is possible to express $p(x) = q(x)^2$ as

$$p(x) = \sum_{j=0}^{2m}\delta_j He_j(x). \quad (143)$$

However, most nonnegative polynomials cannot be expressed as squares of another polynomial, so the order-m SNP family is a strict subset of the order-$2m$ Gram-Charlier family. The order $2m$ SNP family has $2m + 2$ free parameters (the coefficients of an order-$2m$ polynomial plus the location and dispersion parameters a, b, subject to density integrating to one, which in effect removes one parameter); by comparison, the $\mathbf{GC}(a, b; c_1, \ldots, c_{2m})$ distributions have $2m+2$ parameters, but they are not "free" because the resulting density must be nonnegative. It is then not *a priori* clear which of $\mathbf{SNP}(2m)$ or $\mathbf{GC}(a, b; c_1, \ldots, c_{2m})$ would do best. We looked at this problem with the S&P500 annual prices, comparing $\mathbf{SNP}(2)$ and $\mathbf{GC}(a, b; 0, 0, c_3, c_4)$, which both have four parameters. The likelihood function is written the same way in both cases; there are just different restrictions on the parameters. The results for $\mathbf{GC}(a, b; 0, 0, c_3, c_4)$ are given in (138), and those for the $\mathbf{SNP}(2)$ are

$$\begin{aligned} a &= -0.2445, \\ b &= 0.1940, \\ c_1 &= 1.6140, \end{aligned}$$

Table 1: Comparison of ratchet option values computed with the Black-Scholes formula and Gram-Charlier distributions. Rows A to F are based on the four-parameter **GC**$(a, b; 0, 0, c_3, c_4)$, while row G is based on a six-parameter **GC**$(a, b; c_1, c_2, c_3, c_4)$.

	a'	b	c_1	c_2	c_3	c_4	$\bar{s}$	$\bar{k}$	B-E α	$\Delta(A)\%$	V_0 ($\alpha = 0.6$)	$\Delta(A)\%$	V_0 ($\alpha = 0.419$)	$\Delta(A)\%$
A	−0.0042	0.1685	0	0	0	0	0	0	0.419	0	109.26	0	100.00	0
B	−0.0035	0.1685	0	0	−0.1150	0.0360	−0.6898	0.8634	0.443	5.5	107.60	−1.5	98.92	−1.1
C	−0.0034	0.1685	0	0	−0.1749	0.1021	−1.0493	2.4508	0.478	13.9	105.42	−3.5	97.50	−2.5
D	−0.0051	0.1685	0	0	0.1749	0.1021	1.0493	2.4508	0.446	6.3	107.39	−1.7	98.78	−1.2
E	−0.0043	0.1685	0	0	0	0.1667	0	4.0000	0.493	17.5	104.59	−4.3	96.96	−3.0
	a'	b	c_1	c_2	c_3	c_4	$\bar{s}$	$\bar{k}$	B-E α	$\Delta(F)\%$	V_0 ($\alpha = 0.6$)	$\Delta(F)\%$	V_0 ($\alpha = 0.441$)	$\Delta(F)\%$
F	0.0027	0.1595	0	0	0	.0	0	0	0.441	0	107.69	0	100.00	0
G	0.0451	0.1595	−.3054	.09542	−0.12384	0.06120	−0.5437	0.5091	0.438	−0.7	107.90	0.2	100.14	0.14

$$c_2 = 1.1733,$$
$$c_3 = 0.4253,$$
$$c_4 = 0.07320. \tag{144}$$

This translates into the following moments for the **SNP**(2) fitted distribution:

$$\begin{aligned} &\text{mean} && 0.0687 \\ &\text{standard deviation} && 0.1671 \\ &\text{skewness} && -0.6289 \\ &\text{excess kurtosis} && 2.5502. \end{aligned} \tag{145}$$

The excess kurtosis is strikingly far from the data's 0.8903. The likelihood function evaluated at that estimate was lower than the likelihood evaluated at the parameters of the **GC**$(a, b; 0, 0, c_3, c_4)$ distribution. This is a puzzling result that does not mean that **GC**$(a, b; 0, 0, c_3, c_4)$ will always do better than **SNP**(2); it does however show that restricting the search to squares of polynomials does have consequences. León et al. [11] derive formulas for European option prices, taking the martingale restriction into account, based on expression (143); in this paper we do the same calculation for an arbitrary Gram-Charlier distribution. It is easy to see that the SNP class is not closed under convolution; that is, the distribution of the sum of two independent SNP variables is in general not an SNP distribution; the general formulas we derive are essential in order to add independent Gram-Charlier distributed variables (including those in the SNP subset), and therefore the general formulas are needed to define Gram-Charlier processes. León et al. [11] are correct in pointing out that there is an advantage in using $p(x) = q(x)^2$ as far as the nonnegativity constraint is concerned, but we believe that it is no more difficult to estimate the parameters of a general Gram-Charlier distribution, including the non-negativity constraint in the maximization procedure (rather than limiting the parameter space). A full comparison of the numerical and statistical advantages/disadvantages of our approach versus the one in [11] is an interesting avenue for further research.

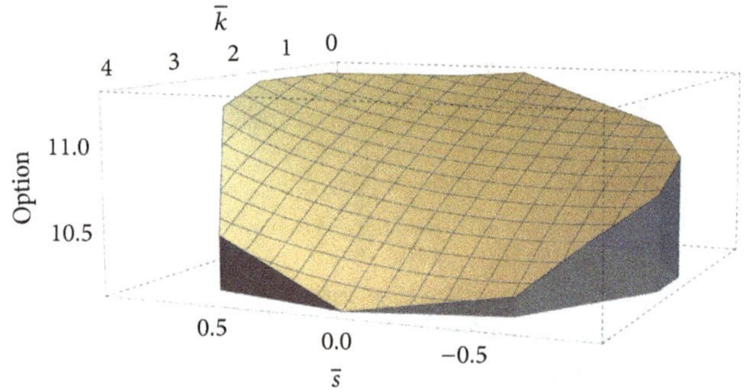

Figure 5: Lookback option prices as a function of skewness and excess kurtosis.

6.2. Lookback Options. Consider a European lookback call option with fixed strike, with payoff $(\bar{S} - K)+$, where $\bar{S}$ is the maximum of the stock price between time 0 and time T. We let the time points used to determine the maximum be equally spaced. There is no closed form formula for the distribution of the maximum price nor for the price of such a discretely monitored lookback option, so simulation is used. We generated normal random variables and used the technique described in Section 2.5. The assumptions are maturity $T = 1$, monthly monitoring, annual volatility 0.1685 (where this number comes from is explained in the next subsection). The interest rate is set to 0, initial stock price is \$100, and strike price is also \$100.

Figure 5 shows the option prices for values of skewness and excess kurtosis that can be achieved with the **GC**$(a, b; 0, 0, c_3, c_4)$ family for monthly rates of return. Roughly speaking, prices decrease when excess kurtosis increases, and also as kurtosis decreases, though the dependence on c_3 and c_4 becomes less linear as $\bar{k}$ increases. When $c_3 = c_4 = 0$ (Gaussian case) the option price is 11.277. The price is a little higher (\$11.286) when $\bar{s} = 0.75$ and $\bar{k} = 1.0$. The lowest price (\$10.163) occurs when $\bar{s} = 0$ and $\bar{k} = 4$. Skewness and kurtosis thus have a significant effect on lookback option prices; the effect will of course vary depending on the parameters. (Observe that all those prices are quite far from the Black-Scholes continuous-monitoring lookback price, \$14.17.)

7. Discussion and Conclusion

This paper tries to provide a framework for applying Gram-Charlier series to option pricing. We have shown that option prices may be significantly affected by varying skewness or kurtosis away from their Gaussian values. The practical consequence is that using the Black-Scholes formula (*i.e.*, zero skewness and excess kurtosis) may distort option prices. Gram-Charlier distributions capture skewness and kurtosis and retain a lot of the tractability of the normal distribution. Estimating parameters remains a challenge for Gram-Charlier distributions involving more than four parameters.

Convergence of Gram-Charlier and other expansions in option pricing and other applied problems is an area for further research. The heuristic derivation of the Gram-Charlier/Edgeworth series of a density $g(\cdot)$ in terms of the normal density and its derivatives (see, e.g., [25], Section 3.2) may give the impression that the inverted series should converge or at least have some asymptotic property; this is not the case in general. Gram-Charlier (or Edgeworth) expansions do not always converge to the true probability distribution, as Section 2.3 indicates; see [26], Section 17, or [25] for more details. In the context of the Central Limit Theorem, the Edgeworth series for the density of the normalized sum or random variables have asymptotic properties, when the number of summands n tends to infinity. Details are given in [25]. Multivariate versions of the Gram-Charlier series exist, for instance, [10, 27, 28], apply such generalizations.

The Gram-Charlier type B series are based on the Poisson distribution, rather than the normal. There are several other expansions that can be used. For instance, series of Laguerre polynomials may be used for densities on the positive half-line (a convergent Laguerre series has some numerical success in pricing Asian options; see [29]). The Jacobi polynomials [30] are orthogonal over the interval $(-1, 1)$, but any other finite interval may be chosen by making the obvious change of variable, and this leads to series expansions for densities with compact support, based on the beta distribution. An application of log-transformed shifted Jacobi polynomials to arbitrary approximation of probability distributions on the positive half-line may be found in [31].

Competing Interests

The authors declare that they have no competing interests.

Acknowledgments

The authors thank the Committee for Knowledge Extension and Research (CKER) of the Society of Actuaries for partially funding this project.

References

[1] P. A. Abken, D. B. Madan, and S. Ramamurtie, "Estimation of risk-neutral and statistical densities by *Hermite polynomial* approximation," Working Paper 96-5, Federal Reserve Bank of Atlanta, 1996.

[2] D. Backus, S. Foresi, K. Li, and L. Wu, "Accounting for biases in Black-Scholes," Working Paper, Stern School of Business, New York University, New York, NY, USA, 1997.

[3] J.-P. D. Chateau, "Marking-to-model credit and operational risks of loan commitments: a Basel-2 advanced internal ratings-based approach," *International Review of Financial Analysis*, vol. 18, no. 5, pp. 260–270, 2009.

[4] J. D. Chateau, "Valuing european put options under skewness and increasing (excess) kurtosis," *Journal of Mathematical Finance*, vol. 4, no. 3, pp. 160–177, 2014.

[5] C. J. Corrado, "The hidden martingale restriction in Gram-Charlier option prices," *Journal of Futures Markets*, vol. 27, no. 6, pp. 517–534, 2007.

[6] E. Jondeau and M. Rockinger, "Gram-Charlier densities," *Journal of Economic Dynamics and Control*, vol. 25, no. 10, pp. 1457–1483, 2001.

[7] E. Jurczenko, B. Maillet, and B. Négrea, "Skewness and kurtosis implied by option prices: a second comment," Discussion Paper of the LSE-FMG 419, 2002.

[8] E. Jurczenko, B. Maillet, and B. Negrea, "A note on skewness and kurtosis adjusted option pricing models under the Martingale restriction," *Quantitative Finance*, vol. 4, no. 5, pp. 479–488, 2004.

[9] J. L. Knight and S. E. Satchell, *Return Distributions in Finance*, Butterworth Heineman, 2001.

[10] E. Schlögl, "Option pricing where the underlying assets follow a Gram/Charlier density of arbitrary order," *Journal of Economic Dynamics & Control*, vol. 37, no. 3, pp. 611–632, 2013.

[11] A. León, J. Mencía, and E. Sentana, "Parametric properties of semi-nonparametric distributions, with applications to option valuation," *Journal of Business & Economic Statistics*, vol. 27, no. 2, pp. 176–192, 2009.

[12] M. Hardy, *Investment Guarantees: Modeling and Risk Management for Equity-Linked Life Insurance*, John Wiley & Sons, New York, NY, USA, 2003.

[13] P. Gaillardetz and X. S. Lin, "Valuation of equity-linked insurance and annuity products with binomial models," *North American Actuarial Journal*, vol. 10, no. 4, pp. 117–144, 2006.

[14] P. Boyle and W. Tian, "The design of equity-indexed annuities," *Insurance: Mathematics & Economics*, vol. 43, no. 3, pp. 303–315, 2008.

[15] S. G. Kou, "Discrete barrier and lookback options," in *Handbooks in Operations Research and Management Science*, vol. 15, chapter 8, pp. 343–373, 2007.

[16] D. E. Barton and K. E. Dennis, "The conditions under which Gram-Charlier and Edgeworth curves are positive definite and unimodal," *Biometrika*, vol. 39, no. 3-4, pp. 425–427, 1952.

[17] H. Cramér, *On Some Classes of Series Used in Mathematical Statistics* , Skandinaviske Matematiker Congres, Copenhagen, Denmark, 1925.

[18] W. Feller, *An Introduction to Probability Theory and Its Applications*, vol. 2, John Wiley & Sons, New York, NY, USA, 1971.

[19] J. Stoyanov, "Krein condition in probabilistic moment problems," *Bernoulli*, vol. 6, no. 5, pp. 939–949, 2000.

[20] F. D. Rouah and G. Vainberg, *Option Pricing Models and Volatility Using Excel-VBA*, John Wiley & Sons, New York, NY, USA, 2007.

[21] L. S. Rompolis and E. Tzavalis, "Retrieving risk neutral densities based on risk neutral moments through a Gram-Charlier series expansion," *Mathematical and Computer Modelling*, vol. 46, no. 1-2, pp. 225–234, 2007.

[22] E. B. Del Brio and J. Perote, "Gram-Charlier densities: maximum likelihood versus the method of moments," *Insurance: Mathematics and Economics*, vol. 51, no. 3, pp. 531–537, 2012.

[23] X. S. Lin and K. S. Tan, "Valuation of equity-indexed annuities under stochastic interest rates," *North American Actuarial Journal*, vol. 7, no. 4, pp. 72–91, 2003.

[24] S. R. Das and R. K. Sundaram, "Of smiles and smirks: a term structure perspective," *Journal of Financial and Quantitative Analysis*, vol. 34, no. 2, pp. 211–239, 1999.

[25] J. E. Kolossa, *Series Approximation Methods in Statistics*, Springer, New York, NY, USA, 3rd edition, 2006.

[26] H. Cramér, *Mathematical Methods of Statistics*, Princeton University, Princeton, NJ, USA, 1946.

[27] E. B. Del Brio, T.-M. Ñíguez, and J. Perote, "Gram-Charlier densities: a multivariate approach," *Quantitative Finance*, vol. 9, no. 7, pp. 855–868, 2009.

[28] E. B. Del Brio, T.-M. Ñíguez, and J. Perote, "Multivariate semi-nonparametric distributions with dynamic conditional correlations," *International Journal of Forecasting*, vol. 27, no. 2, pp. 347–364, 2011.

[29] D. Dufresne, "Laguerre series for Asian and other options," *Mathematical Finance*, vol. 10, no. 4, pp. 407–428, 2000.

[30] N. N. Lebedev, *Special Functions and Their Applications*, Dover, New York, NY, USA, 1972.

[31] D. Dufresne, "Fitting combinations of exponentials to probability distributions," *Applied Stochastic Models in Business and Industry*, vol. 23, no. 1, pp. 23–48, 2007.

21

Stochastic Models for the Infectivity Function in an Infinite Population of Susceptible Individuals

Viswanathan Arunachalam and Liliana Blanco

Department of Statistics, Universidad Nacional de Colombia, Bogotá, Colombia

Correspondence should be addressed to Viswanathan Arunachalam; varunachalam@unal.edu.co

Academic Editor: Chin-Shang Li

Two stochastic models to study the course of the transient behaviour of the total infectivity present in an infinite population of susceptible individuals are developed. The conditional intensity function of the contagion comprises two components: one is due to the external sources only and the other is the contribution of each of the infected persons which is nonstationary in nature. The statistical characteristics of the number of infected individuals at any time are explicitly obtained. Estimation of the model parameters is also indicated.

1. Introduction

Infectious diseases impact public health and consequently a nation's progress in several ways. Diseases such as influenza, gastroenteritis among humans and bird flu, and foot and mouth diseases among animals are a matter of great public concern. In the face of such emerging diseases, time is of the essence in developing and implementing interventions. The ability to predict the transient infectivity function (infected population) imposed by such epidemics is essential for intervention design. Stochastic modeling of such infectious diseases is the only effective tool available for the health planners for optimal intervention design and their implementation.

From the time Kermack and McKendrick [1] published their classic in the proceedings of the Royal Society, there have been various approaches to describe the dynamics of the spread of the infectious diseases. They include stochastic compartment models, conditional intensity models [2, 3], simulation models, inference models [4], Monte Carlo methods [5], and SIR model [6–9]. Recently, Lie algebra approach was used to study susceptible-infected-susceptible (SIS) epidemic models [10, 11].

Kermack and McKendrick [1] model is characterised by the deterministic equation:

$$S'(t) = S(t)\left[\int_0^t S'(t-\tau)\,g(\tau)\,d\tau - f(t)\right], \qquad (1)$$

where $S(t)$ represents the number of susceptible individuals at time t. This model is completely characterized by $S(0)$ and the functions $f(t)$ and $g(t)$. The function $f(t)$ models a possible time dependent outside source of infection or represents the history of the epidemic up to time zero. The function $g(t)$ describes the infectivity of an individual as a function of his age of the infection t, that is, the time elapsed since the moment of his own infection.

Gielen [12] proposed a natural stochastic version of the above model. He observed that his model fills the gap between Bailey's simple stochastic epidemic model [13] (which is obtained as a special case of Kermack and McKendrik's model when the infectivity parameters $f(t)$ and $g(t)$ are constants) and the Reed-Frost model [12] (when the functions $f(t)$ and $g(t)$ are Dirac delta functions). Gielen [14] further proposed an interesting variant of the above model for a small epidemic in a large population of susceptible individuals. This model can be seen to be the direct stochastic version of the Keyfitz [15] deterministic model, governed by the renewal equation:

$$I'(t) = \int_0^t I'(t-\tau)\, g(\tau)\, d\tau + f(t), \tag{2}$$

where $I(t)$ indicates the number of infective individuals at time t in an infinite population of susceptible individuals. Gielen [14] explicitly derived the probability distribution of the number of infective individuals at any time t, using which he arrives at several interesting results. However, the expression for the distribution of the number of infective individuals which involves n-fold integrals can at best be thought of as a formal solution and is not useful for explicit evaluations. Consequently, the statistical characteristics of the number distribution which are very important in the optimal decision making problems cannot be arrived at that in the general case. One of the purposes of the present work is to analyse the stochastic version of the Keyfitz model [15] and obtain explicitly the statistical characteristics of the number of infective individuals in an infinite population.

The traditional epidemic models assumed the contagion rate to depend on the number of infected persons in the population, thereby tacitly assuming that all individuals exhibit the same pattern of infectiousness. However, this being not so in the reality led to the notion of infectivity function. The infectivity function $f(t)$ is the force of infection at any time t which is the probability/unit of time that a susceptible becomes infected. In the dynamic contagion models, the infectivity functions impact endogenous and exogenous factors of the underlying infection. The proposed stochastic infection models in the present paper split the conditional intensity function of infection into endemic and nonstationary epidemic components. The latter, in our models, is the spread of the disease by person to person contact and provides a description preserve at time t caused by each of the infected individuals. The endemic component models imported cases and is contrary to the epidemic component. The number of infected persons is modeled as a point process with its intensity function dependent on the point process itself, thus providing a more effective model to capture the contagion phenomena. The mean number of infected persons obtained from our stochastic epidemic model, for the special case when both the endemic and epidemic components of the intensity function are constants, satisfies Keyfitz's renewal equation (2). Gielen's stochastic version of Keyfitz's model predicts a negative binomial distribution for the number of infected persons (see [14]) with the same mean thus validating our approach. The proposed formulation is broad enough to handle a wide variety of disease progression models for which infectivity varies both within and external to factors, including Ebola virus disease (EVD) which has generated a large epidemic in West Africa and other parts of the world [16, 17].

The present paper is organised as follows. In the next section we propose two stochastic models: the first one is an endemic model in which each person infected from outside source contributes a random amount of infectivity to the force of infectivity (or infectivity function) which decreases at a constant rate over the time. The total infectivity present at any time t is the quantity of interest whose statistical characteristics are obtained. The second model which captures an epidemic has both endemic and epidemic components in the intensity function. The statistical characteristics of the number of infected persons are explicitly obtained. Section 3 presents a discussion on the developed models along with estimation of maximum likelihood parameters of the model.

2. The Models

In this section we propose an endemic and an epidemic stochastic model to describe the dynamics of a disease in a large population of susceptible individuals.

2.1. An Endemic Model. We assume that at the time of origin, there are infinitely many susceptible individuals. A susceptible may be infected at time $t > 0$, after which it is classified as infected. It is assumed that the random variable denoting the time between two successive infections is exponentially distributed and that infections are due to a source of infection from outside of the population. Our interest is analysing the statistical characteristics of the total force of infectivity $f(t)$ present in the population at any time t due to all the infections until time t. In this regard, the contribution to $f(t)$ due to an infection at time t_i ($< t$) is composed of two components, a discrete jump $a(t_i)$ and a nonstationary component $h(t - t_i)$. Thus the force of infectivity at any time t can be written as follows:

$$f(t) = \sum_{i=1}^{N_t} a(t_i)\, h(t - t_i), \tag{3}$$

where N_t is the number of infections until time t.

From our assumptions, N_t is easily seen to be Poisson process. In order to derive the probability frequency functions of the Markov process $f(t)$ we define

$$\pi(y,t)\, dy = P(y < f(t) < y + dy \mid f(0) = 0). \tag{4}$$

The Chapman-Kolmogorov backward equation yields

$$\begin{aligned}\pi(y,t) &= (1 - \lambda\Delta)\, \pi(y, t - \Delta) \\ &\quad + \lambda\Delta\pi(y - ah(t), t - \Delta) + o(\Delta),\end{aligned} \tag{5}$$

where we have assumed $a(t_i) = a$.

The above equation after simplification yields

$$\frac{\partial \pi(y,t)}{\partial t} = -\lambda\pi(y,t) + \lambda\pi(y - ah(t), t) \tag{6}$$

with initial condition

$$\pi(0,t) = 1. \tag{7}$$

Applying the Fourier transform

$$\tilde{\pi}(s,t) = \frac{1}{2\pi} \int_{-\infty}^{\infty} \pi(y,t)\, e^{isy} dy \tag{8}$$

to both sides of (6), we obtain

$$\frac{\partial \tilde{\pi}(s,t)}{\partial t} = -\lambda \tilde{\pi}(s,t) + \lambda e^{isah(t)} \tilde{\pi}(s,t) = -\lambda \left(1 - e^{isah(t)}\right) \tilde{\pi}(s,t). \tag{9}$$

The above partial differential equation with (7) can be solved to give the solution:

$$\tilde{\pi}(s,t) = \exp\left(-\lambda t + \lambda \int_0^t e^{isah(u)} du\right). \tag{10}$$

Now $\pi(y,t)$ could easily be obtained by inverting (6) for specific forms of the function $h(\cdot)$. In the foregoing model, we have assumed that the contribution of each infected individual to the force of infectivity $f(t)$ is the same constant a. However in practice, this may not be so. Thus if we assume the jumps "a" to be a random variable with probability density $p(a)$, then $\pi(y,t)$ in this case can be obtained with minor modifications to the above model. The backward equation (3) now reduces to

$$\frac{\partial \pi}{\partial t} = -\lambda \pi + \lambda \int_a \pi(y - ah(t), t)\, p(a)\, da \tag{11}$$

and the corresponding equation after applying Fourier transform on both sides turns out to be

$$\frac{\partial \tilde{\pi}}{\partial t} = -\lambda \tilde{\pi} + \lambda \int_a e^{isah(t)} \tilde{\pi}(s,t)\, p(a)\, da. \tag{12}$$

Now letting

$$\phi(u) = \int_a e^{iau} p(a)\, da \tag{13}$$

which is nothing but the characteristic function of the random variable a and using (13) in (12) we obtain

$$\frac{\partial \tilde{\pi}}{\partial t} = -\lambda \tilde{\pi} + \lambda \tilde{\pi} \phi(sh(t)). \tag{14}$$

Thus we obtain

$$\tilde{\pi} = \exp\left(-\lambda t + \lambda \int_0^t \phi(sh(\tau))\, d\tau\right). \tag{15}$$

$\pi(y,t)$ can be obtained by inverting (15) for specific forms of the function $h(t)$.

2.2. An Epidemic Model. In the earlier endemic model, it was assumed that the spread of the infection is due to external source only while each infection contributed a nonstationary input to the force of the infection. The present section develops an epidemic model which spreads due to an external source as well as through contagion of the infected population. Instead of the number of infected persons, we use the more realistic concept of the force of infection. Thus the number of infected at any time t is a point process whose conditional intensity function comprises both endemic and epidemic contributions. More specifically, we assume a conditional intensity function of the form

$$f(t) = \lambda + \sum_i b \exp(-a(t - t_i)). \tag{16}$$

The above form is motivated by the following interpretation. The terms on the RHS could be interpreted as follows: the intensity function $f(t)$ which consists of two parts:

(1) The external rate of infection which occurs at a constant rate λ.

(2) The contagion rate due to the already infected persons consists of a constant jump b at the time of infection which decays at a constant rate a so that the contribution to the effective contagion rate at later time t due to an infection at time t_i is $b\exp(-a(t - t_i))$. Thus the net contribution of the infection rate at time t due to all previous infections at $t_1, t_2, \ldots, t_n$ is

$$\sum_i b \exp(-a(t - t_i)). \tag{17}$$

The rate could as well be interpreted as a rate at which the instantaneous jump b decreases due to intervention action.

Let $\pi(y,t)dy$ be the probability that $f(t)$ has a value between y and $y+dy$ at time t. Writing the forward equation,

$$\pi(y, t+\Delta)\, dy = \pi(y - b; t)(y - b)\Delta\, dy + (1 - y\Delta)\pi(y - a(\lambda - y)\Delta; t) + o(\Delta). \tag{18}$$

The above equation is obtained by noting that the rate of decrease of the infection rate is specified by

$$\frac{dy}{dt} = a(\lambda - y). \tag{19}$$

Thus (18) reduces to

$$\frac{\partial \pi}{\partial t} = (y - b)\pi(y - b; t) + (a - y)\pi(y,t) - a(\lambda - y)\frac{\partial \pi}{\partial y}. \tag{20}$$

Letting

$$P(n,t) = \int_0^\infty y^n \pi(y,t)\, dy \tag{21}$$

(20) reduces to

$$\frac{\partial P(n,t)}{\partial t} = an\lambda P(n-1,t) - anP(n,t) + \sum_i \binom{n}{i} b^i P(n - i + 1, t) \tag{22}$$

with initial condition

$$P(0,1) = 1,$$
$$P(n,0) = \lambda^n. \quad (23)$$

Letting $n = 1$ and solving (22), we obtain

$$P(1,t) = \frac{a\lambda}{a-b} - \frac{b\lambda}{a-b}\exp(-(a-b)t). \quad (24)$$

When $b = 0$, so that the contribution of the second term of the contagion rate is not there and only external factor is present, we obtain $P(1,t) = \lambda$ and $E(N(t)) = \lambda t$ which is consistent with the assumptions of the model.

From the solution $P(1,t)$, we are in a position to obtain the expected number of infected persons up to time t. To see this we note that $\gamma_1(t)$, the product density of order one of infections (i.e., probability an infection occurs between t and $t+\Delta$), is straight away given by

$$\gamma_1(t) = P(1,t) = \int_0^\infty y\pi(y,t)\,dy. \quad (25)$$

Thus the expected number of infections in $(0,t)$ is given by

$$E(N(t)) = \int_0^t \gamma_1(u)\,du$$
$$= \frac{1}{a-b}\left\{a\lambda t - \frac{b\lambda}{a-b}\left(1 - e^{-(a-b)t}\right)\right\}. \quad (26)$$

In a similar form, writing the equation satisfied by the conditional density function $\Pi(y_2,t_2 \mid y_1,t_1,\lambda)$ representing the probability that $f(t)$ has a value between y_2 and $y_2 + dy_2$ at time t_2 given that it has a value between y_1 and $y_1 + dy_1$ at time t_1 and initially λ, we obtain the second order product density $\gamma_2(t_1,t_2)$. Then using the relation

$$E(N(t))^2 = \int_0^t \gamma_1(t_1)\,dt_1 + 2\int_0^t\int_0^t \gamma_2(t_1,t_2)\,dt_1dt_2 \quad (27)$$

we obtain explicit expression for the above equation after some calculations, which we do not give as they are lengthy and unwieldy.

3. Discussion

We have proposed two stochastic models to analyse the dynamics of the spread of a disease. The core of the models is the force of the infection. Each of the infected individual contributes a nonstationary term to the total force of the infection. This comprises a discrete jump at the instant of infection which could be deterministic or random and a response function. The statistical characteristics of the number of infected persons are obtained from the conditional intensity function. The developed models subsume some interesting models available in the literature.

In the epidemic model, letting $b = 0$ in (16) so that the contribution to the intensity function is only from external source at a rate λ, we note that the mean and variance of the number of infected persons are equal to λt showing that the infection grows according to Poisson process with rate λ. On the other hand letting $a = 0$ so that the contribution to the intensity function from the external source and infected individuals are constants λ and b, respectively, we are in the domain of a simple stochastic epidemic [13, 14]. In such a scenario, our model predicts

$$E(N(t)) = \frac{1}{b}\left\{\lambda\left(e^{bt} - 1\right)\right\} \quad (28)$$

and Gielen [14] shows the number of infected persons has a negative binomial distribution with the same mean. In this case with probability 1, there are ultimately infinitely many infective individuals. Also the mean number of infective individuals obtained using our model satisfies the renewal equation (2) of Keyfitz's deterministic model [15].

With the usual notations, the equations of the susceptible-infected-recovered (SIR) model are given by

$$\frac{dS}{dt} = -\frac{\beta IS}{N},$$
$$\frac{dI}{dt} = \frac{\beta IS}{N} - \vartheta I, \quad (29)$$
$$\frac{dR}{dt} = \vartheta I.$$

These equations assume that infected people cause infections at rate β and recover at rate ϑ, giving an exponentially distributed infection duration. The process is obviously memoryless. In contrast, in real life situations, for diseases the "age" of an individual's infection affects his/her infectiousness and consequently probability of recovery. Thus our modeling could be visualised as transcending the memoryless regime.

Gielen [14], in his model, considered the intensity function of the form

$$f(t) = a(t) + \sum g(t - t_i). \quad (30)$$

In such case, the total contribution to the infectivity function by the external source and by an infective during the entire infective period, respectively, is given by

$$\int_0^\infty f(t)\,dt = F(\infty),$$
$$\int_0^\infty g(t)\,dt = G(\infty). \quad (31)$$

These two quantities are dimensionless. If $F(\infty) = \infty$, then, with probability 1, there are finally infinitely many defectives which is the case in our model (see page 8 of the paper, [14]). However in Gielen's model, it is possible that $0 < F(\infty) =: A < \infty$ and $G(\infty) = \infty$. In such a scenario, there is a nonzero probability that infection may not start at all so that the probability of infinite infective individuals ultimately is defective with probability $1 - e^{-A}$. In our model we start with $f(0) = \lambda$ which preclude such an eventuality.

In our endemic model, letting $a = 1$ and $h(t) = \delta(t - 1)$ where δ is the Dirac delta function, we note that each infection increases the force of the infectivity by one unit. This is reflected in the final solution as we get

$$\Pi(n,t) = P(f(t) = n) = \frac{e^{-\lambda t}(\lambda t)^n}{n!}, \quad n = 0, 1, \dots. \tag{32}$$

Thus we arrive at a Poisson process of infections. In the endemic model, the contribution to the force of infectivity arises only from the external source. If we further assume that infections also occur due to the already infected persons but such infected individuals are removed from the pool of infected, then the probability of an infection due to the infected population in $(t, t + dt]$ is given by

$$\int_0^\infty y\Pi(y,t)\,dy\,dt. \tag{33}$$

This can be identified with the product density of order one, namely, $\gamma_1(t)$. Hence the expected number of infected individuals at time t is

$$\int_0^t \gamma_1(u)\,du \tag{34}$$

while the expected number of infected persons due to external source is λt.

Finally, we wish to mention the useful tools that a planner needs in designing interventions is the estimated values of the model parameters. In order to estimate the parameters of our model, a and b, we need the function

$$\gamma_n(t_1, t_2, \dots, t_n; a, b) \tag{35}$$

which is the product density of order n. This function which can serve as the likelihood function specifies the probability that infections occur at $t_1, t_2, \dots, t_n$ under the assumed values of a and b. Although such a function could be obtained, it may not be of much value in actual use due to its unwieldy form. Thus we make the reasonable approximation

$$\begin{aligned}&\gamma_n(t_1, t_2, \dots, t_n; a, b)\\&= \gamma_1(t_1; a, b)\,\gamma_1(t_2 - t_1; a, b)\cdots\gamma_1(t_n - t_{n-1}; a, b).\end{aligned} \tag{36}$$

The estimate of the parameters a and b yields the maximum likelihood parameters $\hat{a}$ and $\hat{b}$, by using methods proposed by Rubin [18] and Ozaki [19].

This paper generalizes the stochastic version of Kermack and McKendrick model of Gielen [14] by considering the transient behaviour of the total infectivity present in an infinite population of susceptible individuals. A noteworthy aspect is the explicit analytical expressions for the probability distribution and statistical characteristics of the number of infected individuals. The proposed models can also be solved using Lie algebra approach for finding analytical solutions. The proposed model could be useful in order to clarify theoretical concepts on infectiousness, the force of the infection, in order to compare the infectiousness of different communicable diseases such as Ebola virus disease. The future work is proposed to carry out specific statistical analysis and estimation of parameters using real-life data for epidemic disease.

Competing Interests

The authors declare that they have no competing interests.

Acknowledgments

The authors thank Professor Alagar Rangan for constructive comments and considerable support on this project during his visit to Universidad Nacional de Colombia.

References

[1] W. O. Kermack and A. G. McKendrick, "A contribution to the mathematical theory of epidemics," *Proceedings of the Royal Society of London Series A: Mathematical, Physical and Engineering Sciences*, vol. 115, no. 772, pp. 700–721, 1927.

[2] S. Meyer, J. Elias, and M. Höhle, "A space-time conditional intensity model for invasive meningococcal disease occurrence," *Biometrics. Journal of the International Biometric Society*, vol. 68, no. 2, pp. 607–616, 2012.

[3] A. Dassios and H. Zhao, "A dynamic contagion process," *Advances in Applied Probability*, vol. 43, no. 3, pp. 814–846, 2011.

[4] G. Streftaris and G. J. Gibson, "Bayesian inference for stochastic epidemics in closed populations," *Statistical Modelling*, vol. 4, no. 1, pp. 63–75, 2004.

[5] L. Tierney, "Markov chains for exploring posterior distributions," *The Annals of Statistics*, vol. 22, no. 4, pp. 1701–1762, 1994.

[6] V. Andreasen, "Instability in an SIR-model with age-dependent susceptibility," in *Mathematical Population Dynamics: Analysis of Heterogeneity Volume 1: Theory of Epidemics*, O. Arino, D. Axelrod, M. Kimmel, and M. Langlais, Eds., p. 314, Wuerz Publishing, Winnipeg, Canada, 1995.

[7] T. Britton, "Stochastic epidemic models: a survey," *Mathematical Biosciences*, vol. 225, no. 1, pp. 24–35, 2010.

[8] H. C. Tuckwell and R. J. Williams, "Some properties of a simple stochastic epidemic model of SIR type," *Mathematical Biosciences*, vol. 208, no. 1, pp. 76–97, 2007.

[9] Y. Shang, "Mixed SI (R) epidemic dynamics in random graphs with general degree distributions," *Applied Mathematics and Computation*, vol. 219, no. 10, pp. 5042–5048, 2013.

[10] T. House, "Lie algebra solution of population models based on time-inhomogeneous Markov chains," *Journal of Applied Probability*, vol. 49, no. 2, pp. 472–481, 2012.

[11] Y. Shang, "A lie algebra approach to susceptible-infected-susceptible epidemics," *Electronic Journal of Differential Equations*, vol. 2012, article 233, pp. 1–7, 2012.

[12] J. L. W. Gielen, "Epidemics between Bailey and reed-frost: a stochastic version of the Kermack and McKendrick model," *Journal of Biological Systems*, vol. 5, no. 4, pp. 489–508, 1997.

[13] N. T. J. Bailey, *The Mathematical Theory of Infectious Diseases and Its Applications*, The Griffin, London, UK, 2nd edition, 1975.

[14] J. L. W. Gielen, "A stochastic model for epidemics based on the renewal equation," *Journal of Biological Systems*, vol. 8, no. 1, pp. 1–20, 2000.

[15] N. Keyfitz, *Introduction to the Mathematics of Population*, Addison-Wesley, Reading, Mass, USA, 1968.

[16] G. Chowell and H. Nishiura, "Transmission dynamics and control of Ebola virus disease (EVD): a review," *BMC Medicine*, vol. 12, article 196, 2014.

[17] H. Nishiura and G. Chowell, "Early transmission dynamics of Ebola virus disease (evd), West Africa, march to august 2014," *Eurosurveillance*, vol. 19, no. 36, pp. 1–6, 2014.

[18] I. Rubin, "Regular point processes and their detection," *IEEE Transactions on Information Theory*, vol. 18, pp. 547–557, 1972.

[19] T. Ozaki, "Maximum likelihood estimation of Hawkes' self-exciting point processes," *Annals of the Institute of Statistical Mathematics*, vol. 31, no. 1, pp. 145–155, 1979.

22

A Note on the Waiting-Time Distribution in an Infinite-Buffer $GI^{[X]}/C\text{-}MSP/1$ Queueing System

A. D. Banik,[1] M. L. Chaudhry,[2] and James J. Kim[3]

[1]*School of Basic Sciences, Indian Institute of Technology Bhubaneswar, Permanent Campus Argul, Jatni, Khurda 752 050, Odisha, India*
[2]*Department of Mathematics and Computer Science, Royal Military College of Canada, P.O. Box 17000, STN Forces, Kingston Ont., Canada K7K 7B4*
[3]*Royal Canadian Air Force (RCAF), 34 Tylee Pvt, Ottawa, Canada ON K1V 7M3*

Correspondence should be addressed to A. D. Banik; banikad@gmail.com

Academic Editor: Aera Thavaneswaran

This paper deals with a batch arrival infinite-buffer single server queue. The interbatch arrival times are generally distributed and arrivals are occurring in batches of random size. The service process is correlated and its structure is presented through a continuous-time Markovian service process (*C-MSP*). We obtain the probability density function (p.d.f.) of actual waiting time for the first and an arbitrary customer of an arrival batch. The proposed analysis is based on the roots of the characteristic equations involved in the Laplace-Stieltjes transform (LST) of waiting times in the system for the first, an arbitrary, and the last customer of an arrival batch. The corresponding mean sojourn times in the system may be obtained using these probability density functions or the above LSTs. Numerical results for some variants of the interbatch arrival distribution (Pareto and phase-type) have been presented to show the influence of model parameters on the waiting-time distribution. Finally, a simple computational procedure (through solving a set of simultaneous linear equations) is proposed to obtain the "**R**" matrix of the corresponding $GI/M/1$-type Markov chain embedded at a prearrival epoch of a batch.

1. Introduction

In recent years, analysis of queueing processes with nonrenewal arrival and service processes has been carried out to model data transmission of complex computer and communication networks. The traditional queueing analysis using Poisson processes is not powerful enough to capture the dependence involved in the arrival (service) process. The performance analysis of such bursty and correlated type of the arrival (service) processes may be done through some analytically tractable processes, namely, Markovian arrival process (*C-MAP*) and Markovian service process (*C-MSP*); see Lucantoni et al. [1], Gupta and Banik [2], and references therein. The two processes *C-MAP* and *C-MSP* are convenient representations of a versatile Markovian point process or N-process; see Neuts [3] and Ramaswami [4]. Chaudhry et al. [5] obtain stationary system-length distributions for the infinite-buffer batch arrival $GI/C\text{-}MSP/1$ queues at various epochs using roots. Further, Singh et al. [6] have shown that it is beneficial to use roots rather than the matrix-geometric/matrix-analytic methods in terms of computation time. For further details on the roots method, the readers are referred to the related references in Chaudhry et al. [7].

One may note that Grassmann [8] shows the relation between the characteristics roots and the corresponding eigenvalues of the possible quasi birth-death (QBD) process in connection with the computation of the waiting-time distribution in a $E_n/E_m/1$ queueing model. For further details on the relationships between the characteristics roots of a Markov chain and the corresponding "**G**" matrix of $M/G/1$-type Markov chain or "**R**" matrix of $GI/M/1$-type Markov chain one may see Gail et al. [9].

In this paper we obtain the explicit closed-form expressions for the Laplace-Stieltjes transform (LST) of waiting-time distribution. For analytic purpose, we start our work with associated prearrival epoch probability distribution;

see Chaudhry et al. [5] for details. After that we consider characteristic equations involved in the Laplace-Stieltjes transform's (LST's) waiting-time distribution for the first and an arbitrary customer of an arrival batch. Once the roots of these characteristic equations are found, it becomes easy to obtain the stationary waiting-time distribution for the first and an arbitrary customer of an arrival batch. Also, a simple computational procedure based on the solution of a set of linear simultaneous equations has been proposed to compute the **R** matrix of the corresponding $GI/M/1$-type Markov chain embedded at a prearrival epoch of a batch. It is well-known that the determination of the "**R**" matrix can be done through the minimal nonnegative solution of a nonlinear matrix equation; see Neuts [10] for details. Finally, it may be remarked here that the same queueing model has been dealt with in [5]. But in both the queueing models, corresponding probability density functions of actual waiting time for the first/an arbitrary customer have not been derived. It is needless to mention here that probability density functions of the waiting times are important for obtaining statistical measures of Quality of Service (QoS) requirement of a queueing system. Thus, the present work is a nontrivial extension of [5].

2. Description of the Model: $GI^{[X]}/C\text{-}MSP/1/\infty$ Queue and the Analysis of the Stationary Distribution of Waiting Time

We assume a renewal input single server queueing system where the customers arrive in batches of random size X whose probability mass function is given by $P(X = i) = g_i$, $i = 1, 2, \ldots$, and mean batch size $E(X) = \sum_{i=1}^{\infty} i g_i = \overline{g}$. Let $G(z) = \sum_{i=1}^{\infty} g_i z^i$ be the probability generating function of g_i. The interbatch arrival times are independent identically distributed (i.i.d.) random variables with distribution function (DF) $A(x)$, probability density function $a(x)$ with $x > 0$, and Laplace-Stieltjes transform (LST) $A^*(s)$, where $\Re(s) \geq 0$. Then, $1/\lambda = -A^{*(1)}(0)$; i.e., $1/\lambda$ is the mean interbatch arrival time, where $A^{*(j)}(\theta)$ is the j-th ($j \geq 1$) derivative of $A^*(s)$ at $s = \theta$.

A single server is serving the customers according to a continuous-time Markovian service process ($C\text{-}MSP$) with matrix representation $(\mathbf{L}_0, \mathbf{L}_1)$. For a detailed discussion on $C\text{-}MSP$, the readers are referred to Chaudhry et al. [5]. Let us denote $\mathbf{L}(z) = \mathbf{L}_0 + \mathbf{L}_1 z$ with $\mathbf{L} \equiv \mathbf{L}(1) = \mathbf{L}_0 + \mathbf{L}_1$ being an irreducible infinitesimal generator of the underlying Markov chain $\{J(t)\}$. The mean stationary service rate μ^* of the $C\text{-}MSP$ is given by $\mu^* = \overline{\boldsymbol{\pi}}\mathbf{L}_1\mathbf{e}$, where $\overline{\boldsymbol{\pi}} = [\overline{\pi}_1, \overline{\pi}_2, \ldots, \overline{\pi}_m]$ is the probability row vector which can be computed from $\overline{\boldsymbol{\pi}}\mathbf{L} = \mathbf{0}$ with $\overline{\boldsymbol{\pi}}\mathbf{e} = 1$, where $\mathbf{e}$ is a column vector of ones with an appropriate dimension. The customers are served singly according to a $C\text{-}MSP$ with an average stationary service rate μ^*. The offered load ρ is given by $\rho = \lambda\overline{g}/\mu^*$.

In the following section, we obtain the actual waiting-time distribution for the first customer and an arbitrary customer of an arrival batch. In order to get the actual waiting-time distributions, we first obtain the prearrival epoch probability distribution of system-length at a prearrival epoch. For this, we proceed exactly in the same way as that discussed by Chaudhry et al. [5]. Let $\pi_j^-(n)$ represent the prearrival epoch probability distribution of an arrival batch when there are n (≥ 1) customers in the system and the server is busy in phase j. Also, let $\pi_j^-(0)$ denote the prearrival epoch probability that the system is empty while the server is idle in phase j. Let $\boldsymbol{\pi}^-(n)$ ($n \geq 1$) be the row vector whose j-th ($1 \leq j \leq m$) component is $\pi_j^-(n)$. Also, let us assume $\boldsymbol{\pi}^{-*}(z) = \sum_{n=0}^{\infty} \boldsymbol{\pi}^-(n) z^n$ with $\pi_j^{-*}(z)$ being the j-th $1 \leq j \leq m$ component of $\boldsymbol{\pi}^{-*}(z)$. As presented in [5], $\pi_j^{-*}(z)$ is an analytic function of z for $|z| \leq 1$. Thus, we have

$$\pi_j^{-*}(z) = \sum_{i=1}^{m\hat{r}} \frac{k_{ij}}{1 - \gamma_i z}, \quad 1 \leq j \leq m, \tag{1}$$

where k_{ij} are constants to be determined. Here, γ_i ($1 \leq i \leq m\hat{r}$) are the $m\hat{r}$ roots inside the unit circle of the equation $\det[\mathbf{I}_m - G(z^{-1})\mathbf{S}(z)] = 0$ with $\hat{r}$ being the maximum size of an arrival batch, where $\mathbf{S}(z) = \sum_{n=0}^{\infty} \mathbf{S}_n z^n = A^*(-\mathbf{L}(z))$ with $|z| \leq 1$. The procedure of obtaining $\mathbf{S}_n$ is available in literature; e.g., see Appendix 2 of [5]. One may note that the computation of $\mathbf{S}(z)$ using above relation may be cumbersome. However, the following numerical scheme may be efficient and is given by $\mathbf{S}(z) = \lim_{N \to \infty} \sum_{n=0}^{N} \mathbf{S}_n z^n$, where $\mathbf{S}_n$ may be obtained as proposed in [5]. The equation $\det[\mathbf{I}_m z^{\hat{r}} - (g_1 z^{\hat{r}-1} + g_2 z^{\hat{r}-2} + g_3 z^{\hat{r}-3} + \cdots + g_{\hat{r}})\mathbf{S}(z)] = 0$ has $m\hat{r}$ roots γ_i inside the unit circle. That is, the equation $\det[\mathbf{I}_m - G(z)\mathbf{S}(z^{-1})] = 0$ has $m\hat{r}$ roots $1/\gamma_i$ outside the unit circle. Now, comparing the coefficient of z^n from both sides of (1), one may have the explicit values of $\pi_j^-(n)$ ($1 \leq j \leq m$, $n \geq 0$) in terms of γ_i ($1 \leq i \leq m\hat{r}$) and k_{rj} ($1 \leq i \leq m\hat{r}$, $1 \leq j \leq m$); see Chaudhry et al. [5] for a detailed procedure.

3. Waiting-Time Analysis

In the following we obtain the probability density function (p.d.f.) of actual waiting time for the first and an arbitrary customer of an arrival batch. Let the (i, j)-th entry of $\boldsymbol{\phi}_k(s)$ be the LST of the service times of a total of k (≥ 1) customers and the corresponding phase change of the underlying Markov chain is from i to j. The probability that the service of a customer is completed in the interval $(x, x + dx]$ along with phase changes is given by the matrix $e^{\mathbf{L}_0 x}\mathbf{L}_1 dx + o(dx)$, where $o(x)$ represents a function of x such that $o(x)/x \longrightarrow 0$ as $x \longrightarrow 0$. Since the LST of total service time of k customers $\boldsymbol{\phi}_k(s)$ is the product of their individual LST of each service times $\boldsymbol{\phi}_1(s)$ of those k customers, therefore,

$$\boldsymbol{\phi}_1(s) = \int_0^{\infty} e^{-sx} e^{\mathbf{L}_0 x}\mathbf{L}_1 dx = (s\mathbf{I}_m - \mathbf{L}_0)^{-1}\mathbf{L}_1$$
$$\text{and } \boldsymbol{\phi}_k(s) = \boldsymbol{\phi}_1^k(s), \quad k \geq 2, \tag{2}$$

where $\Re(s) \geq 0$. Let $\mathbf{W}_F(x) = [W_{F1}(x), W_{F2}(x), \ldots, W_{Fm}(x)]$, $x \geq 0$, denote the row vector of order m, where $W_{Fr}(x)$ ($1 \leq r \leq m$) represents the stationary joint probability that the first customer of an arrival batch has to wait up to a maximum of x units of time before getting served with the service process being in phase r. The first customer in an arrival batch waits in the system at most x time units if and only if he sees upon arrival n (≥ 0) customers ahead of him and so he has to wait for $(n+1)$ service completions including itself in the interval $(0, x]$. Similarly, we denote $\mathbf{W}_A(x) = [W_{A1}(x), W_{A2}(x), \ldots, W_{Am}(x)]$, $x \geq 0$, and $\mathbf{W}_L(x) = [W_{L1}(x), W_{L2}(x), \ldots, W_{Lm}(x)]$, $x \geq 0$, as probability vectors whose r-th component $W_{Ar}(x)$ $(1 \leq r \leq m)$ of $\mathbf{W}_A(x)$ denotes the stationary joint probability that an arbitrary customer in an arrival batch has to wait in the system a maximum of x units of time before departing the system with the service process being in phase r. The rth component $W_{Lr}(x)$ $(1 \leq r \leq m)$ of $\mathbf{W}_L(x)$ denotes the stationary joint probability that the last customer in an arrival batch has to wait in the system at most x units of time before getting served and departing the system with the phase of the service process being j. Let $\mathbf{W}_F^*(s) = [W_{F1}^*(s), W_{F2}^*(s), \ldots, W_{Fm}^*(s)]$ be the LST of $\mathbf{W}_F(x)$. Similarly, let $\mathbf{W}_A^*(s) = [W_{A1}^*(s), W_{A2}^*(s), \ldots, W_{Am}^*(s)]$ and $\mathbf{W}_L^*(s) = [W_{L1}^*(s), W_{L2}^*(s), \ldots, W_{Lm}^*(s)]$ be the LSTs of $\mathbf{W}_A(x)$ and $\mathbf{W}_L(x)$, respectively. Hereafter, we need to define a random variable J which denotes the number of customers before an arbitrary customer in a batch. Following Chaudhry and Templeton [7], one may write $g_r^- = P[J = r] = (1/\overline{g}) \sum_{i=r+1}^{\infty} g_i$, $r \geq 0$. Therefore, the LST of the waiting-time distribution after a little algebraic manipulation (see Chaudhry et al. [5]) may be derived as

$$\mathbf{W}_F^*(s) = \left[\sum_{n=0}^{\infty} \boldsymbol{\pi}^-(n)\, \boldsymbol{\phi}_1^{n+1}(s)\right] = \boldsymbol{\pi}^{-*}(\boldsymbol{\phi}_1(s))\, \boldsymbol{\phi}_1(s), \tag{3}$$

$$\mathbf{W}_A^*(s) = \left[\sum_{n=0}^{\infty} \boldsymbol{\pi}^-(n) \sum_{j=0}^{\infty} g_j^- \boldsymbol{\phi}_1^{n+j+1}(s)\right] = \frac{1}{\overline{g}} \cdot \boldsymbol{\pi}^{-*}(\boldsymbol{\phi}_1(s))\, (\mathbf{I}_m - G(\boldsymbol{\phi}_1(s)))\, (\mathbf{I}_m - \boldsymbol{\phi}_1(s))^{-1} \boldsymbol{\phi}_1(s), \tag{4}$$

$$\mathbf{W}_L^*(s) = \left[\sum_{n=0}^{\infty} \boldsymbol{\pi}^-(n) \sum_{j=1}^{\infty} g_j \boldsymbol{\phi}_1^{n+j}(s)\right] = \boldsymbol{\pi}^{-*}(\boldsymbol{\phi}_1(s)) \cdot G(\boldsymbol{\phi}_1(s)), \tag{5}$$

with $\mathfrak{R}(s) \geq 0$. One can find component-wise mean waiting times in the system of the above three types ($\mathbf{W}_F = [W_{F1}, W_{F2}, \ldots, W_{Fm}]$, $\mathbf{W}_A = [W_{A1}, W_{A2}, \ldots, W_{Am}]$, and $\mathbf{W}_L = [W_{L1}, W_{L2}, \ldots, W_{Lm}]$, respectively) by differentiating the above formulae and putting $s = 0$. They are obtained after a little algebraic manipulation; see Chaudhry et al. [5] for explicit formulae of $\mathbf{W}_F$, $\mathbf{W}_A$, and $\mathbf{W}_L$. Using Little's law, we can obtain mean sojourn time as $W_A(LL) = L_s/\lambda\overline{g}$, where L_s is the mean stationary system-length obtained in [5] and one may note that $W_A(LL) = \mathbf{W}_A\mathbf{e}$ while performing numerical computation. Also, one may simplify (3) as stated below. First combine the roots γ_i^n $(1 \leq i \leq m\hat{r})$ and $[\boldsymbol{\phi}_1(s)]^n$, yielding a geometric series, which are then summed over n. As a consequence of the formula regarding geometric series, this sum is $[\mathbf{I}_m - \gamma_i \boldsymbol{\phi}_1(s)]^{-1}$, and this expression is expanded, using the adjoint matrix and determinants (see also [9], (35), and Section A.1 for a proof of formula (6) given below).

$$\begin{aligned}
\mathbf{W}_F^*(s) &= \sum_{n=0}^{\infty} \boldsymbol{\pi}^-(n)\, [\boldsymbol{\phi}_1(s)]^n\, \boldsymbol{\phi}_1(s) \\
&= \sum_{n=0}^{\infty} \left[\sum_{i=1}^{m\hat{r}} k_{i1}\gamma_i^n, \sum_{i=1}^{m\hat{r}} k_{i2}\gamma_i^n, \ldots, \sum_{i=1}^{m\hat{r}} k_{im}\gamma_i^n\right] [\boldsymbol{\phi}_1(s)]^n \\
&\quad \cdot \boldsymbol{\phi}_1(s) = \frac{[k_{11}, k_{12}, \ldots, k_{1m}]\, \mathrm{adj}\,[\mathbf{I}_m - \gamma_1 \boldsymbol{\phi}_1(s)]}{\det[\mathbf{I}_m - \gamma_1 \boldsymbol{\phi}_1(s)]} \\
&\quad \cdot \boldsymbol{\phi}_1(s) + \frac{[k_{21}, k_{22}, \ldots, k_{2m}]\, \mathrm{adj}\,[\mathbf{I}_m - \gamma_2 \boldsymbol{\phi}_1(s)]}{\det[\mathbf{I}_m - \gamma_2 \boldsymbol{\phi}_1(s)]} \\
&\quad \cdot \boldsymbol{\phi}_1(s) + \cdots \\
&\quad + \frac{[k_{m\hat{r}1}, k_{m\hat{r}2}, \ldots, k_{m\hat{r}m}]\, \mathrm{adj}\,[\mathbf{I}_m - \gamma_{m\hat{r}} \boldsymbol{\phi}_1(s)]}{\det[\mathbf{I}_m - \gamma_{m\hat{r}} \boldsymbol{\phi}_1(s)]} \boldsymbol{\phi}_1(s),
\end{aligned} \tag{6}$$

with $\mathfrak{R}(s) \geq 0$. In the following, we develop a numerical scheme for inverting the above LSTs component-wise. Let us first look at the r-th component of the LST given by formula (3) which is further simplified as (6). The right-hand side of (3) involves $\boldsymbol{\pi}^{-*}(\boldsymbol{\phi}_1(s))\boldsymbol{\phi}_1(s)$ which is further simplified as the right-hand side of (6). Both LSTs (3) and (6) are analytic in the region $\mathfrak{R}(s) \geq 0$. Therefore (as $\boldsymbol{\phi}_1(s)$ is the LST of the transition probability matrix of the underlying Markov chain of C-MSP), the r-th $(1 \leq r \leq m)$ component of $\boldsymbol{\pi}^{-*}(\boldsymbol{\phi}_1(s))\boldsymbol{\phi}_1(s)$ may be written as a rational function in s, which is evident from the right-hand side of (6). Now, taking into account the zeros of the denominator of each $m\hat{r}$ terms in the right-hand side of (6), we obtain $m\hat{r}$ characteristic equations; namely,

$$\det[\mathbf{I}_m - \gamma_j \boldsymbol{\phi}_1(s)] = 0, \quad 1 \leq j \leq m\hat{r}. \tag{7}$$

Also, each element of the matrix $\boldsymbol{\phi}_1(s) = (s\mathbf{I}_m - \mathbf{L}_0)^{-1}\mathbf{L}_1 = \mathrm{adj}[s\mathbf{I}_m - \mathbf{L}_0]\mathbf{L}_1/\det[s\mathbf{I}_m - \mathbf{L}_0]$ has denominator $\det[s\mathbf{I}_m - \mathbf{L}_0]$ which has m zeros in the region $\mathfrak{R}(s) < 0$ (see similar proof of Theorem 7.3 of [5]). As a result, $\det[\mathbf{I}_m - \gamma_j \boldsymbol{\phi}_1(s)]$ $(1 \leq j \leq m\hat{r})$ has denominator $\{\det[s\mathbf{I}_m - \mathbf{L}_0]\}^m$ which has m^2 zeros in the region $\mathfrak{R}(s) < 0$ with each zero having multiplicity m. Similarly, the numerator of $\det[\mathbf{I}_m - \gamma_j \boldsymbol{\phi}_1(s)]$ $(1 \leq j \leq m\hat{r})$ also has m^2 zeros in the region $\mathfrak{R}(s) < 0$ for each j $(1 \leq j \leq m\hat{r})$. After canceling the common zeros of the numerator and the denominator of $\det[\mathbf{I}_m - \gamma_j \boldsymbol{\phi}_1(s)] \times \det[s\mathbf{I}_m - \mathbf{L}_0]$ $(1 \leq j \leq m\hat{r})$, one is left with m zeros of the numerator of $\det[\mathbf{I}_m - \gamma_j \boldsymbol{\phi}_1(s)] \times \det[s\mathbf{I}_m - \mathbf{L}_0]$ $(1 \leq j \leq m\hat{r})$ in the region $\mathfrak{R}(s) < 0$. Let us define $\overline{s}_{i,j}$ $(1 \leq i \leq m,\ 1 \leq j \leq m\hat{r})$ as the zeros of the denominator of each component of $\boldsymbol{\pi}^{-*}(\boldsymbol{\phi}_1(s))\boldsymbol{\phi}_1(s)$ in the region $\mathfrak{R}(s) < 0$. Also, it may be noted here that none of these zeros $\overline{s}_{i,j}$ $(1 \leq i \leq m,\ 1 \leq j \leq m\hat{r})$ is equal to a zero of $\det[s\mathbf{I}_m - \mathbf{L}_0]$. Therefore, one may finally note that the denominator of the rth component of the LST $\mathbf{W}_F^*(s)$ given by formula (3) which has $m^2\hat{r}$ zeros: $\overline{s}_{i,j}$ $(1 \leq i \leq m, 1 \leq j \leq m\hat{r})$

with $\Re(s) < 0$. One may finally obtain the rth $(1 \le r \le m)$ component of $\mathbf{W}_F^*(s)$ using partial fractions as follows:

$$W_{Fr}^*(s) = \sum_{i=1}^{m}\sum_{j=1}^{m\hat{r}} \frac{C_{ij}^r}{\left(s - \bar{s}_{ij}\right)}, \tag{8}$$

where C_{ij}^r $(1 \le i \le m,\ 1 \le j \le m\hat{r},\ 1 \le r \le m)$ are constants to be evaluated and $\Re(s) \ge 0$. It may be remarked here that while making partial fractions, we have assumed that the roots $\bar{s}_{ij}$ are distinct. To get the unknown constants, we use the required number of linear equations after substituting any positive values of s ($\Re(s) > 0$) in (8). Finally, we equate the component-wise $\mathbf{W}_F^*(s)$ from (3) with (8) under the same positive values of s. One may note that, as normalizing conditions, the component-wise values of $\mathbf{W}_F^*(0)$ and the components of $\mathbf{W}_F$ must be included in the set of linear equations to get the above constants. These components may be obtained by putting $s = 0$ in (8) and we may use $W_{Fr} = -W_{Fr}^{*(1)}(s)|_{\{s=0\}}$, where $W_{Fr}^{*(1)}(s)$ is the first differentiation of $W_{Fr}^*(s)$ with respect to s. After that, taking inverse Laplace transform of (8), component-wise explicit closed-form expression of probability density function (p.d.f.) $w_{Fr}(x)$ is given by

$$w_{Fr}(x) = \sum_{i=1}^{m}\sum_{j=1}^{m\hat{r}} C_{ij}^r e^{\bar{s}_{ij}x}, \tag{9}$$

$$1 \le i \le m,\ 1 \le j \le m\hat{r},\ 1 \le r \le m,$$

with $x > 0$. Finally, the p.d.f. of waiting time of the first customer may be obtained from $w_F(x) = \sum_{r=1}^{m} w_{Fr}(x)$. From this, we can get $W_{Fr}(u) = \int_0^u w_{Fr}(x)dx$, $(1 \le r \le m)$, and $W_F(u) = \int_0^u w_F(x)dx$.

Similar to the procedure followed earlier, we consider the zeros of the denominator of each element of the matrix $(\mathbf{I}_m - G(\boldsymbol{\phi}_1(s)))$ in (4). As discussed earlier, the denominator of each element of $\boldsymbol{\phi}_1(s)$ has m zeros in the region $\Re(s) < 0$ which may be obtained by solving $\det[s\mathbf{I}_m - \mathbf{L}_0] = 0$. This implies that, given maximum arrival batch size $\hat{r}$, m zeros of the denominator of each element of the matrix $(\mathbf{I}_m - G(\boldsymbol{\phi}_1(s)))$ are repeated up to $\hat{r}$ times. Let the m roots of $\det[s\mathbf{I}_m - \mathbf{L}_0] = 0$ be denoted by $\bar{r}_i$ $(1 \le i \le m)$. Therefore, the denominator of each element of the matrix $(\mathbf{I}_m - G(\boldsymbol{\phi}_1(s)))$ has repeated roots $\bar{r}_i$ $(1 \le i \le m)$ with multiplicity $\hat{r}$. Also, we have to investigate the zeros of the denominator of each element of the matrix $(\mathbf{I}_m - \boldsymbol{\phi}_1(s))^{-1}$. These zeros may be obtained by using $\det[\mathbf{I}_m - \boldsymbol{\phi}_1(s)]$. It can also be shown that there will be m^2 zeros of the numerator of $\det[\mathbf{I}_m - \boldsymbol{\phi}_1(s)]$ in the region $\Re(s) \le 0$ (since $s = 0$ is root of the equation $\det[\mathbf{I}_m - \boldsymbol{\phi}_1(s)] = 0$). Also, there are m^2 zeros of the denominator (which is equal to $\{\det[s\mathbf{I}_m - \mathbf{L}_0]\}^m$) of $\det[\mathbf{I}_m - \boldsymbol{\phi}_1(s)]$ in the region $\Re(s) < 0$ and they are $\bar{r}_i$ $(1 \le i \le m)$ with multiplicity m each. After canceling the common zeros of the numerator and the denominator of $\det[\mathbf{I}_m - \boldsymbol{\phi}_1(s)]$, one is left with $(m-1)$ zeros of the numerator of $\det[\mathbf{I}_m - \boldsymbol{\phi}_1(s)]$ in the region $\Re(s) < 0$. Let these $(m-1)$ roots of the numerator of $\det[\mathbf{I}_m - \boldsymbol{\phi}_1(s)]$ be denoted by $\bar{h}_i$ $(1 \le i \le m-1)$, where $\Re(\bar{h}_i) < 0$. Lastly, the roots of the denominator of each element of the matrix $\boldsymbol{\phi}_1(s)$ in the right-hand side of (4) are also $\bar{r}_i$ $(1 \le i \le m)$. Finally, if we consider all the roots (with $\Re(s) < 0$) of denominators involved in each element of the vector and matrices occurring in the right-hand side of (4), we reach the following conclusion: $\bar{r}_i$'s $(1 \le i \le m)$ roots repeated in the region $\Re(s) < 0$ with multiplicity $(\hat{r} - 1)$ occur in the denominator of each component of $\mathbf{W}_A^*(s)$ in (4). On the other hand $\bar{s}_{ij}$ $(1 \le i \le m,\ 1 \le j \le m\hat{r})$ and $\bar{h}_i$ $(1 \le i \le m-1)$ are the roots in the region $\Re(s) < 0$ that occur in the denominator of each component of $\mathbf{W}_A^*(s)$ in (4). Following this conclusion, after using partial fractions, we may write the r-th $(1 \le r \le m)$ component of $\mathbf{W}_A^*(s)$ (in (4) which also has factor $\boldsymbol{\pi}^{-*}(\boldsymbol{\phi}_1(s))$) as follows:

$$W_{Ar}^*(s) = \sum_{n=1}^{\hat{r}-1}\sum_{i=1}^{m} \frac{\overline{C}_{ri}^n}{\left(s - \bar{r}_i\right)^n} + \sum_{i=1}^{m}\sum_{j=1}^{m\hat{r}} \frac{D_{ij}^r}{\left(s - \bar{s}_{ij}\right)} + \sum_{i=1}^{m-1} \frac{H_i^r}{\left(s - \bar{h}_i\right)},\quad 1 \le r \le m, \tag{10}$$

where $\overline{C}_{ri}^n$ $(1 \le i,\ r \le m,\ 1 \le n \le \hat{r} - 1)$, $D_{i,j}^r$ $(1 \le i,\ r \le m,\ 1 \le i \le m\hat{r})$, and H_i^r $(1 \le r \le m,\ 1 \le i \le m-1)$ are constants to be evaluated and $\Re(s) \ge 0$. It may be remarked here that while making partial fractions, we have assumed that the roots $\bar{h}_i$ and $\bar{s}_{ij}$ are distinct. If they are repeated, a slight modification in partial fraction will be needed. Also, throughout the above discussions of roots, we discard any roots of the denominator of the above LSTs (3)-(5) in the region $\Re(s) \ge 0$. The roots of the denominator with $\Re(s) \ge 0$ must be canceled with numerator as the LSTs (3)-(5) are analytic in the region $\Re(s) \ge 0$.

Now to obtain the unknown constants, we use the required number of linear equations after substituting any positive values of s in (10). Finally, we equate the components of $\mathbf{W}_A^*(s)$ from (4) with the corresponding component of (10) under the same positive values of s. One may note that, as normalizing conditions, the component-wise values of $\mathbf{W}_A^*(0)$ and the components of $\mathbf{W}_A$ must be included in the set of linear equations to get the above constants. These components may be obtained by putting $s = 0$ in (10) and we may use $W_{Ar} = -W_{Ar}^{*(1)}(s)|_{\{s=0\}}$, where $W_{Ar}^{*(1)}(s)$ is the first differentiation of $W_{Ar}^*(s)$ with respect to s. After that, taking inverse Laplace transform of (10), component-wise explicit closed-form expression of probability density function (p.d.f.) $w_{Ar}(x)$ is given by

$$w_{Ar}(x) = \sum_{n=1}^{\hat{r}-1}\sum_{i=1}^{m} \frac{\overline{C}_{ri}^n x^{n-1}}{(n-1)!} e^{\bar{r}_i x} + \sum_{i=1}^{m}\sum_{j=1}^{m\hat{r}} D_{ij}^r e^{\bar{s}_{ij}x} + \sum_{i=1}^{m-1} H_i^r e^{\bar{h}_i x},\quad 1 \le r \le m, \tag{11}$$

with $x > 0$. Finally, the p.d.f. of waiting time of an arbitrary customer may be obtained by $w_A(x) = \sum_{r=1}^{m} w_{Ar}(x)$. Further, let us denote the component-wise and total distribution function of the waiting-time distribution of an arbitrary

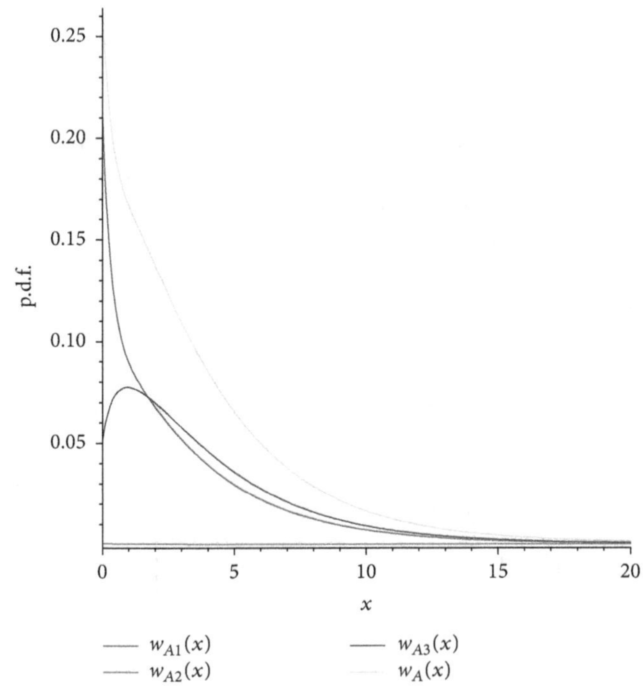

FIGURE 1: x versus p.d.f.

customer of an arrival batch by $W_{Ar}(u) = \int_0^u w_{Ar}(x)dx$, $(1 \leq r \leq m)$, and $W_A(u) = \int_0^u w_A(x)dx$. It may be finally noted that $\lim_{u\to\infty} W_{Ar}(u)$ must be equal to the r-th component of $\mathbf{W}_A^*(s)|_{s=0}$. Exactly similar analysis may be carried out to obtain the component-wise p.d.f. of the waiting time of the last customer in an arrival batch using the LST given by (5).

4. Numerical Results

Using the results formulated in the previous section, some numerical results are presented in this section. Although numerical calculations were carried out with high precision, due to lack of space the results have been reported to 6 decimal places. Since various distributions can be either represented or approximated by PH-distribution, we take interbatch arrival time distribution to be of PH-type having the representation $(\boldsymbol{\alpha}, \mathbf{T})$, where $\boldsymbol{\alpha}$ and $\mathbf{T}$ are of dimension ν. Then using the procedure adopted by Chaudhry et al. [5], $\mathbf{S}(z)$ can be derived as follows:

$$\mathbf{S}(z) = (\mathbf{I}_m \otimes \alpha)(\mathbf{L}(z) \oplus \mathbf{T})^{-1}(\mathbf{I}_m \otimes \mathbf{T}\mathbf{e}_\nu), \quad (12)$$

with $\mathbf{L}(z) \oplus \mathbf{T} = (\mathbf{L}(z) \otimes \mathbf{I}_\nu) + (\mathbf{I}_m \otimes \mathbf{T})$, where $\otimes$ and $\oplus$ are Kronecker product and sum, respectively. For the derivation of $\mathbf{S}(z)$, see Theorem 7.5 in [5]. For more discussion on the derivation of $\mathbf{S}(z)$, see [5]. Numerical computations were performed on a PC having Intel(R) Core i5 processor @1.8 GHz with 4 GB DDR3 RAM using MAPLE 2015.

4.1. The Waiting-Time Analysis of a $PH^{[X]}/C\text{-}MSP/1/\infty$ Queueing System. In Figure 1, we have computed and plotted the component-wise p.d.f. of waiting times in the system of an arbitrary customer in a $PH^{[X]}/C\text{-}MSP/1/\infty$ queueing system using the numerical scheme presented in Section 3. The PH-type interbatch arrival time is given by $\boldsymbol{\alpha} = [0.35 \;\; 0.65]$ and $\mathbf{T} = \begin{bmatrix} -1.183 & 2.423 \\ 1.367 & -2.896 \end{bmatrix}$ with $\lambda = 0.032317$. The batch size distribution is given by $g_1 = 0.0, g_2 = 0.0, g_3 = 0.5$, and $g_4 = 0.5, g_i = 0.0$ $(i \geq 5)$ with mean batch size $\overline{g} = 3.5$. The $C\text{-}MSP$ matrices having positive lag-1 correlation coefficient 0.271908 as

$$\mathbf{L}_0 = \begin{bmatrix} -0.5424095190 & 0.0037279000 & 0.0000000000 \\ 0.0043492170 & -0.0229887200 & 0.0006213170 \\ 0.0000000000 & 0.0012426330 & -2.2696700720 \end{bmatrix}, \quad (13)$$

$$\mathbf{L}_1 = \begin{bmatrix} 0.0205034530 & 0.0000000000 & 0.5181781660 \\ 0.0000000000 & 0.0173968690 & 0.0006213170 \\ 2.2591076880 & 0.0049705340 & 0.0043492170 \end{bmatrix}$$

with stationary mean service rate $\mu^* = 0.500000$ and $\overline{\boldsymbol{\pi}} = [0.464978 \;\; 0.428426 \;\; 0.106596]$ which gives traffic intensity (offered load) $\rho = 0.226216$. The characteristic roots to obtain the system-length distribution at a prearrival epoch are calculated through the nonlinear algebraic equation:

$$\det\left[\mathbf{I}_m z^{\hat{r}} - \left(g_1 z^{\hat{r}-1} + g_2 z^{\hat{r}-2} + g_3 z^{\hat{r}-3} + g_{\hat{r}}\right)\mathbf{S}(z)\right] = 0, \quad (14)$$

where $m = 3, \hat{r} = 4$ and g_i $(i \geq 1)$ are given above and $\mathbf{S}(z)$ may be obtained by (12); see Theorem 7.5 of [5]. The $m\hat{r} = 12$ roots of (14) inside $|z| < 1$ are evaluated and the corresponding k_{ij} $(1 \leq i \leq 12,\ 1 \leq j \leq 3)$ values are calculated using Section 2 where the prearrival epoch probabilities are derived. Details of the characteristic equations and their roots used for the evaluation of waiting-time densities are available with the authors and may be supplied upon specific request to the authors. After computation of the required roots, we construct partial fractions for each component of $\mathbf{W}_A^*(s)$ as given in (10). Finally, using inverse Laplace-transforms we obtain component-wise p.d.f. of waiting times in the system. In Figure 1, we have plotted the above three component-wise densities along with the total waiting-time density function $w_A(x)$ against x. Figure 1 shows that component-wise density of waiting time behaves significantly different from that of $w_A(x)$. Also, in Figure 2, we have plotted the corresponding component-wise distribution function as well as total distribution function against time. Similar behaviour may be observed in the case of DFs, which is what is expected.

4.2. The Waiting-Time Analysis of a $Pareto^{[X]}/C\text{-}MSP/1/\infty$ Queueing System. In Figure 3, we have repeated similar experiment as given in above figures; i.e., we plotted component-wise p.d.f.'s against time in a $Pareto^{[X]}/C\text{-}MSP/1/\infty$

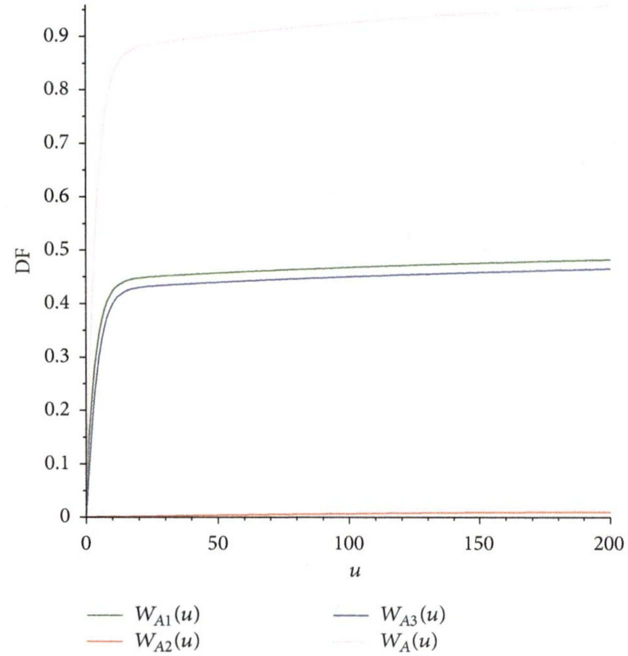

FIGURE 2: u versus distribution function.

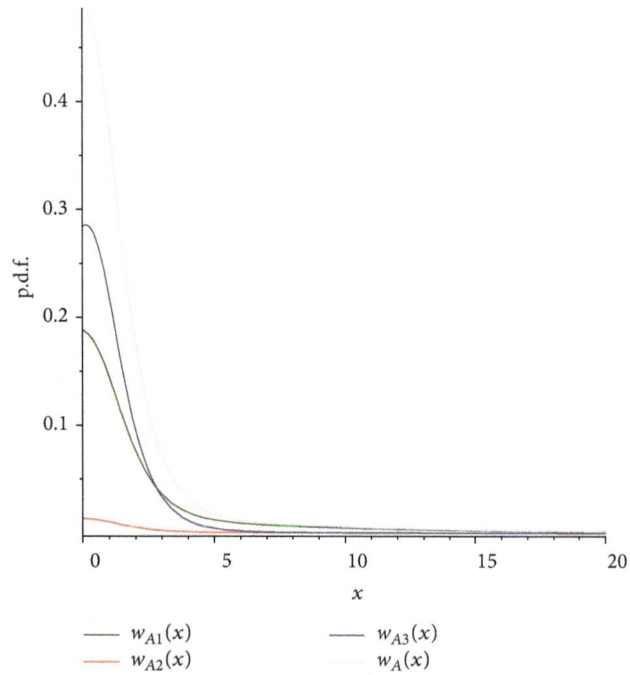

FIGURE 3: x versus p.d.f.

queue, where *Pareto* stands for Pareto distribution whose probability density function and distribution functions are given by $a(x) = \alpha a^{\alpha}/(a+x)^{\alpha+1}$, $\alpha = 0.3$, $a = 2.3$, $x \geq 0$, and $A(x) = 1 - a^{\alpha}/(x+a)^{\alpha}$, $x > 0$, respectively, with shape parameter α and scale parameter a. Batch size distribution is taken arbitrary with probability generating function $G(z) = 0.1z^2 + 0.1z^3 + 0.8z^4$ and mean batch size $\overline{g} = 3.7$. The service matrices having positive lag-1 correlation coefficient 0.018690 are taken as

$$\mathbf{L}_0 = \begin{bmatrix} -2.156750 & 0.007125 & 0.00 \\ 0.149625 & -0.171000 & 0.007125 \\ 0.007125 & 0.007125 & -2.443000 \end{bmatrix},$$
$$\mathbf{L}_1 = \begin{bmatrix} 2.007125 & 0.14250 & 0.00 \\ 0.00 & 0.007125 & 0.007125 \\ 0.00 & 0.00 & 2.42875 \end{bmatrix} \quad (15)$$

with $\mu^* = 1.540543$ and $\overline{\pi} = [0.343750 \ 0.328125 \ 0.328125]$. One may note that the moments do not exist for this case of Pareto interbatch arrival time distribution. From this Pareto distribution function $A(x)$ (through equating $A(x) = 0.5$), we calculate median which is equal 20.882547. Here we assume $\lambda = 0.047887$ so that $\rho = \lambda\overline{g}/\mu^* = 0.115012$. We use geometric transform approximation method (G-TAM); see Shortle et al. [11]. For this, consider $N = 100$ probabilities $y_i = 1 - q^i$ $(1 \leq i \leq 100,\ 0 < q < 1)$. Substituting y_i in the distribution function, i.e., $y_i = 1 - q^i = A(x_i) = 1 - a^{\alpha}/(x_i + a)^{\alpha}$, leads to $x_i = aq^{-i/\alpha} - a$. The weights are $\overline{p}_i = (y_{i+1} - y_{i-1})/2 = (q^{i-1} - q^{i+1})/2$, $\overline{p}_1 = y_1 = 1 - (q + q^2)/2$, and $\overline{p}_{100} = y_{100} = (q^{99} + q^{100})/2$. A binary search on q to match $\sum_{i=1}^{100} \overline{p}_i x_i = 20.882547$ (which is the median of an interbatch arrival time) gives $q = 0.991498$. Thus, the approximate Laplace transform of interbatch arrival time may be taken as

$$A^*(s) = \sum_{i=1}^{100} \overline{p}_i e^{-sx_i}. \quad (16)$$

The Padé approximation [4/5] of (16) is given as follows:

$$A^*(s) \simeq \frac{1.0 - 0.303449s + 104.611371s^2 + 16.029743s^3 + 790.04091s^4}{1.0 + 20.503315s + 189.254931s^2 + 998.209224s^3 + 3069.401520s^4 + 4521.877843s^5}. \quad (17)$$

Now using the above procedure as discussed in previous section, we have computed the component-wise densities of waiting time for the $Pareto^{[X]}/C\text{-}MSP/1/\infty$ using exactly the same procedure as described in Figure 1. Similar to Figure 1, in Figure 3, we have plotted the above three component-wise densities as well as $w_A(x)$ versus x. Here also we have seen

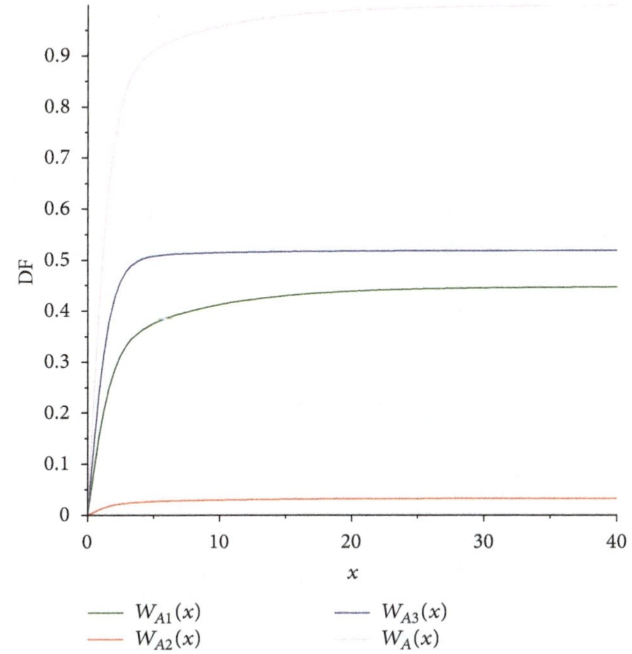

FIGURE 4: x versus distribution function.

TABLE 1: Actual waiting-time (in the system) distribution of a customer.

	Waiting-time distribution function				
u	$W_1(u)$	$W_2(u)$	$W_3(u)$	$W_4(u)$	$W(u)$
0	0.000000	0.000000	0.000000	0.000000	0.000000
0.5	0.497562	0.005344	0.017378	0.002648	0.522932
1.0	0.521051	0.005759	0.035850	0.003488	0.566148
2.5	0.577662	0.006778	0.082479	0.005570	0.672489
3.0	0.593261	0.007059	0.095329	0.006143	0.701792
5.0	0.642893	0.007953	0.136214	0.007968	0.795029
10.5	0.713091	0.009218	0.194041	0.010550	0.926900
⋮	⋮	⋮	⋮	⋮	⋮
90.5	0.752004	0.009919	0.226096	0.011981	0.999999

significant difference among the component-wise densities. Also, we have plotted the corresponding distribution function in component-wise as well as total distribution function in Figure 4.

4.3. Comparison of the Numerical Results Obtained through Roots and Matrix-Geometric Method. In this section we made a thorough numerical comparison of our results with that of Samanta [12] and our computed results are appended in Table 1. For this comparison, we have obtained component-wise waiting-time distribution function (using the procedure discussed in Section 3) of an infinite-buffer $GI/C\text{-}MSP/1$ queueing model with parameters given in Table 2 of [12]. For the same queueing model, we have obtained waiting-time (in system) distribution function (using the procedure discussed in Section 3) which was presented in Table 2 of [12] using matrix-geometric method. As expected, in both the above cases, the results were matched up to six decimal places against Table 2 of [12]. In Table 1, we have presented first seven and the last values of the distribution function. For this experiment, $C\text{-}MSP$ service matrices with positive lag-1 correlation coefficient 0.5381025 are taken as

$$\mathbf{L}_0 = \begin{bmatrix} -15.449000 & 0.116000 & 0.052000 & 0.000000 \\ 0.092000 & -5.069000 & 0.000000 & 0.062000 \\ 0.000000 & 0.213000 & -4.832000 & 4.430000 \\ 0.059000 & 5.033000 & 0.000000 & -5.321000 \end{bmatrix}, \tag{18}$$

$$\mathbf{L}_1 = \begin{bmatrix} 15.068000 & 0.157000 & 0.000000 & 0.056000 \\ 0.123000 & 0.047000 & 4.745000 & 0.000000 \\ 0.084000 & 0.000000 & 0.000000 & 0.105000 \\ 0.051000 & 0.000000 & 0.077000 & 0.101000 \end{bmatrix}$$

with stationary mean service rate $\mu^* = 5.388016$ and $\overline{\boldsymbol{\pi}} = [0.264645\ \ 0.253046\ \ 0.254961\ \ 0.227348]$. The interarrival time is phase-type whose representation was taken as

$$\boldsymbol{\alpha} = [0.22\ \ 0.33\ \ 0.45],$$
$$\mathbf{T} = \begin{bmatrix} -7.423 & 0.000 & 3.812 \\ 8.542 & -9.942 & 1.000 \\ 5.210 & 0.000 & -8.640 \end{bmatrix} \tag{19}$$

with $\lambda = 3.198147$ which gives $\rho = 0.593567$. It may be noted here that for an infinite-buffer $GI/C\text{-}MSP/1$ queue the expression for LST of waiting-time distribution of an arriving customer is the same as $\mathbf{W}_F^*(s) = \mathbf{W}_A^*(s)$ which is given in (3) and for the sake of notational convenience we may denote this as $\mathbf{W}_A^*(s)$.

In Figures 5 and 6, we have plotted the probability densities and distribution functions, respectively, against time for the queueing model discussed in Table 1.

5. Conclusion

In this paper, we analyze the waiting-time distribution of the $GI^{[X]}/C\text{-}MSP/1/\infty$ queue and obtain steady-state probability distribution of the waiting time of the first customer and an arbitrary customer of an arrival batch. The proposed method is based on the roots of a characteristic equation involved in finding probability distribution at a prearrival epoch. After that, we use prearrival epoch probabilities and the roots of a set of characteristic equations involved in the LSTs of the waiting time for the first customer and an arbitrary customer of an arrival batch. Besides the computational advantages in terms of time, this method has other advantages over the matrix-geometric method in the sense that it is accurate even if one does not use all the roots. For example, the values of the distribution function presented in Table 1 of the previous section may be obtained (with an accuracy of up to 6 decimal places) by just using the first 8 roots from the list of $m^2 = 16$ roots. As a consequence, we may obtain good approximate

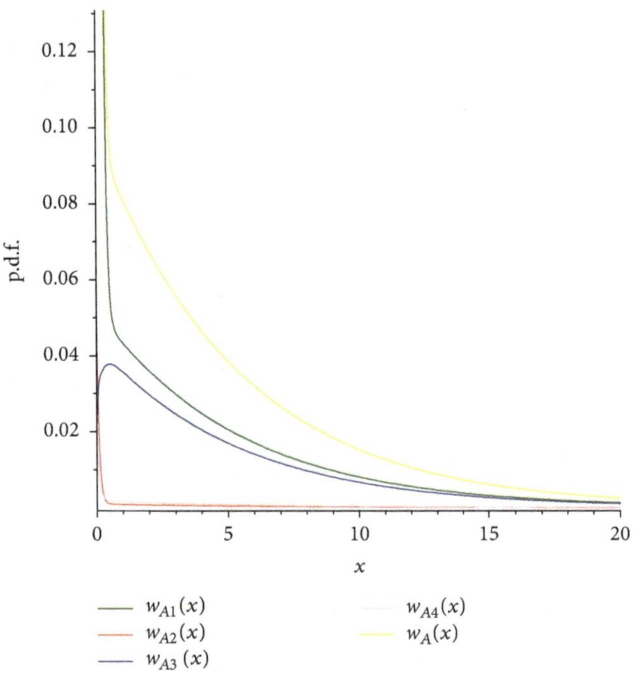

FIGURE 5: x versus p.d.f.

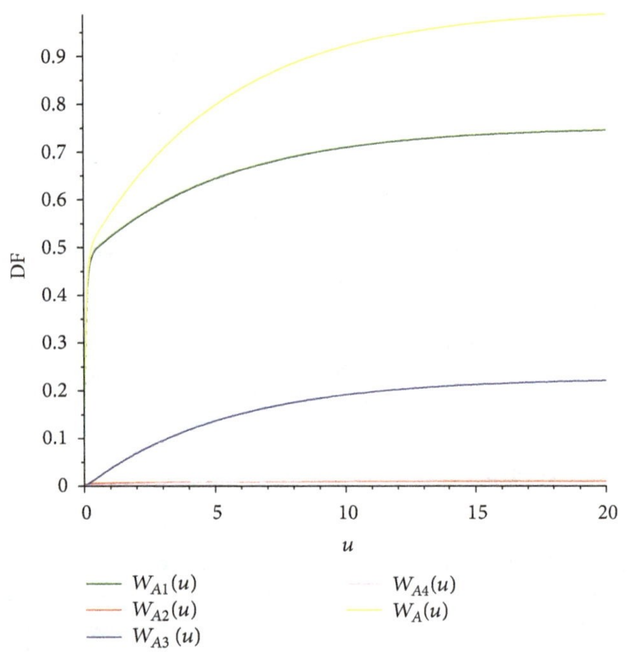

FIGURE 6: u versus distribution function.

waiting-time distribution using less number of roots. Finally, we discuss a simple computational procedure (based on the solution of a set of simultaneous linear equations) which can compute the "R" matrix of the corresponding $GI/M/1$-type Markov chain with a desired level of accuracy; see Section A.2. This procedure may serve as an alternative method for computation of the **R** matrix which is usually computed through the nonnegative solution of a nonlinear matrix equation. Similar procedure may be used to obtain the waiting-time distribution of a $GI/C\text{-}MSP/1$-type queue with set-up time with or without vacation(s). These problems may attract researchers' attention in the future.

Appendix

A. A Proof and Some Additional Numerical Results Related to the Paper Using Roots

A.1. Proof of Formula (6) of the Paper

$$
\begin{aligned}
\mathbf{W}_F^*(s) &= \sum_{n=0}^{\infty} \boldsymbol{\pi}^-(n)\left[\boldsymbol{\phi}_1(s)\right]^n \boldsymbol{\phi}_1(s) \\
&= \sum_{n=0}^{\infty}\left[\sum_{i=1}^{m\hat{r}} k_{i1}\gamma_i^n, \sum_{i=1}^{m\hat{r}} k_{i2}\gamma_i^n, \ldots, \sum_{i=1}^{m\hat{r}} k_{im}\gamma_i^n\right]\left[\boldsymbol{\phi}_1(s)\right]^n \boldsymbol{\phi}_1(s) \\
&= \sum_{n=0}^{\infty}\left[k_{11}\gamma_1^n, k_{12}\gamma_1^n, \ldots, k_{1m}\gamma_1^n\right]\left[\boldsymbol{\phi}_1(s)\right]^n \boldsymbol{\phi}_1(s) \\
&+ \sum_{n=0}^{\infty}\left[k_{21}\gamma_2^n, k_{22}\gamma_2^n, \ldots, k_{2m}\gamma_2^n\right]\left[\boldsymbol{\phi}_1(s)\right]^n \boldsymbol{\phi}_1(s) + \cdots \\
&+ \sum_{n=0}^{\infty}\left[k_{m\hat{r}1}\gamma_{m\hat{r}}^n, k_{m\hat{r}2}\gamma_{m\hat{r}}^n, \ldots, k_{m\hat{r}m}\gamma_{m\hat{r}}^n\right]\left[\boldsymbol{\phi}_1(s)\right]^n \boldsymbol{\phi}_1(s) \\
&= \sum_{n=0}^{\infty}\left[k_{11}, k_{12}, \ldots, k_{1m}\right]\left[\gamma_1\boldsymbol{\phi}_1(s)\right]^n \boldsymbol{\phi}_1(s) \\
&+ \sum_{n=0}^{\infty}\left[k_{21}, k_{22}, \ldots, k_{2m}\right]\left[\gamma_2\boldsymbol{\phi}_1(s)\right]^n \boldsymbol{\phi}_1(s) + \cdots \\
&+ \sum_{n=0}^{\infty}\left[k_{m\hat{r}1}, k_{m\hat{r}2}, \ldots, k_{m\hat{r}m}\right]\left[\gamma_{m\hat{r}}\boldsymbol{\phi}_1(s)\right]^n \boldsymbol{\phi}_1(s) \\
&= \left[k_{11}, k_{12}, \ldots, k_{1m}\right]\left[\mathbf{I}_m - \gamma_1\boldsymbol{\phi}_1(s)\right]^{-1} \boldsymbol{\phi}_1(s) \\
&+ \left[k_{21}, k_{22}, \ldots, k_{2m}\right]\left[\mathbf{I}_m - \gamma_2\boldsymbol{\phi}_1(s)\right]^{-1} \boldsymbol{\phi}_1(s) + \cdots \\
&+ \left[k_{m\hat{r}1}, k_{m\hat{r}2}, \ldots, k_{m\hat{r}m}\right]\left[\mathbf{I}_m - \gamma_{m\hat{r}}\boldsymbol{\phi}_1(s)\right]^{-1} \boldsymbol{\phi}_1(s),
\end{aligned}
\tag{A.1}
$$

which yields the result (6).

*A.2. Demonstration of a Simple Computational Procedure for "**R**" Matrix (Used in Matrix-Geometric Method) Related to the Queueing Model* $\boldsymbol{GI^{[X]}/C\text{-}MSP/1/\infty}$ *of the Paper Using Roots.* As discussed by Neuts [10], the stationary probabilities at an embedded prearrival epoch in a $GI/M/1$-type Markov chain may be obtained in matrix-geometric forms. That is, for a $GI/M/1$-type Markov chain Neuts [10] showed that there exists a nonnegative matrix "**R**" with appropriate dimension such that the stationary probability vectors at an embedded prearrival epoch are given by $\tilde{\boldsymbol{\pi}}(n) = \tilde{\boldsymbol{\pi}}(0)\mathbf{R}^n$, $n \geq 0$, where the matrix "**R**" has spectral radius < 1 in case of a positive recurrent Markov chain and the matrix "**R**" may be computed as a nonnegative solution of a nonlinear matrix equation (see below). Gail et al. [9] proved that the eigenvalues of the "**R**" matrix are related to the characteristic roots of the corresponding $GI/M/1$-type Markov chain. In the following a simple computation procedure for obtaining the "**R**" matrix as a solution of a set of simultaneous linear equations has been proposed for $GI/C\text{-}MSP/1/\infty$ and $GI^{[X]}/C\text{-}MSP/1/\infty$ queueing models.

First, we explain this computational procedure for computing the **R** matrix in the case of $GI/C\text{-}MSP/1/\infty$ queueing model presented in Table 1. Following [10], the stationary probability vectors at an embedded Markov prearrival epoch are given below. Since we have assumed $\rho = \lambda/\mu^* < 1$ and the Markov chain ζ_k is irreducible, there exists the minimal nonnegative solution $\mathbf{R} = [R_{ij}]_{m\times m}$ $(m = 4)$ of the matrix equation $\mathbf{R} = \sum_{k=0}^{\infty} \mathbf{R}^k \mathbf{S}_k$, where **R** is strictly positive and of spectral radius less than one. The matrix **R** can be calculated by an iterative scheme given below:

$$\mathbf{R}(i+1) = \left(\mathbf{S}_0 + \sum_{k=2}^{\infty} \mathbf{R}^k(i)\,\mathbf{S}_k\right)(\mathbf{I} - \mathbf{S}_1)^{-1}, \quad \text{with } \mathbf{R}(0) = \mathbf{0}, \tag{A.2}$$

where $\mathbf{R}(i)$ is the value of **R** at the ith iteration. The invariant probability vector of the Markov chain ζ_k is obtained by

$$\boldsymbol{\pi}^-(k) = \boldsymbol{\pi}^-(0)\,\mathbf{R}^k, \quad k \ge 0, \tag{A.3}$$

where $\boldsymbol{\pi}^-(0)$ is obtained by solving the left eigenvector of the matrix $\mathbf{B}[\mathbf{R}] = \sum_{k=1}^{\infty} \mathbf{R}^{k-1}\mathbf{S}_k^*$ with the normalizing condition $\boldsymbol{\pi}^-(0)[\mathbf{I}_m - \mathbf{R}]^{-1}\mathbf{e} = 1$. Now we obtain the roots of the characteristics equation $\det[\mathbf{I}_m z - \mathbf{S}(z)] = 0$, as $\gamma_1 = 0.180975, \gamma_2 = 0.944434, \gamma_3 = 0.278271 + 0.075281i$, and $\gamma_4 = 0.278271 - 0.075281i$. Thereafter, we get the stationary probability vectors $\boldsymbol{\pi}^-(n), n \ge 0$, at an embedded prearrival epoch using the roots $\gamma_1, \gamma_2, \gamma_3$, and γ_4. One may note that in this case the **R** has dimension 4×4, i.e., 16 unknown elements, and they may be obtained by solving the following 16 simultaneous linear equations derived from (A.3) as follows:

$$\boldsymbol{\pi}^-(1) = \boldsymbol{\pi}^-(0)\,\mathbf{R}, \tag{A.4}$$

$$\boldsymbol{\pi}^-(2) = \boldsymbol{\pi}^-(1)\,\mathbf{R}, \tag{A.5}$$

$$\boldsymbol{\pi}^-(3) = \boldsymbol{\pi}^-(2)\,\mathbf{R}, \tag{A.6}$$

$$\boldsymbol{\pi}^-(4) = \boldsymbol{\pi}^-(3)\,\mathbf{R}. \tag{A.7}$$

Equating $\boldsymbol{\pi}^-(i)$, $i = 1,2,3,4$, and (A.4)-(A.7) componentwise, we obtain the 16 simultaneous equations with 16 elements of the **R** matrix as variables. After solving these 16 simultaneous linear equations with 16 unknown elements of the matrix **R**, we obtain the **R** matrix as follows:

$$\mathbf{R} = \begin{bmatrix} 0.182617 & 0.005479 & 0.004615 & 0.002830 \\ 0.061297 & 0.431499 & 0.157426 & 0.090092 \\ 0.131600 & 0.228110 & 0.590780 & 0.335155 \\ 0.102322 & 0.318496 & 0.209492 & 0.477055 \end{bmatrix}. \tag{A.8}$$

The correctness of the **R** matrix is checked by computing the eigenvalues of **R** given in (A.8) and the eigenvalues are obtained as 0.180975, 0.944434, 0.278271 + 0.075281i, and 0.278271 − 0.075281i which are exactly the same as the characteristic roots. Also, we obtained **R** matrix using iterative scheme (A.2) and found that the **R** matrix is exactly the same as that given in (A.8) after 150 iterations.

Next, we investigate the same computational procedure for computing the **R** matrix in the case of $GI^{[X]}/C\text{-}MSP/1/\infty$ queueing model presented in Section 4.1. It may be noted here that for the case of $GI^{[X]}/C\text{-}MSP/1/\infty$ queueing model as discussed by Neuts [10] (Remarks on bottom of pg. 189), the $\mathbf{R} = [R_{ij}]_{m\hat{r}\times m\hat{r}}$ $(m = 3,\ \hat{r} = 4)$ may be solved by the following set of simultaneous linear equations.

$$\begin{aligned}&[\boldsymbol{\pi}^-(4\times k), \boldsymbol{\pi}^-(4\times k+1), \boldsymbol{\pi}^-(4\times k+2), \\ &\quad \boldsymbol{\pi}^-(4\times k+3)] = [\boldsymbol{\pi}^-(4\times(k-1)), \\ &\quad \boldsymbol{\pi}^-(4\times(k-1)+1), \boldsymbol{\pi}^-(4\times(k-1)+2), \\ &\quad \boldsymbol{\pi}^-(4\times(k-1)+3)]\,\mathbf{R}, \quad 1 \le k \le 12.\end{aligned} \tag{A.9}$$

One may note that for this case the **R** matrix computed by the above procedure (presented below) has spectral radius less than

$$\mathbf{R} = \begin{bmatrix} 0.081741 & 0.001970 & \cdots & 0.000968 \\ 0.009819 & 0.309779 & \cdots & 0.002166 \\ 0.001323 & 0.000346 & \cdots & 0.002384 \\ \vdots & \vdots & \cdots & \vdots \\ 0.026384 & 0.004028 & \cdots & 0.049990 \end{bmatrix}_{12\times 12}. \tag{A.10}$$

Here the eigenvalues of the **R** matrix are not the same as the characteristics roots in this case. The eigenvalues of **R** are given by 0.027661, 0.046758, 0.070065, 0.111779, 0.163563, 0.833729, 0.046207 + 0.023748i, 0.046207 − 0.023748i, 0.067628 + 0.035876i, 0.067628 − 0.035876i, 0.196019 + 0.246307i, and 0.196019 − 0.246307i. Also, we have checked that this **R** satisfies any of the following equations:

$$\begin{aligned}&[\boldsymbol{\pi}^-(4\times k), \boldsymbol{\pi}^-(4\times k+1), \boldsymbol{\pi}^-(4\times k+2), \\ &\quad \boldsymbol{\pi}^-(4\times k+3)] = [\boldsymbol{\pi}^-(0), \boldsymbol{\pi}^-(1), \boldsymbol{\pi}^-(2), \boldsymbol{\pi}^-(3)] \\ &\quad \cdot \mathbf{R}^k, \quad k \ge 0.\end{aligned} \tag{A.11}$$

Therefore, by the above simple computational procedure, one may compute the **R** matrix of the corresponding $GI^{[X]}/M/1$-type using the stationary probabilities (obtained through characteristic roots) at an embedded prearrival epoch.

Conflicts of Interest

The authors declare that they have no conflicts of interest.

Acknowledgments

The second author was supported partially by NSERC under research grant number RGPIN-2014-06604.

References

[1] D. M. Lucantoni, K. S. Meier-Hellstern, and M. F. Neuts, "A single-server queue with server vacations and a class of nonrenewal arrival processes," *Advances in Applied Probability*, vol. 22, no. 3, pp. 676–705, 1990.

[2] U. C. Gupta and A. D. Banik, "Complete analysis of finite and infinite buffer GI/MSP/1 queue—a computational approach," *Operations Research Letters*, vol. 35, no. 2, pp. 273–280, 2007.

[3] M. F. Neuts, "A versatile Markovian point process," *Journal of Applied Probability*, vol. 16, no. 4, pp. 764–779, 1979.

[4] V. Ramaswami, "The N/G/1 queue and its detailed analysis," *Advances in Applied Probability*, vol. 12, no. 1, pp. 222–261, 1980.

[5] M. L. Chaudhry, A. D. Banik, and A. Pacheco, "A simple analysis of the batch arrival queue with infinite-buffer and Markovian service process using roots method: GI[X]/C-MSP/1/∞," *Annals of Operations Research*, vol. 252, no. 1, pp. 135–173, 2017.

[6] G. Singh, U. C. Gupta, and M. L. Chaudhry, "Detailed computational analysis of queueing-time distributions of the BMAP/G/1 queue using roots," *Journal of Applied Probability*, vol. 53, no. 4, pp. 1078–1097, 2016.

[7] M. L. Chaudhry and J. G. C. Templeton, *A first course in bulk queues*, John Wiley & Sons, New York, NY, USA, 1983.

[8] W. K. Grassmann, "A new method for finding the characteristic roots of En/Em/1 queues," *Methodology and Computing in Applied Probability*, vol. 13, no. 4, pp. 873–886, 2011.

[9] H. R. Gail, S. L. Hantler, and B. A. Taylor, "Use of Characteristic Roots for Solving Infinite State Markov Chains," in *Computational Probability*, vol. 24 of *International Series in Operations Research & Management Science*, pp. 205–255, Springer US, Boston, MA, 2000.

[10] M. Neuts, *Matrix-Geometric Solutions in Stochastic Models—An Algorithmic Approach*, The Johns Hopkins University Press, Baltimore, Md, USA, 1981.

[11] J. F. Shortle, P. H. Brill, M. J. Fischer, D. Gross, and D. M. Masi, "An algorithm to compute the waiting time distribution for the M/G/1 queue," *INFORMS Journal on Computing*, vol. 16, no. 2, pp. 152–161, 2004.

[12] S. K. Samanta, "Sojourn-time distribution of the GI/MSP/1 queueing system," *OPSEARCH*, vol. 52, no. 4, pp. 756–770, 2015.

23

A New Class of Distributions Generated by the Extended Bimodal-Normal Distribution

Milton A. Cortés, David Elal-Olivero, and Juan F. Olivares-Pacheco

Departamento de Matemática, Facultad de Ingeniería, Universidad de Atacama, Copiapó, Chile

Correspondence should be addressed to Juan F. Olivares-Pacheco; jfolivar@uda.cl

Academic Editor: Steve Su

In this study, we present a new family of distributions through generalization of the extended bimodal-normal distribution. This family includes several special cases, like the normal, Birnbaum-Saunders, Student's t, and Laplace distribution, that are developed and defined using stochastic representation. The theoretical properties are derived, and easily implemented Monte Carlo simulation schemes are presented. An inferential study is performed for the Laplace distribution. We end with an illustration of two real data sets.

1. Introduction

Although the normal distribution is the most popular probability model in statistics, several random phenomena in nature cannot be described by the normal distribution. In this regard, Azzalini [1] introduced an extension of the normal distribution called skew-normal distribution, where this model shares some properties with the standard normal model; it is mathematically tractable and it has a wide range of the coefficients of skewness and kurtosis. From this work, an important line of research focusing on finding new distributions that offered greater flexibility is generated.

More recently, Elal-Olivero [2] introduced a new class of skew-normal distribution called alpha-skew-normal distribution. In doing so, he first defined a new bimodal-symmetric normal distribution with probability density function given by

$$\zeta(x) = x^2 \phi(x), \quad x \in \mathbb{R}, \tag{1}$$

where $\phi(\cdot)$ is the standard normal density, which is defined as the bimodal-normal (BN) distribution. Furthermore, he studies some properties of this distribution and presents its stochastic representation as the product of two independent random variables $\sqrt{T}$ and V, where $T \sim \chi^2_{(3)}$ and V is a discrete random variable such that $\mathsf{P}(V = \pm 1) = 1/2$; that is, $X = \sqrt{T}V$ has the distribution BN. On the other hand, an extension of the BN density is given by

$$\phi_{\xi}(x) = \left(\frac{1+\xi x^2}{1+\xi}\right)\phi(x), \quad x \in \mathbb{R}, \tag{2}$$

where $\xi \geq 0$ is the shape parameter. Note that this density function is symmetric and is characterized by incorporating bimodality into the normal distribution, which is controlled by the parameter ξ. Elal-Olivero [2] presents this extension as the symmetric-component of the alpha-skew-normal distribution. Furthermore, (2) also can be deduced from the model presented in Elal-Olivero et al. [3]. In this regard, Gui et al. [4] incorporated (2) into the slash distribution, developed its properties, and performed inferential studies, whereas Gómez and Guerrero [5] incorporated (2) into the Birnbaum-Saunders distribution, tested its bimodality, and demonstrated its principal properties.

The objective of this article is to present a new family of distributions through generalization of (2). This generalization can be applied to any density function, thereby producing a more flexible model incorporating a shape parameter. Depending on the density at which we apply this generalization, it is observed that the new model is flexible enough to support uni- and bimodal shapes. Furthermore,

Gui et al. [4] and Gómez and Guerrero [5] are particular cases of the generalization proposal.

This article is organized as follows. In Section 2, we present a generalization of (2) and review some particular cases (normal, Birnbaum-Saunders, Student's t, and Laplace distribution). In Section 3, we develop the basic properties of the cases from Section 2 and study the effects of this new generalization. In Section 4, we study some inferential aspects of the extended Laplace distribution using maximum likelihood estimation and perform a Monte Carlo simulation study. We conclude in Section 5 with a discussion.

2. A General Class of Distributions

This section describes a general class of distributions generated by (2), presents its basic properties, and derives explicit expressions for the normal, Birnbaum-Saunders, Student's t, and Laplace distribution.

2.1. Characterization and Properties

Theorem 1 (general class of distributions). *Let g be a probability density function and h a positive continuous function such that $\mathsf{E}_g[h(Y)] = \kappa < \infty$, where $Y \sim g$. Then,*

$$f(x;\xi) = \left(\frac{1+\xi h(x)}{1+\kappa\xi}\right) g(x), \quad x \in \mathbb{R}, \tag{3}$$

is a probability density function with shape parameter $\xi \geq 0$.

Proof. If we note $f(x;\xi)$ can be represented as a mixture of two densities, then the result follows immediately; that is, $f(x;\xi) = pg(x) + (1-p)(1/\kappa)h(x)g(x)$, where $p = 1/(1+\kappa\xi)$.

Remark 2. On the basis of Theorem 1, we can make the following observations:

(1) If $\xi = 0$, then $f(x;\xi) = g(x)$, $x \in \mathbb{R}$.

(2) If $\xi \longrightarrow \infty$, then $f(x;\xi) \longrightarrow (1/\kappa)h(x)g(x)$, $x \in \mathbb{R}$.

Theorem 3 (stochastic representation). *Let $X_1 \sim g(x)$ and $X_2 \sim (1/\kappa)h(x)g(x)$ be independent random variables. If*

$$X = \begin{cases} X_1, & \textit{with probability } \dfrac{1}{1+\kappa\xi}, \\ X_2, & \textit{with probability } \dfrac{\kappa\xi}{1+\kappa\xi}, \end{cases} \tag{4}$$

then $X \sim f(x;\xi)$.

Proof. Since $f(x;\xi)$ can be represented as a mixture, the result follows immediately.

Remark 4. If $X \sim f(x;\xi)$, then

(1) The cumulative distribution function is given by

$$F(x;\xi) = \frac{1}{1+\kappa\xi}\left(F_1(x;\xi) + \kappa\xi F_2(x;\xi)\right), \tag{5}$$

where $F_1(x;\xi)$ and $F_2(x;\xi)$ are the cumulative distribution functions of $X_1 \sim g(x)$ and $X_2 \sim (1/\kappa)h(x)g(x)$, respectively.

(2) The moment generating function is given by

$$M_X(t) = \frac{1}{1+\kappa\xi}\left(M_1(t) + \kappa\xi M_2(t)\right), \tag{6}$$

where $M_1(t)$ and $M_2(t)$ are the moments generating functions of $X_1 \sim g(x)$ and $X_2 \sim (1/\kappa)h(x)g(x)$, respectively, if both exist.

(3) The r-th moment of the random variable X is given by

$$\mathsf{E}[X^r] = \frac{1}{1+\kappa\xi}\left(\mathsf{E}_g[X^r] + \xi\mathsf{E}_g[X^r h(X)]\right), \quad r = 1, 2, \ldots \tag{7}$$

2.2. Special Cases. In this section, explicit expressions are provided for the probability density function in (3) for the normal, Birnbaum-Saunders, Student's t, and Laplace distribution and different choices of h. These models are selected to show the benefits of the proposed extension, and the choice of the function $h(x)$ is conditioned upon a positive function with finite expectation.

Corollary 5 (normal case). *If $h(x) = x^2$ and $g(x) = \phi(x)$, then X has the probability density function given by*

$$f(x;\xi) = \left(\frac{1+\xi x^2}{1+\xi}\right)\phi(x), \quad x \in \mathbb{R}, \tag{8}$$

and we say that X has an "extended normal distribution," which is denoted as $X \sim EN(\xi)$.

Corollary 6 (Birnbaum-Saunders case). *Let $a_x(\alpha,\beta) = (1/\alpha)(\sqrt{x/\beta} - \sqrt{\beta/x})$. If $h(x) = a_x^2(\alpha,\beta)$ and $g(x) = a_x'(\alpha,\beta)\phi(a_x(\alpha,\beta))$ where $a_x'(\alpha,\beta)$ is the derivative with respect to x, with $\alpha > 0$ and $\beta > 0$, then X has the probability density function given by*

$$f(x;\alpha,\beta,\xi) = \left(\frac{1+\xi a_x^2(\alpha,\beta)}{1+\xi}\right) a_x'(\alpha,\beta)\,\phi(a_x(\alpha,\beta)), \quad x > 0, \tag{9}$$

and we say that X has an "extended Birnbaum-Saunders distribution," which is denoted as $X \sim EBS(\alpha,\beta,\xi)$.

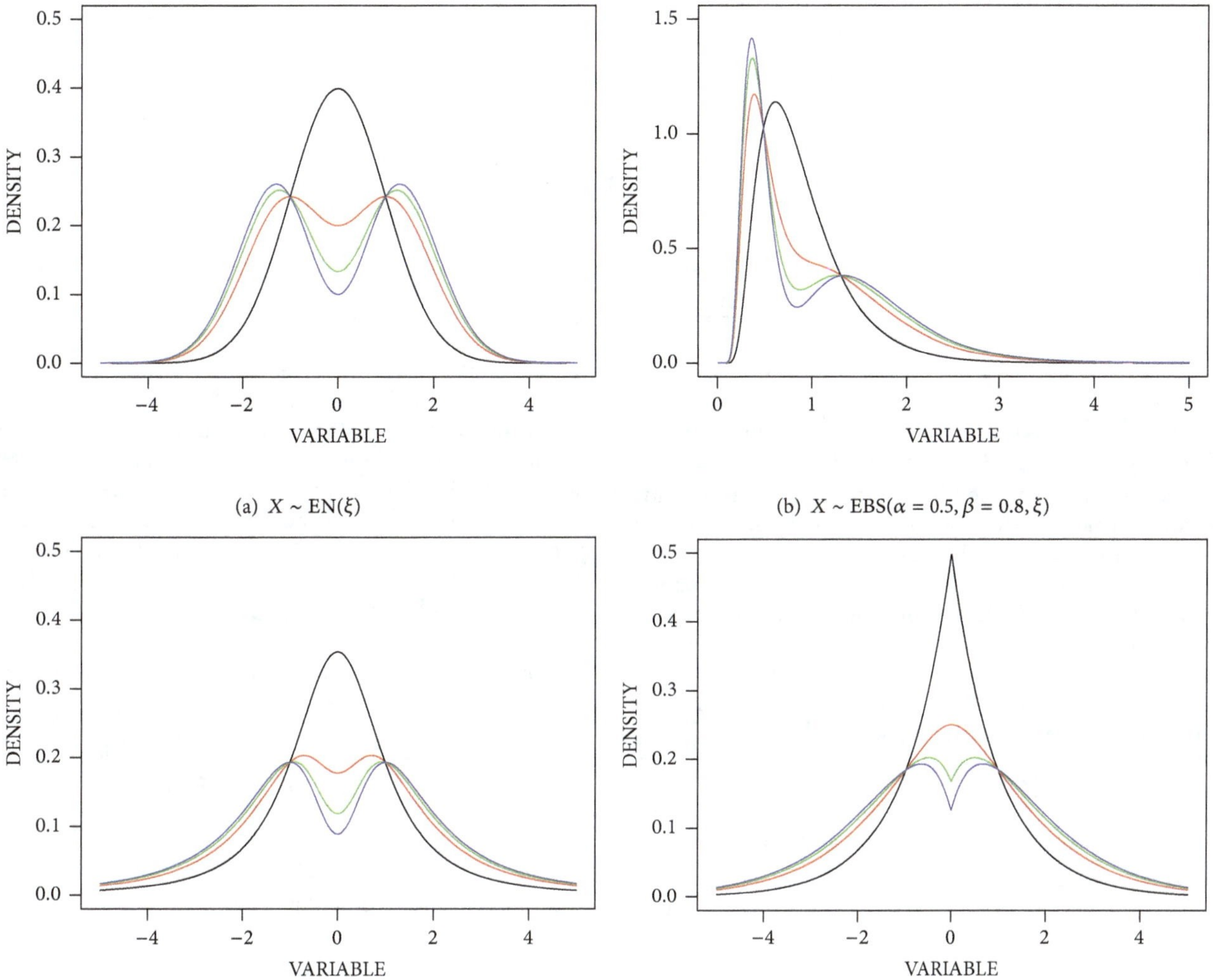

(a) $X \sim \text{EN}(\xi)$

(b) $X \sim \text{EBS}(\alpha = 0.5, \beta = 0.8, \xi)$

(c) $X \sim \text{Et}(\nu = 2, \xi)$

(d) $X \sim \text{EL}(\xi)$

Figure 1: PDF's for the special cases defined in Corollaries 5–8 and several values of ξ. $\xi = 0$ (black line), $\xi = 1$ (red line), $\xi = 2$ (green line), and $\xi = 3$ (blue line).

Corollary 7 (Student's t case). *If $h(x) = x^2(1+\nu)/(x^2+\nu)$ and $g(x) = (\Gamma((1+\nu)/2)/\sqrt{\pi\nu}\Gamma(\nu/2))(1+x^2/\nu)^{-(1+\nu)/2}$ where $\nu > 0$, then X has the probability density function given by*

$$f(x;\nu,\xi) = \left(\frac{\nu + (1+\xi+\xi\nu)x^2}{(\nu+x^2)(1+\xi)}\right) \cdot \frac{\Gamma((1+\nu)/2)}{\sqrt{\pi\nu}\Gamma(\nu/2)}\left(1+\frac{x^2}{\nu}\right)^{-(1+\nu)/2}, \quad x \in \mathbb{R}, \tag{10}$$

and we say that X has an "extended Student's t distribution," which is denoted as $X \sim Et(\nu, \xi)$.

Corollary 8 (Laplace case). *If $h(x) = |x|$ and $g(x) = (1/2)e^{-|x|}$, then X has the probability density function given by*

$$f(x;\xi) = \left(\frac{1+\xi|x|}{2(1+\xi)}\right)e^{-|x|}, \quad x \in \mathbb{R}, \tag{11}$$

and we say that X has an "extended Laplace distribution," which is denoted as $X \sim EL(\xi)$.

As we notice, in the Corollaries 5–8 and Figure 1, when the function $g(x)$ is a symmetric density, the effect of the extension is that the model supports uni- and bimodal shapes. On the other hand, if the model has positive support, the bimodality depends on the choice of parameters, as seen in the Birnbaum-Saunders distribution case.

3. Some Results of the Special Cases

In this section, we develop some properties associated with the models defined in Corollaries 5–8. The cumulative distribution function, moment, and stochastic representation will be presented when they correspond to the cases at hand. Some proofs are straightforward and are, therefore, omitted.

3.1. Extended Normal Distribution. The extended normal distribution is the basis for the development of the specific

Table 1: Alternative stochastic representation of special cases.

Case	Representation	Variables
EN	$\sqrt{\frac{1}{1+\xi}}Z + \sqrt{\frac{\xi}{1+\xi}}X$	$Z \sim \mathrm{N}(0,1)$ and $X \sim \mathrm{BN}$
EBS	$\beta\left(\frac{\alpha}{2}Y + \sqrt{\left(\frac{\alpha}{2}Y\right)^2 + 1}\right)^2$	$Y \sim \mathrm{EN}(\xi)$
Et	$\frac{Y}{\sqrt{V/\nu}}$	$Y \sim \mathrm{EN}(\xi)$ and $V \sim \chi^2_{(\nu)}$
EL	$\sqrt{2WY}$	$W \sim \exp(1)$ and $Y \sim \mathrm{EN}(\xi)$

cases discussed previously. If $g(x) = \phi(x)$ and $h(x) = x^2$, then, from Theorem 3, we know that $X_1 \sim \phi(x)$ and $X_2 \sim x^2\phi(x)$. Note that the distribution of X_2 corresponds to the bimodal-normal distribution, for which the stochastic representation was presented in Section 1.

The stochastic representation of $X \sim \mathrm{EN}(\xi)$ is obtained through Theorem 3. Table 1 shows an alternative way to generate random variables $X \sim \mathrm{EN}(\xi)$. Furthermore, the stochastic representation has a form that is similar to the representation given in Henze [6] for the skew-normal distribution presented in Azzalini [1].

Remark 9. If $X \sim \mathrm{EN}(\xi)$, then

(1) The cumulative distribution function is given by

$$F(x;\xi) = \Phi(x) - \frac{\xi x}{1+\xi}\phi(x), \quad x \in \mathbb{R}. \tag{12}$$

(2) The r-th moment of the random variable X is given by

$$\mathsf{E}[X^r] = (1+(-1)^r)\frac{2^{-1+r/2}(1+\xi+r\xi)\Gamma((1+r)/2)}{\sqrt{\pi}(1+\xi)}, \quad r = 1,2,\ldots \tag{13}$$

(3) The expected value and variance of the random variable X is given by

$$\mathsf{E}[X] = 0, \qquad \mathsf{V}(X) = \frac{1+3\xi}{1+\xi}. \tag{14}$$

3.2. Extended Birnbaum-Saunders distribution. The Birnbaum-Saunders (BS) distribution (see Birnbaum and Saunders [7, 8]) describes the lifetime of components exposed to fatigue caused by cyclical stress and tension. Since 1969, the number of studies that have investigated this distribution and discussed the development of both its theoretical properties and its applications has increased dramatically. Because of its significance, this distribution has been extended in a variety of manners to relax its behavior and thus make it applicable to a wide range of situations. For example, see Birnbaum and Saunders [7, 8], Mann et al. [9], Desmond [10, 11], Chang and Tang [12], Díaz-García and Leiva-Sánchez [13], Gómez et al. [14], and Olmos et al. [15]. The BS distribution with parameters $\alpha > 0$ and $\beta > 0$ has density function given by

$$g(x) = a'_x(\alpha,\beta)\,\phi(a_x(\alpha,\beta)), \quad x > 0, \tag{15}$$

where $a_x(\alpha,\beta)$ is defined in Corollary 6 and is denoted as $\mathrm{BS}(\alpha,\beta)$. If $X_1 \sim \mathrm{BS}(\alpha,\beta)$ and $h(x) = a_x^2(\alpha,\beta)$, then

$$X_2 \sim \frac{1}{\kappa}h(x)\,g(x) = a_x^2(\alpha,\beta)\,a'_x(\alpha,\beta)\,\phi(a_x(\alpha,\beta)), \quad x > 0, \tag{16}$$

with a stochastic representation given by

$$X_2 = \beta\left(\frac{\alpha}{2}Y + \sqrt{\left(\frac{\alpha}{2}Y\right)^2 + 1}\right)^2, \tag{17}$$

where $Y \sim \mathrm{BN}$. From Theorem 1, the extended Birnbaum-Saunders distribution has density (9) and from Theorem 3 we can generate random variables $X \sim \mathrm{EBS}(\alpha,\beta,\xi)$. An alternative way to generate this random variable can be seen in Table 1.

Theorem 10. *Let $X \sim EBS(\alpha,\beta,\xi)$ with $\alpha > 0$, $\beta > 0$ and $\xi \geq 0$. Then*

(1) *$aX \sim EBS(\alpha, a\beta, \xi)$, for $a > 0$.*

(2) *$X^{-1} \sim EBS(\alpha, \beta^{-1}, \xi)$.*

Proof. The proofs are immediate from the theorem of the change of variable.

Remark 11. Like the Birnbaum-Saunders distribution observes that the property (1) established in Theorem 10 implies that the EBS distribution belongs to the scale family, whilst the property (2) implies that it also belongs to the family of random variables closed under reciprocation; see Saunders [16]. Furthermore, based on properties (1) and (2), we can have the two-parameter EBS distribution: $\alpha X/\beta \sim \mathrm{EBS}(\alpha,\alpha,\xi)$.

Remark 12. If $X \sim \text{EBS}(\alpha, \beta, \xi)$, then

(1) The cumulative distribution function is given by

$$F(x;\alpha,\beta,\xi) = F_{BS}(x;\alpha,\beta) - \frac{\xi a_x(\alpha,\beta)}{1+\xi}\phi(a_x(\alpha,\beta)), \quad x > 0, \tag{18}$$

where $F_{BS}(\cdot)$ is the cumulative distribution function of the Birnbaum-Saunders distribution.

(2) The r-th moment of the random variable X is given by

$$\mathsf{E}[X^r] = \beta^r \sum_{j=0}^{r} \binom{2r}{2j} \sum_{m=0}^{j} \binom{j}{m} \left(\frac{\alpha}{2}\right)^{2(r+m-j)} \mathsf{E}\left[Y^{2(r+m-j)}\right], \quad r = 1,2,\ldots, \tag{19}$$

where $Y \sim \text{EN}(\xi)$.

(3) The expected value and variance of the random variable X is given by

$$\mathsf{E}[X] = \beta + \frac{\beta(1+3\xi)\alpha^2}{2(1+\xi)},$$
$$\mathsf{V}(X) = \frac{\alpha^2\beta^2\left(4(1+\xi)(1+3\xi) + (5+3\xi(10+7\xi))\alpha^2\right)}{4(1+\xi)^2}. \tag{20}$$

3.3. Extended Student's t-Distribution. The Student's t-distribution serves as a robust alternative when it is desired to model data sets with atypical values and with a coefficient of kurtosis that is greater than of the normal distribution. The Student's t-distribution with parameter $\nu > 0$ has a density function given by

$$g(x) = \frac{\Gamma((1+\nu)/2)}{\sqrt{\pi\nu}\Gamma(\nu/2)}\left(1+\frac{x^2}{\nu}\right)^{-(1+\nu)/2}, \quad x \in \mathbb{R}. \tag{21}$$

If $X_1 \sim g(x)$ and $h(x) = x^2(1+\nu)/(x^2+\nu)$, then,

$$X_2 \sim \frac{1}{\kappa}h(x)g(x) = \left(\frac{x^2(1+\nu)}{x^2+\nu}\right)\frac{\Gamma((1+\nu)/2)}{\sqrt{\pi\nu}\Gamma(\nu/2)}\left(1+\frac{x^2}{\nu}\right)^{-(1+\nu)/2}, \quad x \in \mathbb{R}, \tag{22}$$

with a stochastic representation given by $X_2 = Y/\sqrt{W/\nu}$, where $Y \sim \text{BN}$ and $W \sim \chi^2_{(\nu)}$ are independent random variables. From Theorem 1, the extended Student's t-distribution has density (10) and from Theorem 3 we can generate random variables $X \sim \text{Et}(\nu,\xi)$. An alternative way to generate this random variable can be seen in Table 1.

Remark 13. If $X \sim \text{Et}(\nu, \xi)$, then

(1) The cumulative distribution function is given by

$$F(x;\nu,\xi) = F_T(x;\nu) - \frac{\xi x}{1+\xi}f_T(x;\nu), \quad x \in \mathbb{R}, \tag{23}$$

where $F_T(\cdot)$ and $f_T(\cdot)$ are the cumulative distribution function and probability density function of the t-Student distribution, respectively.

(2) The r-th moment of the random variable X is given by

$$\mathsf{E}[X^r] = (1+(-1)^r)\frac{\nu^{r/2}(1+\xi+r\xi)}{2\sqrt{\pi}(1+\xi)\Gamma(\nu/2)}\Gamma\left(\frac{1+r}{2}\right) \cdot \Gamma\left(\frac{-r+\nu}{2}\right), \quad r = 1,2,\ldots, \tag{24}$$

where $\nu > r$.

(3) The expected value and variance of the random variable X is given by

$$\mathsf{E}[X] = 0, \quad \nu > 1,$$
$$\mathsf{V}(X) = \frac{(1+3\xi)\nu}{(1+\xi)(-2+\nu)}, \quad \nu > 2. \tag{25}$$

3.4. Extended Laplace Distribution. The Laplace (L) or double exponential distribution, which was originally published by Pierre Laplace in 1774, is a symmetric distribution with density function given by

$$g(x) = \frac{1}{2}e^{-|x|}, \quad x \in \mathbb{R}. \tag{26}$$

If $X_1 \sim \text{L}$ and $h(x) = |x|$, then,

$$X_2 \sim \frac{1}{\kappa}h(x)g(x) = \frac{|x|}{2}e^{-|x|}, \quad x \in \mathbb{R}, \tag{27}$$

with a stochastic representation given by $X_2 = \sqrt{2W}Y$, where $W \sim \exp(1)$ and $Y \sim \text{BN}$ are independent random variables. From Theorem 1, the extended Laplace distribution has density (11) and from Theorem 3 we can generate random variables $X \sim \text{EL}(\xi)$. An alternative way to generate this random variable can be seen in Table 1.

Remark 14. If $X \sim \text{EL}(\xi)$, then

(1) The cumulative distribution function is given by

$$F(x;\xi) = \begin{cases} \frac{1}{2}e^{x}\left(1-\frac{\xi x}{1+\xi}\right), & \text{if } x < 0, \\ 1-\frac{1}{2}e^{-x}\left(1+\frac{\xi x}{1+\xi}\right), & \text{if } x \geq 0. \end{cases} \tag{28}$$

(2) The r-th moment of the random variable X is given by

$$\mathsf{E}[X^r] = (1+(-1)^r)\frac{r(1+\xi+r\xi)}{2(1+\xi)}\Gamma(r), \quad r = 1,2,\ldots \tag{29}$$

(3) The expected value and variance of the random variable X are given by

$$\mathrm{E}[X] = 0,$$
$$\mathrm{V}(X) = \frac{2(1+3\xi)}{1+\xi}. \tag{30}$$

Table 1 shows an alternative way to generate random variables for the special cases defined in Corollaries 5–8. We can see that the extended normal distribution is the basis for the development of the specific cases discussed previously.

4. Inferential Aspects of the EL Distribution

In this section, we will study some inferential properties of the extended Laplace distribution defined in Corollary 8 . We will explore maximum likelihood estimators and Monte Carlo simulation and will apply there results to two real data sets, comparing the fit with the Laplace distribution using the likelihood ratio and the Akaike Information Criterion (AIC).

4.1. Maximum Likelihood Estimator. In practice, it is common to work with a location and scale transformation $X = \mu + \sigma Y$, where $\mu \in \mathbb{R}$, $\sigma > 0$, and $Y \sim \mathrm{EL}(\xi)$ with $\xi \geq 0$. Hence, the density for the random variable X, denoted as $X \sim \mathrm{EL}(\mu, \sigma, \xi)$, is

$$f(x; \mu, \sigma, \xi) = \left(\frac{\sigma + \xi|x-\mu|}{2\sigma^2(1+\xi)}\right)e^{-|x-\mu|/\sigma}, \quad x \in \mathbb{R}. \tag{31}$$

Assume that $\boldsymbol{x} = (x_1, x_2, \ldots, x_n)^\top$ is a random sample of size n from an $\mathrm{EL}(\mu, \sigma, \xi)$ distribution. From (31), the log-likelihood function is

$$\ell(\boldsymbol{\theta}) \propto -2n\log(\sigma) - n\log(1+\xi) + \sum_{i=1}^{n}\log(\sigma + \xi|x_i-\mu|) - \frac{1}{\sigma}\sum_{i=1}^{n}|x_i-\mu|, \tag{32}$$

where $\boldsymbol{\theta} = (\mu, \sigma, \xi)^\top$, which is a continuous function in each parameter, but it is not differentiable at $x_i = \mu$, for $i = 1, \ldots, n$. Thus, by assuming $x_i \neq \mu$, for $i = 1, \ldots, n$, we have that elements of the score vector are $\mathbf{S}(\boldsymbol{\theta}) = (S_\mu, S_\sigma, S_\xi)^\top$, where $S_a = \partial\ell/\partial a$, given by

$$\begin{aligned}
S_\mu &= -\xi\sum_{i=1}^{n}\frac{\mathrm{sgn}(x_i-\mu)}{\sigma+\xi|x_i-\mu|} + \frac{1}{\sigma}\sum_{i=1}^{n}\mathrm{sgn}(x_i-\mu),\\
S_\sigma &= -\frac{2n}{\sigma} + \frac{1}{\sigma^2}\sum_{i=1}^{n}|x_i-\mu| + \sum_{i=1}^{n}\frac{1}{\sigma+\xi|x_i-\mu|},\\
S_\xi &= -\frac{n}{1+\xi} + \sum_{i=1}^{n}\frac{|x_i-\mu|}{\sigma+\xi|x_i-\mu|},
\end{aligned} \tag{33}$$

where $\mathrm{sgn}(\cdot)$ denotes the sign function.

Hence, the maximum likelihood estimator $\widehat{\boldsymbol{\theta}}$ solves the score equations $\mathbf{S}(\boldsymbol{\theta}) = \mathbf{0}$. Which must be obtained through a numerical method. A lot of software, including optimization toolbox, can be used for obtaining the maximum likelihood estimates. To achieve the maximization of log-likelihood function, we used the function `optim` on `R` (see R Core Team [17]), the specific method being `Nelder-Mead` (see Nelder and Mead [18]), that uses only function values and is robust but relatively slow. It will work reasonably well for nondifferentiable functions.

For obtaining the standard errors of the maximum likelihood estimates one should compute the information matrix $\boldsymbol{F}(\boldsymbol{\theta})$. It is well known that the elements of $\boldsymbol{F}(\boldsymbol{\theta})$ are given by

$$\boldsymbol{F}(\boldsymbol{\theta}) = -\mathrm{E}\left[\frac{\partial^2\ell}{\partial\theta_i\partial\theta_j}\right], \quad i, j = 1, 2, 3 \text{ and } \boldsymbol{\theta} = (\mu, \sigma, \xi)^\top. \tag{34}$$

Since expectation over EL distribution is not straightforward, numerical methods should be performed to obtain the explicit form of the information matrix. This matrix can be approximated by the observed information matrix $\boldsymbol{I}(\widehat{\boldsymbol{\theta}})$, which is defined as minus the Hessian matrix evaluated at $\widehat{\boldsymbol{\theta}}$; that is

$$\boldsymbol{I}(\widehat{\boldsymbol{\theta}}) \approx -\left(\frac{\partial^2\ell}{\partial\theta_i\partial\theta_j}\right)_{i,j}, \tag{35}$$

where the second derivatives are given below:

$$\begin{aligned}
\frac{\partial^2\ell}{\partial\mu^2} &= -\xi^2\sum_{i=1}^{n}\frac{1}{(\sigma+\xi|x_i-\mu|)^2},\\
\frac{\partial^2\ell}{\partial\mu\partial\sigma} &= -\frac{1}{\sigma^2}\sum_{i=1}^{n}\mathrm{sgn}(x_i-\mu) + \xi\sum_{i=1}^{n}\frac{\mathrm{sgn}(x_i-\mu)}{(\sigma+\xi|x_i-\mu|)^2},\\
\frac{\partial^2\ell}{\partial\mu\partial\xi} &= -\sigma\sum_{i=1}^{n}\frac{\mathrm{sgn}(x_i-\mu)}{(\sigma+\xi|x_i-\mu|)^2},\\
\frac{\partial^2\ell}{\partial\sigma^2} &= \frac{2n}{\sigma^2} - \sum_{i=1}^{n}\frac{1}{(\sigma+\xi|x_i-\mu|)^2} - \frac{2}{\sigma^3}\sum_{i=1}^{n}|x_i-\mu|,\\
\frac{\partial^2\ell}{\partial\sigma\partial\xi} &= -\sum_{i=1}^{n}\frac{|x_i-\mu|}{(\sigma+\xi|x_i-\mu|)^2},\\
\frac{\partial^2\ell}{\partial\xi^2} &= \frac{n}{(1+\xi)^2} - \sum_{i=1}^{n}\frac{(x_i-\mu)^2}{(\sigma+\xi|x_i-\mu|)^2}.
\end{aligned} \tag{36}$$

Thus, we use the observed information matrix for computing the standard errors in the rest of the paper. Note that this approximation of the observed information matrix is obtained under a less stringent supposition, this is, assuming that the density function is absolutely continuous, as is the case with the Laplace distribution (see Kotz et al. [19], remark 2.6.1).

4.2. Numerical Study. We shall use Monte Carlo simulation to evaluate the finite sample performance of the maximum

TABLE 2: Empirical means and standard error for maximum likelihood estimators.

n	Parameters			Estimations		
	μ	σ	ξ	$\hat{\mu}$ (SE)	$\hat{\sigma}$ (SE)	$\hat{\xi}$ (SE)
50	0	1	0.5	0.0204 (0.4820)	1.0389 (0.3676)	0.5309 (1.0881)
			1	0.0242 (0.3654)	1.0280 (0.2816)	1.0336 (1.5642)
			3	0.0427 (0.1831)	0.9960 (0.1650)	3.1850 (3.7918)
	5	2	0.5	5.1040 (0.8495)	2.0082 (0.6631)	0.6770 (1.1768)
			1	5.1001 (0.6149)	2.0170 (0.5393)	1.2138 (1.7345)
			3	5.0971 (0.3587)	1.9945 (0.3280)	3.2898 (3.9105)
100	0	1	0.5	0.0204 (0.2839)	1.0221 (0.2745)	0.5318 (0.8100)
			1	0.0252 (0.2173)	1.0146 (0.1887)	1.0345 (1.0549)
			3	0.0248 (0.1206)	0.9968 (0.1203)	3.1760 (2.8356)
	5	2	0.5	5.0733 (0.5707)	1.9895 (0.4675)	0.6398 (0.7921)
			1	5.0679 (0.3964)	1.9881 (0.3951)	1.1650 (1.2693)
			3	5.0555 (0.2397)	1.9874 (0.2444)	3.2888 (2.9978)
250	0	1	0.5	0.0145 (0.1416)	1.0051 (0.1676)	0.5347 (0.5004)
			1	0.0213 (0.1290)	1.0011 (0.1171)	1.0335 (0.6633)
			3	0.0065 (0.0699)	0.9963 (0.0732)	3.1530 (1.6987)
	5	2	0.5	5.0293 (0.2936)	1.9586 (0.2806)	0.6131 (0.4630)
			1	5.0441 (0.2255)	1.9774 (0.2034)	1.1208 (0.6059)
			3	5.0255 (0.1387)	1.9922 (0.1469)	3.2357 (1.7630)
500	0	1	0.5	0.0107 (0.0806)	0.9960 (0.1048)	0.5341 (0.3108)
			1	0.0149 (0.0856)	0.9992 (0.0721)	1.0362 (0.3982)
			3	0.0067 (0.0488)	0.9944 (0.0497)	3.1315 (1.1299)
	5	2	0.5	5.0058 (0.1794)	1.9592 (0.1822)	0.5861 (0.2886)
			1	5.0205 (0.1587)	1.9802 (0.1348)	1.0922 (0.3904)
			3	5.0165 (0.0954)	1.9908 (0.0985)	3.2127 (1.1524)

likelihood estimator. The number of Monte Carlo replications was 1000 from simulated samples of the EL distribution for several samples sizes. Each sample was generated using the stochastic representation of the EL distribution, described above. For each generated sample, we obtain the maximum likelihood estimates using the function `optim` on R, the specific method being `Nelder-Mead`.

In order to analyze the point estimation results, we computed, for each sample size and for each estimator, the standard error from the observed information matrix defined in (35). The result can be seen in Table 2. From the results, we can see that the estimates are quite stable and estimates are asymptotically unbiased as expected, that is, it is observed that the bias becomes smaller as the sample size n increases.

4.3. Data Illustration. In this section we shall examine the application of the EL distribution to two real data sets. The first data set is related to the project WHO MONICA (World Health Organization Multinational Monitoring of Trends and Determinants in Cardiovascular Disease). This data set has been previously analyzed and studied in Kuulasmaa et al. [20], Kulathinal et al. [21], and de Castro et al. [22] and corresponds to the average annual rate of occurrence of cardiovascular mortality or the presence of coronary disease. The data are as follows: −5.6, −3.1, 0.3, −1.9, −1.6, 2.0, −0.9, −3.1, −7.4, −7.2, −4.6, 1.2, −2.3, −3.5, −4.9, −1.6, −3.5, −0.8, −4.1, −2.2, 0.9, 1.8, −6.7, −1.0, −2.3, 0.6, −1.8, 0.8, 2.2, −1.3, −4.9, −5.4, 0.7, −2.9, −6.1, −0.6, −5.9, 2.3, −5.8, −4.3, −4.7, −0.4, 4.0, −2.9, 4.9, −0.9, −5.4, −8.7, −1.6, −5.9, −3.9, −7.0, −0.1, −5.9, −0.3, 2.4, 1.4, −5.1, 8.4, 4.8, −6.1, −1.1, 0.0, −8.5, −1.7, 7.5, 0.5, −0.1, 0.4, −5.4, −4.0, −3.3, −1.8, 4.3. The second data set consists of the heights in inches of 126 students from University of Pennsylvania. This data set has been previously analyzed and studied by Hassan and Hijazi [23] and Gui et al. [4]. The data are as follows: 55.00, 60.00, 60.25, 61.00, 61.75, 62.25, 62.25, 62.63, 62.75, 63.00, 63.25, 63.25, 63.25, 63.25, 63.38, 64.00, 64.25, 64.25, 64.50, 64.75, 64.75, 65.00, 65.00, 65.13, 65.13, 65.17, 65.25, 65.25, 65.25, 65.25, 65.25, 65.50, 65.75, 66.00, 66.00, 66.25, 66.25, 66.25, 66.50, 66.75, 66.75, 66.75, 66.75, 67.00, 67.13, 67.25, 67.38, 67.50, 67.50, 67.75, 67.75, 67.75, 68.13, 68.75, 69.00, 69.00, 69.25, 69.50, 69.88, 70.00, 70.00, 70.25, 70.38, 71.00, 71.00, 71.25, 71.75, 65.50, 65.75, 66.00, 66.25, 66.75, 67.00, 67.13, 67.25, 68.00, 68.13, 68.75, 68.75, 69.13, 69.25, 69.25, 69.38, 69.75, 70.00, 70.13, 70.50, 70.75, 70.75, 71.00, 71.00, 71.50, 71.50, 71.63, 71.75, 71.75, 72.00, 72.00, 72.00, 72.00, 72.00, 72.00, 72.25, 72.25, 72.25, 72.38, 72.50, 72.50, 72.88, 73.00, 73.00, 73.00, 73.25, 73.25, 73.63, 74.00, 74.63, 75.00, 75.00, 75.05, 76.00, 77.00, 77.13, 77.50, 77.50, 78.50. Table 3 shows a descriptive summary of the data sets analyzed.

From these data sets, the maximum likelihood estimators of the parameters associated with the Laplace ($L(\mu,\sigma)$) and the extended Laplace ($EL(\mu,\sigma,\xi)$) distributions are obtained; the results of the comparison are summarized in Table 4 and

TABLE 3: Descriptive summary.

Data set	n	$\overline{x}$	s_x	Asymmetry	Kurtosis
WHO MONICA	74	−1.9014	3.6037	0.4724	3.0893
HEIGHT	126	68.5462	4.1578	−0.0497	3.0528

TABLE 4: Maximum likelihood estimation and standard error.

Parameter estimates	WHO MONICA		HEIGHT	
	L	EL	L	EL
μ	−1.7000 (0.0580)	−2.1310 (0.1889)	68.7500 (0.0353)	68.4236 (0.1469)
σ	3.3319 (0.4563)	1.6204 (0.1876)	3.7162 (0.3628)	1.7988 (0.1308)
ξ	–	3.5025 (2.9071)	–	7.9689 (5.1001)
Log-likelihood	−204.0124	−199.5330	−368.2153	−354.6919
AIC	412.0249	405.0661	740.4306	715.3838

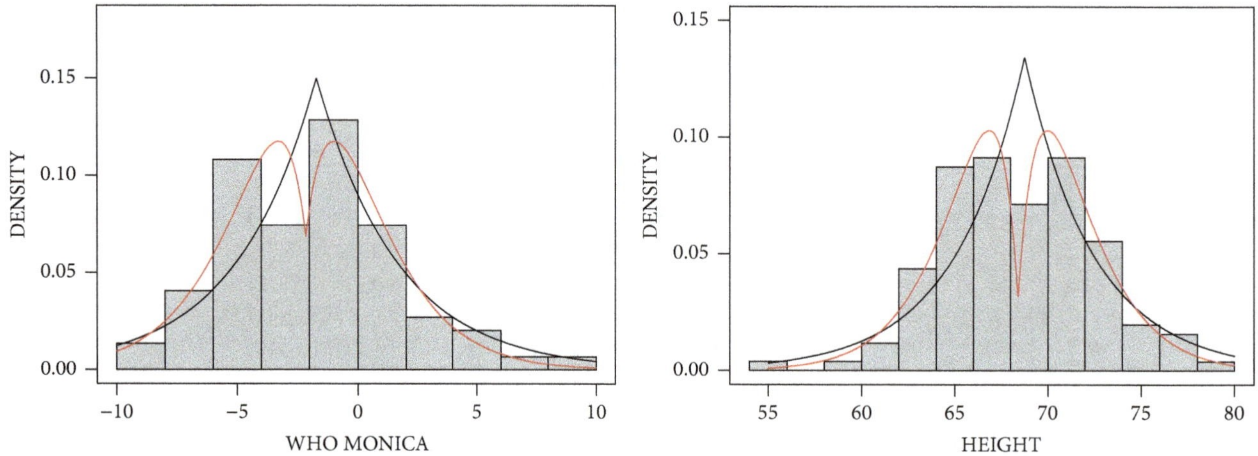

(a) WHO MONICA data set

(b) HEIGHT data set

FIGURE 2: Histograms and the fitted curves. Extended Laplace distribution (red line) and Laplace distribution (black line).

Figure 2 shows the histograms and the fitted curves for the data set.

The AIC is used to compare the estimated models. As one can see, our model with the smallest values of AIC is preferable. In addition, we can use the likelihood ratio (LR) test statistic to confirm our claim. To do this, we consider the following hypotheses:

$$H_0 : \xi = 0 \ \left(L(\mu, \sigma)\right) \text{ vs. } H_1 : \xi \neq 0 \ \left(EL(\mu, \sigma, \xi)\right) \quad (37)$$

The value of the LR test statistic for the data WHO MONICA is 7.22 and for the data set of the height of students is 26.96 and comparing this quantity with $\chi^2_{(1)} = 3.83$, the null hypothesis is rejected.

Figure 3 shows the graphs of the QQ-plots for the WHO MONICA and HEIGHT data sets calculated with the Laplace and extended Laplace models fitted with the maximum likelihood estimates of the parameters. These also show the good agreement of the EL distribution for the two data sets.

5. Conclusion and Final Comments

We have presented a generalization of the extended bimodal-normal distribution that depends on a shape parameter that controls the effect of bimodality when the density is symmetric. But, generally speaking, it produces a more flexible model in terms of asymmetry and kurtosis coefficients. Additionally, we have demonstrated its basic properties and stochastic representation, the latter of which played a significant role in the development of this work. The family of distributions includes a large number of distributions. For example, four of them were presented as corollaries, leaving the Laplace distribution for last, and used to develop some inferential aspects and Monte Carlo simulation schemes, which were facilitated by the use of stochastic representation in the generation of random variables. Finally, using two real data sets, we demonstrated that the proposed model resulted in better behavior relative to the standard Laplace model. Moreover, in the statistical literature, there are a variety of extensions of the Laplace distribution, in order to achieve greater flexibility, but without the effect of bimodality that

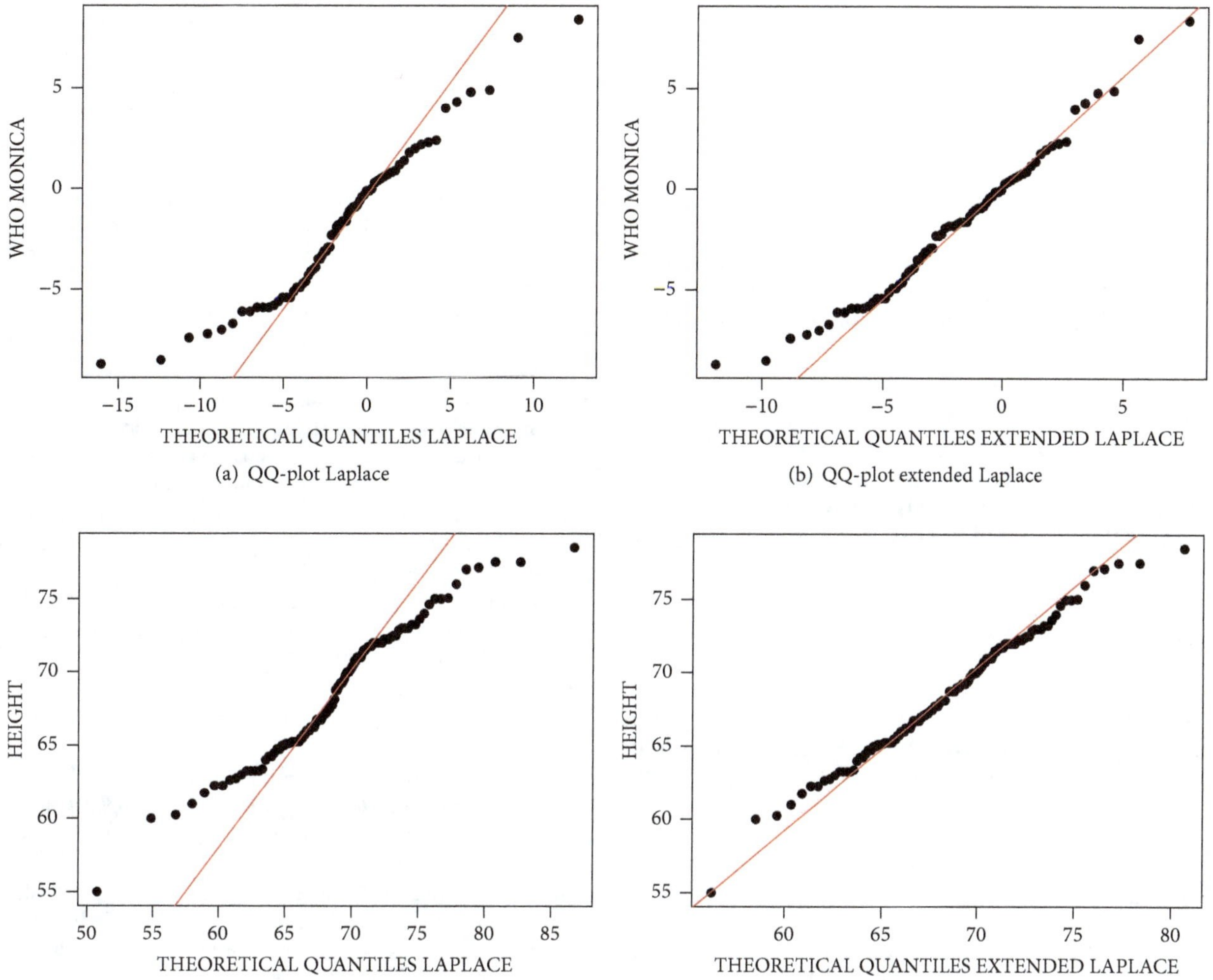

(a) QQ-plot Laplace

(b) QQ-plot extended Laplace

(c) QQ-plot Laplace

(d) QQ-plot extended Laplace

FIGURE 3: QQ-plot for the WHO MONICA and HEIGHT data sets.

fitted the data analyzed. Although bimodality can be achieved through a mixture of distributions, the proposed model is more parsimonious in terms of the number of parameters. It is important to emphasize that Theorem 1 can be extended, as demonstrated below.

Theorem 15. *Let g_1 and g_2 be a probability density functions and h a positive continuous function such that $\mathsf{E}_{g_2}[h(Y)] = \kappa < \infty$, where $Y \sim g_2$. Then*

$$f(x;\xi) = \left(\frac{1}{1+\kappa\xi}\right) g_1(x) + \left(\frac{\xi h(x)}{1+\kappa\xi}\right) g_2(x), \quad (38)$$

is a probability density function with shape parameter $\xi \geq 0$.

Note that when $g_1(x) = g_2(x) = g(x)$ we have Theorem 1. Furthermore, this new extension has stochastic representation given by the following theorem.

Theorem 16. *Let $X_1 \sim g_1(x)$ and $X_2 \sim (1/\kappa)h(x)g_2(x)$ be independent random variables. If*

$$X = \begin{cases} X_1, & \text{with probability } \dfrac{1}{1+\kappa\xi}, \\ X_2, & \text{with probability } \dfrac{\kappa\xi}{1+\kappa\xi}, \end{cases} \quad (39)$$

then $X \sim f(x;\xi)$.

Proof. Since $f(x;\xi)$ is a mixture, the result follows immediately.

This new extension includes, for examples, the case of the slash distribution Rogers and Tukey [24], Mosteller and Tukey [25], and Kafadar [26], if we consider the following definitions: (i) $g_1(x) = q\int_0^1 u^q \phi(xu)du$, (ii) $g_2 = (q+2)\int_0^1 u^{q+2}\phi(xu)du$, and (iii) $h(x) = qx^2/(q+2)$, with $x \in \mathbb{R}$ and $q > 2$, with density given by

$$f(x;\xi) = \left(\frac{1}{1+\xi}\right) q \int_0^1 u^q \phi(xu)\,du + \left(\frac{\xi}{1+\xi}\right) qx^2 \int_0^1 u^{2+q} \phi(xu)\,du. \quad (40)$$

Disclosure

Preliminary results of this manuscript were presented as a poster in "Flexible Statistical Models for a Skewed World of Data: A Workshop in Honor of Reinaldo B. Arellano-Valle's 65th Birthday" 2017.

Conflicts of Interest

The authors declare that there are no conflicts of interest regarding the publication of this paper.

Funding

The research of M. A. Cortés was supported in part by Grant DIUDA 221230. The research of D. Elal-Olivero was supported in part by Grant DIUDA 22287. The research of J. F. Olivares-Pacheco was supported in part by Grant DIUDA 22257.

References

[1] A. Azzalini, "A class of distributions which includes the normal ones," *Scandinavian Journal of Statistics*, vol. 12, no. 2, pp. 171–178, 1985.

[2] D. Elal-Olivero, "Alpha-skew-normal distribution," *Proyecciones Journal of Mathematics*, vol. 29, no. 3, pp. 224–240, 2010.

[3] D. Elal-Olivero, H. W. Gómez, and F. A. Quintana, "Bayesian modeling using a class of bimodal skew-elliptical distributions," *Journal of Statistical Planning and Inference*, vol. 139, no. 4, pp. 1484–1492, 2009.

[4] W. Gui, P.-H. Chen, and H. Wu, "A symmetric component alpha normal slash distribution: properties and inferences," *Journal of Statistical Theory and Applications*, vol. 12, no. 1, pp. 55–66, 2013.

[5] H. W. Gómez and G. A. Guerrero, *Bimodal Birnbaum-Saunders Distribution*, COMCA 2014: Congreso de Matemática Capricornio, Copiapó, Chile, 2014.

[6] N. Henze, "A probabilistic representation of the 'skew-normal' distribution," *Scandinavian Journal of Statistics*, vol. 13, no. 4, pp. 271–275, 1986.

[7] Z. W. Birnbaum and S. C. Saunders, "A new family of life distributions," *Journal of Applied Probability*, vol. 6, no. 2, pp. 319–327, 1969.

[8] Z. W. Birnbaum and S. C. Saunders, "Estimation for a family of life distribution with applications to fatigue," *Journal of Applied Probability*, vol. 6, no. 2, pp. 328–347, 1969.

[9] N. R. Mann, R. E. Schafer, and N. D. Singpurwalla, *Methods for Statistical Analysis of Reliability and Lifetime Data*, John Wiley & Sons, New York, NY, USA, 1974.

[10] A. Desmond, "Stochastic models of failure in random environments," *Canadian Journal of Statistics*, vol. 13, no. 2, pp. 171–183, 1985.

[11] A. F. Desmond, "On the Relationship Between Two Fatigue-Life Models," *IEEE Transactions on Reliability*, vol. 35, no. 2, pp. 167–169, 1986.

[12] D. S. Chang and L. C. Tang, "Graphical analysis for Birnbaum-Saunders distribution," *Microelectronics Reliability*, vol. 34, no. 1, pp. 17–22, 1994.

[13] J. A. Díaz-García and V. Leiva-Sánchez, "A new family of life distributions based on the elliptically contoured distributions," *Journal of Statistical Planning and Inference*, vol. 128, no. 2, pp. 445–457, 2005.

[14] H. W. Gómez, J. F. Olivares-Pacheco, and H. Bolfarine, "An extension of the generalized Birnbaum-Saunders distribution," *Statistics & Probability Letters*, vol. 79, no. 3, pp. 331–338, 2009.

[15] N. M. Olmos, G. Martínez-Flórez, and H. Bolfarine, "Bimodal Birnbaum-Saunders distribution with applications to non negative measurements," *Communications in Statistics—Theory and Methods*, vol. 46, no. 13, pp. 6240–6257, 2017.

[16] S. C. Saunders, "A family of random variables closed under reciprocation," *Journal of the American Statistical Association*, vol. 69, no. 346, pp. 533–539, 1974.

[17] R Core Team, "R: A Language and Environment for Statistical Computing," R Foundation for Statistical Computing, Vienna, Austria, https://www.R-project.org/, 2018.

[18] J. A. Nelder and R. Mead, "A simplex algorithm for function minimization," *The Computer Journal*, vol. 7, no. 4, pp. 308–313, 1965.

[19] S. Kotz, T. J. Kozubowski, and K. Podgórski, *The Laplace Distribution and Generalizations: A Revisit with Applications to Communications, Economic, Engineering and Finance*, Birkhäuser-Springer, Boston, Mass, USA, 2001.

[20] K. Kuulasmaa, H. Tunstall-Pedoe, A. Dobson et al., "Estimation of contribution of changes in classic risk factors to trends in coronary-event rates across the WHO MONICA Project populations," *The Lancet*, vol. 355, no. 9205, pp. 675–687, 2000.

[21] S. B. Kulathinal, K. Kuulasmaa, and D. Gasbarra, "Estimation of an errors-in-variables regression model when the variances of the measurement errors vary between the observations," *Statistics in Medicine*, vol. 21, no. 8, pp. 1089–1101, 2002.

[22] M. de Castro, M. Galea, and H. Bolfarine, "Hypothesis testing in an errors-in-variables model with heteroscedastic measurement errors," *Statistics in Medicine*, vol. 27, no. 25, pp. 5217–5234, 2008.

[23] M. Y. Hassan and R. H. Hijazi, "A bimodal exponential power distribution," *Pakistan Journal of Statistics*, vol. 26, no. 2, pp. 379–396, 2010.

[24] W. H. Rogers and J. W. Tukey, "Understanding some long-tailed symmetrical distributions," *Statistica Neerlandica. Journal of the Netherlands Society for Statistics and Operations Research*, vol. 26, no. 3, pp. 211–226, 1972.

[25] F. Mosteller and J. Tukey, *Data Analysis and Regression: A Second Course in Statistics*, Addison-Wesley Pub. Co., 1977.

[26] K. Kafadar, "A biweight approach to the one-sample problem," *Journal of the American Statistical Association*, vol. 77, no. 378, pp. 416–424, 1982.

Permissions

All chapters in this book were first published in JPS, by Hindawi Publishing Corporation; hereby published with permission under the Creative Commons Attribution License or equivalent. Every chapter published in this book has been scrutinized by our experts. Their significance has been extensively debated. The topics covered herein carry significant findings which will fuel the growth of the discipline. They may even be implemented as practical applications or may be referred to as a beginning point for another development.

The contributors of this book come from diverse backgrounds, making this book a truly international effort. This book will bring forth new frontiers with its revolutionizing research information and detailed analysis of the nascent developments around the world.

We would like to thank all the contributing authors for lending their expertise to make the book truly unique. They have played a crucial role in the development of this book. Without their invaluable contributions this book wouldn't have been possible. They have made vital efforts to compile up to date information on the varied aspects of this subject to make this book a valuable addition to the collection of many professionals and students.

This book was conceptualized with the vision of imparting up-to-date information and advanced data in this field. To ensure the same, a matchless editorial board was set up. Every individual on the board went through rigorous rounds of assessment to prove their worth. After which they invested a large part of their time researching and compiling the most relevant data for our readers.

The editorial board has been involved in producing this book since its inception. They have spent rigorous hours researching and exploring the diverse topics which have resulted in the successful publishing of this book. They have passed on their knowledge of decades through this book. To expedite this challenging task, the publisher supported the team at every step. A small team of assistant editors was also appointed to further simplify the editing procedure and attain best results for the readers.

Apart from the editorial board, the designing team has also invested a significant amount of their time in understanding the subject and creating the most relevant covers. They scrutinized every image to scout for the most suitable representation of the subject and create an appropriate cover for the book.

The publishing team has been an ardent support to the editorial, designing and production team. Their endless efforts to recruit the best for this project, has resulted in the accomplishment of this book. They are a veteran in the field of academics and their pool of knowledge is as vast as their experience in printing. Their expertise and guidance has proved useful at every step. Their uncompromising quality standards have made this book an exceptional effort. Their encouragement from time to time has been an inspiration for everyone.

The publisher and the editorial board hope that this book will prove to be a valuable piece of knowledge for researchers, students, practitioners and scholars across the globe.

List of Contributors

Livio Fenga
University of California San Diego (UCSD), San Diego, CA, USA

K. Perera
Department of Engineering Mathematics, Faculty of Engineering, University of Peradeniya, Peradeniya, Sri Lanka

Kobkun Raweesawat, Yupaporn Areepong and Saowanit Sukparungsee
Department of Applied Statistics, Faculty of Applied Science, King Mongkut's University of Technology North Bangkok, Bangkok 10800,Thailand

Katechan Jampachaisri
Department of Mathematics, Faculty of Science, Naresuan University, Phitsanulok 65000,Thailand

Hare Krishna and Neha Goel
Department of Statistics, Ch. Charan Singh University, Meerut, India

Maxwell Akwasi Boateng
Faculty of Engineering, Ghana Technology University College, Accra, Ghana

Akoto Yaw Omari-Sasu, Richard Kodzo Avuglah and Nana Kena Frempong
Department of Mathematics, Kwame Nkrumah University of Science and Technology, Kumasi, Ghana

N. A. Rather and T. A. Rather
Department of Mathematics, University of Kashmir, Srinagar 190006, India

Jimmy Reyes and Héctor W. Gómez
Departamento de Matemáticas, Facultad de Ciencias Básicas, Universidad de Antofagasta, Antofagasta, Chile

Osvaldo Venegas
Departamento de Ciencias Matemáticas y Físicas, Facultad de Ingeniería, Universidad Católica de Temuco, Temuco, Chile

Freddy H. Marín Sánchez and Verónica M. Gallego
Department of Mathematical Sciences, Universidad Eafit, Carrera 49 No. 7 Sur 50, Medellin, Colombia

Suddhendu Biswas
Department of Statistics, Amity Institute of Applied Science (AIAS), Amity University, Noida 201301, India

Kalpana Singh
Department of Statistics, Amity Institute of Applied Science (AIAS), Amity University, Noida 201301, India
Public Health Foundation of India, Sector 44, Plot 47, Gurgaon, Haryana 122002, India

Oluyemi Oyeniran
Manufacturing, Toxicology and Applied Statistical Sciences, Janssen Research & Development, Spring House, PA 19002, USA

Hanfeng Chen
Department of Mathematics and Statistics, Bowling Green State University, Bowling Green, OH 43403, USA

M. Ganji and F. Gharari
Department of Statistics, University of Mohaghegh Ardabili, Ardabil, Iran

Daya K. Nagar, Edwin Zarrazola and Luz Estela Sánchez
Instituto de Matemáticas, Universidad de Antioquia, Calle 67, No. 53–108, Medellín, Colombia

Rada Somkhuean, Sa-aat Niwitpong and Suparat Niwitpong
Department of Applied Statistics, Faculty of Applied Science, King Mongkut's University of Technology North Bangkok, Bangkok 10800,Thailand

Masood Anwar and Jawaria Zahoor
Department of Mathematics, COMSATS Institute of Information Technology, Park Road, Chak Shahzad, Islamabad, Pakistan

Tamás F. Móri
Department of Probability Theory and Statistics, Eötvös Loránd University, Pázmány P. s. 1/C, Budapest 1117, Hungary

Meredith A. Ray
Division of Epidemiology, Biostatistics, and Environmental Health, University of Memphis, 220 Robison Hall, Memphis, TN 38152, USA

Jian Kang
Department of Biostatistics, University of Michigan, 1415 Washington Heights, Ann Arbor, MI 48109, USA

Hongmei Zhang
Division of Epidemiology, Biostatistics, and Environmental Health, University of Memphis, 224 Robison Hall, Memphis, TN 38152, USA

Wolf-Dieter Richter and Kay Schicker
Institute of Mathematics, University of Rostock, Ulmenstr. 69, Haus 3, 18057 Rostock, Germany

Jaime A. Londoño
Departamento de Matemáticas y Estadística, Universidad Nacional de Colombia, Manizales, Colombia

Lionel Cucala
Institut Montpelliérain Alexander Grothendieck, 34000 Montpellier, France

Jean-Pierre Chateau
Faculty of Business Administration, University of Macau, Macau

Daniel Dufresne
Montreal, QC, Canada

Viswanathan Arunachalam and Liliana Blanco
Department of Statistics, Universidad Nacional de Colombia, Bogotá, Colombia

A. D. Banik
School of Basic Sciences, Indian Institute of Technology Bhubaneswar, Permanent Campus Argul, Jatni, Khurda 752 050, Odisha, India

M. L. Chaudhry
Department of Mathematics and Computer Science, Royal Military College of Canada, STN Forces, Kingston Ont., Canada K7K 7B4

James J. Kim
Royal Canadian Air Force (RCAF), 34 Tylee Pvt, Ottawa, Canada ON K1V 7M3

Milton A. Cortés, David Elal-Olivero and Juan F. Olivares-Pacheco
Departamento de Matemática, Facultad de Ingeniería, Universidad de Atacama, Copiapó, Chile

Index